Geomorphic Impacts of Earthquakes

Geomorphic Impacts of Earthquakes

Post-Seismic Landscape Evolution

Edited by

TIM DAVIES
School of Earth and Environment, University of Canterbury, Christchurch, New Zealand

ALEXANDER DENSMORE
Department of Geography, Durham University, Durham, United Kingdom

XUANMEI FAN
State Key Laboratory of Geohazard Prevention and Geoenvironment Protection, Chengdu University of Technology, Chengdu, Sichuan, P.R. China

ELSEVIER

Elsevier
Radarweg 29, PO Box 211, 1000 AE Amsterdam, Netherlands
125 London Wall, London EC2Y 5AS, United Kingdom
50 Hampshire Street, 5th Floor, Cambridge, MA 02139, United States

ISBN: 978-0-443-40375-0

For Information on all Elsevier publications
visit our website at https://www.elsevier.com/books-and-journals

Publisher: Peter Llewellyn
Acquisitions Editor: Jennette McClain
Editorial Project Manager: Prachi Mittal
Production Project Manager: Punithavathy Govindaradjane
Cover Designer: Mark Rogers

Working together to grow libraries in developing countries
www.elsevier.com • www.bookaid.org

Typeset by MPS Limited, Chennai, India

Contents

List of contributors

Kate E. Allstadt
U.S. Geological Survey, Geologic Hazards Science Center, Golden, CO, United States

Sarah J. Boulton
School of Geography, Earth and Environmental Sciences, University of Plymouth, Plymouth, United Kingdom

Marie-Luce Chevalier
Institute of Geology, Chinese Academy of Geological Sciences, Beijing, P.R. China

Kristen L. Cook
IRD, ISTerre, Universite Grenoble Alpes, Grenoble, France

Lanxin Dai
State Key Laboratory of Geohazard Prevention and Geoenvironment Protection, Chengdu University of Technology, Chengdu, Sichuan, P.R. China

Tim Davies
School of Earth and Environment, University of Canterbury, Christchurch, New Zealand

Alexander Densmore
Department of Geography, Durham University, Durham, United Kingdom

Benjamin Edwards
School of Environmental Sciences, University of Liverpool, Liverpool, United Kingdom

Xuanmei Fan
State Key Laboratory of Geohazard Prevention and Geoenvironment Protection, Chengdu University of Technology, Chengdu, Sichuan, P.R. China

Tristram C. Hales
School of Earth and Environmental Sciences, Cardiff University, Cardiff, United Kingdom

Erin L. Harvey
Department of Geography, Durham University, Durham, United Kingdom

Farnaz Kamranzad
School of Environmental Sciences, University of Liverpool, Liverpool, United Kingdom

Mark Kincey
School of Geography, Politics and Sociology, Newcastle University, Newcastle upon Tyne, United Kingdom

Gen K. Li
Department of Earth Science, University of California, Santa Barbara, CA, United States

Sally F. McGill
Department of Geological Sciences, California State University, San Bernardino, CA, United States

Mark Quigley
School of Geography, Earth and Atmospheric Sciences, University of Melbourne, Parkville, VIC, Australia

Tom R. Robinson
School of Earth and Environment, University of Canterbury, Christchurch, New Zealand

Nick J. Rosser
Department of Geography, Institute of Hazard Risk and Resilience, Durham University, Durham, United Kingdom

Eric M. Thompson
U.S. Geological Survey, Geologic Hazards Science Center, Golden, CO, United States

Maximillian Van Wyk de Vries
Department of Geography, University of Cambridge, Cambridge, United Kingdom; Department of Earth Sciences, University of Cambridge, Cambridge, United Kingdom

Gonghui Wang
Research Center for Landslide Risk Cognition and Reduction, DPRI, Kyoto University, Uji, Kyoto, Japan

Jin Wang
State Key Laboratory of Loess Science, Institute of Earth Environment, Chinese Academy of Sciences, Xi'an, P.R. China

Alexander C. Whittaker
Department of Earth Science and Engineering, Royal School of Mines, Imperial College, London, United Kingdom

Foreword

On the flanks of Mount Parnassus, high above the plains of Chrysso, where olive trees grow for a thousand years, lies ancient Delphi. It is said that the Pythia, the high priestess of the temple of Apollo there, would retreat into a cavity to connect with her god and return with prophesies. This spectacle turned Delphi into a cultural, financial and political hub of ancient Greece. Today, a sign at the sprawling archaeological site locates the Pythia's inner sanctum in an inconspicuous hollow at the base of a steep outcrop. An active normal fault bounds the cliff. Landslide debris fills the hollow.

The interaction of tectonic and geomorphic processes is a primordial characteristic of our planet. It shapes Earth's topography, drives matter fluxes and has guided the trajectory of the Earth's surface system and the life it harbours. This interaction may be steady in a geodynamic sense, governed by rates of crustal deformation and the gross climatic conditions. On a large scale, it marks plate boundaries with long-lived mountain ranges and escarpments, which direct the flow of air, water and ice across continents. However, the dialogue of tectonics and Earth surface processes is punctuated by the elemental manifestations of its grand drivers on smaller scales, but with major consequences. Earthquakes are perhaps the best example of this mechanism. They can cause abrupt, local displacements of the Earth's surface and are accompanied by strong ground motion radiating from an epicentre. As a result, distinct features form in the landscape – fault scarps, waterfalls, terraces, abandoned river courses – that can be read for evidence of past seismic activity. Faulting and shaking can fragment rock masses, creating or cutting off pathways for groundwater flow, and distending, sometimes even liquefying near-surface materials. And, the temporary, cyclic changes in the orientation and strength of mechanical forces acting on topography due to seismic shaking can affect slope stability, sediment production, weathering and the integrity of ecosystems and the built environment.

Travellers and chroniclers have long observed and recorded earthquakes and their effects at Earth's surface. The resulting image is, predominantly, a toxic mix of unpredictability and damaging impacts associated with processes that can't be stopped. To address this, and for lack of prevention, earthquake research has a century-old history of asking why, what, where, when, how much and with what consequences. The scientific evaluation of seismically triggered surface processes has followed in this track, spurred on by the fact that landslides, liquefaction and other companion processes are a major cause of damage and loss in many earthquakes. While this work can build on the laws of physics, it is made difficult by geological, topographic and process complexity and inaccessibility of relevant structures and strata at depth below the surface. Systematic geophysical and geomorphological documentation and analysis of larger continental earthquakes in recent decades have helped to find general, predictable features of seismically triggered processes. Importantly, these case studies have also found that earthquakes can have longer-lasting, cascading effects within the Earth's surface system. These effects set dynamic conditions for slope stability, sediment flux and hazard risk not only during but also after earthquakes, and they can propagate deep into ecosystems and biogeochemical cycles.

In a field that has evolved rapidly, this volume takes stock of our current insight into the geomorphic impacts of earthquakes. It brings together fundamentals of seismic and geomorphic processes, the effects of earthquakes on near-surface materials and individual surface process systems, and the specific and systemic features of some well-studied major seismo-geomorphic events. The ensemble captures a detailed, but bigger, picture of the natural dynamics of seismically active landscapes. Humanity has chosen to live in these landscapes: they offer shelter and an exceptional abundance of resources as a counter to their inherent hazards, much like Apollo carried both poisoned arrows and the power of healing. To safeguard life and socio-economic activity in the proximity of earthquakes, knowledge and understanding of their immitigable geomorphic effects enables protection by informed anticipation, planning and preparation, as well as targeted efficiency of disaster management. This volume sets out to contribute to this high aim, with the clear voice of science rather than the riddles of prophecy. Beyond this, it may also serve to highlight the fascination of seismic geomorphology and its role in the Earth sciences and to encourage engagement with the substantial remaining challenges and emerging opportunities.

Niels Hovius, *Potsdam*, April 2026
GFZ Helmholtz-Zentrum für Geoforschung, Telegrafenberg, Potsdam, Germany;
Institute of Geosciences, University of Potsdam, Potsdam, Germany

CHAPTER 1

Introduction: tectonics, earthquakes, landforms and society

Tim Davies[1], Alexander Densmore[2], and Xuanmei Fan[3]
[1]School of Earth and Environment, University of Canterbury, Christchurch, New Zealand,
[2]Department of Geography, Durham University, Durham, United Kingdom,
[3]State Key Laboratory of Geohazard Prevention and Geoenvironment Protection, Chengdu University of Technology, Chengdu, Sichuan, P.R. China

1.1 Introduction

In regions that experience major earthquakes, these events, and the subsequent long-term response of the landscape to them, influence the whole geomorphology of the region at all spatiotemporal scales. Given the fundamental ways in which geomorphology influences and constrains the natural and societal systems and processes that occupy Earth's surface, the importance of understanding seismic impacts on geomorphology is self-evident. Nevertheless, the geomorphic impacts of earthquakes are relatively little reported, and this volume is intended to spark and set some basis for better recognition and understanding of this topic.

1.2 Background

Tectonic plate motion generates the crustal stresses that cause earthquakes, but it also causes the slow, long-term crustal deformation that results in mountain ranges. Close to a plate-boundary fault, much, if not all, of that deformation occurs during earthquakes, so the fault-proximal geomorphology can be expected to clearly reflect the occurrence of earthquakes. Farther from the fault, where ground shaking and displacement from plate-boundary earthquakes are less, a larger proportion of tectonic deformation can be more or less continuous (aseismic), and so the geomorphology becomes (in principle) less dominated by earthquakes themselves. Thus characterisation of the geomorphic consequences of earthquakes (i.e. 'seismic geomorphology') requires setting some limitations to distinguish it within the more general topic of tectonic geomorphology, or the geomorphology of tectonically active regions. That is one of the aims of this volume.

The operative characteristics of earthquakes that result in landscape change are (1) short-term intense ground shaking radiating from the rupturing fault surface and (2) rapid short-term displacement of the ground resulting in permanent topographic displacement close to the fault. These are the direct triggers of the geomorphic processes following the earthquake and of the resulting landforms, many of which are explored in

Geomorphic Impacts of Earthquakes
DOI: https://doi.org/10.1016/B978-0-443-40375-0.00009-8

the present volume. Other aseismic tectonic deformation phenomena, such as slow earthquakes and tectonic uplift, depression and translation, are longer-term intermittent or continuous drivers of geomorphology that are not specifically considered herein. Still other, less direct triggers, such as seismically induced changes in surface or groundwater flow, form part of the postseismic geomorphic cascade, parts of which are considered in the chapters that follow.

The distinction between seismic and tectonic drivers becomes fuzzy, however, when considering not just the geomorphic impacts of large earthquakes on plate-boundary faults, but also those of lesser earthquakes which occur more frequently in faulted regions and whose cumulative effect may be appreciable. The distinction between 'characteristic' earthquakes (in which a single fault or segment ruptures more or less regularly) and earthquakes described by the Gutenberg-Richter power-law distribution (in which, across a region, small earthquakes are consistently more numerous than large earthquakes) is important here; in reality, however, earthquake occurrence at a specific location is a combination of both types - for example, close to an active plate-boundary fault large-magnitude ruptures and widespread shaking result from characteristic behaviour, while in the adjacent region Gutenberg-Richter behaviour occurs with more frequent, spatially distributed ruptures of widely varied magnitudes causing similarly widespread but more evenly distributed shaking. While the present volume is not explicitly restricted to 'large' earthquakes, its contents probably are biased in this direction simply because large earthquakes are dramatic and societally important events, and it is thus easier to fund their investigation (e.g. Caputo, 2005; Robinson & Davies, 2013) than it is for smaller events.

Both seismology and geomorphology have long and distinguished histories, and both are currently highly active fields of research, in particular in the context of the widespread emphasis on disaster risk reduction that has burgeoned since the Sendai Framework for Disaster Risk Reduction 2015–30 was signed by most nations on Earth. Given the obvious impacts of earthquakes on landscapes in the form of surface ruptures and coseismic landslides, one would have expected a similarly enthusiastic focus on earthquake geomorphology in general, but this is not the case; most publications involving both disciplines have focused on specific faults, or described the impacts of specific earthquakes, or related earthquakes directly to hazards and societal impacts. The present volume attempts to start to fill this research gap with a set of contributions that explore in detail how landscapes respond in the longer-term to the triggering of geomorphic processes by intense shaking and ground displacement.

One of the reasons for the relative neglect of seismic geomorphology, even in highly seismically active locations, is that earthquakes are generally perceived as occasional phenomena that, from time to time, perturb the even tenor of life, but most of the time have no impact, so can be ignored. Rare large events can certainly have a huge impact on 'business as usual', but they may not be recognised as inevitable events that can continue to affect everyday life once damage has been repaired. Indeed, even in earthquake-prone societies, governments see the need to continually remind people by means of publicity that earthquakes and their potentially devastating consequences can occur at any time. Earthquakes are easily neglected.

An analogous situation can be seen in professions focused on land use and commerce in earthquake-prone regions: everyday plans and considerations often ignore the potential for earthquakes to disturb, perhaps for long periods, the economic processes and functioning of society. Neither can science itself be absolved from the same charge; for example, it is common to find river engineers and land-use planners designing modifications to rivers and uses of the landscape that will be severely affected by earthquakes, but with no or inadequate consideration of these impacts.

These apparent blind spots result largely from the long time intervals (usually centuries or greater) between major earthquakes; this can mean that for much of the interseismic period, no living memory exists of the last major event, which in any case probably affected a very different society. Furthermore, because all major development decisions are politically and economically driven, and both electoral and economic planning time-frames are typically years to at most a decade, there may be little societal appetite for serious consideration of an event that might not occur for a hundred years, no matter how devastating it might be and how long-term its impacts. Thus with little interest among the populace or its leaders, there is correspondingly small incentive for Earth scientists to be concerned with the impacts of earthquakes on landscapes and the way they function. This makes an interesting contrast with flood hazards, which are commonly experienced, have different forms and types of scientific and lay knowledge, and whose mitigation many societies take for granted as an obvious necessity.

Nevertheless, in much the same way for both society and science, significant future benefits can result from taking earthquakes seriously as an influence on many aspects of both everyday life and the Earth's surface in general. As an example, the widespread coseismic landsliding associated with a large earthquake in mountainous terrain has obvious and immediate impacts on society; what is largely unconsidered, however, is that it can also result in slowly developing, widespread, long-term and high-magnitude aggradation in river systems that flow from the source area across societally-occupied floodplains (e.g. Blagen et al., 2022; Fan et al., 2019). This aggradation phase may continue for several decades, during which river management and flood control may severely test local authorities, and until it ends, it will cause serious problems of infrastructure maintenance in the affected areas. The time scales over which landsliding and consequent aggradation develop are much longer than the typical postearthquake response times mandated by national and international organisations (e.g. Inter-Agency Standing Committee, 2015; Williams et al., 2018), meaning that it can be structurally difficult or even impossible for that response to be informed by an awareness of the evolving hazard (e.g. Rosser et al., 2021). The science underlying such an event, the development of societal responses, and the relationships between them scarcely feature in current literature, but are crucial to elaborate and understand if the situation is to be planned and managed so that economic damage is limited.

1.3 Outline

Herein, we take a broad look at the landforms that result directly or indirectly from land displacement and ground shaking in earthquakes. In many cases, similar landforms can result from nonseismic causes – coseismic rockfalls, for example, often result from loosening and displacement of a rock mass so that gravity can mobilise it in much the same way that it mobilises rainfall-triggered rockfalls. By contrast, large coseismic landslides are often the result of formation and completion of a deep-seated failure surface due to prolonged (and perhaps repeated) seismic shaking (Bazgard et al., 2011; Davies & Moretti, 2022), whereas aseismic weathering would not have generated a deep-seated failure surface; thus such a failure surface, typified by a bowl-shaped headscarp, is a distinctive signature of an earthquake origin (Turnbull & Davies, 2006). Another characteristic signature of earthquake activity is the long linear features formed by fault traces, and their equivalents in offset features such as river terraces. Other aspects of landform assemblages can also indicate a seismic origin; for example, coseismic landslides and rockfalls often occur in widespread contemporaneous clusters, whereas landslides and rockfalls triggered by rainfall and flooding tend to be less dramatically clustered in time and space, and those triggered by stress corrosion tend to be isolated events. Even though the individual events in many cases have largely similar characteristics, their geomorphic consequences can be significantly different depending on the number and spatial distribution of simultaneous events that occur in the vicinity.

The contents of this volume are intended to provide both geomorphologists and seismologists with an adequate background in both topics so that the more focused material on specific aspects of seismic geomorphology can be appreciated. A set of case studies then demonstrates practical applications of this knowledge, and illustrates both what we as a community have learned from recent earthquakes and what outstanding knowledge gaps still remain. Thus Chapters 2 to 5 provide background information on earthquakes (Tim Davies), ground motion (Benjamin Edwards and Farnaz Kamranzad), rock fracture and ground failure processes (Kate Allstadt and Eric Thompson), and Earth surface processes and geomorphology (Tim Davies), and are aimed at readers who have not specialised in these areas. Following this, Chapters 6 to 12 deal in detail with fault trace processes (Marie-Luce Chevalier); coseismic landslides and rockfalls (Wang Gonghui); liquefaction and lateral spreading (Mark Quigley); fluvial responses to earthquake deformation and to coseismic sediment inputs (Sarah Bolton, Alex Whittaker and TC Hales); glacial processes and impacts (Max van Wyk de Vries); postseismic sediment and geochemical fluxes (Jin Wang and Gen Li); and geomorphic indicators of past earthquakes, including palaeoseismology (Sally McGill). Three case studies of the geomorphic impacts of recent earthquakes are then outlined: the 1999 Chi-Chi earthquake in Taiwan (Kristen Cook), the 2008 earthquake in China (TC Hales, Erin Harvey, Dai Lanxin and Fan Xuanmei), and the 2015 Gorkha earthquake in Nepal (Mark Kincey and Nick Rosser). The volume continues with a chapter by Tom Robinson on the implications of seismic geomorphology for hazard and risk analysis, impacts and mitigation, and concludes with a summary and review of future needs and prospects in the field by the editors and Niels Hovius.

While we have attempted to span the full extent of an extremely broad topic, some significant topics remain unaddressed herein. This is in part due to the intense demands that professional employment makes on experts in modern times, so that a number of invitations to expert scientists to contribute to this volume were unable to be accepted. Nevertheless, we are hopeful that this volume will provide both background and inspiration for future researchers – there remains plenty of scope!

References

Bazgard, A., Davies, T., and Cundall, P. (2011) Dynamic analysis of mountain response using PFC3D and FLAC3D. In Continuum and distinct element modelling in geomechanics. Itasca, Minneapolis, MN, USA, pp.1-7.

Blagen, J. R., Davies, T. R. H., Wells, A., & Norton, D. A. (2022). Post-seismic aggradation history of the West Coast, South Island, Aotearoa/New Zealand: Dendrogeomorphological evidence and disaster recovery implications. *Natural Hazards, 114*, 2545–2570.

Caputo, R. (2005). Ground effects of large morphogenic earthquakes. *Journal of Geodynamics, 40*, 113–118.

Davies, T. R. H., & Moretti, D. (2022). Geomorphic precursors of large landslides: Seismic preconditioning and slope-top benches. In *Landslide hazards, risks, and disasters*. Elsevier, 641–666.

Fan, X., Scaringi, G., Korup, O., West, A. J., van Westen, C. J., Tanyas, H., Hovius, N., Hales, T. C., Jibson, R. W., Allstadt, K. E., Zhang, L., Evans, S. G., Xu, C., Li, G., Pei, X., Xu, Q., & Huang, R. (2019). Earthquake-induced chains of geologic hazards: Patterns, mechanisms, and impacts. *Reviews of Geophysics, 57*, 421–503. https://doi.org/10.1029/2018RG000626.

Inter-Agency Standing Committee (2015) Multi-Sector Initial Rapid Assessment (MIRA) Guidance. https://reliefweb.int/node/1155511.

Robinson, T. R., & Davies, T. R. H. (2013). Potential geomorphic consequences of a future great (M_w = 8.0+) Alpine Fault earthquake, South Island, New Zealand. *Natural Hazards and Earth System Sciences, 13*, 2279–2299.

Rosser, N., Kincey, M., Oven, K., Densmore, A., Robinson, T., Pujara, D. S., Shrestha, R., Smutny, J., Gurung, K., Lama, S., & Dhital, M. R. (2021). Changing significance of landslide hazard and risk after the 2015 M_w 7.8 Gorkha, Nepal earthquake. *Progress in Disaster Science, 10*. https://doi.org/10.1016/j.pdisas.2021.100159.

Turnbull, J. M., & Davies, T. R. H. (2006). A mass movement origin for cirques. *Earth Surface Processes and Landforms, 31*, 1129–1148.

Williams, J. G., Rosser, N. J., Kincey, M. E., Benjamin, J., Oven, K. J., Densmore, A. L., Milledge, D. G., Robinson, T. R., Jordan, C. A., & Dijkstra, T. A. (2018). Satellite-based emergency mapping: landslides triggered by the 2015 Nepal earthquake. *Natural Hazards and Earth System Sciences, 18*, 185–205. https://doi.org/10.5194/nhess-18-185-2018.

CHAPTER 2

Earthquakes

Tim Davies
School of Earth and Environment, University of Canterbury, Christchurch, New Zealand

2.1 Introduction

Earthquakes only occur occasionally in any given location, and last at most for a few minutes, but they can apply large forces to the surface materials of the landscape which can exceed the static forces that act in the absence of earthquakes, such as weight and friction. These seismic forces can cause the ground itself to lose much of its strength, and to move rapidly and substantially, and ground displacements can be many metres; ground accelerations in earthquake shaking can exceed that of gravity. Thus earthquakes can result in significant ground material failures and displacements, which, together with consequent and later land surface processes, can generate significant geomorphological features. In seeking to understand these geomorphic consequences of earthquakes, it is important to know where earthquakes can occur, with what intensity, and how frequently, so that associations between seismicity and landforms can be identified and investigated.

2.2 What causes an earthquake? Fault creation and evolution

Earthquakes are the result of elastic strain energy being stored in deforming crustal rock as stress in the crust builds up at a given location due to the relative motion of adjoining tectonic plates. In some places, the crust readjusts more or less continuously in response to the accumulating tectonic stresses by slow creep along rock cracks, so the stress does not build up to a sufficiently high level that major ruptures occur. In other locations, or at other times, the frictional resistance to motion on the joints is sufficient to resist the tendency to creep, and the joints are said to be 'locked'; in this situation, the tectonic stress applied to the crust increases over time. Earthquakes are initiated when the shear stress applied to a jointed zone of crustal rock exceeds its static strength; this results in the rock failing by shearing, so that a (set of) joint(s) is reactivated (or is initiated as a (set of) crack(s)) within the rock mass. The rock on either side of the released joint set then rebounds elastically in response to the accumulated stress, and stored strain energy is released to generate (1) a pervasive all-frequency high-stress elastic body-wave field as a result of intense rock fracture as the crack propagates and rock is crushed by shearing; (2) lower-frequency body waves and surface waves caused by the elastic rebound of the

Geomorphic Impacts of Earthquakes
DOI: https://doi.org/10.1016/B978-0-443-40375-0.00001-3

initially-deformed rock; and (3) irreversible deformation of the land surface. Conventionally, the fracture-generated energy (1) is conventionally assumed to be absorbed as surface energy and lost to the situation, but recently Davies et al. (2019, 2020) showed both experimentally and conceptually that this energy loss does not, in fact, occur; this correction becomes quantitatively important when the ultra-fine (submicron) nature of much of the fault gouge (rock in the fractured zone that is pervasively crushed during shear) is recognised (Wilson et al., 2005).

For such a rupture to develop into a significant earthquake, the initial rock failure must release sufficient strain energy that the transfer of stress results in the yield strength of adjacent material being exceeded. Rupture then occurs in these adjacent zones, and the rupture propagates (i.e. the crack extends) through the crust. This extension often takes place along a fault (a narrow region of rock previously shattered by earlier earthquakes and perhaps strengthened subsequently by gradual chemical recementation), and can propagate at a rate of several kilometres per second. Rupture propagation can cease when the rupture front encounters a much weaker area of rock where the strain energy released by the advancing crack tip decreases, and also when the rupture front encounters a stronger area of crust. Cessation of rupture propagation causes energy radiation to fall to zero, and the earthquake stops. A major earthquake on a fault several hundred km long may continue to emit seismic energy for the order of 10 to 100 s.

Although some earthquakes are 'first-time' ruptures through intact rock (e.g. Wilson et al., 2005), many, if not most, occur in locations where previous ruptures have occurred. The process of crack development and extension, and the subsequent relative motion of the rock on either side of the crack, results in a linear zone of severely crushed and sheared rock called a fault zone, whose mechanical and hydraulic properties differ from those of the surrounding rock. Thus when, after an appropriate time has elapsed after the previous earthquake, the ongoing tectonic plate motion again results in high stresses on rock in the vicinity of the fault, the most likely location of the next rupture is somewhere along the fault, and the rupture is likely to propagate along the fault again. Therefore in identifying future earthquake locations, it is useful to look for existing faults; however, knowledge of fault locations is rarely complete – for example, in New Zealand it is estimated that '… about half of all historical earthquakes $M_w \geq 7.0$ ruptured faults that, based on today's state of knowledge of active-fault locations, would not have been identified as active prior to the event' (Nicol et al., 2016) – so it is not unusual for earthquakes to occur unexpectedly in regions where seismic hazard was unrecognised or underestimated (England & Jackson, 2011).

Particularly where many earthquakes have occurred along the same fault, the repeated intense seismic wave field radiating from the fault causes an extensive zone of damaged rock to develop adjacent to the fault. This may extend for up to a kilometre from the fault (e.g. Dor et al., 2006), and is often asymmetric, being wider on the hanging-wall side (i.e. above a dipping fault) than on the footwall side below the fault.

This damage zone, comprising fractured rock in the zone of most intense ground motion, can be the preferred location for initiation of earthquake-generated landsliding (Massey et al., 2020).

During the time between fault ruptures, the highly comminuted fault 'gouge', that is the result of rupture propagation and subsequent rapid high-stress slip, undergoes physical and chemical changes as a result of the large surface area of chemically-active ultra-fine particles that are created and the seepage of fluids containing minerals in solution. The processes of compaction, cementation, solution and dissolution can result in changes to the permeability (which generally decreases) and strength (which generally increases) of the fault over time following an earthquake. It has been suggested that these processes tend to homogenise material strength along the fault over time, resulting eventually in a fault of relatively uniform strength along which a rupture can propagate uninterruptedly for a long distance (Davies et al., 2012).

2.3 Earthquake triggers

The major contributor to the applied shear stress causing an earthquake is usually the gradually increasing regional tectonic stress resulting from plate motion or glacio-isostatic loading and unloading. However, as this stress approaches the local shear strength in a given location, other short-term applied stresses are increasingly likely to cause the local rock strength to be exceeded. These temporary stresses can include seismic stresses from nearby earthquakes, pore fluid stresses (which may also reduce shear strength temporarily) resulting from seismicity or hydrological processes, mass redistribution caused, for example, by large landslides or lake filling and draining, and stresses caused by volcanic eruptions or (more rarely) bolide impacts (Karlstrom et al., 2025). In the absence of any such distinct trigger, a rupture will occur where and when the local rock strength is first exceeded by the increasing local tectonic stress. Whether or not it develops into a significant earthquake depends on the stress and strength distributions along the fault and in the adjacent crustal region.

2.4 Spatial distribution of earthquakes

Since tectonic stress is usually the result of relative tectonic plate motion, the global distribution of earthquakes is distinctly related to the major tectonic plate boundaries where relative motion is greatest (Fig. 2.1), with the vast majority occurring close to these boundaries. While major earthquakes can occur at significant distances from plate boundaries, this is unusual, and we expect that the geomorphic consequences of large earthquakes will be most apparent where the events are most frequent. Note, however, that several regions (e.g. the Himalaya, the eastern Mediterranean, and Arabia) show high seismicity over large areas, indicating that plate boundary influences can extend in

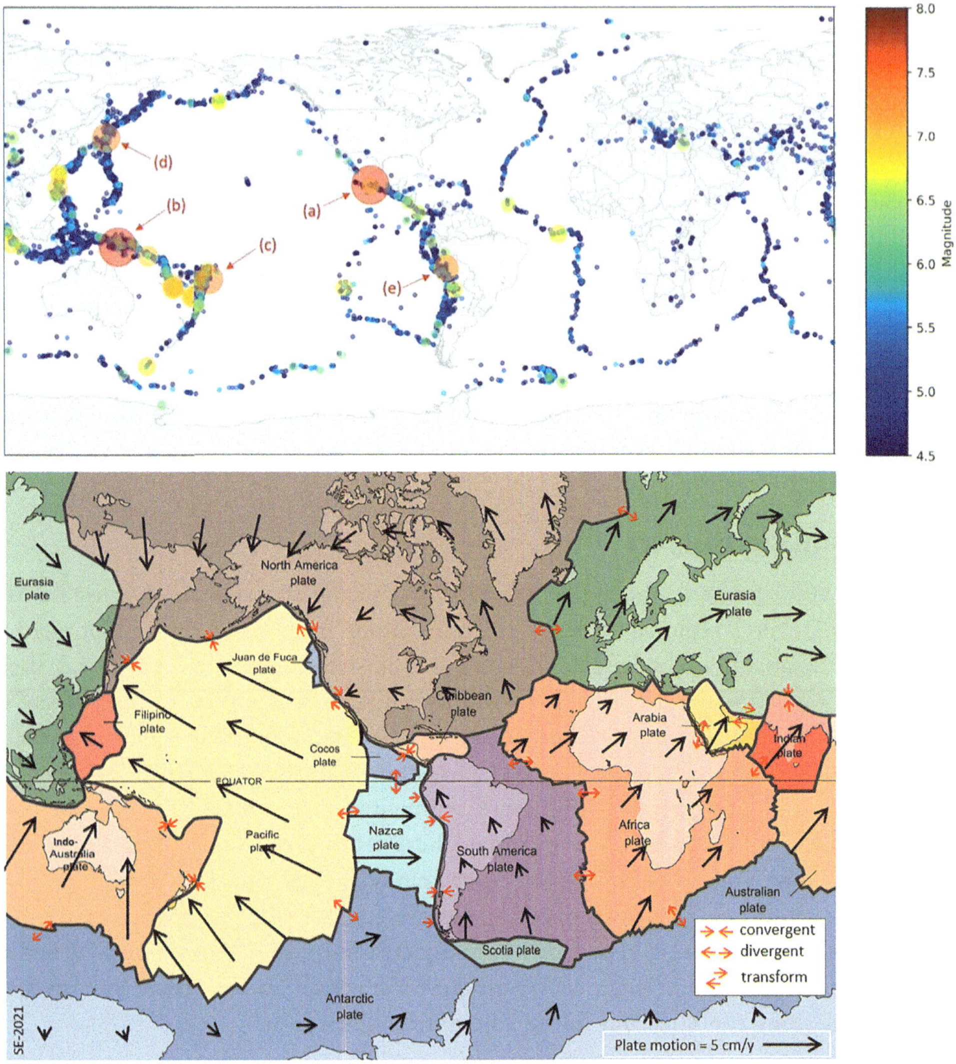

Figure 2.1 Upper: worldwide map of earthquakes in 2022. Points (a) - (e) indicate the largest five earthquakes. Lower: Tectonic plates with motions (https://environmental-geology-dev.pressbooks.tru.ca/chapter/plate-tectonics/). Note that upper and lower maps have different projections (Iacoletti, 2023). Note: Map lines delineate study areas and do not necessarily depict accepted national boundaries. *Modified from Iacoletti, G., (2023). An unpredictable peril; why earthquake should always be on the risk radar. https://axaxl.com/fast-fast-forward/an-unpredictable-peril-why-earthquake-should-always-be-on-the-risk-radar.*

the form of active crustal deformation for several hundreds of kilometres. Furthermore, several large earthquakes are known to have occurred far from plate boundaries (e.g. the 1811–12 New Madrid earthquake sequence in Missouri, central USA, which has also been linked to deglaciation-induced isostatic stresses: Zoback & Grollimund, 2001) and have caused distinct geomorphic consequences (e.g. Guccione et al., 2000; McEwan et al., 2023, 2025).

Tectonic stress can also undergo long-term changes that are not the result of relative plate motion. For example, parts of Scandinavia are today undergoing rock uplift as a result of crustal unloading as the last-glacial ice sheet (which was up to 3 km thick) finally disappears after about 20,000 years of melting; nevertheless, seismicity in Scandinavia is today considered low. This isostatic unloading is known to have caused several major faults to rupture to full crustal depth in the mid-Holocene (Eldholm & Bungum, 2021) when rock uplift was faster. Similar deglaciation-related seismicity may be associated with a well-defined peak in postglacial landsliding in Europe (e.g. Ballantyne et al., 2014a, b; Darnault et al., 2012; McColl et al., 2012). Such postglacial seismicity is also likely to have contributed to supraglacial coseismic landsliding, resulting in the formation of large terminal moraines (Reznichenko et al., 2011).

2.5 Local seismicity

The degree to which earthquakes can affect geomorphology is related to the frequency of occurrence and magnitude of earthquakes in any specific locality. These characteristics also form the basis of seismic hazard mapping, which is in widespread use for quantifying the potential damage to society due to, for example, building and infrastructure collapse, tsunami or landslide occurrence; thus seismic hazard assessments also form a useful basis for understanding the extent to which seismicity contributes to geomorphology.

Seismic hazard assessments are derived from empirical data on recorded earthquakes and from investigations of prehistoric seismicity on specific faults using trenching and local geomorphology, with all data being used to calibrate regional, national and even global models for earthquake magnitude and frequency (e.g. Gerstenberger et al., 2020; Silva et al., 2020). However, it is important to realise that modern data acquisition technology means that the known distribution of seismic hazard can change rapidly as knowledge increases. This is illustrated dramatically in New Zealand, where the distribution of seismic hazard changed remarkably between 2010 and 2022, largely as a result of the Canterbury earthquake sequence of 2010–12 and the 2016 M_w 7.8 Kaikōura earthquake (Fig. 2.2; chapter 16).

2.6 Energy radiation and earthquake magnitude

During rupture propagation, the intensity of the body-wave field carrying the released tectonic strain energy decreases with distance from the fault as friction turns part of the energy field to heat, with higher-frequency components attenuating more rapidly than

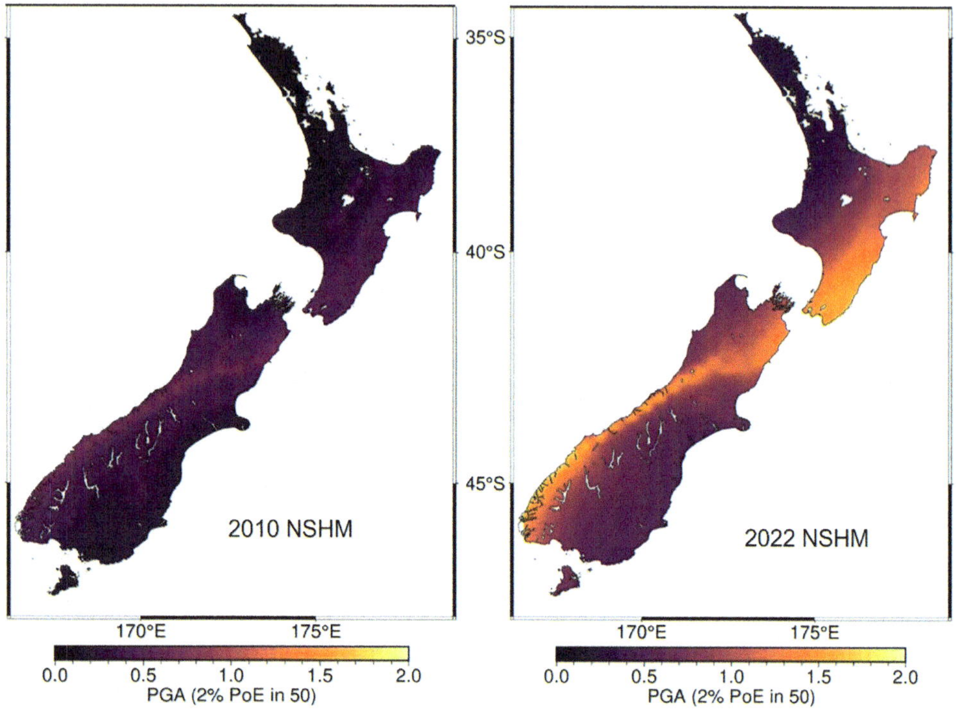

Figure 2.2 Probabilistic seismic hazard maps for peak ground acceleration (PGA) in New Zealand, with 2% probability of exceedance in 50 years in 2010 (left) and 2022 (right). (https://www.gns.cri.nz/assets/Research-projects/NSHM/NSHM-Presentation-slides.pdf). Note: Map lines delineate study areas and do not necessarily depict accepted national boundaries.

lower frequencies. The spatial distribution of wave-field intensity is also affected by variations in crustal lithology, other faults and discontinuities, topography, and the direction of movement of the rupture locus, but in general nearby earthquakes are experienced as high-amplitude shaking at a wide range of frequencies (with maximum energy usually associated with the 1–10 Hz range), while more distant earthquakes are felt as lower-amplitude, lower-frequency (but still potentially damaging) shaking.

The quantitative magnitude of an earthquake is determined by the total energy it releases, or 'seismic moment'; this is equal to the integral over time and space of the stress reduction at failure multiplied by the area of crust that failed (usually assumed to be the fault rupture length times the rupture depth). That can then be scaled to yield the moment magnitude M_W for comparison between earthquakes and with other magnitude scales. Moment magnitude is measured on a logarithmic scale so that the radiated energy increases by a factor of 31.6 for a unit increase in M_W. The energy released from a long fault rupture will tend to be greater than that from a short fault rupture, and fault length generally varies more (from less than one to thousands of kilometres) than fault depth and stress drop. Generally, the very largest earthquakes (up to M_W 9.5) are those generated in plate boundary subduction zones, mainly because these zones contain the longest faults with the deepest active slip surfaces. Major active plate boundaries can also generate large non-subduction earthquakes (up to M_W 8+). For present purposes, a 'large' earthquake could be defined as one that releases enough energy to significantly modify the landscape, which is usually an M_W 6 or greater. However, the cumulative effects of a series of smaller earthquakes can also give rise to significant geomorphic consequences.

2.7 Fault rupture types and seismic waves

Depending on the direction of the applied stress and the nature of the crustal structure, fault ruptures can take several forms (Fig. 2.3). Purely lateral shear is called 'strike-slip' faulting, and pure strike-slip faults are aligned vertically or subvertically. In contrast, shear on 'dip-slip' faults is parallel to the dip direction, and these faults tend to dip at more moderate angles (typically 60 degrees or less). Dip-slip faults can be further divided into two types: 'normal' faults are the result of tensile crustal stress, and 'thrust' or 'reverse' faults result from compressive crustal stress. Combinations of strike-slip and dip-slip can occur. Faults may not rupture the ground surface, in which case they are termed 'blind' faults, and may form gentle ridges or scarps.

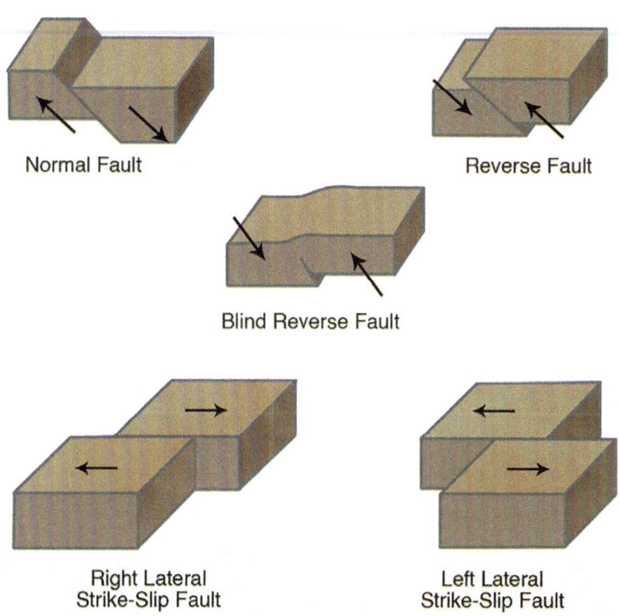

Figure 2.3 Types of fault.

The local displacement of the crustal rock that comprises the travelling fault rupture generates a variety of body waves that propagate through the adjacent crust; the major components of these waves, that can travel long distances and affect the land surface hundreds of kilometres from the fault, are compression waves or P waves, shear waves or S waves, and surface (Rayleigh or Love) waves (Fig. 2.4).

The rough orders of magnitude of amplitude, velocity, frequency and attenuation with distance of these seismic wave types are shown in Table 2.1. In terms of landscape disturbance and geomorphic effects, surface waves with the greatest amplitude and low attenuation with distance are the most significant, but both S and surface waves are in the frequency range that may be able to excite the resonant frequencies of mountain edifices, resulting in the substantial displacements that lead to the development of large-scale (and often deep-seated) failure surfaces. Brain et al. (2017) used geotechnical laboratory testing to simulate earthquake loading of hillslopes and found that different sequences of ground shaking influence hillslope stability prior to, during, and following an earthquake mainshock. Brain et al. (2015) showed in addition that seismic waves can affect the runout dynamics of coseismic landslides.

The impact of an earthquake at a specific site in a landscape determines its potential geomorphic consequences there, and is obviously not solely dependent on the earthquake moment magnitude because of seismic energy dissipation between source and site and the nature of the surface and subsurface material and topography at the site.

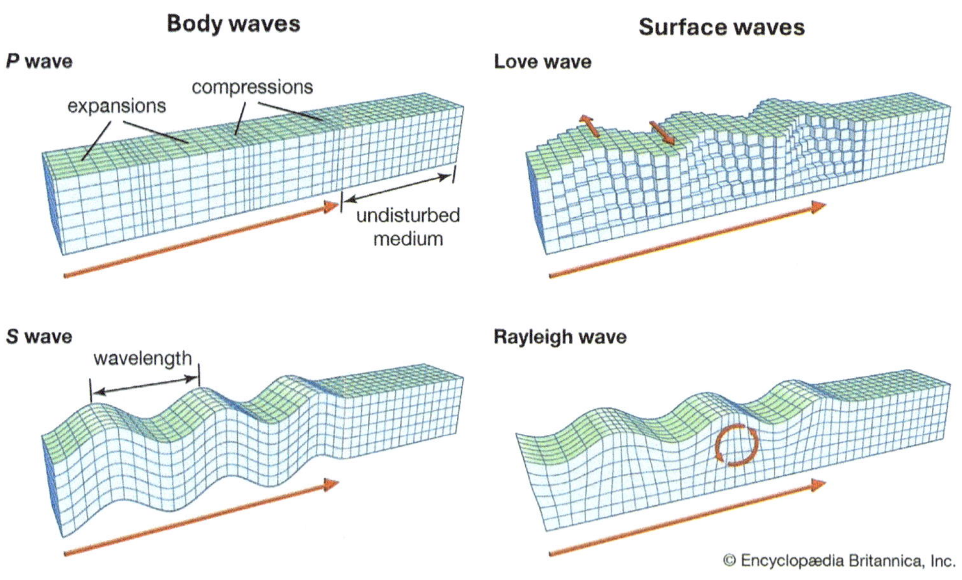

Figure 2.4 Different types of seismic waves. Redrawn from https://open.oregonstate.education/app/uploads/sites/7/2019/06/Page_035_Figure_1-1-1.jpg. Living With Earthquakes In The Pacific Northwest, Subtitle: A Survivor's Guide, by Robert S. Yeats, Fig. 3.10a

Table 2.1 Seismic wave types and characteristics: orders of magnitude.

Wave type	Velocity km/s	Frequency Hz	Amplitude	Attenuation
P	5–8	5–30	mm–cm	Low
S	3–4	1–5	cm–dm	High
Surface	0.5–2.5	0.1–1	dm–m	Low

Figure 2.5 Shattered ridge-top of Little Red Hill, a minor glaciated edifice in the South Island, New Zealand. Buech et al. (2010) reported up to 12-fold amplification of ground acceleration in minor earthquakes at this site. Author's photo.

Local shaking is described by parameters such as the peak ground acceleration (PGA, m/s^2), peak ground velocity (m/s) and Modified Mercalli Intensity, the latter scale being defined in relation to the effects of shaking on infrastructure but also relevant to geomorphic impacts (see Chapter 3). Close to a fault rupture, ground accelerations in an intense earthquake can exceed 1 g (~10 m/s^2), which can obviously exert exceptionally large forces on land surface materials and cause potentially significant impacts such as slips and liquefaction. Ground motions can be amplified or attenuated by local geology, weathering pattern and topography. In particular, the topography of ridge-tops can amplify ground accelerations several-fold (Buech et al., 2010; Faccioli, 1991) compared with flat ground nearby; one geomorphic consequence of the latter effect is the presence of shattered bedrock on ridge-tops (e.g. Fig. 2.5).

Earthquakes can also cause permanent long-term displacement of the land surface by the relative movement that takes place across the fault. The surface displacement pattern can be quite complex, as shown by the 1983 M_w 6.9 Borah Peak earthquake in the western USA (Stein & Barrientos, 1985). While in an individual earthquake displacement is at most of the order of metres vertically and perhaps a few tens of metres horizontally, the displacements that accumulate over the course of many fault ruptures can cause horizontal displacements of hundreds of kilometres and vertical displacements of kilometres. For example, the transpressive plate-boundary Alpine fault in South Island, New Zealand, has accumulated a total horizontal offset of about 450 km (Wellman, 1955) in its 25-million-year history; rock in the Southern Alps to its east has experienced total uplift of about 20 km (Kamp et al., 1989), but ongoing erosion has limited the total relief to about 3 km. Each individual earthquake on the Alpine fault can cause up to about 8 m of horizontal and 2–3 m of vertical displacement, and these are unequivocally seismically-generated; however, the longer-term displacements, which are the result of many earthquakes, are also tectonic in origin.

Many faults and fault segments move more or less continuously by creeping, rather than suddenly in large earthquakes, so offset land features along a fault do not necessarily reflect the occurrence of earthquakes. However, creeping faults may in fact generate large numbers of very small earthquakes, and some faults may exhibit both creep and large earthquakes: '… well-recorded creeping fault earthquakes of up to magnitude 6.6 that have occurred in shallow continental regions produce similar fault-surface rupture areas and similar peak ground shaking as their locked fault counterparts of the same earthquake magnitude' (Harris, 2017; see also Chen & Bürgmann, 2017).

It is clear that earthquakes will tend to occur frequently in locations where tectonic stress buildup is rapid. Thus in identifying the geomorphic consequences of earthquakes, there is a need to distinguish the effects of gradual tectonic stress buildup from those of the instantaneous release of tectonic strain energy upon fault rupture. For example, increasing tectonic stress, evidenced by increasing frequency of minor tremors, may cause an increase in rockfall and landslide frequencies (McSaveney et al., 2015), and local land surface elevations may alter due to a gradual crustal response to tectonic stresses.

2.8 Earthquake frequency of occurrence

There are two common representations of the temporal occurrence of earthquakes. Across a seismically active region containing many faults, there is strong evidence that, like other geophysical phenomena such as landslides, the magnitude-frequency relationship that describes the occurrence of earthquakes is well-represented by a negative power-law. Fig. 2.6 shows this relationship for global earthquakes in 2022, and Fig. 2.1 shows the spatial distribution of these earthquakes. This is the well-known

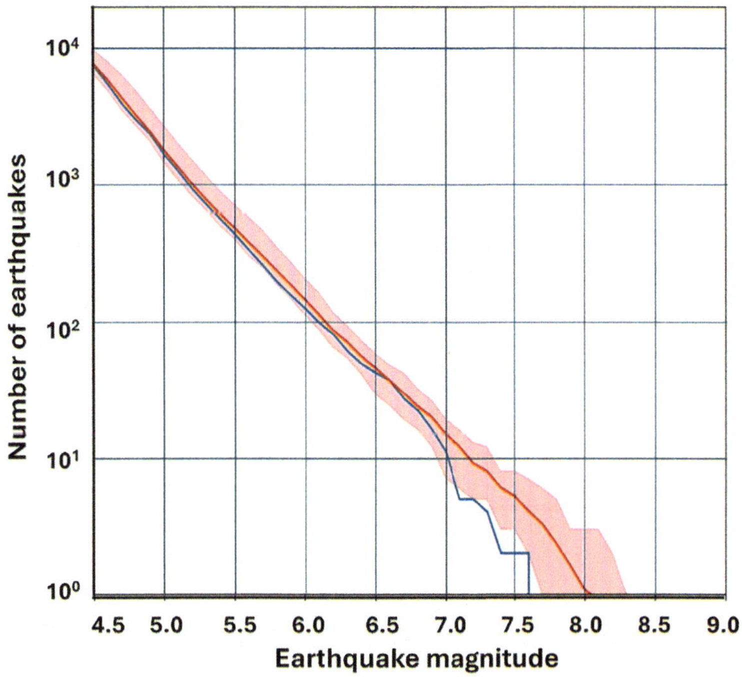

Figure 2.6 Global earthquake magnitude-frequency distribution (blue line) for 2022, along with the mean (red line) and minimum–maximum range (light red band) of all years between 2011 and 2021 (Iacoletti, 2023). *Redrawn from Iacoletti, G. (2023). An unpredictable peril; why earthquake should always be on the risk radar. https://axaxl.com/fast-fast-forward/an-unpredictable-peril-why-earthquake-should-always-be-on-the-risk-radar.*

Gutenberg-Richter law (Gutenberg and Richter, 1955). On average, there is about one $M_W > 8$ event somewhere in the world every year, and about 12 earthquakes between M_W 7 and 8. There is no well-defined physical explanation for this relationship, aside from the intuitive idea that in the nature of things, small events occur more frequently than large ones. Recent suggestions that such behaviour is based on the emergent scaling and cascading behaviour of complex dynamic systems (e.g. Turcotte & Malamud, 2002; Yakovlev et al., 2005), however, align with the trends outlined in Chapter 5 herein and appear to be a promising area of investigation. This statistical relationship, while it displays about an order of magnitude of scatter, may be useful in understanding regional seismically-generated geomorphology because a geomorphic feature is often not the result of a single earthquake, but rather the cumulative result of many past earthquakes of a variety of magnitudes, possibly sourced on many different faults.

On a specific fault, or a segment of a long fault, earthquakes occur when sufficient tectonic strain has accumulated since the previous earthquake that the stress on the fault is sufficient to cause a rupture. Assuming that the rate of change of tectonic stress is steady, this process would result in earthquakes with similar magnitudes occurring at well-defined intervals or return periods, which is known as 'characteristic' behaviour. Such behaviour has been demonstrated on a number of plate-boundary faults (e.g. the Alpine fault in New Zealand; Berryman et al., 2012), and although average return intervals can be well-defined, any specific interseismic interval can vary considerably (Howarth et al., 2021). This concept implies that the annual probability of occurrence and the magnitude of the next rupture are related to the time since the last rupture, leading to emphasis on the identification of 'seismic gaps' (Jackson & Kagan, 2021) as locations where, in spite of unusually low recorded earthquake occurrence, seismic hazard may in fact be locally higher than elsewhere. For example, Fig. 2.7 shows the distribution of historic seismicity in the Himalaya Range and the current slip potential that has accumulated since the most recent event (Bilham, 2019). Characteristic behaviour is relevant for geomorphology in that it anticipates the repeated (but not necessarily periodic) occurrence of both a particular magnitude of shaking in the rock mass and a magnitude of permanent displacement of the surface. Thus a series of earthquakes that produce no obvious geomorphic outcome can precondition a hillslope, increasing the probability that it will fail in a subsequent earthquake (Dadson et al., 2004; Marc et al., 2019; Parker et al., 2015).

Li et al. (2024) provided an interesting commentary on these two different processes in the context of the so-called 'dragon king' theory, which conceptualises the

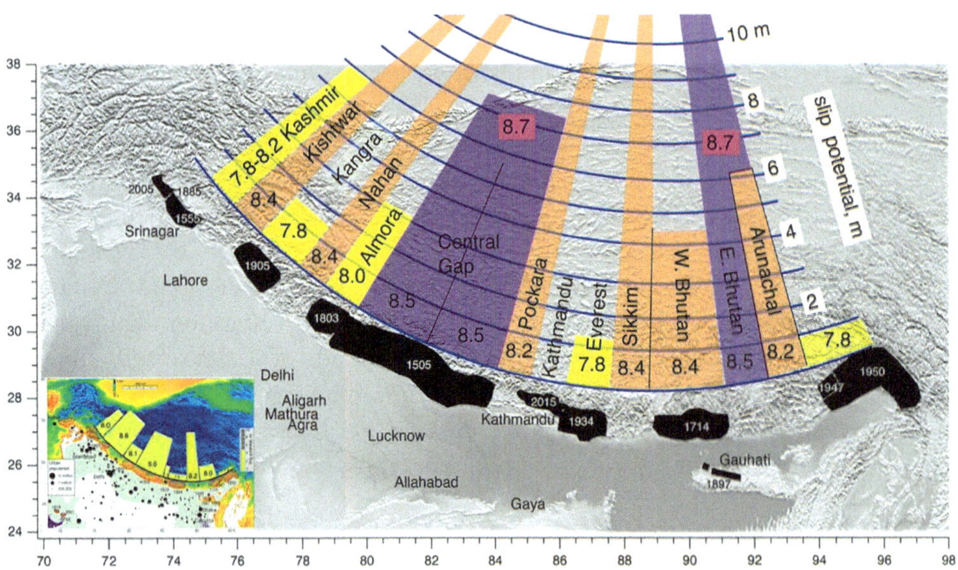

Figure 2.7 Historical earthquakes and accumulated slip along the Himalaya range (Bilham, 2019).

occurrence of unexpectedly large events that appear to exceed statistical predictions of system behaviour.

2.9 Aftershocks

All major earthquakes are followed by a series of aftershocks caused by the sudden change of crustal stress on the fault, which is then transmitted rapidly through the volume of rock around the rupture patch, altering the stress distribution over larger areas and causing other faults (or parts of the same fault) to rupture. Stresses are also transmitted through crustal rocks by changes in groundwater pressures; these can cause distant stresses to change more slowly, because groundwater diffuses via rock cracks that allow pressure changes to be transmitted relatively slowly (Mohr et al., 2012; Ge and Garven, 1992).

While all aftershock sequences are different in their spatiotemporal distribution and evolution, even for similar-magnitude earthquakes, there are some general tendencies. Large earthquakes can have hundreds to thousands of instrumentally-detectable aftershocks, which decrease in magnitude and frequency over time according to a consistent pattern: Båth's law (Båth, 1965) states that the largest aftershock is about 1.1–1.2 units of magnitude less energetic than the mainshock, and Omori's law (Omori, 1894) describes the decay of aftershock frequency with time, with the frequency of aftershocks decreasing roughly with the reciprocal of time after the mainshock. Typically, aftershock epicentres are found up to a distance equal to the mainshock rupture length away from the fault plane, as demonstrated, for example, in the 1992 M_W 7.3 Landers earthquake in southern California (Liu et al., 2003).

The spatial distribution of aftershocks also varies widely between earthquakes, based on the sense of motion on faults and the nature of the surrounding rock masses. An example is the temporal and spatial distribution of aftershocks following the M_W 7.1 Greendale earthquake in New Zealand in September 2010 (Kaiser et al., 2012); it is notable that the aftershocks following each of the three major 'main' shocks were clustered but showed distinctly different spatial distributions, probably indicating the locations of stresses imposed by the preceding fault rupture (e.g. Steacy et al., 2014). The full aftershock sequence from the Greendale earthquake involved over 10,000 events and lasted over a year.

2.10 Conclusions

This chapter has outlined the main features of earthquake occurrence that can affect the geomorphology of an area. At a given location, the potential geomorphic significance of earthquakes can be estimated from probabilistic seismic hazard models, allowing for the fact that seismicity is likely to vary with time. The occurrence of an earthquake can cause metre-scale horizontal and/or vertical local ground displacements due to fault movement, as well as short-term high-intensity shaking of all frequencies. The ground displacements will reduce to zero within at most tens of kilometres from the fault, while the shaking will more gradually become lower in frequency and intensity with increasing distance from the fault; perceptible ground motion can be felt several hundred of kilometres from a major earthquake epicentre. Complex three-dimensional ground shaking affects the properties and behaviour of near-surface materials, and ground displacement alters local stress distributions. While large earthquakes have the most obvious influence on geomorphology, sequences of smaller events impose repeated impacts that can cumulatively precondition a landscape for a major response to a relatively minor event. Many of these matters are explored further in the rest of this volume.

References

Ballantyne, C. K., Sandeman, G. F., Stone, J. O., & Wilson, P. (2014a). Rock-slope failure following Late Pleistocene deglaciation on tectonically stable mountainous terrain. *Quaternary Science Reviews, 86*, 144–157.

Ballantyne, C. K., Wilson, P., Gheorghiu, D., & Rodés, À. (2014b). Enhanced rock-slope failure following ice-sheet deglaciation: Timing and causes. *Earth Surface Processes and Landforms, 39*(7), 900–913.

Båth, M (1965). Lateral inhomogeneities of the upper mantle. *Tectonophysics, 2*(6), 483–514.

Berryman, K. R., Cochran, U. A., Clark, K. J., Biasi, G. P., Langridge, R. M., & Villamor, P. (2012). Major earthquakes occur regularly on an isolated plate boundary fault. *Science (New York, N.Y.), 336*(6089), 1690–1693.

Bilham, R. (2019). Himalayan earthquakes: A review of historical seismicity and early 21st century slip potential. *Geological Society, London, Special Publications, 483*(1), 423–482.

Brain, M. J., Rosser, N. J., Sutton, J., Snelling, K., Tunstall, N., & Petley, D. N. (2015). The effects of normal and shear stress wave phasing on coseismic landslide displacement. *Journal of Geophysical Research: Earth Surface, 120*(6), 1009–1022.

Brain, M. J., Rosser, N. J., & Tunstall, N. (2017). The control of earthquake sequences on hillslope stability. *Geophysical Research Letters, 44*(2), 865–872.

Buech, F., Davies, T. R. H., & Pettinga, J. R. (2010). The little red hill seismic experimental study: Topographic effects on ground motion at a bedrock-dominated mountain edifice. *Bulletin of the Seismological Society of America, 100*(5A), 2219–2229.

Chen, K. H., & Bürgmann, R. (2017). Creeping faults: Good news, bad news? *Reviews of Geophysics, 55*(2), 282–286.

Dadson, S. J., Hovius, N., Chen, H., Dade, W. B., Lin, J. C., Hsu, M. L., Lin, C. W., Horng, M. J., Chen, T. C., Milliman, J., & Stark, C. P. (2004). Earthquake-triggered increase in sediment delivery from an active mountain belt. *Geology, 32*(8), 733–736.

Darnault, R., Rolland, Y., Braucher, R., Bourlès, D., Revel, M., Sanchez, G., & Bouissou, S. (2012). Timing of the last deglaciation revealed by receding glaciers at the Alpine-scale: Impact on mountain geomorphology. *Quaternary Science Reviews, 31*, 127–142.

Davies, T. R. H., McSaveney, M. J., & Boulton, C. J. (2012). Elastic strain energy release from fragmenting grains: Effects on fault rupture. *Journal of Structural Geology, 38*, 265–277.

Davies, T. R. H., McSaveney, M. J., & Reznichenko, N. V. (2019). What happens to fracture energy in brittle fracture? Revisiting the Griffith assumption. *Solid Earth, 10*(4), 1385–1395.

Davies, T. R. H., Reznichenko, N. V., & McSaveney, M. J. (2020). Energy budget for a rock avalanche: Fate of fracture-surface energy. *Landslides, 17*(1), 3–13.

Dor, O., Ben-Zion, Y., Rockwell, T. K., & Brune, J. (2006). Pulverized rocks in the Mojave section of the San Andreas Fault Zone. *Earth and Planetary Science Letters, 245*(3-4), 642–654.

Eldholm, O., & Bungum, H. (2021). Potential secondary events caused by early Holocene paleoearthquakes in Fennoscandia – A climate-related review. *Norwegian Journal of Geology 101, 202108*. https://doi.org/10.17850/njg101-2-3.

England, P., & Jackson, J. (2011). Uncharted seismic risk. *Nature Geosciences, 4*, 348–349. https://doi.org/10.1038/ngeo1168.

Faccioli, E. (1991). Seismic amplification in the presence of geological and topographic irregularities. In *Second international conference on recent advances in geotechnical earthquake engineering & soil dynamics* (Paper No. SOA7, pp. 1779–1797). St. Louis, Missouri.

Ge, S. and Garven, G (1992). *Hydromechanical modeling of tectonically driven groundwater flow with application to the Arkoma foreland basin. Journal of Geophysical Research: Solid Earth, 97*(B6), 9119–9144.

Gerstenberger, M. C., Marzocchi, W., Allen, T., Pagani, M., Adams, J., Danciu, L., Field, E. H., Fujiwara, H., Luco, N., Ma, K. F., & Meletti, C. (2020). Probabilistic seismic hazard analysis at regional and national scales: State of the art and future challenges. *Reviews of Geophysics, 58*(2), e2019RG000653.

Guccione, M. J., Van Arsdale, R. B., & Hehr, L. H. (2000). Origin and age of the Manila high and associated Big Lake "sunklands" in the New Madrid seismic zone, northeastern Arkansas. *Geological Society of America Bulletin, 112*(4), 579–590.

Gutenberg, B. and Richter, C.F (1955). Magnitude and energy of earthquakes. *Nature, 176*(4486), 795.

Harris, R. A. (2017). Large earthquakes and creeping faults. *Reviews of Geophysics, 55*(1), 169–198.

Howarth, J. D., Barth, N. C., Fitzsimons, S. J., Richards-Dinger, K., Clark, K. J., Biasi, G. P., Cochran, U. A., Langridge, R. M., Berryman, K. R., & Sutherland, R. (2021). Spatiotemporal clustering of great earthquakes on a transform fault controlled by geometry. *Nature Geoscience, 14*(5), 314–320.

Iacoletti, G. (2023). An unpredictable peril; why earthquake should always be on the risk radar. https//:axaxl.com/fast-fast-forward/an-unpredictable-peril-why-earthquake-should-always-be-on-the-risk-radar.

Jackson, D. D., & Kagan, Y. Y. (2021). Characteristic earthquakes and seismic gaps. In *Encyclopedia of solid earth geophysics*. Cham: Springer International Publishing, 53–56.

Kamp, P. J. J., Green, P. F., & White, S. H. (1989). Fission track analysis reveals character of collisional tectonics in New Zealand. *Tectonics, 8*, 169–195.

Kaiser, A., Holden, C., Beavan, J., Beetham, D., Benites, R., Celentano, A., Collett, D., Cousins, J., Cubrinovski, M., Dellow, G., & Denys, P. (2012). The M_w 6.2 Christchurch earthquake of February 2011: Preliminary report. *New Zealand Journal of Geology and Geophysics, 55*(1), 67–90.

Karlstrom, K. E., Baisan, C. H., Kring, D. A., Hereford, R., Turney, C., Hogg, A., Norman, L. M., O'Brien, P., Palmer, J. G., Rittenour, T. M., & Ballensky, J. (2025). Grand Canyon landslide-dam and paleolake triggered by the Meteor Crater impact at 56 ka. *Geology*.

Li, J., Sornette, D., Wu, Z., & Li, H. (2024). New horizon in the statistical physics of earthquakes: Dragon-king theory and dragon-king earthquakes. arXiv:2408.10857.

Liu, J., Sieh, K., & Hauksson, E. (2003). A structural interpretation of the aftershock "cloud" of the 1992 M w 7.3 Landers earthquake. *Bulletin of the Seismological Society of America, 93*(3), 1333–1344.

Marc, O., Behling, R., Andermann, C., Turowski, J. M., Illien, L., Roessner, S., & Hovius, N. (2019). Long-term erosion of the Nepal Himalayas by bedrock landsliding: The role of monsoons, earthquakes and giant landslides. *Earth Surface Dynamics*, 7(1), 107–128.

Massey, C. I., Townsend, D. T., Lukovic, B., Morgenstern, R., Jones, K., Rosser, B., & de Vilder, S. (2020). Landslides triggered by the MW 7.8 14 November 2016 Kaikōura earthquake: An update. *Landslides, 17*, 2401–2408.

McColl, S. T., Davies, T. R. H., & McSaveney, M. J. (2012). The effect of glaciation on the intensity of seismic ground motion. *Earth Surface Processes and Landforms, 37*(12), 1290–1301.

McEwan, E., Stahl, T., Howell, A., Langridge, R., & Wilson, M. (2023). Coseismic river avulsion on surface rupturing faults: Assessing earthquake-induced flood hazard. *Science Advances, 9*(18), eadd2932.

McEwan, E., Stahl, T., Langridge, R., Davies, T. R. H., Howell, A., & Wilson, M. (2025). Seismic hazard and shifting channels: Exploring coseismic river response. *Earth-Science Reviews, p*105042.

McSaveney, M., Cox, S., & Hancox, G. (2015). Increasing rock-avalanche frequency correlates with increasing seismic moment release in New Zealand's Southern Alps. *In EGU General Assembly Conference Abstracts,* 8074.

Mohr, C. H., Montgomery, D. R., Huber, A., Bronstert, A., & Iroumé, A. (2012). Streamflow response in small upland catchments in the Chilean coastal range to the MW 8.8 Maule earthquake on 27 February 2010. *Journal of Geophysical Research: Earth Surface, 117*(F2).

Nicol, A., Van Dissen, R. J., Stirling, M. W., & Gerstenberger, M. C. (2016). *Seismological Research Letters, 87*(6), 1299–1310. https://doi.org/10.1785/0220160088.

Omori, F (1894). *Journal of the College of Science, Imperial University, Tokyo, Japan, Vol.* 7, 111–200.

Parker, R. N., Hancox, G. T., Petley, D. N., Massey, C. I., Densmore, A. L., & Rosser, N. J. (2015). Spatial distributions of earthquake-induced landslides and hillslope preconditioning in the northwest South Island, New Zealand. *Earth Surface Dynamics, 3*(4), 501–525.

Reznichenko, N.V., Davies, T.R.H. and Alexander, D.J (2011). Effects of rock avalanches on glacier behaviour and moraine formation. *Geomorphology, 132*(3-4), 327–338.

Silva, V., Amo-Oduro, D., Calderon, A., Costa, C., Dabbeek, J., Despotaki, V., Martins, L., Pagani, M., Rao, A., Simionato, M., & Viganò, D. (2020). Development of a global seismic risk model. *Earthquake Spectra, 36*(1_suppl), 372–394.

Steacy, S., Jiménez, A., & Holden, C. (2014). Stress triggering and the Canterbury earthquake sequence. *Geophysical Journal International, 196,* 473–480. https://doi.org/10.1093/gji/ggt380.

Stein, R. S., & Barrientos, S. E. (1985). Planar high-angle faulting in the Basin and Range: Geodetic analysis of the 1983 Borah Peak, Idaho, earthquake. *Journal of Geophysical Research: Solid Earth, 90*(B13), 11355–11366.

Turcotte, D. L., & Malamud, B. D. (2002). Earthquakes as a complex system. In *In* International geophysics, *81,* Academic Press, 209 cp1).

Wellman, H. W. (1955). New Zealand quaternary tectonics. *Geologische Rundschau, 43*(1), 248–257.

Wilson, B., Dewers, T., Reches, Z. E., & Brune, J. (2005). Particle size and energetics of gouge from earthquake rupture zones. *Nature, 434*(7034), 749–752.

Yakovlev, G., Newman, W. I., Turcotte, D. L., & Gabrielov, A. (2005). An inverse cascade model for self-organized complexity and natural hazards. *Geophysical Journal International, 163*(2), 433–442.

Zoback, M. D., & Grollimund, B. (2001). Impact of deglaciation on present-day intraplate seismicity in eastern North America and western Europe. *Comptes Rendus de l'Académie des Sciences-Series IIA-Earth and Planetary Science, 333*(1), 23–33.

CHAPTER 3

Coseismic ground motion

Benjamin Edwards and Farnaz Kamranzad
School of Environmental Sciences, University of Liverpool, Liverpool, United Kingdom

3.1 Introduction

Elastic strain energy, released suddenly during an earthquake, is transmitted through the Earth's crust and near-surface in the form of seismic waves. The arrival and passage of these waves at the surface manifests as coseismic ground motion, which represents the primary mechanism by which earthquake energy is delivered to the geomorphic systems of the landscape (e.g. Scholz, 2019). This ground motion is the critical link between tectonic fault rupture and the often-dramatic modifications this imparts on surface and near-surface topography and processes (Robinson & Davies, 2013). The resulting coseismic ground shaking and displacement are, therefore, the direct initiators and drivers for the spectrum of geomorphic phenomena discussed throughout the remainder of this book.

Coseismic ground motion can be broadly categorised into two distinct but related forms, both of which are fundamental mechanisms of geomorphic change. The first is dynamic ground motion, or transient shaking, which causes oscillating stresses that can cyclically weaken or fracture rock and soil masses, leading to widespread landslides, rockfall and liquefaction (Hovius & Meunier, 2012; Keefer, 1984). The second is static ground displacement, which represents the permanent, inelastic deformation of the ground surface, including surface rupture and regional uplift or subsidence (Bürgmann et al., 2000). This permanent displacement directly alters the form of the landscape, for example, by creating fault scarps, changing river gradients and even vertically shifting coastal tracts relative to sea level. The interplay of these dynamic and static motions initiates complex cascades of Earth surface processes that can reshape landscapes for decades or more following an event (Fan et al., 2019).

This chapter provides the essential physical background required to understand the character and variability of seismic wave propagation and coseismic ground motion. This background should allow the reader to understand the link between tectonic fault rupture and the dynamic inputs on the Earth's near-surface as a driver of subsequent geomorphic consequences, the focus of later chapters. We begin by reviewing the fundamental principles of seismic wave propagation, examining the different wave types generated by an earthquake. Subsequently, we discuss the methods used to quantify ground motion, from historical, qualitative intensity scales to modern instrumental

Geomorphic Impacts of Earthquakes
DOI: https://doi.org/10.1016/B978-0-443-40375-0.00011-6

parameters that measure acceleration, velocity and permanent displacement. The chapter concludes by exploring the principal controls on the nature of ground motion at a given location, including the effects of the earthquake source, the travel path of the seismic waves and modification by local site conditions.

3.2 Seismic wave propagation

The seismic energy released during an earthquake originates at the hypocentre and is radiated progressively as the rupture propagates across the fault plane, generating a complex three-dimensional wavefield. For the purposes of understanding the resultant ground motions at and near the surface, it is convenient to simplify this wavefield into distinct types of seismic waves, each with characteristic particle motions and velocities. These waves travel from the earthquake source to a site on the surface via various routes, or ray paths, through the Earth's interior and along the surface. The angle at which these ray paths arrive at the surface is a key control on the polarisation of the resulting ground motion (i.e. the balance between vertical and horizontal shaking components). A simplified illustration of body-wave ray paths from a crustal earthquake source is shown in Fig. 3.1.

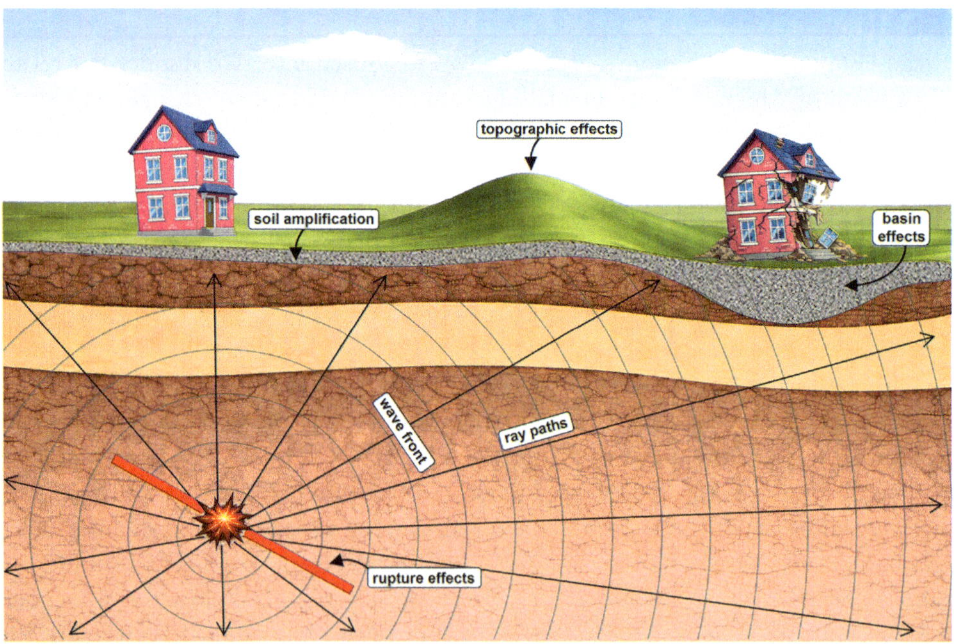

Figure 3.1 A simplified cross-section showing ray paths from a shallow crustal earthquake source to the surface, highlighting source rupture mechanisms alongside topographic and near-surface site effects. The layers are shown schematically, with colour used to illustrate the layered structure of the Earth.

The principal wave types are divided into two main groups: body waves, which travel through the Earth's interior, and surface waves, which are trapped near the free surface.

3.2.1 Body waves

Body waves (Aki & Richards, 2002; Shearer, 2019; Stein & Wysession, 2009) propagate through the volume of the Earth and are the first to arrive at a given location. There are two fundamental types of body waves:

P-waves (primary waves) are compressional (sometimes referred to as longitudinal) waves in which the particle motion is parallel to the direction of wave propagation. Analogous to sound waves, they can travel through solids, liquids and gases. P-waves have the highest velocity of all seismic waves and are therefore the first to be recorded on a seismogram.

S-waves (secondary waves) are shear, or transverse, waves where the particle motion is perpendicular to the direction of wave propagation. S-waves can only propagate through solid materials, as liquids and gases cannot transmit shear stresses, and they travel more slowly than P-waves (Fig. 3.2).

For earthquakes occurring at some distance from a recording site, the ray paths of body waves travel through the deeper, higher-velocity crust before being refracted upwards as they enter the shallower, lower-velocity crust. This reduction in seismic velocity towards the surface causes the wave trajectory to become progressively steeper, resulting in a near-vertical angle of incidence at the surface. Consequently, the arriving P-wave energy produces particle motion that is predominantly in the vertical direction, while the S-wave energy results in predominantly horizontal ground motion.

The dominant horizontal motion associated with S-waves is typically more damaging to the built environment (Fig. 3.3) and a more significant driver of slope failure than the motion from P-waves, particularly for shallow failures and in far-field conditions. However, in near-fault settings, on steep slopes or for deep-seated failures, strong vertical ground motion associated with P-waves can also play an important contributory role.

3.2.2 Surface waves

Surface waves (Aki & Richards, 2002; Shearer, 2019; Stein & Wysession, 2009) are generated by the constructive interference of body waves that are reflected and refracted near the Earth's surface and in the shallow crust. Their energy is confined to the near-surface, and consequently, they attenuate more slowly with distance than body waves. This can result in surface waves having the largest amplitudes and longest durations in a seismogram, particularly at significant distances from the earthquake epicentre.

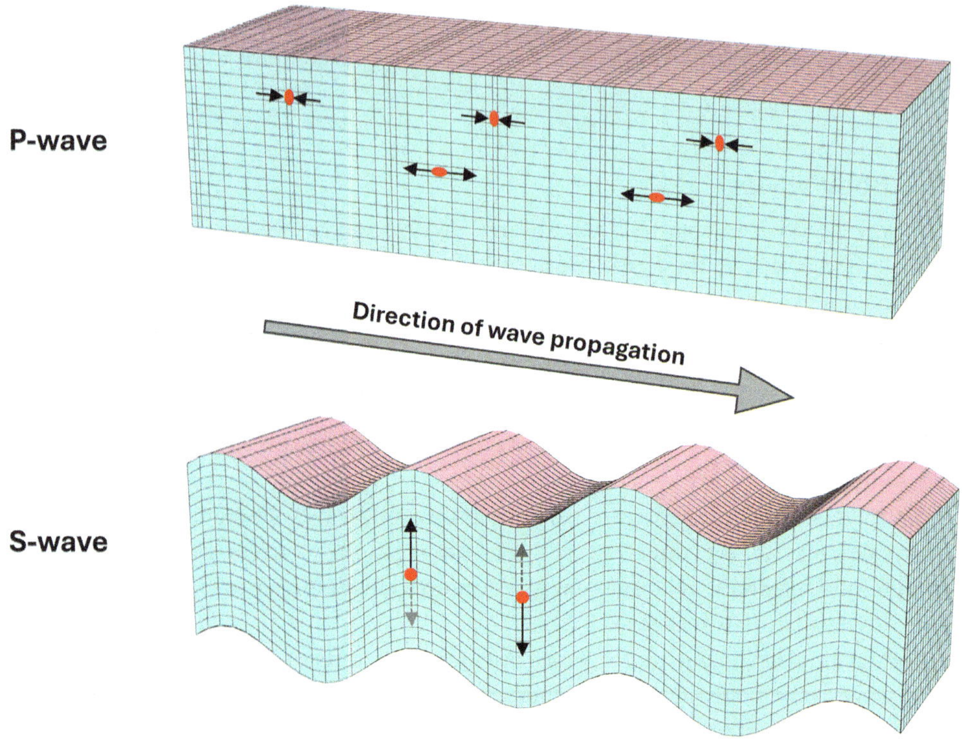

Figure 3.2 An illustration of the particle motion of P-waves (compressional) and S-waves (shear, here shown polarised vertically, denoted SV) relative to their direction of propagation (Note: S-waves can oscillate in any direction orthogonal to the direction of propagation).

Surface waves manifest as two types: Love and Rayleigh waves. Love waves consist of horizontally polarised shear (SH) motion. Their particle motion is entirely horizontal and perpendicular to the direction of propagation. They are a result of S-waves being trapped in a low-velocity layer near the surface, such as the crust. The large-amplitude transverse motion associated with Love waves makes them particularly destructive for structures and potentially significant in triggering ground failure.

At the surface, Rayleigh waves exhibit a retrograde elliptical particle motion in the vertical plane, similar to the motion of waves on the surface of water. They involve both vertical and horizontal components of motion and are generated by the interactions of incident and reflected P- and SV- (vertically polarised shear) waves at the free surface (Fig. 3.4). The long-period high-displacement motion of Rayleigh waves can be particularly effective at disturbing near-surface materials and contributing to processes such as liquefaction (Jiang et al., 2022) and settlement. Interaction of surface waves and body waves, for example at basin-edges, can lead to strong amplification of ground shaking (Kawase, 1996; see also Chapter 4).

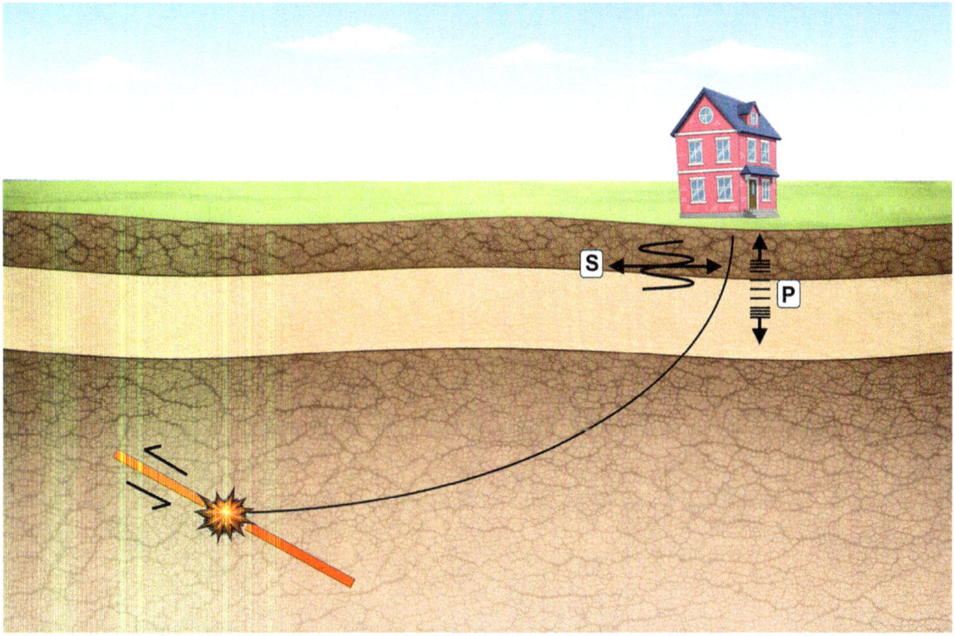

Figure 3.3 A diagram illustrating the predominant particle motion at the ground surface resulting from near-vertical incidence of P- and S-waves. The red line shows a cross-section of the fault rupture plane, with arrows indicating the sense of relative motion during the earthquake.

3.2.3 Illustrative examples

The destructive potential of surface waves at large distances was tragically demonstrated by the 1985 M_w 8.0 Michoacán earthquake in Mexico. The epicentre was located over 350 km from Mexico City, yet the capital suffered catastrophic damage. This was because long-period surface waves generated by the earthquake travelled into the basin where Mexico City is located, which is underlain by soft, water-saturated lakebed sediments. These sediments dramatically amplified the ground motion, particularly at a period of around 2–5 seconds, corresponding to the fundamental resonant period of the basin (Fäh et al., 1994). The correspondence of this basin-amplified shaking with the natural resonant frequency of many mid-rise buildings contributed to their collapse (Seed et al., 1988).

While horizontal shaking from S-waves and surface waves is responsible for the majority of earthquake damage, strong vertical motion associated with P-waves can be a major contributing factor in the near-source region or at basin edges (where waves may undergo conversion from high-amplitude SV to P). The 1994 M_w 6.7 Northridge, California, and 2011 M_w 6.2 Christchurch, New Zealand, earthquakes both generated exceptionally strong vertical accelerations (Bradley & Cubrinovski, 2011;

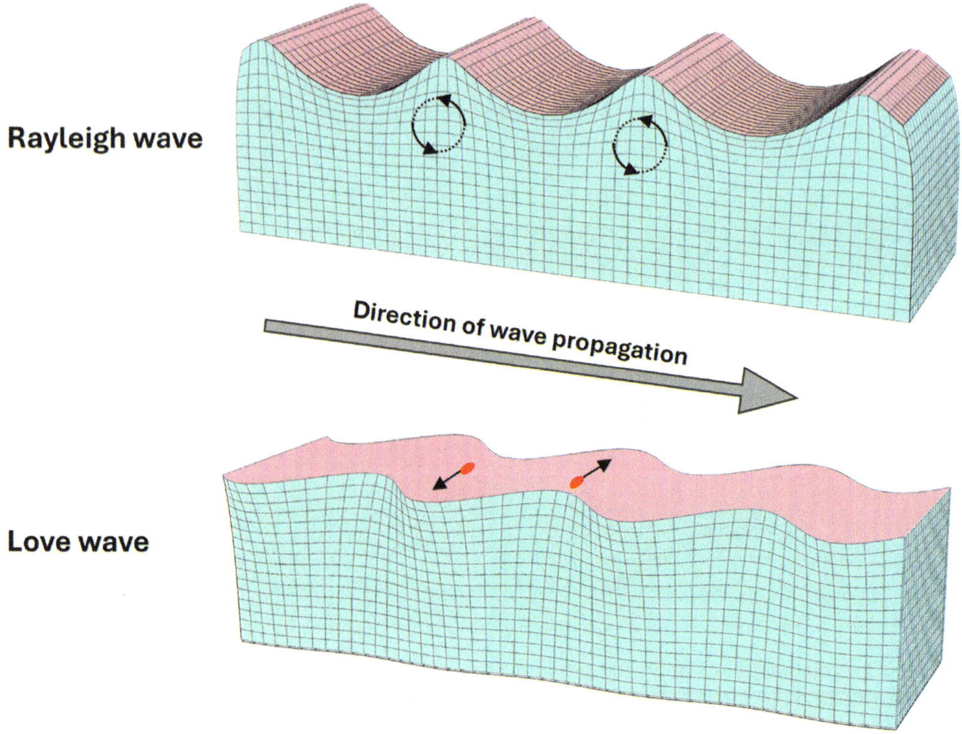

Figure 3.4 An illustration of the particle motion of Rayleigh waves (retrograde elliptical at the surface) and Love waves (horizontal transverse) relative to their direction of propagation.

Stewart et al., 1996) in some cases exceeding the acceleration of gravity. Such high vertical accelerations are typically confined to areas close to the fault rupture and may often exceed horizontal accelerations, particularly in the case of soil non-linearity (i.e. when the stress–strain relationship becomes non-linear under strong shaking) and liquefaction (Bradley & Cubrinovski, 2011; see Chapter 8).

3.3 Quantification of ground motion

The ground motion at the Earth's surface during an earthquake is a complex, three-dimensional time history of shaking characterised by ground surface displacement, velocity and acceleration (Fig. 3.5). While a full time history provides a complete description of the shaking, it is often simplified into a set of scalar or vector parameters for engineering design, hazard assessment and analysis of geomorphic response. This section reviews the primary methods for quantifying ground motion, from qualitative scales based on observed effects to quantitative measures derived from instrumental recordings.

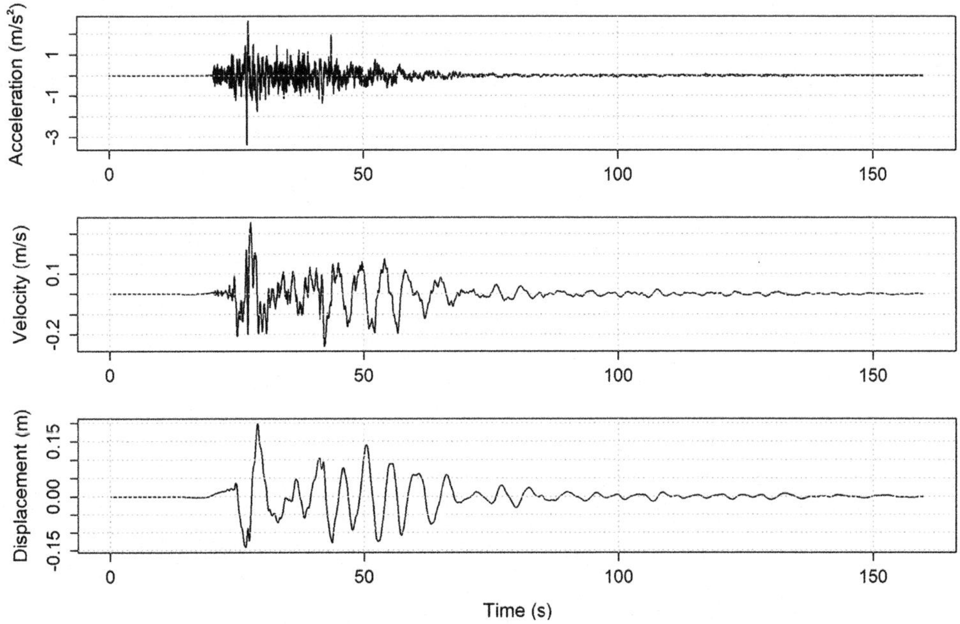

Figure 3.5 An example of a strong-motion recording, showing the time-histories of acceleration, velocity and displacement for a single component of motion, recorded for the 1999 M_W 7.6 Chi-Chi earthquake in Taiwan (see Chapter 13).

3.3.1 Intensity

Before the advent of modern seismographs, the only way to quantify an earthquake was to observe its effects. This led to the development of seismic intensity scales, which provide a qualitative, location-dependent measure of the strength of ground shaking. These scales classify the severity of shaking based on observed damage to structures, effects on people and impacts on the natural environment. These scales are still used today, providing complementary information to digital instrumentation. Several scales have been developed for use in different regions, most notably the Modified Mercalli Intensity scale, widely used in the United States, and the European Macroseismic Scale (EMS-98) (Grünthal, 1998), which is an update of the earlier Medvedev–Sponheuer–-Karnik scale used in Europe and elsewhere. These scales typically use twelve levels of intensity, denoted by Roman numerals (e.g. I for not felt, to XII for catastrophic destruction), with key effects summarised in Table 3.1.

While subjective, intensity data are invaluable for studying historical earthquakes (and related secondary effects) that predate instrumental records and for communicating earthquake impacts to the public. In the modern era, the collection of observational data has been revolutionised by internet-based systems. The U.S. Geological Survey's 'Did You Feel It?' programme, for example, uses online questionnaires to rapidly gather

Table 3.1 A summary of seismic intensity scales Modified Mercalli Intensity (USGS (United States Geological Survey), 2026) and EMS-98 (Swiss Seismological Service (SED), 2026), showing selected intensity levels and their corresponding descriptions of shaking, human perception and typical damage to buildings of varying vulnerability.

Intensity Level	MMI	EMS-98
I	Not felt except by a very few under especially favorable conditions.	Not felt
II	Felt only by a few persons at rest, especially on upper floors of buildings.	Felt indoors by a few people. People at rest feel a swaying or light trembling.
III	Felt quite noticeably by persons indoors, especially on upper floors of buildings. Many people do not recognize it as an earthquake. Standing motor cars may rock slightly. Vibrations similar to the passing of a truck.	
IV	Felt indoors by many, outdoors by few during the day. At night, some awakened. Dishes, windows, doors disturbed; walls make cracking sound. Sensation like heavy truck striking building. Standing motor cars rocked noticeably.	Felt indoors by many people, outdoors by very few. A few people are awakened. Windows, doors and dishes rattle.
V	Felt by nearly everyone; many awakened. Some dishes, windows broken. Unstable objects overturned. Pendulum clocks may stop.	Felt indoors by most, outdoors by few. Many sleeping people awake. A few are frightened. Buildings tremble throughout. Hanging objects swing considerably. Small objects are shifted. Doors and windows swing open or shut.
VI	Felt by all, many frightened. Some heavy furniture moved; a few instances of fallen plaster. Damage slight.	Many people are frightened and run outdoors. **Some objects fall. Many houses suffer slight** non-structural damage like hair-line cracks and fall of small pieces of plaster.
VII	Damage negligible in buildings of good design and construction; slight to moderate in well-built ordinary structures; considerable damage in poorly built or badly designed structures; some chimneys broken.	Most people are frightened and run outdoors. Furniture is shifted and objects fall from shelves in large numbers. Many well built **ordinary buildings suffer moderate damage:** small cracks in walls, fall of plaster, parts of chimneys fall down; older buildings may show large cracks in walls and failure of fill-in walls.
VIII	Damage slight in specially designed structures; considerable damage in ordinary substantial buildings with partial collapse. Damage great in poorly built structures. Fall of chimneys, factory stacks, columns, monuments, walls. Heavy furniture overturned.	**Many people find it difficult to stand. Many** houses have large cracks in walls. A few well built ordinary buildings show serious failure of walls, while weak older structures may collapse.
IX	Damage considerable in specially designed structures; well-designed frame structures thrown out of plumb. Damage great in substantial buildings, with partial **collapse. Buildings shifted off** foundations.	General panic. Many weak constructions collapse. Even well built ordinary buildings show very heavy damage: serious failure of walls and partial structural failure.
X / X+	Some well-built wooden structures destroyed; most masonry and frame structures destroyed with foundations. Rails bent.	Most ordinary well built buildings collapse, even some with good earthquake resistant design are destroyed.

MMI modified from USGS (United States Geological Survey). Retrieved February 2026, from https://www.usgs.gov/programs/earthquake-hazards/modified-mercalli-intensity-scale. EMS-98 modified from Swiss Seismological Service (SED) at ETH Zurich. Retrieved February 2026, from Swiss Seismological Service website.

shaking and damage reports from the public immediately following an earthquake (Wald, Quitoriano, Dengler, et al., 1999). This 'citizen science' approach allows for the collection of a vast number of data points, leading to more detailed and robust community intensity maps. These observational data can be integrated with instrumental ground motion data (Wald, Quitoriano, Heaton, et al., 1999) to produce rapid estimates of the intensity distribution, known as 'ShakeMaps'.

3.3.2 Instrumental parameters

Modern seismic networks use highly sensitive instruments to record ground motion in three orthogonal directions (e.g. north-south, east-west and vertical). Seismometers are designed to measure ground velocity, while accelerometers (or 'strong-motion sensors') directly measure ground acceleration. Instruments can be deployed at the surface, in boreholes or even on the seafloor to capture a complete record of the shaking. Of course, displacement, velocity and acceleration can all be determined from any seismic record (whether from a seismometer or accelerometer) through differentiation or integration. However, recording natively in a given unit has many advantages: for example, seismometer records can record very low-frequency motion, while accelerometer records tend not to 'clip' (i.e. the recorded signal does not saturate during strong shaking) even at very high levels of ground shaking (Edwards et al., 2025). For these reasons, it is not unusual to have multi-instrument sites within a seismic network.

From the resulting time-histories (typically from the horizontal components, but also sometimes from the vertical), key parameters are derived to characterise the ground motion (Baker et al., 2021):

- Peak ground acceleration (PGA) is the maximum acceleration recorded during an earthquake at a site. It is often expressed as a fraction of the acceleration of gravity (g). PGA is a widely used parameter in engineering and is often correlated with damage to short, stiff structures and with the initiation of some types of landslides, including rock falls and debris slides (e.g. Keefer, 1984; see Chapter 7).
- Peak ground velocity (PGV) is the maximum absolute velocity recorded, typically reported in units of cm/s or m/s. PGV is often better correlated with damage to more flexible, medium-to-high-rise structures than PGA. It is also considered a robust indicator of the potential for liquefaction and landslide triggering (Keefer, 1984).
- Pseudo-spectral acceleration (PSA) describes the maximum acceleration experienced by a simple, single-degree-of-freedom oscillator with a specific natural period and damping, when subjected to the recorded ground motion. By calculating PSA for a range of periods, a response spectrum is generated, which provides a powerful tool for assessing the likely response of different structures and landscape elements to shaking.

Other parameters are also used to characterise the duration and energy content of shaking, such as Arias intensity (a measure of the total energy of the ground motion) and cumulative absolute velocity, but the peak and spectral parameters remain the most common in practice (Douglas, 2003).

3.3.3 Static displacement (permanent ground deformation)

Distinct from the transient, (mostly) elastic deformation associated with the passage of seismic waves, static displacement is the permanent, inelastic shift of the ground surface that remains after an earthquake. This deformation is a direct expression of the slip on the underlying fault. It can be measured using a variety of geodetic techniques:

- Global Navigation Satellite Systems, such as GPS and Galileo, use networks of permanent ground stations to measure their positions with millimetre accuracy. By comparing pre- and post -earthquake positions, the full three-dimensional static displacement field can be mapped.
- Interferometric Synthetic Aperture Radar is a satellite-based remote sensing technique that can map ground deformation over wide areas with centimetre-level precision. By comparing radar images taken before and after an earthquake, an interferogram is created that reveals the pattern of ground displacement along the satellite's line-of-sight (Elliott et al., 2016).

Static displacement can also be derived from seismological data, for example, by performing a double integration of strong-motion accelerograph records. This provides a crucial link between the high-frequency dynamic shaking and the final permanent deformation at a single point. The 2016 M_w 7.8 Kaikōura, New Zealand, earthquake provides a good example of complex static displacement, with multiple faults rupturing to produce intricate patterns of horizontal and vertical ground motion, including over 8 m of coastal uplift in some locations (Hamling et al., 2017).

3.4 Controls on ground motion

The character of ground motion experienced at any point on the Earth's surface is the complex product of three main contributing factors: the kinematics of the earthquake source, the modification of the seismic waves as they propagate along the path from the source to the site and the influence of local geological and topographical conditions at the site (see Chapter 4). This source–path–site framework is fundamental to seismology and engineering seismology. Understanding these factors is critical for explaining the spatial patterns of coseismic landscape response, such as the distribution of landslides or the variability of liquefaction.

In practice, the physical processes governing strong ground motion are highly complex and cannot be described in full detail for every earthquake. Instead, ground-motion models (GMMs) adopt a simplified but physically motivated representation in which the combined effects of the source, wave propagation and site conditions are expressed through a limited number of key variables (Douglas & Edwards, 2016). These models aim to predict a measure of ground-motion intensity at a given site while retaining a link to the underlying physics. Most empirical and semiempirical GMMs – sometimes referred to as attenuation models or ground-motion prediction equations (GMPEs) – can therefore be written in a general compact logarithmic form as:

$$\ln Y = f(M) + f(R) + f(F) + f(S) + f(B) + \varepsilon \tag{3.1}$$

where:

- Y =a ground-motion intensity measure, such as PGA, PGV or PSA at a given period;
- M =the earthquake magnitude (usually the moment magnitude, M_w);
- R =source-to-site distance metric, capturing the effects of geometric spreading and attenuation as seismic waves propagate along the path from the source to the site;
- F =faulting mechanism and source geometry, accounting for differences between normal, reverse and strike-slip earthquakes, as well as rupture progression along the fault plane;
- S =local site conditions, including near-surface geology classification, shear-wave velocity (e.g. characterised by the upper 30 m travel-time-average shear-wave velocity, V_{S30}), and local topography, all of which strongly influence site amplification;
- B =basin or sedimentary depth effects, which can further modify ground motion, particularly at longer periods;
- ε =aleatory variability, representing random fluctuations due to unmodelled effects.

This additive logarithmic formulation shows that ground motion intensity is not controlled by a single parameter, but rather by the combined influence of earthquake size, wave propagation effects and local site response. As such, it provides a clear and intuitive framework for interpreting observed ground-motion variability and forms the basis of modern GMPEs used in seismic hazard analysis.

3.4.1 Source effects

The physical process of fault rupture is the origin of all subsequent ground motion, and consequently, the nature of the rupture – its size, geometry, and dynamics – exerts a primary control on the resulting wavefield. The most basic of these controls is the earthquake's size, often quantified by the moment magnitude (M_w). This scale is directly related to the seismic moment (M_0), a physical measure of the 'work done' in rupturing the fault, and is closely related to the energy radiated. M_0 is calculated as the product of the fault rupture area, the average slip, and the shear modulus (strength) of the rock. Both amplitude (Boore et al., 2014) and duration (Kempton & Stewart, 2006) of ground shaking generally increase with magnitude; for every unit increase in moment magnitude, there is a ~32-fold increase in energy release, leading to significantly stronger and longer-lasting ground motion. An important characteristic of amplitude-duration scaling is that as earthquake magnitude increases beyond around M_w ~7, the increase of peak amplitude slows, with increased energy release instead expressed as longer shaking durations (Fig. 3.6). These durations are often quantified using the significant duration, defined as the time interval over which the central portion of the shaking energy is released, typically between 5% and 95% of the cumulative energy.

In addition to earthquake magnitude, the kinematics of the faulting process are also critical. The style of faulting (normal, reverse or strike-slip) influences the radiation pattern of seismic energy, with the amplitude of waves varying systematically with direction from the fault. Furthermore, the stress drop on the fault affects the high-frequency content of the ground motion. Higher stress drop events release more stored elastic strain energy and tend to generate stronger shaking for a given magnitude.

The dynamics of the rupture itself also impart a strong signature on the ground motion. For example, the ground motion at a site is strongly influenced by the direction

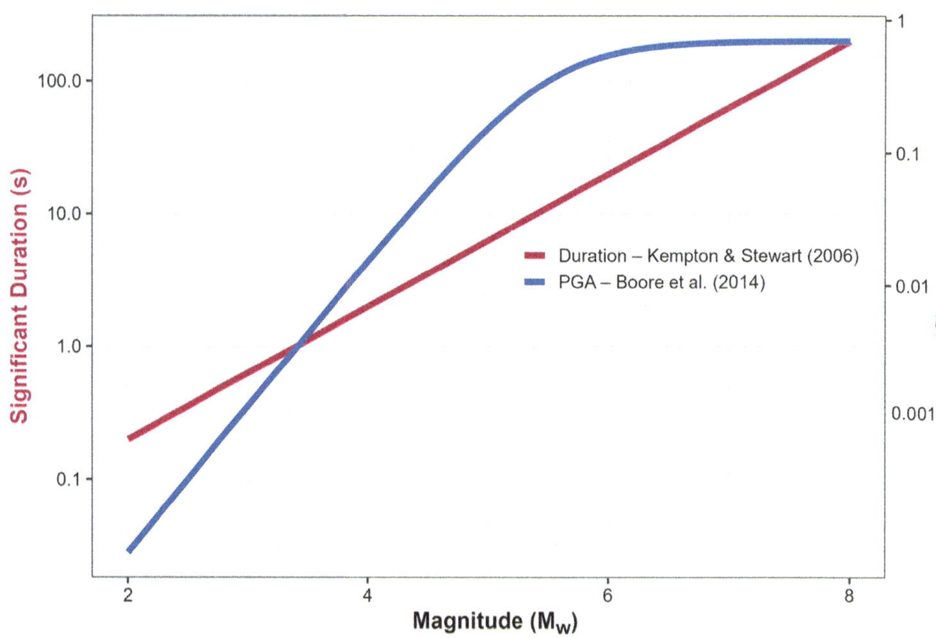

Figure 3.6 Plot showing the scaling of peak ground acceleration (PGA) and significant shaking duration with moment magnitude (Boore et al., 2014; Kempton & Stewart, 2006), illustrating how duration becomes dominant for large-magnitude events.

of fault rupture propagation relative to the site. If the rupture propagates towards a site, seismic energy arrives in a compressed timeframe, resulting in a larger-amplitude, shorter-duration packet of shaking. This phenomenon, known as forward directivity, can produce exceptionally high peak ground velocities and is a major driver of damage in the near-fault region (Bayless et al., 2025; Somerville et al., 1997). Conversely, sites located in the direction opposite to rupture propagation experience lower amplitude, but longer-duration shaking, which may be more impactful in some situations - such as liquefaction triggering.

The three-dimensional geometry of the fault is also important. For dipping faults, sites located on the hanging wall (the block above an inclined fault plane) often experience stronger shaking than sites at an equivalent distance on the footwall block below the fault (Abrahamson & Somerville, 1996; Chang et al., 2004). This is primarily because the hanging wall is located closer to a larger area of the energy-radiating fault plane, an effect that is particularly pronounced for shallow-dipping reverse faults. Given these complexities, particularly for large-magnitude earthquakes where the fault rupture can extend for tens to hundreds of kilometres, simple distance measures such as epicentral distance become inadequate. Instead, distance metrics that account for the finite size of the fault (along with models for hanging-wall effects) are required. An example of this is rupture distance (R_{rup}), defined as the closest distance from a site to any part of the fault rupture plane. Another commonly used metric is the Joyner–Boore distance (R_{JB}), defined as the closest horizontal distance from a site to the surface projection of the rupture plane (Fig. 3.7).

3.4.2 Path effects

As seismic waves propagate away from the source, their characteristics are modified by the properties of the crust through which they travel (Boore, 2003). This modification is controlled by two primary mechanisms. The first, geometric spreading, describes how energy is spread over an increasingly larger area as waves radiate from the source. This is a primary reason for the decay of wave amplitude with distance, with body waves (spreading spherically) decaying faster than surface waves (spreading cylindrically). The second mechanism is attenuation, where seismic energy is lost through both scattering off heterogeneities in the crust and anelastic absorption (material damping), where vibrational energy is converted into heat. The efficiency of this attenuation is frequency-dependent (high-frequency waves attenuate more rapidly) and varies significantly between different tectonic regions. For example, the old, cold and stable crust of central and eastern North America is less attenuating than the younger, hotter and more fractured crust of active regions such as California; the former therefore allows seismic waves to travel much farther with damaging amplitude (Atkinson & Boore, 2006).

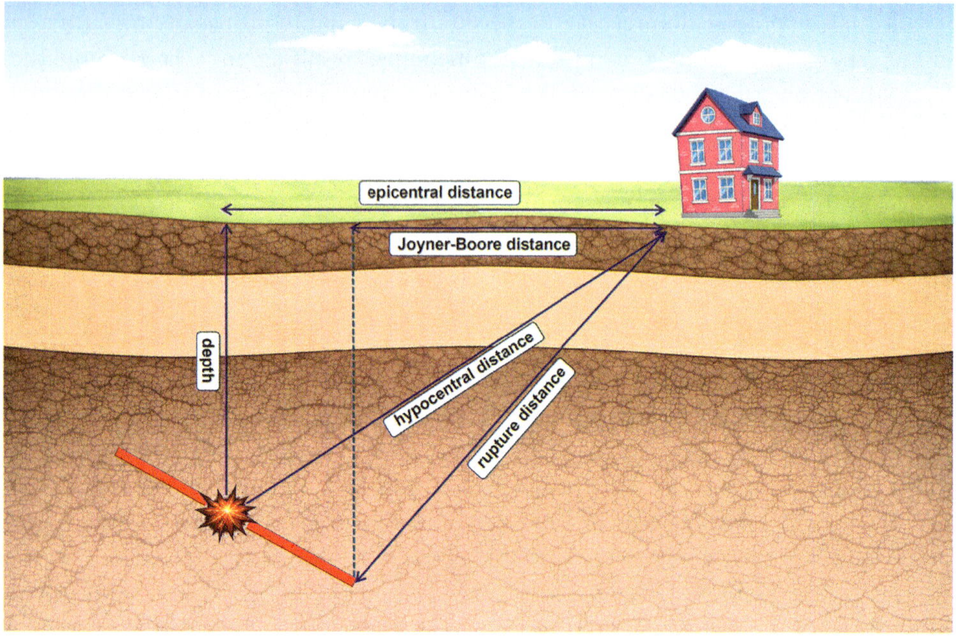

Figure 3.7 A schematic diagram illustrating different source-to-site distance metrics for an extended fault rupture, including epicentral distance, hypocentral distance, rupture distance (R_{rup}), and Joyner–Boore distance (R_{JB}), shown for a finite fault plane.

Regional variability can be significant even over smaller scales where major geological provinces meet (Kotha et al., 2020). For instance, the complex geology of Switzerland, which transitions from the crystalline rocks of the Alps to the deep sedimentary basins of the foreland, results in distinctly different attenuation characteristics between the two regions. This necessitates the use of separate GMMs for the Alpine and foreland areas to accurately predict the decay of shaking with distance (Edwards & Fäh, 2013).

3.4.3 Local site effects

The final, and often most dramatic, modification to the ground motion occurs in the uppermost tens to hundreds of metres of the Earth's surface. When seismic waves travel from a harder, denser material (e.g. intact bedrock) into a softer, less dense material (e.g. alluvial sediments), their propagation velocity decreases. To conserve energy, the amplitude of the waves must therefore increase. This phenomenon controlled by the acoustic impedance contrast between the materials, can lead to significant amplification of ground motion on soil sites compared to nearby rock sites (Abercrombie, 1997; Edwards & Michel, Poggi, et al., 2013).

The amplification at a site is described by a site amplification function, which quantifies the ratio of surface ground motion to that on a reference rock site (or buried horizon) across a range of frequencies. For a simple layer of soft soil with thickness H and average shear-wave velocity V_s overlying bedrock, this function has a characteristic peak at the site's fundamental frequency, $f_0 \approx V_s/4H$, corresponding to the natural resonant period of the soil column (Kramer, 2024). In practice, site conditions are complex, and empirical proxies are needed to characterise site effects, for example, in GMPEs. The most widely used proxy is V_{s30}, the time-averaged shear-wave velocity in the upper 30 m of the ground. While a simplification, it has proven to be a robust parameter for predicting average seismic site amplification (Boore et al., 2014; see Chapter 4).

Although soils typically amplify low-to-moderate levels of shaking, they can behave non-linearly during strong shaking. The high strains induced in the soil can cause a reduction in its stiffness and an increase in its damping, which can de-amplify the peak accelerations but potentially amplify motions at longer periods.

The surface topography itself can also focus or defocus seismic energy. Ridges and hilltops often experience amplified shaking compared to valleys, as seismic waves become trapped and constructively interfere near the convex landform (Buech et al., 2010; Geli et al., 1988; Hough et al., 2010). This effect is a critical factor in the spatial distribution of coseismic landslides, which are often concentrated on ridges (see Chapter 7), and is particularly significant where seismic frequencies match natural frequencies, giving rise to resonant response and correspondingly enhanced ground motion intensity.

3.5 Conclusions

Coseismic ground motion constitutes the mechanism through which elastic strain energy released during fault rupture is transmitted through the crust and manifested at the Earth's surface. It therefore provides the essential physical link between tectonic processes at depth and geomorphic response at the surface. Two components are fundamental: (1) dynamic ground motion imposes transient stresses that may exceed material strength and initiate slope failure, liquefaction and deformation of near-surface materials; (2) static ground displacement represents permanent inelastic deformation associated directly with fault slip, expressed as surface rupture, uplift or subsidence, and results in enduring modification of topography.

The characteristics of this motion are governed by seismic wave propagation. Body waves and surface waves differ in particle motion, velocity and attenuation behaviour. At local distances, near-vertical incidence of refracted body waves results in predominantly horizontal ground motion from S-waves and strong vertical components from P-waves. At regional distances, dispersive surface waves, particularly Love and Rayleigh waves, often dominate the amplitude and duration of shaking. Source properties, including rupture geometry, stress drop and rupture directivity, further control the amplitude, frequency content and duration of motion, especially within the near-field of the source.

Quantification of ground motion requires both observational (macroseismic) and instrumental approaches. Intensity scales describe the severity of shaking at different locations based on observed effects. Instrumental parameters such as PGA, PGV, response spectra and duration measures provide simplified but physically meaningful representations of complex time-histories. Static displacement derived from geodetic and remote sensing techniques defines the permanent deformation field associated with the earthquake source.

Ground motion at any site reflects the combined influence of earthquake source characteristics, wave propagation and path effects, and local geological and topographic conditions. This source–path–site framework provides a physically coherent basis for interpreting the spatial variability of shaking and deformation. The concepts outlined in this chapter establish the necessary foundation for understanding the geomorphic processes triggered by earthquakes. Ongoing research aims to address limitations of GMMs, particularly in terms of their reliance on ergodic assumptions in their derivation, which can increase model uncertainty or lead to biases. GMMs derived in a nonergodic framework (Lavrentiadis et al., 2023), with intensity modelled as source–path–site specific, rather than magnitude-distance-Vs30, must overcome a lack of data. Current research aims to bridge the gap between physical understanding of earthquake processes and simulations thereof, with advanced statistical-empirical approaches.

References

Abercrombie, R. E. (1997). Near-surface attenuation and site effects from comparison of surface and deep borehole recordings. *Bulletin of the Seismological Society of America, 87*(3), 731–744.

Abrahamson, N. A., & Somerville, P. G. (1996). Effects of the hanging wall and footwall on ground motions recorded during the Northridge earthquake. *Bulletin of the Seismological Society of America, 86*(1B), S93–S99.

Aki, K., & Richards, P. G. (2002). *Quantitative seismology*. MIT Press.

Atkinson, G. M., & Boore, D. M. (2006). Earthquake ground-motion prediction equations for eastern North America. *Bulletin of the Seismological Society of America, 96*(6), 2181–2205.

Baker, J., Bradley, B., & Stafford, P. (2021). *Seismic hazard and risk analysis*. Cambridge University Press.

Bayless, J., Abrahamson, N. A., & Somerville, P. G. (2025). A rupture directivity adjustment model and its application in seismic hazard. *Earthquake Spectra, 41*(1), 753–781.

Boore, D. M. (2003). Simulation of ground motion using the stochastic method. *Pure and Applied Geophysics, 160*(3), 635–676.

Boore, D. M., Stewart, J. P., Seyhan, E., & Atkinson, G. M. (2014). NGA-West2 equations for predicting PGA, PGV, and 5% damped PSA for shallow crustal earthquakes. *Earthquake Spectra, 30*(3), 1057–1085.

Bradley, B. A., & Cubrinovski, M. (2011). Near-source strong ground motions observed in the 22 February 2011 Christchurch earthquake. *Seismological Research Letters, 82*(6), 853–865.

Buech, F., Davies, T. R., & Pettinga, J. R. (2010). The little red hill seismic experimental study: Topographic effects on ground motion at a bedrock-dominated mountain edifice. *Bulletin of the Seismological Society of America, 100*(5A), 2219–2229.

Bürgmann, R., Rosen, P. A., & Fielding, E. J. (2000). Synthetic aperture radar interferometry to measure Earth's surface topography and its deformation. *Annual Review of Earth and Planetary Sciences, 28*(1), 169–209.

Chang, T. Y., Cotton, F., Tsai, Y. B., & Angelier, J. (2004). Quantification of hanging-wall effects on ground motion: Some insights from the 1999 Chi-Chi earthquake. *Bulletin of the Seismological Society of America, 94*(6), 2186–2197.

Douglas, J. (2003). Earthquake ground motion estimation using strong-motion records: A review of equations for the estimation of peak ground acceleration and response spectral ordinates. *Earth-Science Reviews, 61*(1-2), 43–104.

Douglas, J., & Edwards, B. (2016). Recent and future developments in earthquake ground motion estimation. *Earth-Science Reviews, 160*, 203–219.

Edwards, B., & Fäh, D. (2013). A stochastic ground-motion model for Switzerland. *Bulletin of the Seismological Society of America, 103*(1), 78–98.

Edwards, B., Michel, C., Poggi, V., & Fäh, D. (2013). Determination of site amplification from regional seismicity: Application to the Swiss National Seismic Networks. *Seismological Research Letters, 84*(4), 611–621.

Edwards, B., Ntinalexis, M., Vanneste, K., & van Elk, J. (2025). Reconstruction of clipped time-history records, with application to ground motions from the 1992 ML 5.8 roermond, netherlands, earthquake. *Seismological Research Letters*. https://doi.org/10.1785/0220250139.

Elliott, J. R., Walters, R. J., & Wright, T. J. (2016). The role of space-based observation in understanding and responding to active tectonics and earthquakes. *Nature Communications*, 7(1), 13844.

Fäh, D., Suhadolc, P., & Panza, G. F. (1994). A hybrid method for the estimation of ground motion in sedimentary basins: Quantitative modeling for Mexico City. *Bulletin of the Seismological Society of America, 84*(2), 383–399.

Fan, X., Scaringi, G., Korup, O., West, A. J., Van Westen, C. J., Tanyas, H., Hovius, N., Hales, T. C., Jibson, R. W., Allstadt, K. E., & Zhang, L. (2019). Earthquake-induced chains of geologic hazards: Patterns, mechanisms, and impacts. *Reviews of Geophysics, 57*(2), 421–503.

Geli, L., Bard, P. Y., & Jullien, B. (1988). The effect of topography on earthquake ground motion: A review and new results. *Bulletin of the Seismological Society of America, 78*(1), 42–63.

Grünthal (1998). *European macroseismic scale (1998)*. Luxembourg City, Luxembourg: European Centre for Geodynamics and Seismology, Cahiers du Centre Européen de Geodynamique et de Seismologie, 15.

Hamling, I. J., Hreinsdóttir, S., Clark, K., Elliott, J., Liang, C., Fielding, E., Litchfield, N., Villamor, P., Wallace, L., Wright, T. J., & D'Anastasio, E. (2017). Complex multifault rupture during the 2016 M_w 7.8 Kaikōura earthquake, New Zealand. *Science, 356*(6334) p.eaam7194.

Hough, S. E., Altidor, J. R., Anglade, D., Given, D., Janvier, M. G., Maharrey, J. Z., Meremonte, M., Mildor, B. S. L., Prepetit, C., & Yong, A. (2010). Localized damage caused by topographic amplification during the 2010 M 7.0 Haiti earthquake. *Nature Geoscience, 3*(11), 778–782.

Hovius, N., & Meunier, P. (2012). Earthquake ground motion and patterns of seismically induced landsliding. *Landslides*, 24–36 edited by: Clague, J. J. and Stead, D.

Jiang, M., Kamura, A., & Kazama, M. (2022). Numerical study on liquefaction characteristics of granular materials under Rayleigh-wave strain conditions using 3D DEM. *Soils and Foundations, 62*(4), 101176.

Kawase, H. (1996). The cause of the damage belt in Kobe:"The basin-edge effect," constructive interference of the direct S-wave with the basin-induced diffracted/Rayleigh waves. *Seismological Research Letters, 67*(5), 25–34.

Keefer, D. K. (1984). Landslides caused by earthquakes. *Geological Society of America Bulletin, 95*(4), 406–421.

Kempton, J. J., & Stewart, J. P. (2006). Prediction equations for significant duration of earthquake ground motions considering site and near-source effects. *Earthquake Spectra, 22*(4), 985–1013.

Kotha, S. R., Weatherill, G., Bindi, D., & Cotton, F. (2020). A regionally-adaptable ground-motion model for shallow crustal earthquakes in Europe. *Bulletin of Earthquake Engineering, 18*(9), 4091–4125.

Kramer, S. L. (2024). *Geotechnical earthquake engineering*. CRC Press.
Lavrentiadis, G., Abrahamson, N. A., Nicolas, K. M., Bozorgnia, Y., Goulet, C. A., Babič, A., Macedo, J., Dolšek, M., Gregor, N., Kottke, A. R., & Lacour, M. (2023). Overview and introduction to development of non-ergodic earthquake ground-motion models. *Bulletin of Earthquake Engineering, 21*(11), 5121–5150.
Robinson, T. R., & Davies, T. R. H. (2013). Potential geomorphic consequences of a future great (M_w=8.0+) Alpine Fault earthquake, South Island, New Zealand. *Natural Hazards and Earth System Sciences, 13*(9), 2279–2299.
Scholz, C. H. (2019). *The mechanics of earthquakes and faulting*. Cambridge University Press.
Seed, H. B., Romo, M. P., Sun, J. I., Jaime, A., & Lysmer, J. (1988). The Mexico earthquake of September 19, 1985—Relationships between soil conditions and earthquake ground motions. *Earthquake Spectra, 4*(4), 687–729.
Shearer, P. M. (2019). *Introduction to seismology*. Cambridge University Press.
Somerville, P. G., Smith, N. F., Graves, R. W., & Abrahamson, N. A. (1997). Modification of empirical strong ground motion attenuation relations to include the amplitude and duration effects of rupture directivity. *Seismological Research Letters, 68*(1), 199–222.
Stein, S., & Wysession, M. (2009). *An introduction to seismology, earthquakes, and earth structure*. John Wiley & Sons.
Stewart, J. P., Seed, R. B., & Bray, J. D. (1996). Incidents of ground failure from the 1994 Northridge earthquake. *Bulletin of the Seismological Society of America, 86*(1B), S300–S318.
Swiss Seismological Service (SED) at ETH Zurich. Retrieved February 2026. https://www.seismo.ethz.ch/en/knowledge/faq/what-does-ems-98-mean.
USGS (United States Geological Survey). Retrieved February 2026. https://www.usgs.gov/programs/earthquake-hazards/modified-mercalli-intensity-scale.
Wald, D. J., Quitoriano, V., Dengler, L. A., & Dewey, J. W. (1999). Utilization of the Internet for rapid community intensity maps. *Seismological Research Letters, 70*(6), 680–697.
Wald, D. J., Quitoriano, V., Heaton, T. H., & Kanamori, H. (1999). Relationships between peak ground acceleration, peak ground velocity, and modified Mercalli intensity in California. *Earthquake Spectra, 15*(3), 557–564.

CHAPTER 4

Ground failure resulting from earthquake ground motions☆

Kate E. Allstadt and Eric M. Thompson
U.S. Geological Survey, Geologic Hazards Science Center, Golden, CO, United States

4.1 Introduction

One of the most immediate geomorphic consequences of an earthquake is ground failure, a term that is inclusive of landslides, liquefaction, cyclic softening and lateral spreading (Fig. 4.1). Coseismic ground failure can manifest as a wide variety of phenomena and consequences over many scales, but all involve permanent ground deformation and share a single cause: the interaction of earthquake shaking with shallow subsurface materials. The objective of this chapter is to describe this interaction.

After briefly introducing the different types and processes of coseismic ground failure, we build on concepts introduced in Chapter 3 by detailing how earthquake ground motions are altered by geological structures and topography near and at the surface, and how surface materials susceptible to ground failure respond to that altered shaking. We then describe different ways of approximating the complex relationship between ground motions and ground failure, with particular emphasis on geospatial modelling of global and regional patterns.

4.1.1 Ground failure types and processes

The term 'ground failure' includes a variety of physical processes (Fig. 4.1). All types of coseismic ground failure can be hazardous. However, the consequences vary dramatically, depending on both the characteristics of failure and proximity to people and the built environment.

4.1.1.1 Ground failure on sloped ground

When ground failure occurs on sloped ground, deformation is predominantly in the downslope direction, and the feature is referred to as a landslide. This is a broad term encompassing phenomena ranging from small rock falls that may impede traffic on a road to the collapse of entire mountainsides, which can fill valleys and block the flow of rivers.

☆ Any use of trade, firm, or product names is for descriptive purposes only and does not imply endorsement by the U.S. Government

Geomorphic Impacts of Earthquakes
DOI: https://doi.org/10.1016/B978-0-443-40375-0.00008-6

Most landslide classification schemes are based on material, style of movement and speed (e.g. Hungr et al., 2014; Varnes, 1978). These general classification schemes are applicable to coseismic landslides, but earthquakes trigger some types more readily than others. Chapter 7 describes these classification schemes and the most common types of coseismic landslides in detail. For the purposes of this chapter, following Keefer (1984),

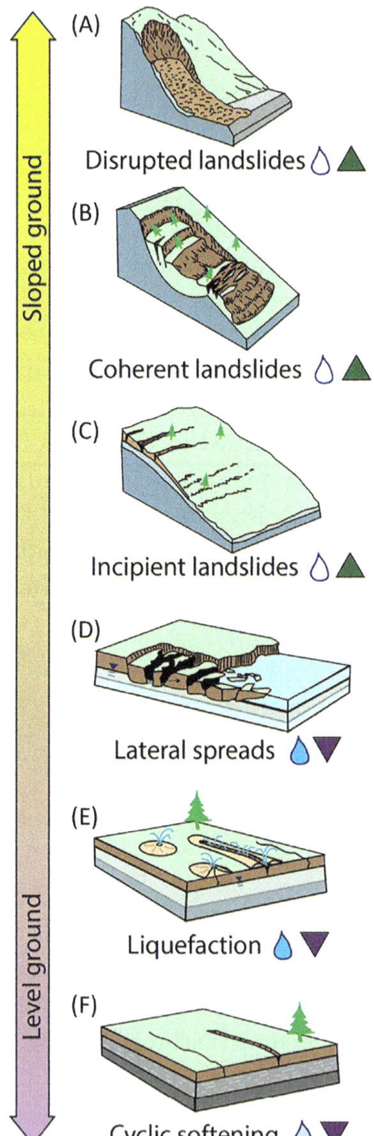

we group coseismic landslides into two major categories: disrupted and coherent (Fig. 4.1A,B). Either type can originate in rock or unconsolidated sediment, and volumes range over many orders of magnitude. In disrupted landslides, during the failure process, the original source volume breaks into many smaller blocks and fragments that move chaotically and reach high speeds (>1 m/s) by falling, sliding, toppling, avalanching, flowing or slumping (Hungr et al., 2014; Keefer, 1984). In contrast, coherent landslides tend to move more slowly (~0.1 mm/s to <5 m/s) and less chaotically, typically over shorter distances and mainly by sliding or slumping (Hungr et al., 2014; Keefer, 1984). They also largely retain their internal structure, though they may break into a few larger blocks, and some disrupted movement may occur on the margins. Coherent slides tend to be relatively thick (>3 to >100 m), whereas disrupted landslides can be both deep (>3 m thick) or shallow (<3 m thick) (Keefer, 1984). Some landslides are complex and may have both coherent and disrupted components.

A hillslope is unstable when the shear stress exceeds the shearing resistance (Kramer & Stewart, 2024). Earthquakes can trigger ground failure on sloped ground both by increasing shear stresses within a hillslope (e.g. inertial force due to ground acceleration) and by reducing the effective shear strength resisting downslope motion (e.g. fracturing, pore-pressure increases) (Varnes, 1978). Increases in shear stress caused by shaking (also known as earthquake loading) are transient and time varying because the ground shaking is temporary and, during the shaking, the loading will not always be imposed in a direction that increases shear stresses (e.g. when the acceleration direction is upslope). Strength reduction, however, can be permanent (e.g. via fracturing) or transient (e.g. via an increase in pore pressure). Unlike many other types of landslides, earthquake-triggered landslides can be entirely dry (Keefer, 1984), though the presence of water can make failure more likely by reducing effective shear strength. Many slopes, especially those composed of ductile materials, can tolerate brief exceedances of shear strength and may require an accumulation of damage or strain over several cycles of ground motion

Figure 4.1 Illustrations of the major categories of coseismic ground failure discussed in this chapter. The solid blue water-drop icon indicates the material must be saturated, light blue indicates that the material is typically saturated but does not have to be and white indicates that the phenomenon can occur under any saturation conditions (though the presence of water may make failure more likely). A green upward-arrow icon indicates that the primary mechanism of failure is typically due to the shear resistance of the slope being exceeded by increased shear stresses from earthquake loading, whereas a purple downward-arrow icon indicates that the primary mechanism is typically soil strength reduction, though both mechanisms may contribute to all types of ground failure. The light beige colour indicates a saturated, liquefiable layer in illustrations (D) and (E), the light grey colour with horizontal line segments indicates a plastic (i.e. clay-rich) layer in illustration (F). Dark grey indicates nonliquefiable layers (Highland, 2004). *Illustrations A-D modified from Highland, L. M. (2004).* Landslide types and processes *(Fact Sheet No. 2004–3072). U.S. Geological Survey. https://doi.org/10.3133/fs20043072.*

to fail substantially (Jibson, 2011; Newmark, 1965). Therefore the duration of ground motions can be equally as important as the amplitude in determining the extent and severity of landslides. Some slopes may begin to deform but do not fully mobilise before the shaking ceases. This phenomenon, known as an incipient landslide, is characterised by ground cracks and minor deformations (Fig. 4.1C), and forms a subcategory of coherent landslides that has been widespread following many major landslide-triggering earthquakes and has important implications for future landslide hazards (e.g. Görüm et al., 2023; Jibson et al., 2018; Petley et al., 2010; Yin et al., 2009).

4.1.1.2 Ground failure on level ground

Ground failure that occurs primarily on level ground is classified by the physical process that causes a loss of soil strength that ultimately results in ground deformation. Liquefaction is the most well known and well studied of these processes (Fig. 4.1E; see also Chapter 8). Liquefaction occurs when shaking increases pore pressure in loose, saturated, noncohesive (i.e. nonplastic) soils. The increase in pore pressure is caused by the soil matrix contracting, thereby transferring interparticle stresses that were previously supported by the soil to the water (Kramer & Stewart, 2024). Sandy soils are most susceptible to this process, although liquefaction is also possible in silty and gravelly soils in some conditions (e.g. Boulanger & Idriss, 2006; Rollins et al., 2021). Although liquefaction rarely occurs in clayey materials due to their cohesion, surface manifestations similar to those caused by liquefaction can occur in clay-rich materials due to cyclic softening. Cyclic softening (Fig. 4.1F) refers to a loss of strength and the resulting deformation due to shaking that occurs in finer-grained moist or saturated soils, which exhibit high cohesion and tend to deform plastically (e.g. Boulanger & Idriss, 2007). A special case of cyclic softening occurs in sensitive clays, also known as quick clays. Quick clays are marine clays leached by freshwater, resulting in a fragile structure that can collapse when strained (e.g. by earthquake shaking). This substantial loss in strength can result in soil that behaves like a viscous fluid (e.g. Crawford, 1968; Updike et al., 1988). Similar to landslides, the duration of the ground motions is an important factor for ground failure on level ground (Kramer & Stewart, 2024).

Vertical settlement generally dominates ground failure on level ground, though horizontal deformations also occur. The severity of the deformations varies substantially, and small deformations may not be easily detected when they occur in locations where structures or infrastructure are not present. All types of ground failure on level ground can result in ground cracking without phenomena-specific manifestations, so identifying the physical process and soil type responsible for the ground failure can be challenging (Idriss & Boulanger, 2008). The presence of sand boils is one discriminating factor (Fig. 4.1E). Sand boils may form during liquefaction as slurries of pressurised water and sediment are ejected to the surface through new or preexisting weaknesses in an overlying nonliquefiable crust, depositing sediment on the surface around a vent (e.g. Idriss & Boulanger, 2008). Sand boils can form from a single vent or along fissures, and the ejection process can continue for tens of minutes after the shaking has ceased and can shoot water and sediment up to several feet in the air (Ambraseys & Sarma, 1969). However, sand boils are not always present and can occasionally occur in the absence of liquefaction (Kramer et al., 2015). If liquefaction, cyclic softening or cyclic degradation of sensitive clay occurs near an open face (e.g. a stream bank) or on a gentle slope, horizontal displacements may become substantial and the feature is then called a lateral spread (Fig. 4.1D). Lateral spreads are also considered the third major category of landslides by Keefer (1984) and are a transition point in the spectrum of ground-failure behaviour between landslides and level-ground processes, as illustrated in Fig. 4.1.

Though liquefaction evidence can sometimes be subtle and transient, deposits of sand at the surface and subsurface dikes and sills of injected sediment can persist in the geologic record and can be used, for example, to characterise past earthquake activity (e.g. Obermeier et al., 2001). More detail about liquefaction and lateral spreading can be found in Chapter 8.

4.2 Interaction of ground motion and the surface

In this section, we detail how the seismic wavefield generated by an earthquake interacts with and is altered near the ground surface in ways that are relevant to ground failure. As detailed in Chapter 3, the characteristics of earthquake shaking that arrive at a site of interest on the surface of the Earth can be generally attributed to three sequential factors: (1) source effects, (2) path effects and (3) site effects (e.g. Boore, 2004; Borcherdt, 1970) (Fig. 4.2). Source effects refer to the characteristics of the wavefield determined by the fault rupture process directly, such as magnitude, faulting mechanism, rupture geometry, propagation direction and slip distribution (see Chapters 2 and 3). The wavefield at the source, initially composed primarily of body waves, propagates through the Earth's structure, which modifies the waves as they spread out, attenuate and interact with subsurface structures. These modifications are referred to as path effects (detailed in Chapter 3).

Site effects can be defined as the final portion of the path effects in the vicinity of the location of interest, where the ground motions are altered by near-surface geological materials (Kramer & Stewart, 2024). Site effects are generally defined as the ratio of the ground-motion amplitude relative to a reference site (Boore, 2004). The reference site can be defined as something observable, such as a site where flat bedrock reaches the surface (e.g. Fig. 4.2B) or something more abstract, such as an empirical prediction of ground motion or a median ground motion for a group of similar sites. For example, Fig. 4.3 shows the ground motions of the 2014 M_W 6.0 Napa earthquake recorded on the Mission Peak landslide in California compared with a reference site nearby on stable ground during a study on landslide site effects by Hartzell et al. (2017).

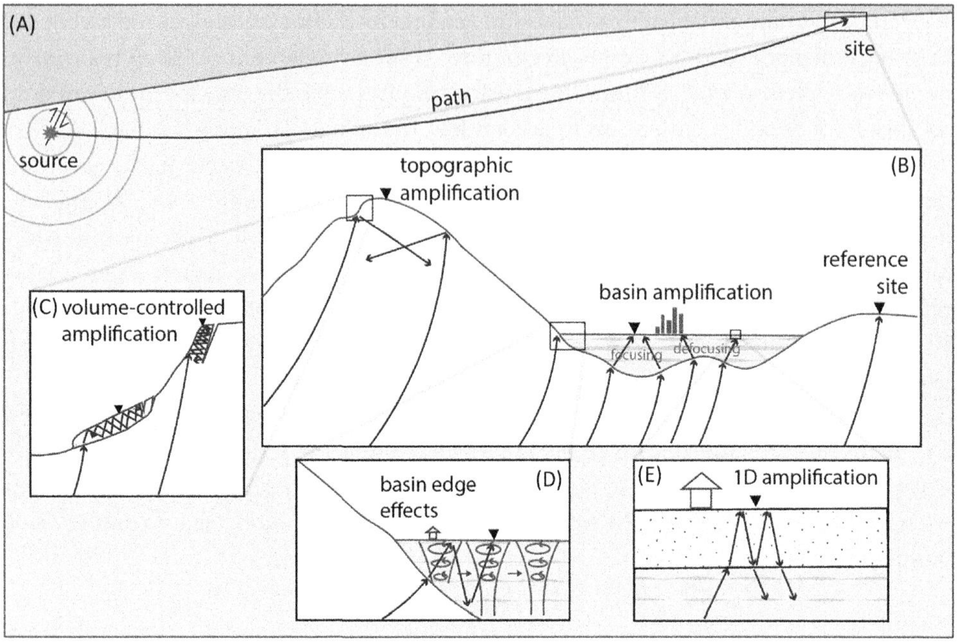

Figure 4.2 Schematic of source, path and site effects, showing (A) simplified ray paths (arrows) of seismic waves emitted by an earthquake arriving at different theoretical sites (inverted triangles) relative to a "reference site" on flat bedrock shown in panel b. Inset panels illustrate the different types of site effects including (B) 2D or 3D amplification within sedimentary basins and within topography due to focusing/defocusing and constructive/destructive interference of reflected and refracted waves at specific wavelengths; (C) volume-controlled amplification within a body bounded by open fractures (e.g., a rock column) or by boundaries of high impedance contrast (e.g., a preexisting landslide of low-velocity sediment over bedrock), which traps energy, resulting in resonance of the body at specific frequencies; (D) the conversion of body waves to surface waves at sedimentary basin edges that can result in high amplitude surface waves that then travel across the basin, illustrated by cones representing a decrease in amplitude with depth and ellipses showing the retrograde particle motion of Rayleigh waves; and (E) 1D amplification within softer soil layers overlying bedrock due to trapped energy constructively and deconstructivity interfering at specific frequencies.

Although all three factors, from source to site, ultimately determine the characteristics of ground shaking, site effects are most intimately connected to ground failure. They can be so intertwined that some site effects only occur because a preexisting weakness related to prior ground failure exists, such as amplification within a preexisting landslide body. Before we detail different types of site effects and their relevance for ground failure, we first briefly review a few basic seismological concepts required to understand the rest of this chapter. We refer the reader to Chapter 3 and to Kramer and Stewart (2024) and Shearer (2009) for more detailed fundamentals.

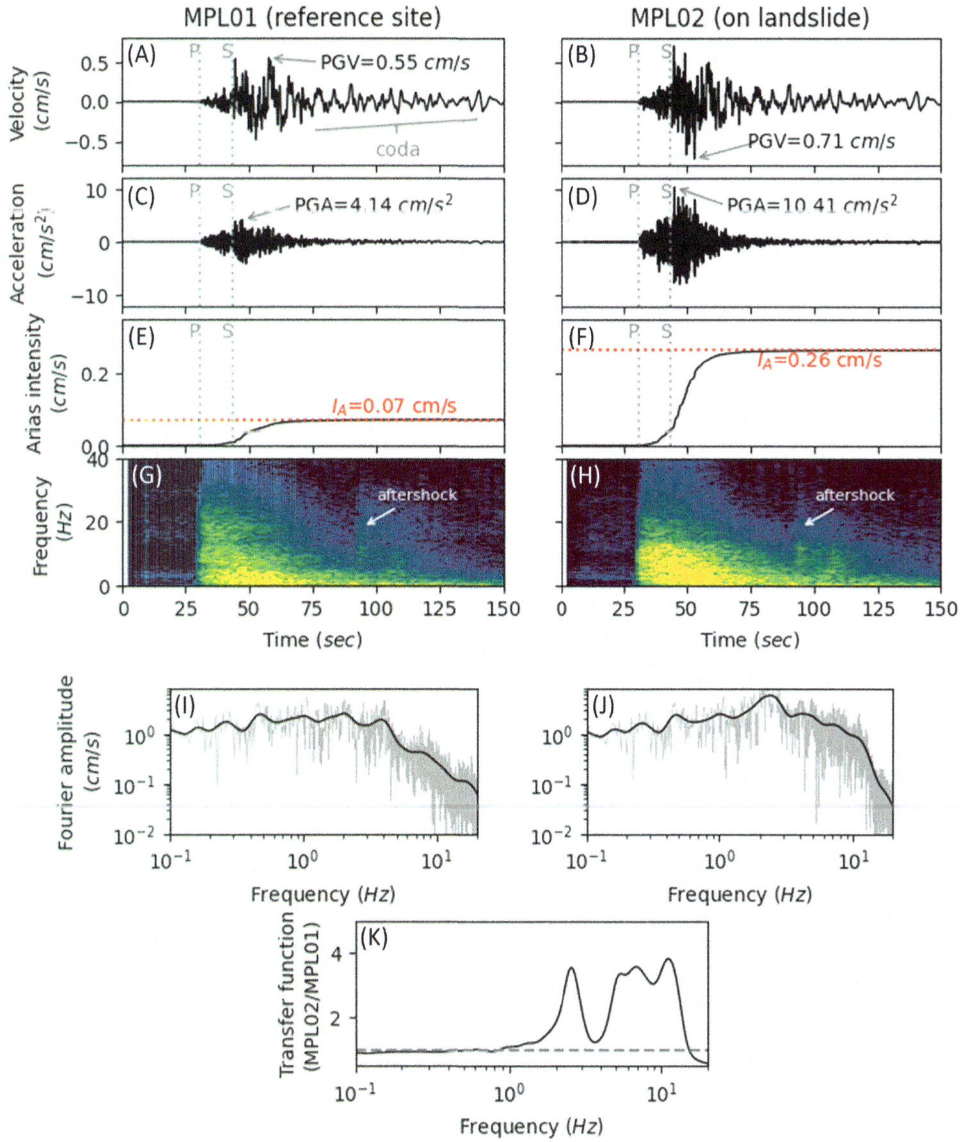

Figure 4.3 Illustration of ground motion concepts discussed in the text using recordings 85 km from the 2014 M_w 6.0 Napa, California, earthquake on the north component of a seismometer located on the body of the Mission Peak landslide complex (right panels, MPL02) compared to a reference site off the landslide about 100 m to the north (left panels, MPL01). (A, B) The ground velocity recordings with peak ground velocity (PGV) labelled. (C, D) The same signal differentiated to show acceleration with peak ground acceleration (PGA) labelled. (E, F) The cumulative integration process used to compute Arias intensity (I_A), labelled in red. (G, H) The spectrogram of the acceleration records. (I, J) Fourier amplitude spectra (grey), which have been smoothed (black) to compute the ratio of MPL02 to MPL01 shown in (K), the transfer function between the two stations. The transfer function indicates that below 1 Hz, MPL02 is not amplified relative to MPL01 (ratio ~1, dashed grey line), but is amplified by a factor of ~3.6 at 2.6 Hz and also amplified from ~5–12 Hz. Hartzell et al. (2017) concluded that this amplification is primarily due to resonance of the landslide deposits. Approximate P- and S-wave arrivals and the coda are labelled in grey.

4.2.1 Basic concepts

An earthquake ground-motion recording can be characterised by three components of motion: one vertical and two orthogonal horizontal directions (Douglas, 2003). Each component records a time series, typically of ground velocity or acceleration, depending on the instrument type. For example, Fig. 4.3C and D show a landslide and reference rock recording from the 2014 M_w 6.0 Napa Valley, California, earthquake. These waveforms are recorded on seismometers that measure velocity; acceleration is subsequently computed by differentiating the velocity record (Fig. 4.3A, B). As is apparent in Fig. 4.3, earthquake ground motions exhibit a complex evolution of both frequency (i.e. number of wave oscillations per second) and amplitude over time as different types of waves (P, S, surface waves) arrive from different parts of the fault as it ruptures (Stein & Wysession, 2003).

The amplitude spectrum of the signal (Fig. 4.3I, J), commonly computed using the Fourier transform (e.g. Bracewell, 2000), shows the distribution of amplitude as a function of frequency. When the amplitude spectrum is computed over smaller sliding time windows to compute a 2D view of the waveform called a spectrogram, one can visualise the time evolution of both frequency and amplitude, which in our example (Fig. 4.3G, H) reveals an aftershock that was not readily visible in the original time-series waveform. Both amplitude and frequency are important considerations for ground-failure triggering, as is wavelength, because seismic waves are propagating both in time and space. Wavelength (λ) is related to the wave's frequency (f) and the seismic velocity (V) of the material it is propagating through: $\lambda = V/f$.

Seismic velocities vary spatially and, therefore, as a wavetrain propagates, wavelengths change as the energy passes into a new layer with different seismic velocities. The waves will also refract (bend) and/or reflect (bounce) by varying amounts determined by the ratio of the velocities of the two layers and the incidence angle. The incident, reflected and refracted waves travelling through a volume can interfere with each other, causing constructive and destructive interference in differing amounts and locations dependent on wavelength (and thus also dependent on frequency and seismic velocity). Repeated reflections, such as between the perfectly reflective free-surface boundary and a large change of impedance ($I = \rho V$, where ρ is density and V is seismic velocity), such as between hard rock and soft sediments, can essentially trap some of the energy, building up high-amplitude resonances. Postcritical reflections (trapped energy) can also result in longer-duration motions.

Such wave-propagation effects in the heterogeneous subsurface are the root cause of many of the types of site effects we will discuss. The relevant subsurface properties, such as shear-wave velocity and density, are controlled by geological processes that influence material properties, such as grain-size distribution, density, the degree of cementation in sediments and the level of fracturing and weathering in rock. A common simplification of subsurface velocity structure that is used for estimating some types of site effects is V_{S30}, the the time-averaged shear-wave velocity in the upper 30 m of the ground (Boore et al., 1993). V_{S30} also correlates strongly with geology and thus where detailed knowledge of subsurface conditions is unavailable, V_{S30} and, by extension, site effects, can be estimated from geology maps (Boore et al., 1993; Seyhan et al., 2014; Wills & Silva, 1998). For global earthquake monitoring, the geology-based method is not feasible, and this led Wald and Allen (2007) to develop correlations of V_{S30} with topographic slope. This first-order approximation is rooted in the idea that V_{S30} is linked to geomorphic features. For example, steep slopes tend to be correlated with mountains and thus hard rock and large V_{S30} values, and within alluvial fans grain size (and thus V_{S30}) and slope tend to decrease with distance from the source of the sediments. Subsequently, Yong et al. (2012) developed a similar method that uses topographic convexity and roughness as well as slope. These correlations break down in some specific settings such as glaciated terrains or bedrock mesas. Heath et al. (2020) compiled many sources of V_{S30} maps across the globe, most of which make use of surface geology.

4.2.2 Ground motion parameters

As detailed in Chapter 3, earthquake waveforms are often simplified to parameters commonly known as intensity measures (IMs) for many types of statistical or spatial analysis, including ground-failure modelling. The simplest IMs are based on the maximum absolute value of amplitude, commonly quantified by peak ground acceleration (PGA) and peak ground velocity (PGV), illustrated in Fig. 4.3A–D. These and other IMs, are typically computed using horizontal components, which are generally larger in amplitude than vertical, and also of the greatest engineering importance (Joyner & Boore, 1988).

The fundamental limitation of these peak values is that they do not capture the frequency or duration of shaking, both of which can be important for ground failure (Kayen & Mitchell, 1997; Youd & Idriss, 2001). Earthquake magnitude is traditionally used as a proxy for duration of shaking in liquefaction models (Youd & Idriss, 2001). There is substantial evidence that IMs based on integrals of the acceleration waveform are better for predicting liquefaction and landslides (e.g. Bullock et al., 2019; Cho & Rathje, 2020; Jibson, 2007; Kayen & Mitchell, 1997; Kramer & Mitchell, 2006). These 'energy-based' metrics are therefore a function of both the amplitude and duration of shaking. One of the best-known is Arias intensity (I_A; Arias, 1970), which is defined as:

$$I_A = \frac{\pi}{2g} \int_0^{t_{\max}} [a(t)]^2 dt \tag{4.1}$$

where g is the acceleration due to gravity, $t_{\max}$ is the duration of the time series, and $a(t)$ is the ground acceleration at each time step, t (Fig. 4.3E, F). Cumulative absolute velocity (CAV; Electrical Power Research Institute EPRI, 1988) is a similar metric which integrates over the absolute value of acceleration rather than its square (Campbell & Bozorgnia, 2012). Despite their potential usefulness for ground-failure modelling, I_A and CAV are not as commonly computed or reported as other IMs, so the available datasets and models are more limited.

Frequency content can be parameterised using IMs such as the mean period, T_m (Rathje et al., 1998), which is computed from the Fourier amplitude spectrum, and has been used as a measure of frequency content for landslide studies (e.g., Jibson & Tanyaş, 2020; Saygili & Rathje, 2008). Similar to I_A and CAV, T_m is less commonly adopted than PGA or PGV and datasets are rarer. A more commonly used type of IM that reflects the frequency content and is used for many earthquake-engineering applications (though less often for ground failure) is based on the elastic response spectrum (Chopra, 2007). The most common response-spectrum quantity is the spectral acceleration $SA(T)$ which is the maximum acceleration of a single-degree-of-freedom system oscillator with natural period T (1/frequency), in which the damping is commonly assumed to be 5% (Rezaeian et al., 2014).

A key benefit of standard IMs is that empirical ground-motion models (Bozorgnia et al., 2022; Douglas, 2003) are available to estimate the mean and standard deviation of IM values based on simple explanatory variables such as earthquake magnitude, mechanism, V_{S30} and site-to-source distance. Although these models are conceptually simple, implementation is prone to error and it is useful to use well-vetted open-source implementations, e.g. OpenQuake (Pagani et al., 2014). As we discuss in detail in 4.3.3, geospatial estimates of event-specific IMs are needed for most regional geospatial ground-failure models, and the U.S. Geological Survey's (USGS) ShakeMap product is one of the most commonly used sources. For real events, maps of standard IMs (PGA, PGV, Intensity, SA) are available from the ShakeMap system within minutes of the earthquake origin time and these parameters can also be computed for scenarios (Wald et al., 2022). ShakeMap uses the OpenQuake library of ground-motion models and incorporates available ground-motion records and 'Did you feel it (DYFI)?' data from citizen reports (Quitoriano & Wald, 2020) to constrain the IMs at locations throughout the region that experienced shaking from the event (Engler et al., 2022; Worden et al., 2018). ShakeMaps are updated over the subsequent hours and days as additional ground motion constraints and information on the fault-rupture geometry become available. Estimates of the rupture geometry are not automatically created, and this is an important factor that constrains the IM estimates for larger earthquakes (larger than ~M_w 6). Thus extra uncertainty can be present for larger magnitude earthquakes when no rupture geometry is available. This uncertainty is mitigated if the ShakeMap is constrained by dense instrumentation. Fig. 4.4 shows the influence of depth and rupture geometry on the PGA contours and the resulting landslide model from the USGS Ground Failure product (Allstadt et al., 2022) for a hypothetical earthquake. Point source

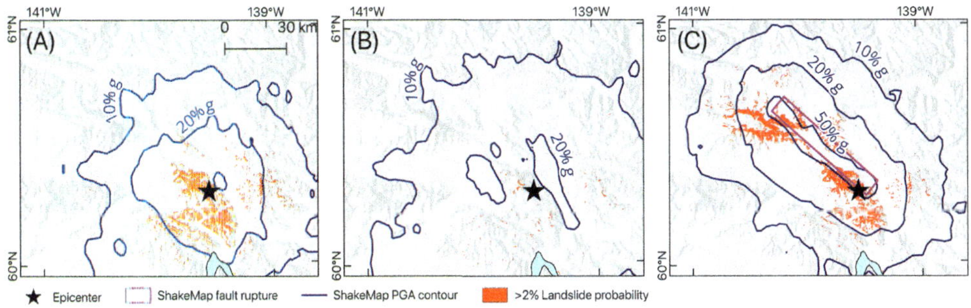

Figure 4.4 Variability in the spatial distribution of ShakeMap shaking estimates (shown as peak ground acceleration contours, in blue) and resulting landslide probability estimates from the U.S. Geological Survey Ground Failure product (using the Nowicki Jessee et al. (2018) model, with areas of >2% landslide-probability shown in red) for a hypothetical M_w 7 earthquake modelled as (A) a point source with a depth of 5 km, (B) a point source with a depth of 50 km and (C) with a rupture plane starting at 3.3 km depth dipping towards the northeast to 21.2 km depth.

representations are sensitive to the estimated depth (hypocentral depth is 5 km in Fig. 4.4A and 50 km in 4.4b), which can change as more data become available, and the addition of the rupture geometry (Fig. 4.4C) substantially alters the shaking estimates and resulting landslide estimates. Refined ShakeMaps that incorporate additional scientific findings are ultimately released in the ShakeMap Atlas (Marano et al., 2024) and these should be used whenever possible for research.

ShakeMaps, which can be downloaded from the USGS comprehensive catalogue (USGS Earthquake Hazards Program, 2019), are used as the ground-motion input for many modern ground-failure research studies (e.g., Kaynia et al., 2010; Nowicki Jessee et al., 2018; Tanyaş et al., 2019a; Zhu et al., 2017). However, because some of the most useful IMs for ground failure are not included in ShakeMap (e.g., I_A, CAV, T_m), and ShakeMap does not account for all types of site effects, directly computing IMs from waveforms can also be useful. We summarise some of the available resources here. The Centre for Engineering Strong Motion (ESM) Data (California Geological Survey, U.S. Geological Survey, 2005) provides access to ground-motion records and many IMs from past earthquakes. The ESM Database (Luzi et al., 2020) provides waveforms and IMs for past earthquakes in the European and Mediterranean regions and the Middle East. Many international seismic data centres also distribute their waveforms through the Federation of Digital Seismograph Networks web services. In Japan and Taiwan, the National Research Institute for Earth Science and Disaster Resilience (Aoi et al., 2004) and the Central Weather Administration (Central Weather Administration, 2012) provide public access to an extensive ground-motion dataset. The USGS provides software to facilitate access to these data centres and compute IMs from the waveforms (Thompson et al., 2024).

4.2.3 Site effects

Site effects can be quantified using the ratio of the spectrum of the site of interest to the spectrum of a reference site, referred to as a transfer function; this is illustrated for the 2014 Napa Valley, California, earthquake example in Fig. 4.3K. The transfer function captures how the ground motion is modified as a function of frequency. Empirically, site effects can be estimated using a variety of techniques, and Field and Jacob (1995) provided quantitative comparisons between many of these techniques.

Broadly speaking, site effects can be categorised into linear and nonlinear effects. Nonlinear effects refer to differences in site effects that occur as shaking intensification induces changes in soil material properties. In other words, the assumption that site amplification is constant at a single location breaks down. This is a complicated topic, the details of which are outside the scope of this chapter, but generally nonlinear effects result in decreased amplification relative to site effects that may be measured or modelled under weak ground motion (Kramer & Stewart, 2024). Ground failure itself is a nonlinear process, analogous to nonlinear site response that results from large strains created when shaking intensifies. For example, liquefaction and cyclic softening tend to shift energy towards lower frequencies, sometimes causing an abrupt shift at the time of liquefaction onset that can even be used to detect liquefaction below a seismic station (e.g. Kramer et al., 2015; Youd & Carter, 2005).

In the following sections, we describe four types of site effects that are relevant to the patterns of ground failure triggered by individual earthquakes and their cumulative geomorphic signatures, illustrated schematically in Fig. 4.2: (1) 1D effects due to sediment layering (depth-controlled amplification); (2) 2D or 3D effects due to subsurface lateral heterogeneities, such as sedimentary basins (basin effects); (3) 2D or 3D amplification due to variability in surface elevation (topographic amplification); and (4) 3D amplification within bounded structures (volume-controlled amplification). For simplicity, we discuss these types from a linear perspective, but note that all types can be altered by nonlinear effects if shaking is strong enough to cause large strains, though this is most common in softer sediments.

4.2.4 Depth-controlled amplification of ground motion (1D)

Amplification due to soil layering is referred to as 1D amplification or depth-controlled amplification. These effects are strongest when lateral variations are gradual enough that they can be neglected (Fig. 4.2E).

Two key factors that influence 1D site effects are (1) amplification due to changes in material impedance and (2) resonances due to reflections and refractions in the soil column. Impedance effects arise because the energy in a wave is proportional to velocity and the square of amplitude (Kramer & Stewart, 2024). As a wave enters materials with a lower seismic velocity, it slows down, and so, to conserve energy, the amplitude must increase to compensate for the change in velocity. This effect is widely approximated based on the square root of the impedance ratio, with a correction for attenuation (Anderson & Hough, 1984; Boore, 2013; Joyner et al., 1981). In contrast, resonances are amplifications at specific frequencies due to multiple arrivals, including from multiple reflections within a layer, that constructively interfere (e.g. Fig. 4.5B). If we consider a single layer of soil with height H and a shear-wave velocity V_S, a simple way

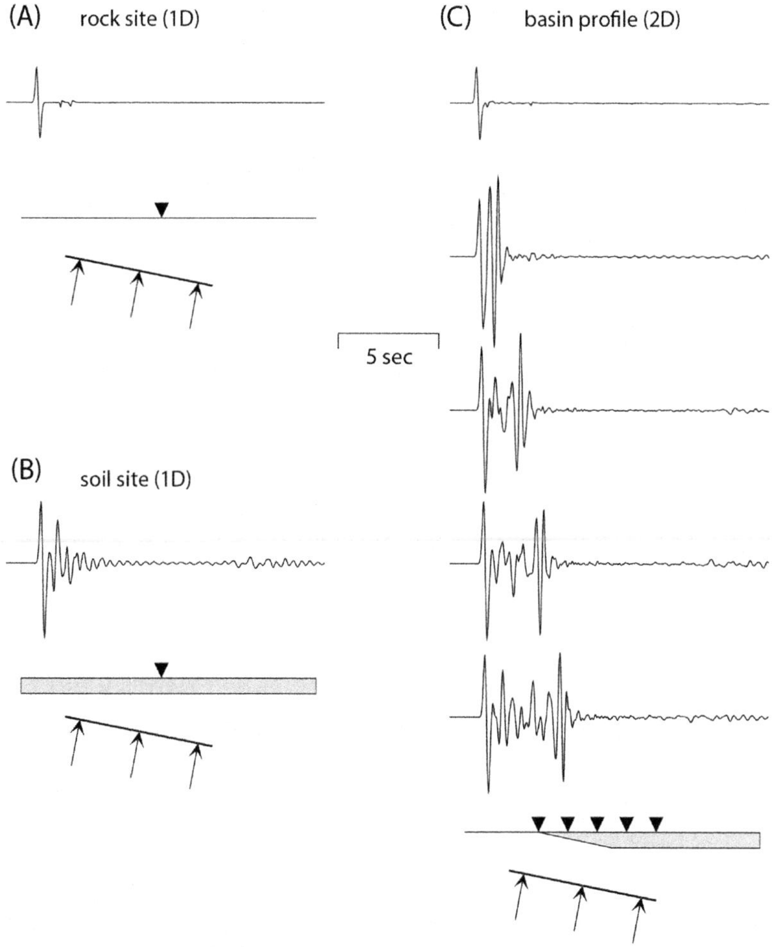

Figure 4.5 Illustration of basin effects with simulated ground motions. In each panel, the simulated waveform is displayed above an illustration of the site conditions. The locations of the seismic stations are labelled as inverted triangles, and the softer sediments are shaded grey. (A) rock site, (B) 1D site response amplification, (C) transect of sites moving into the basin from the rock at the basin edge. *Reproduced from Graves, R. W. (1993). Modeling three-dimensional site response effects in the Marina District Basin, San Francisco, California. 83(4), 1042–1063. https://doi.org/10.1785/BSSA0830041042.*

to think about resonance is as a standing wave with a node (i.e. no displacement) at the base of the layer and an antinode (maximum displacement) at the top. In this case, the fundamental mode shape is one-quarter of a wavelength spanning the soil layer. The period of the fundamental mode (T_0), at which amplification is typically highest, is thus four times the travel time of an S wave: $T_0 = 4H/V_S$, and the degree of amplification depends on how attenuating the materials are (Kramer & Stewart, 2024). Additional amplification peaks will occur at the higher modes, decreasing in amplitude as the period of the mode becomes shorter (frequencies become higher). Multiple layers can result in more complex amplification patterns. Fig. 4.3K is an example of a transfer function showing primarily 1D amplification. At this site, Hartzell et al. (2017) determined that the 2.6 Hz peak that shows amplification by a factor of ~4 relative to a nearby reference site is the fundamental mode of resonance of landslide deposits overlying a higher-velocity substrate. The more complex amplification pattern at higher frequencies may represent higher modes and/or the effect of additional layering within the landslide deposits.

One-dimensional amplification is approximated in ground-motion models by using V_{S30} as a proxy for site conditions (Boore et al., 1993). Boore and Joyner (1997) found that ground motions for soft-rock sites are, on average, amplified by less than a factor of 1.8, peaking between 2 and 5 Hz, whereas for soil sites, ground motions are amplified by more than a factor of 2.0 over 0.4 to 4 Hz. ShakeMap incorporates this V_{S30} term for approximating site effects using the global compilation of V_{S30} maps by Heath et al. (2020), which defaults to the slope-based V_{S30} where regional maps are not available, as described in 4.2.1.

4.2.5 Basin amplification (2D or 3D)

Waves propagate in three dimensions and interact with heterogeneities at a wide range of scales. Observations from past earthquakes, such as the 1985 M_w 8.0 Michoacan earthquake in Mexico (Esteva, 1988), the 1989 M_w 6.9 Loma Prieta earthquake (Graves, 1993) and 1994 M_w 6.7 Northridge earthquake (Graves et al., 1998) in the United States and the 1995 M_w 6.9 Kobe earthquake in Japan (Kawase, 1996), have indicated that sedimentary basins (large depressions in the crust where layers of softer sediments and sedimentary rocks have accumulated) are one of the most important sources of path and site effects that cannot be modelled in 1D. The lower impedance of the sediments within the basin not only amplifies the shaking due to 1D effects, but the sediments can also function as a lens, focusing the energy of the incident waves (Fig. 4.2B). Surface waves are generated at basin edges due to the interaction of incoming body waves with the 3D structure and then propagate across the basin (Fig. 4.2D) (Graves et al., 1998; Joyner, 2000; Vidale & Helmberger, 1988). Additionally, the body waves within the basin can become trapped when the incidence angle of the downgoing wave is greater than the critical angle (the angle at which all energy is reflected) and no energy leaves the basin due to refraction. This phenomenon was numerically illustrated by Graves (1993) using a 2D idealisation of a sedimentary basin (Fig. 4.5) with the finite-difference methodology for simulating linear wave propagation. Fig. 4.5A shows the simulated waveform for a site without a shallow soil layer. The waveform in Fig. 4.5B is identical except that a shallow soil layer is present, which increases the amplitude and duration of shaking due to the additional arrivals at the site due to multiple reflections within the soil layer, which is the 1D amplification discussed in section 4.2.4. Fig. 4.5C shows a transect of waveforms extending from just outside the basin, which is essentially identical to the waveform with no soil layer, into the basin. The trapped waves cause the amplitude of shaking to remain large for an extended duration, greater than that of the 1D soil layer, and the duration increases for sites that are farther into the basin. Surface waves that form on the edges of sedimentary basins can be of sufficiently high amplitude to be of engineering concern, especially for processes sensitive to lower frequencies (0.1–0.3 Hz; Joyner & Boore, 1988). Because basin effects are inherently 3D, they are sensitive to the direction from which the waves are arriving and can vary substantially depending on where the earthquake source is located relative to the basin (Wirth et al., 2019).

Basin effects are more clearly relevant for liquefaction and other level-ground processes, but they can also be an important consideration for landslides because sedimentary basins can be large features with substantial topography within them. For example, the city of Seattle, in the northwestern United States, overlies a large sedimentary basin that is draped with unconsolidated glacial deposits that form steep slopes throughout the city, posing a substantial coseismic landslide hazard that is sensitive to basin effects (Allstadt et al., 2013). Many ground-motion models approximate the basin effects with a parameter defined as the depth to a 1 or 2.5 km/s horizon (Campbell & Bozorgnia, 2014; Chiou & Youngs, 2014; Day et al., 2008). One-dimensional site response is difficult to separate from basin effects in statistical models because these velocity-horizon terms are correlated with V_{S30} (Boore & Atkinson, 2008), and so some basin effects are inevitably captured by V_{S30}. Basin-depth parameters are not currently used in ShakeMap.

4.2.6 Topographic amplification

When ground motions propagate into areas of topographic relief, incoming waves can be reflected, refracted and scattered within structures like ridges or mountains, constructively and destructively interfering with other wavefronts (Fig. 4.2B). This focusing, defocusing and trapping of energy can result in complex frequency dependent and spatially variable patterns of amplification and deamplification, referred to collectively as topographic amplification (Ashford et al., 1997; Bard, 1982; Boore, 1972). Amplification factors as high as 3 to 10 at specific frequencies have been observed at topographic highs (e.g., Massa et al., 2014), though PGA tends to show lower amplification factors of 1.2 to 1.8 (Molina et al., 2019). Topographic amplification is commonly cited as an explanation for observations of concentrated and localised structural damage on ridges and hilltops (e.g., Çelebi, 1987; Hough et al., 2010; McCrink et al., 2010), though structures on slopes can have many additional vulnerabilities that can also contribute to this pattern (Roshan & Pal, 2023). Topographic amplification has also long been investigated as an explanatory factor for observed clustering of coseismic landslides at certain aspects (e.g. Harp et al., 1978, 2014) and as a control on the locations of the largest landslides (Dunham et al., 2022). However, numerical simulations struggle to reproduce amplifications as high as many historical observations, perhaps in part because it can be challenging to isolate topographic amplification from other sources of near-surface amplification like soil layering, weathering and fracturing (e.g. Burjánek et al., 2014; Geli et al., 1988; Hartzell et al., 2013, 2017; Massa et al., 2014; Molina et al., 2019; Stone et al., 2022).

Despite challenges with *a priori* quantitative prediction in real settings, decades of observational, experimental, analytic and numerical studies have identified some general qualitative patterns (e.g. Bard, 1982; Boore, 1972; Geli et al., 1988; Hartzell et al., 2013; Massa et al., 2014). Topographic amplifications at the surface are typically higher:

- at topographic highs,
- on horizontal components of ground motion,
- at frequencies corresponding to wavelengths having about the same dimension as the topographic feature (i.e. matching the feature's resonance frequencies),
- where the topographic feature is steeper,
- where the feature has a higher height-to-width ratio and/or a sharper ridge crest,
- where the incidence angle of the incoming waves is small (closer to vertical) and
- in the direction transverse to the major axis of elongated features.

Wavelengths shorter than the topographic feature may activate higher resonance modes and produce alternating patterns of amplification and deamplification along a slope flank, whereas wavelengths much longer than the dimension of the topography may have minimal effect (Massa et al., 2014).

Realistic topography may be composed of nested features of many scales that are all interacting with and modifying the wavefield at different wavelengths simultaneously. Analytic solutions for topographic amplification, which typically idealise topography into simple 2D geometries like a triangular hill or a cliff edge (e.g. Bouchon, 1973; Paolucci, 2002; Sánchez-Sesma, 1990), have limited usefulness in such settings. Numerical simulations give more flexibility (e.g. Ashford & Sitar, 1997; Bouchon et al., 1996; Geli et al., 1988), but it can be computationally intensive and require detailed seismic-velocity models to get to a three-dimensional simulation of the high frequencies which are of interest for landslide triggering. Numerical studies can also best highlight azimuthal controls on amplification because, like basin amplification, topographic amplification has been shown to be sensitive to which direction the seismic waves are coming from (direction to the source) (e.g. Dunham et al., 2022; Harp et al., 2014; Massa et al., 2014; Maufroy et al., 2014; Stone et al., 2022).

The complexities of topographic amplification are difficult to simplify. However, some engineering solutions have been proposed to approximate the degree of amplification. Most approaches are frequency dependent and intended for modification of response spectral accelerations based on topographic curvature or relative elevation (Maufroy et al., 2014; Rai et al., 2017). Molina et al. (2019), however, included an equation for PGA based on regressions of a dataset of reported amplifications compiled from the literature:

$$A_h(PGA) = 1.00 + 0.96(H/L) \tag{4.2}$$

where $A_h\,(PGA)$ is amplification factor for PGA, and H/L is the shape ratio equal to the height of the feature over half its base width. A few seismic building codes account for topographic amplification even more simply by multiplying ground motions by a factor of 1.1 to 1.4 (Molina et al., 2019). Ground-motion models do not generally account for topographic amplification, and thus ShakeMap does not incorporate topographic-amplification models into its ground-motion estimates.

On a regional scale, the scattering of body waves, and especially surface waves passing through mountainous areas, can reduce ground motions on the other side (Lee et al., 2009; Ma et al., 2007), though in some configurations, basin surface waves may also be reradiated as body waves into surrounding mountains (e.g. Lee et al., 2009; Vidale & Helmberger, 1988). Ground-motion durations within regions of complex topography may also be extended due to wave scattering (Bard, 1982).

4.2.7 Volume-controlled amplification of ground motion (2D and 3D)

Another category of amplification occurs when seismic energy is trapped and amplified at resonant frequencies (i.e. eigenfrequencies) within a 2D or 3D volume at the surface that itself could be a potential failure mass (Fig. 4.2C). Examples include a rock pillar or an unstable rock slope bounded by deep open fractures (e.g. Burjánek et al., 2012; Lévy et al., 2010; Moore et al., 2011). This phenomenon, referred to by Kleinbrod et al. (2019) as volume-controlled amplification, is primarily considered for the triggering of rock falls and rock slides in rugged terrain, but can become important for any volume of material that is bounded by impedance contrasts and/or discontinuities near the surface. In fact, in some settings, it may be the most important type of amplification for landslides: modelling by Gischig et al. (2015) suggests that volume-controlled amplification can enhance downslope coseismic slope displacements by a factor of 10–100, substantially more than topographic or depth-controlled amplification, for which they estimated factors of 1.1 to 1.6 and 1 to 5, respectively. The coherency of volume-controlled amplification may promote failure more than depth-controlled amplification because the entire mass is responding as one. Typical observed amplification factors of weak ground motions are around ~7–10, though occasionally much higher (e.g. 80) relative to nearby stable sites, but generally only over very narrow frequency bands corresponding to the fundamental and higher-mode resonant frequencies (Burjánek et al., 2019; Kleinbrod et al., 2019). Resonant frequencies are controlled by the volume's geometry as well as its elastic properties (i.e. density, stiffness), contact stiffnesses along boundaries and the internal fracture-network geometry (Kleinbrod et al., 2019).

As for most types of site effects, dimension is a major controlling factor, so the fundamental frequency (eigenfrequency) tends to be lower for larger volumes. For example, in a series of investigations of unstable slopes, Kleinbrod et al. (2019) reported eigenfrequencies of ~1–5 Hz for features with volumes in the millions of m^3 whereas for smaller features (~1000 m^3), frequencies ranged from ~6–15 Hz. In this way, volume-dependent amplification of fracture-bounded rock structures can be similar to the response of buildings to shaking and is sometimes modelled using similar approaches (e.g. Gischig et al., 2015; Lévy et al., 2010; Moore et al., 2011). Volume-controlled amplification is typically directional (i.e. polarised), and for unstable slopes, is typically highest perpendicular to the dominant crack direction (e.g. Kleinbrod et al., 2019; Moore et al., 2011).

During an earthquake, the response of a given resonant volume to the incoming ground motions depends on the alignment of the frequency content of the incoming seismic energy with the resonant frequencies of the volume and also the alignment of the directionality of incoming peak ground motions relative to the volume's peak amplification-response direction (and, for landslides, potential sliding direction), although the relationship between these factors and resulting displacement is complex (Del Gaudio et al., 2013; Gischig et al., 2015). Additionally, the response can evolve during a long-duration earthquake or an earthquake sequence (e.g. mainshock and subsequent aftershocks). The alteration of material properties due to nonlinear effects (i.e. as strains build and/or material cracks and dilates as it approaches failure) and increased wavefield scattering in newly fractured material could diminish, augment, shift or otherwise alter the amplification transfer function, or new volume-controlled amplification can be introduced as cracks develop (e.g. Gischig et al., 2015).

Slow-moving or dormant landslides can also trap and amplify ground motions within the landslide body if it contrasts with surrounding undamaged material. Volume-controlled amplification may influence the reactivation or acceleration of preexisting instabilities that have been observed during many earthquakes, sometimes at great distances from the fault rupture or during small-magnitude earthquakes (e.g. Bozzano et al., 2011; Dellow et al., 2017; Görüm et al., 2023; Martino, 2016). Whether the observed amplification is better approximated as depth-controlled (1D) or volume-controlled (2D/3D) depends on the geometry and properties of the mass. In natural, complex and heterogeneous settings, multiple types of amplification can occur concurrently and can be challenging to separate (Gischig et al., 2015; Hartzell et al., 2017; Yeh et al., 2025).

4.2.8 Other factors

Though we focus on site effects in this chapter, source effects are also important for ground failure, especially where the distance from the site of interest to the source is small. In ground-motion models, these factors are often referred to as 'near-source' effects and can be difficult to quantify because ground motion records within about 10–20 km of the source are uncommon. Abrahamson and Somerville (1996) showed that PGA can be up to 50% larger on the hanging wall of a dipping fault compared to the footwall. The hanging-wall effect has been explained in terms of the relative difference in mass between the hanging wall and footwall, and the trapping of waves within the hanging wall (Oglesby et al., 2000). Another important near-source effect is directivity, which describes the interaction between the rupture propagation direction and the site of interest and can result in spatial variations in amplitude and duration, notably the focusing of energy in areas towards which the fault ruptures (Somerville et al., 1997). This is analogous to the Doppler effect but is complicated by the earthquake-source radiation pattern.

4.3 Coseismic ground-failure patterns

Each earthquake leaves a unique fingerprint of ground failure, shaped by the complex combination of spatially varying shaking and ground-failure susceptibility across the landscape. This fingerprint is often captured by a ground-failure inventory, which documents the locations and characteristics of ground failure triggered by a given earthquake (e.g. Harp et al., 2011; Schmitt et al., 2017; Tanyaş et al., 2017). There is a long tradition of describing the general spatial patterns of ground-failure occurrences with the goals of hazard mitigation, emergency planning and situational awareness while responding to an earthquake. In the remainder of this chapter, we build on the factors that control ground motions introduced in the previous sections to summarise controls on the regional patterns of ground failure.

4.3.1 Basic observations

As long as susceptible terrain is available, the first-order patterns of where ground failure tends to occur, and how intensively, generally correlate with ground-motion patterns (e.g. Meunier et al., 2007; Tatard & Grasso, 2013). For example, landslides tend to be larger and more numerous on the hanging wall of the fault than the footwall and tend to be concentrated in a narrower band along the fault for strike-slip earthquakes than for dip-slip events (Fan et al., 2019; Gorum et al., 2014; Meunier et al., 2007; Qin et al., 2026). The underlying reason for the higher rate of occurrence on the hanging wall can be difficult to determine because numerous factors affecting landslide occurrence vary between the hanging wall and footwall. This includes the ground-motion intensity, frequency content, the direction of displacement and the susceptibility of the materials. Directivity of unilateral ruptures may cause more substantial ground failure at the end towards which the fault ruptured (e.g. Jibson et al., 2004). Deeper earthquakes tend to have weaker ground motions and thus trigger less intense ground failure than shallow crustal earthquakes (e.g. Jibson et al., 2020), but can trigger ground failure at larger distances (e.g. Havenith et al., 2016; Tatard & Grasso, 2013). The biggest earthquakes on Earth, megathrust subduction-zone earthquakes, trigger ground failure over the largest areas because they have the largest fault ruptures (Keefer, 1984). The limited megathrust ground-failure inventory datasets that are available indicate that landslides may differ in style and may not be as intensively triggered as for crustal earthquakes in terms of number or density (Grant et al., 2022; Serey et al., 2019), but the widespread exposure to intense, long duration, low-frequency-rich ground motions has resulted in some of the largest and most disastrous landslides and dramatic liquefaction cases in recent history (e.g. Hansen, 1965; Plafker et al., 1971). Their documentation may be incomplete, however, because evidence, especially of coastal liquefaction, may be washed away by subsequent tsunami runup (e.g. Yamaguchi et al., 2012) and much of the area exposed to strong shaking may lie below sea level.

Landslide inventories tend to be more complete and are more numerous than inventories of ground failure on level ground, so more quantitative statistics about overall patterns are available. Most landslides occur within 100 km of the fault rupture, with about a third within 25 km (Tanyaş et al., 2017), and blind ruptures tend to trigger landslides at larger distances than those that rupture to the surface (Tatard & Grasso, 2013). Disrupted landslides tend to occur at the greatest distances from the fault, whereas coherent landslides and lateral spreads tend to only be triggered closer in (Keefer, 1984). There tend to be many more smaller landslides than larger landslides: frequency-area statistics follow power-law scaling, though the relationship diverges for the smallest landslides (less than ~10,000 m^2) (Malamud et al., 2004; Tanyaş et al., 2019b). Smaller disrupted slides like rock falls and rock slides are most common in the historical record, whereas coherent soil and rock slides and larger disrupted slides like rock avalanches are less common (Keefer, 1984; see also Chapter 7). The largest landslides tend to be more commonly triggered by larger-magnitude earthquakes, which also happen to have longer duration shaking and more low-frequency energy (Jibson & Tanyaş, 2020). The largest landslides also tend to occur closer to the fault trace (Valagussa et al., 2019). Ground motions are higher closer to the fault trace, but other factors can also contribute to this pattern. For example, near-fault damage zones may already be more susceptible to ground failure because of increased fracture density, mineral changes along the fault zone, such as the development of clayey fault gouges and the topographic relief produced by long-term tectonics (Choi et al., 2016; Ehteshami-Moinabadi, 2022). Strains caused by permanent tectonic deformation of the near-surface during the earthquake may also contribute to landslide triggering along the fault-rupture zone (e.g. Gorum et al., 2013; Massey et al., 2018; Tatard & Grasso, 2013). The spatial clustering of landslides tends to be patchy and irregular in shape, with dense areas interspersed with areas of little to no landsliding even within steep terrain (Keefer, 1984). Spatial clustering indicates second-order control by a combination of geological susceptibility factors and heterogeneities of shaking patterns, such as topographic or volume-controlled amplification, that is challenging to account for.

Spatial heterogeneities in shaking patterns, for example, caused by depth-dependent amplification and basin effects, are also important controls for liquefaction. Additionally, liquefaction can only occur in soils that are saturated and is more likely to occur as the density of the soil decreases, when the level of cementation of the soil particles is minimal and as the amount of fine-grained particles (clays and silt) decreases; thus the geological/geomorphic features that correspond to these factors strongly influence the spatial pattern of liquefaction occurrences. Youd and Perkins (1978) provided guidelines on the geomorphic factors that control liquefaction susceptibility, which are primarily the type of deposit (e.g. alluvial fan, sand dunes), the age of deposition (typically Holocene) and the overburden (loading) history (see also Chapter 8). The deposit type controls the grain size distribution, the age controls the level of cementation and the overburden history affects the density. These factors mean that liquefaction occurrence tends to cluster near water bodies and on flat ground. Youd and Perkins (1978) categorised river-channel and delta deposits younger than 500 years and uncompacted artificial fill as the most susceptible to liquefaction when saturated. Human alteration of subsurface conditions can result in liquefaction in unexpected locations. For example, one of the deadliest cases of liquefaction in recent history, the lateral spreads and flow slides caused by the 2018 M_w 7.5 Palu, Indonesia, earthquake, were the result of artificially-high water tables caused by rice cultivation in gently sloping areas normally not susceptible to liquefaction (Bradley et al., 2019).

4.3.2 Thresholds

Some early basic rules of thumb for coseismic ground failure still effectively hold true. Based on records available at the time, Keefer (1984) concluded that M_w 4 is the smallest earthquake one might expect to trigger landslides, and then only rarely and in small numbers, whereas widespread landsliding might only be expected for M_w >6. The threshold is higher for liquefaction: a recent study by Green and Bommer (2019) supported early conclusions that the minimum magnitude that might be expected to trigger lateral spreading and liquefaction is M_w 5 (Keefer, 1984; Kuribayashi & Tatsuoka, 1975). However, there are some historical instances of landslides and liquefaction in highly susceptible locations at lower magnitudes than these thresholds (e.g. Galli & Meloni, 1993; Green & Bommer, 2019; Keefer, 1984).

Prior to the wide availability of detailed geospatial shaking estimates and high-quality digital ground-failure inventories, early researchers published relationships to define the boundary of the farthest distance from the epicentre (or fault distance, when available) within which ground failure might be expected from a given magnitude earthquake (Galli, 2000; Keefer, 1984; Papadopoulos & Lefkopoulos, 1993; Rodríguez et al., 1999; Wilson & Keefer, 1985). Equations based on epicentral distance are less useful for larger-magnitude earthquakes with fault ruptures that can extend up to hundreds of km from the epicentre (e.g. Fig. 4.4C). Fault rupture distance equations are more informative; they are generally just a simple proxy for an average ground-motion-intensity threshold.

In search of a minimum shaking threshold for landslide triggering, Jibson and Harp (2016) sought out the most distant landslides they could find from six earthquakes, finding they occurred where ShakeMap PGA estimates were as low as 0.02 to 0.04 g. However, these most-distant landslides were so minor they may typically be ignored (e.g. $<1\ m^3$ rock falls in highly susceptible road cuts). The vast majority of inventoried coseismic landslides (i.e. those of sufficient size and/or consequence to warrant mapping) have occurred in areas where PGA shaking estimates exceed 0.1 g (Tanyaş et al., 2017). Arias intensity (I_A) datasets are more limited, but Wilson and Keefer (1985) suggested minimum thresholds likely to trigger landslides of $I_A > 0.5$ m/s for coherent landslides and $I_A > 0.15$ m/s for disrupted landslides. Liquefaction typically has a higher threshold; it has been observed in highly-susceptible soils at PGA levels as low as ~0.1 g and I_A as low as 0.8 m/s (Brandenberg et al., 2020; Kayen & Mitchell, 1997; Santucci de Magistris et al., 2014). However, explicit ground motion minimum thresholds should be interpreted carefully because, in many cases, the ground-motion estimates are based on imperfect or interpolated IM estimates that may not accurately take into account site effects or depth variations. Additionally, the edge cases that define the thresholds typically represent extremely susceptible conditions that may be rare. Simple thresholds also do not account for the spatial distribution of susceptible terrain in relation to the shaking. To do this, one generally would turn to geospatial models.

4.3.3 Modelling ground failure

Although our understanding of the basics of ground failure triggering predates the digital era, our ability to model this relationship has improved substantially. The quality and quantity of ground motion, susceptibility and ground-failure-inventory datasets have increased in recent years with advances to ShakeMap, satellite-derived products, numerical modelling and openly-shared data (e.g. Ahern, 2003; Schmitt et al., 2017; Suárez et al., 2008; Thompson et al., 2024; Wilkinson et al., 2016).

Regional-scale ground failure patterns can be approximated anywhere in the world from maps of shaking intensities and simple, globally available proxies for susceptibility. This approach is necessary because detailed geotechnical or geological susceptibility inputs are not available globally. For example, the USGS Ground Failure product (Allstadt et al., 2022), produces maps of liquefaction and landslide likelihood based on five models trained on global shaking and ground-failure inventory datasets (Godt et al., 2008; Nowicki et al., 2014; Nowicki Jessee et al., 2018; Zhu et al., 2015, 2017). The product includes an alert level based on aggregate statistics of area exposed to hazard and the population in that area (Fig. 4.6). The Ground Failure product is initially released automatically within minutes of most earthquakes greater than M_w 5 worldwide, and improves as more information on shaking constraints becomes available, especially for larger earthquakes with extended fault ruptures (Fig. 4.7). Qualitative and quantitative comparisons with ground-failure inventories (Allstadt et al., 2022; Asadi et al., 2024; Burrows et al., 2021; Rashidian & Baise, 2020) show that, if the ShakeMap is well constrained, the Ground Failure product generally assigns appropriate alert levels and identifies areas of heightened hazard. However, the models tend to overestimate the probabilities and extent, especially for liquefaction, whereas the models occasionally also underestimate probabilities largely due to the coarseness of scale and limitations in accuracy of global inputs.

Other globally applicable models for rapidly characterising first-order patterns or aggregate statistics (e.g. total area, total volume, total number) of coseismic landslides have been proposed and demonstrated with hindsight case studies but have not yet been implemented into routine monitoring operations (e.g. Havenith et al., 2016; Keefer, 1994; Kritikos et al., 2015; Marc et al., 2016; Tanyaş et al., 2019a). At this time, few alternative global geospatial models for liquefaction have been proposed. An exception is the work of Sanger et al. (2025), which includes additional geospatial predictor variables, integrates geospatial proxies with subsurface geotechnical data, and uses machine- learning methods. Some authors have also proposed updating global models to improve their performance based on remote sensing information as it becomes available (Burrows et al., 2021; Robinson et al., 2017; Xu et al., 2022) or by updating global models for a specific region using local geotechnical data (Engler et al., 2025).

Where the necessary datasets are available, region-specific modelling approaches generally can be run at higher resolution and perform better than global models (e.g., Kramer & Stewart, 2024; Sanger et al., 2025; Wang & Rathje, 2015). Switzerland, New Zealand and Taiwan have incorporated near-real-time, region-specific models for landslides into routine earthquake-monitoring procedures (Cauzzi et al., 2018; Ke et al., 2025; Massey et al., 2022) and Japan has operationalised region-specific models for both

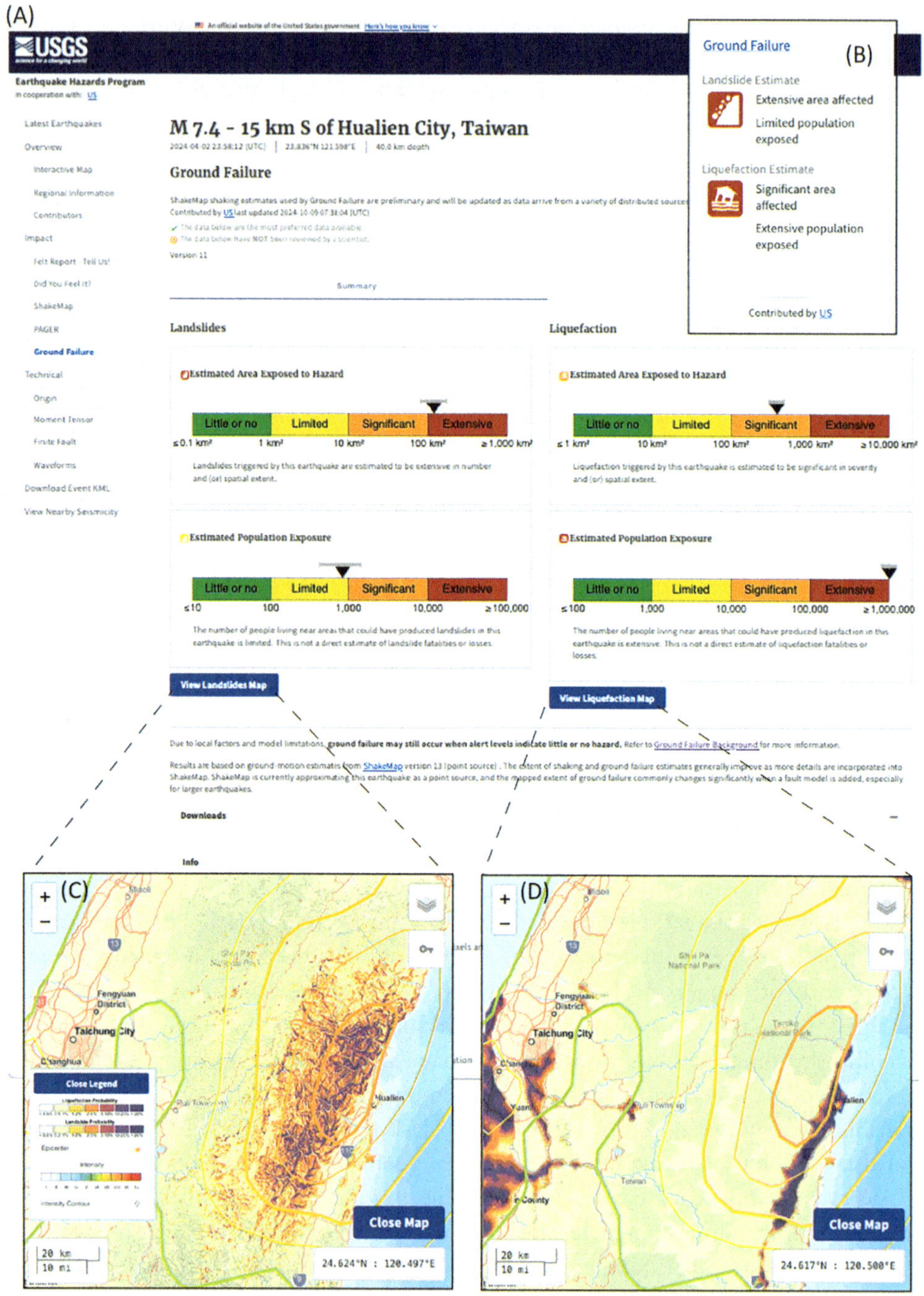
(A)
USGS
Earthquake Hazards Program
M 7.4 - 15 km S of Hualien City, Taiwan
Ground Failure
Landslides
Estimated Area Exposed to Hazard
Little or no
Limited
Significant
Extensive
Estimated Population Exposure
Liquefaction
View Landslides Map
View Liquefaction Map
(B)
Ground Failure
Landslide Estimate
Extensive area affected
Limited population exposed
Liquefaction Estimate
Significant area affected
Extensive population exposed
Contributed by US
(C)
Close Map
24.624°N : 120.497°E
(D)
Close Map
24.617°N : 120.500°E

liquefaction and landslides (Iwahashi et al., 2024; Kamiya et al., 2014). Like many of the aforementioned global models, many region-specific models are statistical models trained on ground-failure inventories, shaking estimates and more detailed region-specific susceptibility proxies using logistic regression (e.g. Asadi et al., 2024; Bozzoni et al., 2023; Chang et al., 2007; Dai et al., 2022; Lin et al., 2021; Massey et al., 2018) or newer machine-learning approaches (e.g., Geyin et al., 2022; Jena et al., 2023; Lee & Evangelista, 2006; Todorovic & Silva, 2022). However, as is true for all statistical models, the meaning of the outputs and quality of the results depend heavily on the class balance of ground-failure occurrence and nonoccurrence data points used (e.g. Oommen et al., 2011) and the quality and completeness of the available training and susceptibility datasets. The challenge of not knowing the true class balance due to data incompleteness results in models that output probabilities with no clear physical meaning. The susceptibility proxies available to be used as input may also vary from region to region. For these reasons, the outputs of different statistical models are rarely directly comparable (Allstadt et al., 2017).

Mechanistic models, based on simplifications of site-specific geotechnical or constitutive approaches, can also be feasible for specific regions if the necessary inputs are available or can be compiled, though they still may require substantial simplifying assumptions (e.g. uniform failure thickness, full saturation). The most common mechanistic regional approach for landslides is based on simplifications of the Newmark method that assumes a block sliding on an infinite slope subjected to shaking and treats the resulting displacements as an index of catastrophic failure likelihood (Allstadt, 2013; Jibson et al., 2000; Jibson, 2011; Kaynia et al., 2010; Peng et al., 2009; Wang & Rathje, 2015). A few approaches add in additional failure modes (Grant et al., 2016; Saade et al., 2016). The most common mechanistic approach for regional liquefaction-hazard mapping makes use of the liquefaction potential index (Chen et al., 2016; Holzer et al., 2006; Lenz and Baise, 2007). However, mechanistic regional models tend to be less common because, even at regional scales, the required

Figure 4.6 Screenshots of the U.S. Geological Survey Ground Failure product webpage for the 2024 M_w 7.4 Hualien earthquake in Taiwan (accessed 2 January 2026). (A) The main Ground Failure product page summarises hazard and population exposure for each type of ground failure and their uncertainties. Note the downloads section at the bottom, where the geospatial files can be obtained. (B) Screenshot of the Ground Failure summary, which appears on the event overview page and describes the overall level of concern for each type of ground failure at a glance, using coloured alert levels and short descriptions. (C) Screenshot of the interactive map that opens when clicking 'View Landslides Map' on the Ground Failure product webpage and (D) same as (C) for liquefaction. Note: Map lines delineate study areas and do not necessarily depict accepted national boundaries. *The Ground Failure product shown here can be accessed at https://earthquake.usgs.gov/earthquakes/eventpage/us7000m9g4/ground-failure/summary.*

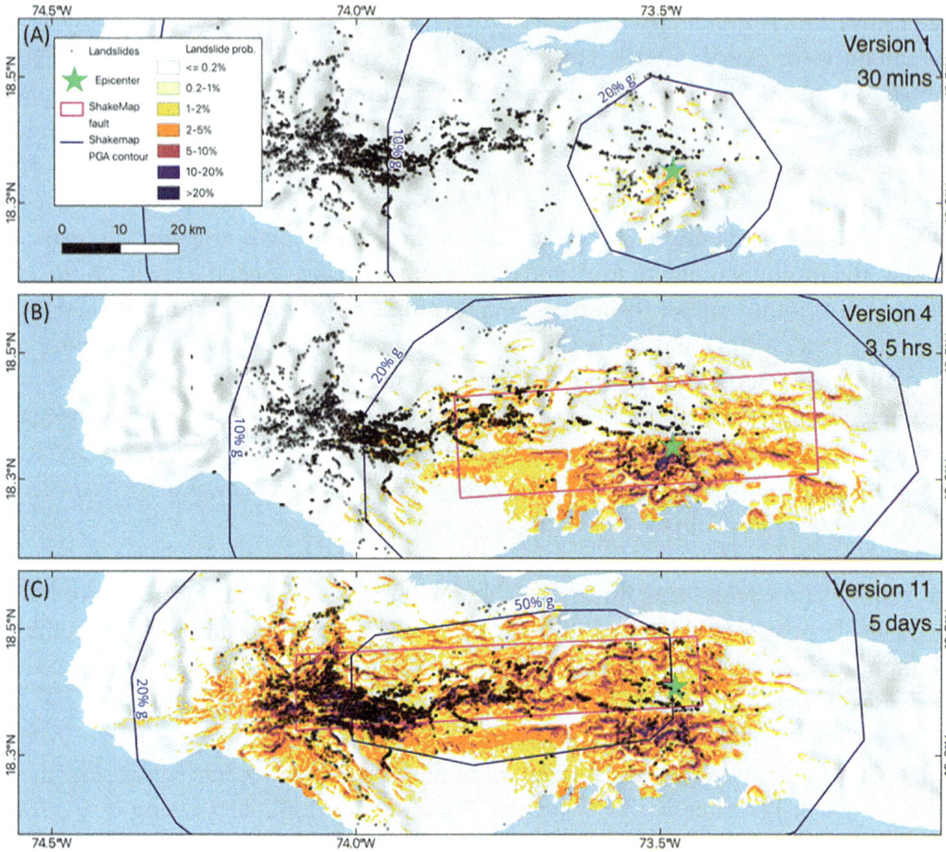

Figure 4.7 Evolution of the primary landslide-hazard model (Nowicki Jessee et al., 2018) from the U.S. Geological Survey Ground Failure (GF) product as the ShakeMap shaking estimates improved over time for the M7.2 Nippes, Haiti, earthquake in 2021 compared to the landslide inventory by Martinez et al. (2021) for (A) the first version of ShakeMap based on a point source to model ground shaking, the GF product was available within 30 min of the earthquake, (B) 3.5 h after the earthquake when the approximate fault geometry was first added to the ShakeMap (version 4) shaking estimates based on teleseismic data, (C) 5 days after the earthquake with the addition of satellite-imagery constraints on fault-rupture geometry and some ground-motion data to the ShakeMap shaking estimates (version 11). Note: Map lines delineate study areas and do not necessarily depict accepted national boundaries. *ShakeMap and GF versions accessed from Comcat (USGS Earthquake Hazards Program, 2019).*

susceptibility inputs may not be available or sufficiently well constrained and mechanistic models are sensitive to these values (e.g. Allstadt et al., 2017; Dreyfus et al., 2013). Some authors have proposed methods of regional estimation of geotechnical parameters, such as through back-analysis of ground-failure inventories (Gallen et al., 2015; Gong et al., 2023; Medwedeff et al., 2025), statistical analysis of topography (Olsen et al., 2015) or sophisticated interpolation of borehole datasets and other point measurements (Engler et al., 2025; Greenfield & Grant, 2020).

One limitation of all modelling approaches that use geospatial shaking estimates as the earthquake-loading input is that not all site effects are accounted for, yet those site effects may be a major control on any given instance of ground failure and the overall occurrence patterns observed in the landscape. For example, ShakeMap does not explicitly estimate topographic amplification or volume-controlled amplification, and thus the effects of these processes will only be represented if an instrument recorded that effect nearby, which is a rare occurrence. Additionally, the spatial scale at which ShakeMap applies the V_{S30} and basin-depth metrics that roughly approximate basin amplification and 1D site amplification is coarse and may not highlight finer geological features that are crucial for ground failure. In fact, the global V_{S30} map used by ShakeMap is largely based on slope angle rather than highlighting specific geological units. However, Bullock et al. (2019) suggested that reference rock IMs perform better for liquefaction modelling than IMs that account for site response. More complex methods of estimating the earthquake forcing require substantial computing power and detailed site information and are generally only feasible for hindsight research studies or scenario studies (e.g. Allstadt et al., 2013; Dunham et al., 2022; Harp et al., 2014; Peng et al., 2009; Stone et al., 2022), not for operational use.

4.4 Conclusions

Coseismic ground failure is caused by the interaction of earthquake ground motions with shallow subsurface materials. Ground failure can occur on both sloped and level ground, and occurs when earthquake loading exceeds material strength or degrades material strength, or a combination of both factors occurs. The characteristics of shaking at a site susceptible to ground failure reflect the sequential contributions of source, path and site effects. In this chapter, we focused primarily on site effects, which refer to modifications to ground-motion characteristics (e.g. amplitude, frequency and duration) caused by the geological and geotechnical conditions at the site of interest. The site effects of most relevance to ground failure are (1) 1D depth-controlled amplification within shallow sediment layering, (2) 2D or 3D amplification due to subsurface geometry, specifically within sedimentary basins, (3) 2D or 3D amplification due to resonance within topographic relief and (4) 3D volume-controlled amplification, describing resonance of bounded structures. In addition, earthquake-source factors such as the depth and extent of the fault that ruptured, rupture directivity and the hanging-wall effect, can also exert a strong influence on ground motions and the resulting ground failure, especially near the fault. As long as susceptible terrain is present, the factors that control the ground motions are first-order controls on ground-failure occurrence patterns. Geological, geomorphical and geotechnical factors control ground-failure susceptibility. Geospatial modelling approaches applicable regionally or globally, some for operational near-real-time use, generally take into account maps of estimated ground motions (e.g. from ShakeMap) and proxies for susceptibility such as slope angle, distance from water bodies and geological unit and are trained on inventories of past ground failure from historical earthquakes.

References

Abrahamson, N. A., & Somerville, P. G. (1996). Effects of the hanging wall and footwall on ground motions recorded during the Northridge earthquake. *Bulletin of the Seismological Society of America, 86*, S93–S99. https://doi.org/10.1785/BSSA08601B0S93.

Ahern, T. K. (2003). The FDSN and IRIS data management system: Providing easy access to terabytes of information. *International Geophysics, 81*(B), 1645–1655. https://doi.org/10.1016/S0074-6142(03)80298-X.

Allstadt, K., Vidale, J. E., & Frankel, A. D. (2013). A scenario study of seismically induced landsliding in Seattle using broadband synthetic seismograms. *103*(6), 2971–2992. https://doi.org/10.1785/0120130051

Allstadt, K. E. (2013). *Surficial seismology: Landslides, glaciers, and volcanoes in the Pacific Northwest through a seismic lens* (Doctoral dissertation). University of Washington, Seattle, WA.

Allstadt, K. E., Thompson, E. M., Hearne, M., Jessee, M. A. N., Zhu, J., Wald, D. J., & Tanyas, H. (2017). *Integrating landslide and liquefaction hazard and loss estimates with existing USGS real-time earthquake information products*. Paper presented at the 16th World Conference on Earthquake Engineering.

Allstadt, K. E., Thompson, E. M., Jibson, R. W., Wald, D. J., Hearne, M., Hunter, E. J., Fee, J. M., Schovanec, H., Slosky, D., & Haynie, K. L. (2022). The US geological survey ground failure product: Near-real-time estimates of earthquake-triggered landslides and liquefaction. *Earthquake Spectra,* 1–32. https://doi.org/10.1177/87552930211032685.

Ambraseys, N., & Sarma, S. (1969). Liquefaction of soils induced by earthquakes. *59*(2), 651–664. https://doi.org/10.1785/BSSA0590020651

Anderson, J. G., & Hough, S. E. (1984). A model for the shape of the Fourier amplitude spectrum of acceleration at high frequencies. *74*(5), 1969–1993. https://doi.org/10.1785/BSSA0740051969

Aoi, S., Kunugi, T., & Fujiwara, H. (2004). Strong-motion seismograph network operated by nied: K-Net and Kik-Net. *Journal of JAEE (commonly referring to the Journal of the Japan Association for Earthquake Engineering). 4*, 65–74. https://doi.org/10.5610/jaee.4.3_65.

Arias, A. (1970). A measure of earthquake intensity. *Seismic Design of Nuclear Power Plants,* 438–483.

Asadi, A., Sanon, C., Cakir, E., Zhan, W., Shirzadi, H., Baise, L. G., Cetin, K. Ö., & Moaveni, B. (2024). Geospatial liquefaction modeling of the 2023 Türkiye earthquake sequence by an ensemble of global, continental, regional, and event-specific models. https://doi.org/10.1785/0220230287

Ashford, S. A., & Sitar, N. (1997). Analysis of topographic amplification of inclined shear waves in a steep coastal bluff. *Bulletin of the Seismological Society of America, 87*, 692–700. https://doi.org/10.1785/BSSA0870030692.

Ashford, S. A., Sitar, N., Lysmer, J., & Deng, N. (1997). Topographic effects on the seismic response of steep slopes. *87*(3), 701–709. https://doi.org/10.1785/BSSA0870030701.

Bard, P.-Y. (1982). Diffracted waves and displacement field over two-dimensional elevated topographies. *Geophysical Journal International, 71*, 731–760. https://doi.org/10.1111/j.1365–246X.1982.tb02795.x.

Boore, D. M. (1972). A note on the effect of simple topography on seismic SH waves*. *62*(1), 275–284. https://doi.org/10.1785/BSSA0620010275

Boore, D. M. (2004). Can site response be predicted? *Journal of Earthquake Engineering, 08*, 1–41. https://doi.org/10.1142/S1363246904001651.

Boore, D. M. (2013). The uses and limitations of the square-root-impedance method for computing site amplification. *03*(5), 2356–2368. https://doi.org/10.1785/0120120283

Boore, D. M., & Atkinson, G. M. (2008). Ground-motion prediction equations for the average horizontal component of PGA, PGV, and 5%-Damped PSA at Spectral Periods between 0.01 s and 10.0 s. *Earthquake Spectra, 24*, 99–138. https://doi.org/10.1193/1.2830434.

Boore, D. M., & Joyner, W. B. (1997). Site amplifications for generic rock sites. *Bulletin of the Seismological Society of America, 87*, 327–341. https://doi.org/10.1785/BSSA0870020327.

Boore, D. M., Joyner, W. B., & Fumal, T. E. (1993). *Estimation of response spectra and peak accelerations from western North American earthquakes: An interim report* (Open-File Report No. 93–509). U.S. Geological Survey. https://doi.org/10.3133/ofr93509

Borcherdt, R. D. (1970). Effects of local geology on ground motion near San Francisco Bay*. *60*(1), 29–61. https://doi.org/10.1785/BSSA0600010029

Bouchon, M. (1973). Effect of topography on surface motion. *63*(2), 615–632. https://doi.org/10.1785/BSSA0630020615

Bouchon, M., Schultz, C. A., & Toksöz, M. N. (1996). Effect of three-dimensional topography on seismic motion. *Journal of Geophysical Research: Solid Earth, 101*, 5835–5846. https://doi.org/10.1029/95JB02629.

Boulanger, R. W., & Idriss, I. M. (2006). Liquefaction susceptibility criteria for silts and clays. *Journal of Geotechnical and Geoenvironmental Engineering, 132*, 1413–1426. https://doi.org/10.1061/(ASCE)1090-0241(2006)132:11(1413).

Boulanger, R. W., & Idriss, I. M. (2007). Evaluation of cyclic softening in silts and clays. *Journal of Geotechnical and Geoenvironmental Engineering, 133*, 641–652. https://doi.org/10.1061/(ASCE)1090-0241(2007)133:6(641).

Bozorgnia, Y., Abrahamson, N. A., Ahdi, S. K., Ancheta, T. D., Atik, L. A., Archuleta, R. J., Atkinson, G. M., Boore, D. M., Campbell, K. W., Chiou, B. S.-J., Contreras, V., Darragh, R. B., Derakhshan, S., Donahue, J. L., Gregor, N., Gulerce, Z., Idriss, I., Ji, C., Kishida, T., et al. (2022). NGA-Subduction research program. *38*(2), 783–798. https://doi.org/10.1177/87552930211056081

Bozzano, F., Lenti, L., Martino, S., Paciello, A., & Scarascia Mugnozza, G. (2011). Evidences of landslide earthquake triggering due to self-excitation process. *International Journal of Earth Sciences, 100*, 861–879. https://doi.org/10.1007/s00531-010-0514-5.

Bozzoni, F., Furiosi, A., & Lai, C. G. (2023). Probabilistic assessment of the earthquake-induced soil liquefaction hazard at national scale: Macrozonation of the Italian territory. *Natural Hazards, 115*, 2237–2255. https://doi.org/10.1007/s11069-022-05636-w.

Bracewell, R. (2000). *The Fourier transform & its applications*. Boston: McGraw-Hill Science/Engineering/Math.

Bradley, K., Mallick, R., Andikagumi, H., Hubbard, J., Meilianda, E., Switzer, A., Du, N., Brocard, G., Alfian, D., Benazir, B., Feng, G., Yun, S.-H., Majewski, J., Wei, S., & Hill, E. M. (2019). Earthquake-triggered 2018 Palu Valley landslides enabled by wet rice cultivation. *Nature Geoscience, 12*, 935–939. https://doi.org/10.1038/s41561-019-0444-1.

Brandenberg, S. J., Zimmaro, P., Stewart, J. P., Kwak, D. Y., Franke, K. W., Moss, R. E., Çetin, K. Ö., Can, G., Ilgac, M., Stamatakos, J., Weaver, T., & Kramer, S. L. (2020). Next-generation liquefaction database. *36*(3), 939–959. https://doi.org/10.1177/8755293020902477

Bullock, Z., Dashti, S., Liel, A. B., Porter, K. A., & Karimi, Z. (2019). Assessment supporting the use of outcropping rock evolutionary intensity measures for prediction of liquefaction consequences. *35*(4), 1899–1926. https://doi.org/10.1193/041618EQS094M

Burjánek, J., Edwards, B., & Fäh, D. (2014). Empirical evidence of local seismic effects at sites with pronounced topography: A systematic approach. *Geophysical Journal International, 197*, 608–619. https://doi.org/10.1093/gji/ggu014.

Burjánek, J., Kleinbrod, U., & Fäh, D. (2019). Modeling the seismic response of unstable rock mass with deep compliant fractures. *Journal of Geophysical Research: Solid Earth, 124*, 13039–13059. https://doi.org/10.1029/2019JB018607.

Burjánek, J., Moore, J. R., Yugsi Molina, F. X., & Fäh, D. (2012). Instrumental evidence of normal mode rock slope vibration. *Geophysical Journal International, 188*, 559–569. https://doi.org/10.1111/j.1365–246X.2011.05272.x.

Burrows, K., Milledge, D., Walters, R. J., & Bellugi, D. (2021). Integrating empirical models and satellite radar can improve landslide detection for emergency response. *Natural Hazards and Earth System Sciences, 21*, 2993–3014. https://doi.org/10.5194/nhess-21–2993-2021.

California Geological Survey, & U.S. Geological Survey. (2005). *Center for Engineering Strong Motion Data (CESMD)*. https://doi.org/10.5066/P13HTZQS

Campbell, K. W., & Bozorgnia, Y. (2012). A comparison of ground-motion prediction equations for Arias intensity and cumulative absolute velocity developed using a consistent database and functional form. *28*(3), 931–941. https://doi.org/10.1193/1.4000067

Campbell, K. W., & Bozorgnia, Y. (2014). NGA-West2 ground motion model for the average horizontal components of PGA, PGV, and 5% damped linear acceleration response spectra. *Earthquake Spectra, 30*, 1087–1115. https://doi.org/10.1193/062913EQS175M.

Cauzzi, C., Fäh, D., Wald, D. J., Clinton, J., Losey, S., & Wiemer, S. (2018). ShakeMap-based prediction of earthquake-induced mass movements in Switzerland calibrated on historical observations. *Natural Hazards, 92*, 1211–1235. https://doi.org/10.1007/s11069-018-3248-5.

Çelebi, M. (1987). Topographical and geological amplifications determined from strong-motion and aftershock records of the 3 March 1985 Chile earthquake. *Bulletin of the Seismological Society of America, 77*, 1147–1167. https://doi.org/10.1785/BSSA0770041147.

Central Weather Administration & U.S. Geological Survey. (2012). *Central Weather Administration Seismographic Network*. https://doi.org/10.7914/SN/T5

Chang, K.-T., Chiang, S.-H., & Hsu, M.-L. (2007). Modeling typhoon- and earthquake-induced landslides in a mountainous watershed using logistic regression. *Geomorphology, 89*, 335–347. https://doi.org/10.1016/j.geomorph.2006.12.011.

Chen, Q., Wang, C., & Hsein Juang, C. (2016). CPT-Based Evaluation of Liquefaction Potential Accounting for Soil Spatial Variability at Multiple Scales. *Journal of Geotechnical and Geoenvironmental Engineering, 142*(2). 10.1061/(asce)gt.1943-5606.0001402.

Chiou, B. S.-J., & Youngs, R. R. (2014). Update of the chiou and youngs NGA model for the average horizontal component of peak ground motion and response spectra. *Earthquake Spectra, 30*, 1117–1153. https://doi.org/10.1193/072813EQS219M.

Cho, Y., & Rathje, E. M. (2020). Displacement hazard curves derived from slope-specific predictive models of earthquake-induced displacement. *Soil Dynamics and Earthquake Engineering, 138*, 106367. https://doi.org/10.1016/j.soildyn.2020.106367.

Choi, J.-H., Edwards, P., Ko, K., & Kim, Y.-S. (2016). Definition and classification of fault damage zones: A review and a new methodological approach. *Earth-Science Reviews, 152*, 70–87. https://doi.org/10.1016/j.earscirev.2015.11.006.

Chopra, A. K. (2007). Elastic response spectrum: A historical note. *Earthquake Engineering and Structural Dynamics, 36*, 3–12. https://doi.org/10.1002/eqe.609.

Crawford, C. B. (1968). Quick clays of eastern Canada. Engineering Geology*, 2*, 239–265. https://doi.org/10.1016/0013–7952(68)90002-1.

Dai, Y., Bai, X., Nie, G., & Huangfu, G. (2022). A rapid assessment method for earthquake-induced landslide casualties based on GIS and logistic regression model. *Geomatics, Natural Hazards and Risk, 13*, 222–248. https://doi.org/10.1080/19475705.2021.2017022.

Day, S. M., Graves, R., Bielak, J., Dreger, D., Larsen, S., Olsen, K. B., Pitarka, A., & Ramirez-Guzman, L. (2008). Model for basin effects on long-period response spectra in southern California. *Earthquake Spectra, 24*, 257–277. https://doi.org/10.1193/1.2857545.

Del Gaudio, V., Wasowski, J., & Muscillo, S. (2013). New developments in ambient noise analysis to characterise the seismic response of landslide-prone slopes. *Natural Hazards and Earth System Sciences, 13*, 2075–2087. https://doi.org/10.5194/nhess-13–2075-2013.

Dellow, S., Massey, C., Cox, S., Archibald, G., Begg, J., Bruce, Z., Carey, J., Davidson, J., Pasqua, F. D., Glassey, P., Hill, M., Jones, K., Lyndsell, B., Lukovic, B., McColl, S., Rattenbury, M., Read, S., Rosser, B., Singeisen, C., ... Little, M. (2017). Landslides caused by the Mw7.8 Kaikōura earthquake and the immediate response. *Bulletin of the New Zealand Society for Earthquake Engineering, 50*, 106–116. https://doi.org/10.5459/bnzsee.50.2.106-116.

Douglas, J. (2003). Earthquake ground motion estimation using strong-motion records: A review of equations for the estimation of peak ground acceleration and response spectral ordinates. *61*(1–2), 43–104. https://doi.org/10.1016/S0012-8252(02)00112-5

Dreyfus, D., Rathje, E. M., & Jibson, R. W. (2013). The influence of different simplified sliding-block models and input parameters on regional predictions of seismic landslides triggered by the Northridge earthquake. Engineering Geology, *163*, 41–54. https://doi.org/10.1016/j.enggeo.2013.05.015.

Dunham, A. M., Kiser, E., Kargel, J. S., Haritashya, U. K., Watson, C. S., Shugar, D. H., Hughes, A., & DeCelles, P. G. (2022). Topographic control on ground motions and landslides from the 2015 gorkha earthquake. *Geophysical Research Letters, 49*, e2022GL098582. https://doi.org/10.1029/2022GL098582.

Ehteshami-Moinabadi, M. (2022). Properties of fault zones and their influences on rainfall-induced landslides, examples from Alborz and Zagros ranges. *Environmental Earth Sciences, 81*, 168. https://doi.org/10.1007/s12665-022-10283-2.

Electrical Power Research Institute (EPRI). (1988). *A criterion for determining exceedance of the operating basis earthquake* (EPRI Report No. NP-5980). Palo Alto, CA: EPRI.

Engler, D. T., Thompson, E. M., Maurer, B. W., Geyin, M., Bürgi, P. M., Allstadt, K. E., & Jaiswal, K. S. (2025). Updating regional-scale geospatial liquefaction models with locally available geotechnical data. *Seismological Research Letters, 96*, 3365–3378. https://doi.org/10.1785/0220250134.

Engler, D. T., Worden, C. B., Thompson, E. M., & Jaiswal, K. S. (2022). Partitioning ground-motion uncertainty when conditioned on station data. *112*(3), 1060–1079. https://doi.org/10.1785/0120210177

Esteva, L. (1988). The Mexico earthquake of September 19, 1985—Consequences, lessons, and impact on research and practice. *4*(3), 413–426. https://doi.org/10.1193/1.1585482

Fan, X., Scaringi, G., Korup, O., West, A. J., van Westen, C. J., Tanyas, H., Hovius, N., Hales, T. C., Jibson, R. W., Allstadt, K. E., Zhang, L., Evans, S. G., Xu, C., Li, G., Pei, X., Xu, Q., & Huang, R. (2019). Earthquake-induced chains of geologic hazards: Patterns, mechanisms, and impacts. *Reviews of Geophysics, 57*, 421–503. https://doi.org/10.1029/2018RG000626.

Field, E. H., & Jacob, K. H. (1995). A comparison and test of various site-response estimation techniques, including three that are not reference-site dependent. *85*(4), 1127–1143. https://doi.org/10.1785/BSSA0850041127

Gallen, S. F., Clark, M. K., & Godt, J. W. (2015). Coseismic landslides reveal near-surface rock strength in a high-relief, tectonically active setting. *Geology, 43*, 11–14. https://doi.org/10.1130/G36080.1.

Galli, P. (2000). New empirical relationships between magnitude and distance for liquefaction. *Tectonophysics, 324*, 169–187. https://doi.org/10.1016/S0040-1951(00)00118-9.

Galli, P., & Meloni, F. (1993). Nuovo catalogo nazionale dei processi di liquefazione avvenuti in occasione dei terremoti storici in Italia. *Il Quaternario Italian Journal of Quaternary Science, 6*(2), 271–292.

Geli, L., Bard, P.-Y., & Jullien, B. (1988). The effect of topography on earthquake ground motion: A review and new results. *78*(1), 42–63. https://doi.org/10.1785/BSSA0780010042

Geyin, M., Maurer, B. W., & Christofferson, K. (2022). An AI-driven, mechanistically grounded geospatial liquefaction model for rapid response and scenario planning. *159*, 107348. https://doi.org/10.1016/j.soildyn.2022.107348

Gischig, V. S., Eberhardt, E., Moore, J. R., & Hungr, O. (2015). On the seismic response of deep-seated rock slope instabilities—Insights from numerical modeling. Engineering Geology, *193*, 1–18. https://doi.org/10.1016/j.enggeo.2015.04.003.

Godt, J. W., Sener, B., Verdin, K. L., Wald, D. J., Earle, P. S., Harp, E. L., & Jibson, R. W. (2008). Rapid assessment of earthquake-induced landsliding. In *Proceedings of the First World Landslide Forum* (pp. 392–395), Tokyo, Japan, November 18–21, 2008. International Programme on Landslides.

Gong, W., Zekkos, D., & Clark, M. (2023). A pseudo-3D methodology for regional-scale back-analysis of earthquake-induced landslides. Engineering Geology, *325*, 107277. https://doi.org/10.1016/j.enggeo.2023.107277.

Gorum, T., Korup, O., van Westen, C. J., van der Meijde, M., Xu, C., & van der Meer, F. D. (2014). Why so few? Landslides triggered by the 2002 Denali earthquake, Alaska. *Quaternary Science Reviews, 95*, 80–94. https://doi.org/10.1016/j.quascirev.2014.04.032.

Görüm, T., Tanyas, H., Karabacak, F., Yılmaz, A., Girgin, S., Allstadt, K. E., Süzen, M. L., & Bürgi, P. (2023). Preliminary documentation of coseismic ground failure triggered by the February 6, 2023 Türkiye earthquake sequence. *327*, 107315. https://doi.org/10.1016/j.enggeo.2023.107315

Gorum, T., van Westen, C. J., Korup, O., van der Meijde, M., Fan, X., & van der Meer, F. D. (2013). Complex rupture mechanism and topography control symmetry of mass-wasting pattern, 2010 Haiti earthquake. *Geomorphology, 184*, 127–138. https://doi.org/10.1016/j.geomorph.2012.11.027.

Grant, A., Wartman, J., & Abou-Jaoude, G. (2016). Multimodal method for coseismic landslide hazard assessment. Engineering Geology, *212*, 146–160. https://doi.org/10.1016/j.enggeo.2016.08.005.

Grant, A. R. R., Struble, W. T., & LaHusen, S. R. (2022). Limits to coseismic landslides triggered by Cascadia Subduction Zone earthquakes. *Geomorphology, 418*, 108477. https://doi.org/10.1016/j.geomorph.2022.108477.

Graves, R. W. (1993). Modeling three-dimensional site response effects in the Marina District Basin, San Francisco, California. *83*(4), 1042–1063. https://doi.org/10.1785/BSSA0830041042

Graves, R. W., Pitarka, A., & Somerville, P. G. (1998). Ground-motion amplification in the Santa Monica area: Effects of shallow basin-edge structure. *Bulletin of the Seismological Society of America, 88*, 1224–1242. https://doi.org/10.1785/BSSA0880051224.

Green, R. A., & Bommer, J. J. (2019). What is the smallest earthquake magnitude that needs to be considered in assessing liquefaction hazard? *Earthquake Spectra, 35*, 1441–1464. https://doi.org/10.1193/032218EQS064M.

Greenfield, M. W., & Grant, A. (2020). Probabilistic regional-scale liquefaction triggering modeling using 3D Gaussian processes. *Soil Dynamics and Earthquake Engineering, 134*, 106159. https://doi.org/10.1016/j.soildyn.2020.106159.

Hansen, W. R. (1965). *Effects of the earthquake of March 27, 1964, at Anchorage, Alaska* (Professional Paper No. 542-A). U.S. Government Printing Office. https://doi.org/10.3133/pp542A

Harp, E. L., Hartzell, S. H., Jibson, R. W., Ramirez-Guzman, L., & Schmitt, R. G. (2014). Relation of landslides triggered by the Kīholo Bay earthquake to modeled ground motion. *104*(5), 2529–2540. https://doi.org/10.1785/0120140047

Harp, E. L., Keefer, D. K., Sato, H. P., & Yagi, H. (2011). Landslide inventories: The essential part of seismic landslide hazard analyses. Engineering Geology, *122*, 9–21. https://doi.org/10.1016/j.enggeo.2010.06.013.

Harp, E. L., Wilson, R. C., Wieczorek, G. F., & Keefer, D. K. (1978). Landslides from the February 4, 1976 Guatemala earthquake: Implications for seismic hazard reduction in the Guatemala City area. In *Proceedings of the second international conference on microzonation* (pp. 353–366).

Hartzell, S., Leeds, A. L., & Jibson, R. W. (2017). Seismic response of soft deposits due to landslide: The mission peak, California, landslide. *Bulletin of the Seismological Society of America, 107*, 2008–2020. https://doi.org/10.1785/0120170033.

Hartzell, S., Meremonte, M., Ramírez-Guzmán, L., & McNamara, D. (2013). Ground motion in the presence of complex topography: Earthquake and ambient noise sources. *104*(1), 451–466. https://doi.org/10.1785/0120130088

Havenith, H.-B., Torgoev, A., Braun, A., Schlögel, R., & Micu, M. (2016). A new classification of earthquake-induced landslide event sizes based on seismotectonic, topographic, climatic and geologic factors. 3, 6. https://doi.org/10.1186/s40677-016-0041-1

Heath, D. C., Wald, D. J., Worden, C. B., Thompson, E. M., & Smoczyk, G. M. (2020). A global hybrid VS30 map with a topographic slope–based default and regional map insets. *Earthquake Spectra, 36*, 1570–1584. https://doi.org/10.1177/8755293020911137.

Highland, L. M. (2004). *Landslide types and processes* (Fact Sheet No. 2004–3072). U.S. Geological Survey. https://doi.org/10.3133/fs20043072

Holzer, T. L., Bennett, M. J., Noce, T. E., Padovani, A. C., & Tinsley, J. C. (2006). Liquefaction Hazard Mapping with LPI in the Greater Oakland, California, Area. *Earthquake Spectra, 22*(3), 693–708. 10.1193/1.2218591.

Hough, S. E., Altidor, J. R., Anglade, D., Given, D., Janvier, M. G., Maharrey, J. Z., Meremonte, M., Mildor, B. S.-L., Prepetit, C., & Yong, A. (2010). Localized damage caused by topographic amplification during the 2010 M 7.0 Haiti earthquake. *Nature Geoscience, 3*, 778–782. https://doi.org/10.1038/ngeo988.

Hungr, O., Leroueil, S., & Picarelli, L. (2014). The varnes classification of landslide types, an update. *Landslides, 11*, 167–194. https://doi.org/10.1007/s10346-013-0436-y.

Idriss, I. M., & Boulanger, R. W. (2008). *Soil liquefaction during earthquakes* (EERI Publication No. MNO-12). Earthquake Engineering Research Institute.

Iwahashi, J., Endo, R., & Nakano, T. (2024). Seismic ground disaster assessment system (SGDAS) of GSI and its scope. *Japanese Geotechnical Society Special Publication, 10*, 2361–2366. https://doi.org/10.3208/jgssp.v10.P2-02.

Jena, R., Pradhan, B., Almazroui, M., Assiri, M., & Park, H.-J. (2023). Earthquake-induced liquefaction hazard mapping at national-scale in Australia using deep learning techniques. *Geoscience Frontiers, 14*, 101460. https://doi.org/10.1016/j.gsf.2022.101460.

Jibson, R. W. (2007). Regression models for estimating coseismic landslide displacement. Engineering Geology, *91*, 209–218. https://doi.org/10.1016/j.enggeo.2007.01.013.

Jibson, R. W. (2011). Methods for assessing the stability of slopes during earthquakes—A retrospective. Engineering Geology, *122*, 43–50. https://doi.org/10.1016/j.enggeo.2010.09.017 The next generation of research on earthquake-induced landslides.

Jibson, R. W., Allstadt, K. E., Rengers, F. K., & Godt, J. W. (2018). *Overview of the geologic effects of the November 14, 2016, Mw 7.8 Kaikōura, New Zealand, earthquake* (Scientific Investigations Report No. 2017–5146). U.S. Geological Survey. https://doi.org/10.3133/sir20175146

Jibson, R. W., Grant, A. R. R., Witter, R. C., Allstadt, K. E., Thompson, E. M., & Bender, A. M. (2020). Ground failure from the anchorage, Alaska, earthquake of 30 November 2018. *Seismological Research Letters, 91*, 19–32. https://doi.org/10.1785/0220190187.

Jibson, R. W., & Harp, E. L. (2016). Ground motions at the outermost limits of seismically triggered landslides. *Bulletin of the Seismological Society of America, 106*, 708–719. https://doi.org/10.1785/0120150141.

Jibson, R. W., Harp, E. L., & Michael, J. A. (2000). A method for producing digital probabilistic seismic landslide hazard maps. Engineering Geology, *58*, 271–289. https://doi.org/10.1016/S0013-7952(00)00039-9.

Jibson, R. W., Harp, E. L., Schulz, W., & Keefer, D. K. (2004). Landslides triggered by the 2002 Denali fault, Alaska, earthquake and the inferred nature of the strong shaking. *Earthquake Spectra, 20*, 669–691. https://doi.org/10.1193/1.1778173.

Jibson, R. W., & Tanyaş, H. (2020). The influence of frequency and duration of seismic ground motion on the size of triggered landslides—A regional view. Engineering Geology, *273*, 105671. https://doi.org/10.1016/j.enggeo.2020.105671.

Joyner, W., & Boore, D. M. (1988). Measurement, characterization, and prediction of strong ground motion. In *Earthquake engineering and soil dynamics II—Recent advances in ground-motion evaluation: Proceedings of the specialty conference* (pp. 43–102).

Joyner, W. B. (2000). Strong motion from surface waves in deep sedimentary basins. *Bulletin of the Seismological Society of America, 90*, S95–S112. https://doi.org/10.1785/0120000505.

Joyner, W. B., Warrick, R. E., & Fumal, T. E. (1981). The effect of Quaternary alluvium on strong ground motion in the Coyote Lake, California, earthquake of 1979. *71*(4), 1333–1349. https://doi.org/10.1785/BSSA0710041333

Kamiya, I., Koarai, M., Otoi, K., & Nakano, T. (2014). Construction of seismic ground disaster assessment system (SGDAS). *Journal of the Geospatial Information Authority of Japan, 126*, 61–66.

Kawase, H. (1996). The cause of the damage belt in kobe: "The basin-edge effect," constructive interference of the direct S-wave with the basin-induced diffracted/rayleigh waves. *Seismological Research Letters, 67*, 25–34. https://doi.org/10.1785/gssrl.67.5.25.

Kayen, R. E., & Mitchell, J. K. (1997). Assessment of liquefaction potential during earthquakes by Arias intensity. *Journal of Geotechnical and Geoenvironmental Engineering, 123*, 1162–1174. https://doi.org/10.1061/(ASCE)1090-0241(1997)123:12(1162).

Kaynia, A. M., Skurtveit, E., & Saygili, G. (2010). Real-time mapping of earthquake-induced landslides. *Bulletin of Earthquake Engineering, 9*, 955–973. https://doi.org/10.1007/s10518-010-9234-2.

Ke, S., Hsu, C., Yang, C., Wu, C. C., & Huang, M. (2025). Leveraging big data for earthquake disaster response: Insights from Taiwanese decision-support systems from 1999 to 2024. *Bulletin of the Seismological Society of America, 116*, 162–175. https://doi.org/10.1785/0120250188.

Keefer, D. K. (1984). Landslides caused by earthquakes. *GSA Bulletin, 95*, 406–421. https://doi.org/10.1130/0016-7606(1985)96%3C1093:LCBEDA%3E2.0.CO;2.

Keefer, D. K. (1994). The importance of earthquake-induced landslides to long-term slope erosion and slope-failure hazards in seismically active regions. *Geomorphology, 10*, 265–284. https://doi.org/10.1016/0169-555X(94)90021-3.

Kleinbrod, U., Burjánek, J., & Fäh, D. (2019). Ambient vibration classification of unstable rock slopes: A systematic approach. *249*, 198–217. https://doi.org/10.1016/j.enggeo.2018.12.012

Kramer, S. L., Astaneh Asl, B., Ozener, P., & Sideras, S. S. (2015). Effects of liquefaction on ground surface motions. In A. Ansal & M. Sakr (Eds.), *Perspectives on earthquake geotechnical engineering: In honour of Prof. Kenji Ishihara* (pp. 285–309). Springer. https://doi.org/10.1007/978-3-319-10786-8_11

Kramer, S. L., & Mitchell, R. A. (2006). Ground motion intensity measures for liquefaction hazard evaluation. *Earthquake Spectra, 22*, 413–438. https://doi.org/10.1193/1.2194970.

Kramer, S. L., & Stewart, J. P. (2024). *Geotechnical earthquake engineering* (2nd ed.). Boca Raton, FL. https://doi.org/10.1201/9781003512011

Kritikos, T., Robinson, T. R., & Davies, T. R. H. (2015). Regional coseismic landslide hazard assessment without historical landslide inventories: A new approach. *Journal of Geophysical Research: Earth Surface, 120*. https://doi.org/10.1002/2014JF003224.

Kuribayashi, E., & Tatsuoka, F. (1975). Brief review of liquefaction during earthquakes in Japan. *The Japanese Geotechnical Society, 15*(4), 81–92.

Lee, S., & Evangelista, D. G. (2006). Earthquake-induced landslide-susceptibility mapping using an artificial neural network. *Natural Hazards and Earth System Sciences, 6*, 687–695. https://doi.org/10.5194/nhess-6-687-2006.

Lee, S.-J., Komatitsch, D., Huang, B.-S., & Tromp, J. (2009). Effects of topography on seismic-wave propagation: An example from northern Taiwan. *Bulletin of the Seismological Society of America, 99*, 314–325. https://doi.org/10.1785/0120080020.

Lenz, J. A., & Baise, L. G. (2007). Spatial variability of liquefaction potential in regional mapping using CPT and SPT data. *Soil Dynamics and Earthquake Engineering, 27*(7), 690–702. 10.1016/j.soildyn.2006.11.005.

Lévy, C., Baillet, L., Jongmans, D., Mourot, P., & Hantz, D. (2010). Dynamic response of the Chamousset rock column (Western Alps, France). *Journal of Geophysical Research: Earth Surface, 115*. https://doi.org/10.1029/2009JF001606.

Lin, A., Wotherspoon, L., Bradley, B., & Motha, J. (2021). Evaluation and modification of geospatial liquefaction models using land damage observational data from the 2010–2011 Canterbury Earthquake Sequence. Engineering Geology*, 287*, 106099. https://doi.org/10.1016/j.enggeo.2021.106099.

Luzi, L., Lanzano, G., Felicetta, C., D'Amico, M. C., Russo, E., Sgobba, S., Pacor, F., & ORFEUS Working Group 5. (2020). *Engineering Strong Motion Database (ESM), version 2.0*. https://doi.org/10.13127/ESM.2

Ma, S., Archuleta, R. J., & Page, M. T. (2007). Effects of large-scale surface topography on ground motions, as demonstrated by a study of the San Gabriel Mountains, Los Angeles, California. *Bulletin of the Seismological Society of America, 97*, 2066–2079. https://doi.org/10.1785/0120070040.

Malamud, B. D., Turcotte, D. L., Guzzetti, F., & Reichenbach, P. (2004). Landslide inventories and their statistical properties. *Earth Surface Processes and Landforms, 29*, 687–711. https://doi.org/10.1002/esp.1064.

Marano, K. D., Hearne, M., Jaiswal, K. S., Thompson, E. M., Worden, C. B., & Wald, D. J. (2024). ShakeMap Atlas 4.0 and AtlasCat: An archive of the recent and the historical earthquake ShakeMaps, and impacts for global hazard analyses and loss model calibration. *Seismological Research Letters*, 95, 879-899. https://doi.org/10.1785/0220220324

Marc, O., Hovius, N., Meunier, P., Gorum, T., & Uchida, T. (2016). A seismologically consistent expression for the total area and volume of earthquake-triggered landsliding. *Journal of Geophysical Research: Earth Surface, 121*(4), 640–663. https://doi.org/10.1002/2015JF003732.

Martinez, S. N., Allstadt, K. E., Slaughter, S. L., Schmitt, R. G., Collins, E., Schaefer, L. N., & Ellison, S. (2021). *Landslides triggered by the August 14, 2021, magnitude 7.2 Nippes, Haiti, earthquake* (Open-File Report No. 2021–1112). U.S. Geological Survey. https://doi.org/10.3133/ofr20211112

Martino, S. (2016). Earthquake-induced reactivation of landslides: Recent advances and future perspectives. In S. D'Amico (Ed.). *Earthquakes and their impact on society* (pp. 291–322). Cham: Springer International Publishing. https://doi.org/10.1007/978-3-319-21753-6_10.

Massa, M., Barani, S., & Lovati, S. (2014). Overview of topographic effects based on experimental observations: Meaning, causes and possible interpretations. *Geophysical Journal International, 197*, 1537–1550. https://doi.org/10.1093/gji/ggt341.

Massey, C., Lukovic, B., & Dellow, S. (2022). A Prototype earthquake-induced landslide forecast tool for New Zealand. In I. Towhata, G. Wang, Q. Xu, & C. Massey (Eds.). *Coseismic landslides: Phenomena, long-term effects and mitigation* (pp. 617–631). Singapore: Springer Nature. https://doi.org/10.1007/978-981-19-6597-5_17.

Massey, C., Townsend, D., Rathje, E., Allstadt, K. E., Lukovic, B., Kaneko, Y., Bradley, B., Wartman, J., Jibson, R. W., Petley, D. N., Horspool, N., Hamling, I., Carey, J., Cox, S., Davidson, J., Dellow, S., Godt, J. W., Holden, C., Jones, K., ... Villeneuve, M. (2018). Landslides triggered by the 14 November 2016 Mw 7.8 Kaikōura Earthquake, New Zealand. *Bulletin of the Seismological Society of America, 108*, 1630–1648. https://doi.org/10.1785/0120170305.

Maufroy, E., Cruz-Atienza, V. M., Cotton, F., & Gaffet, S. (2014). Frequency-scaled curvature as a proxy for topographic site-effect amplification and ground-motion variability. *105*(1), 354–367. https://doi.org/10.1785/0120140089

McCrink, T. P., Wills, C. J., Real, C. R., & Manson, M. W. (2010). Effects of topographic position and geology on shaking damage to residential wood-framed structures during the 2003 San Simeon Earthquake, Western San Luis Obispo County, California. *Earthquake Spectra, 26*, 779–802. https://doi.org/10.1193/1.3459160.

Medwedeff, W. G., Clark, M. K., & Zekkos, D. (2025). Regional back-analysis of earthquake triggered landslide inventories: A 2D method for estimating rock strength from remote sensing data. *Journal of Geophysical Research: Earth Surface, 130*, e2023JF007471. https://doi.org/10.1029/2023JF007471.

Meunier, P., Hovius, N., & Haines, A. J. (2007). Regional patterns of earthquake-triggered landslides and their relation to ground motion. *Geophysical Research Letters, 34*. https://doi.org/10.1029/2007GL031337.

Molina, S., Lang, D. H., Singh, Y., & Meslem, A. (2019). A period-dependent topographic amplification model for earthquake loss estimation. *17*, 3709–3725. https://doi.org/10.1007/s10518-019-00608-1

Moore, J. R., Gischig, V., Burjánek, J., Loew, S., & Fäh, D. (2011). Site effects in unstable rock slopes: Dynamic behavior of the Randa instability (Switzerland). *101*(6), 3110–3116.

Newmark, N. M. (1965). Effects of earthquakes on dams and embankments. *Geotechnique, 15*, 139–159.

Nowicki Jessee, M. A., Hamburger, M. W., Allstadt, K. E., Wald, D. J., Robeson, S. M., Tanyas, H., Hearne, M., & Thompson, E. M. (2018). A global empirical model for near-real-time assessment of seismically induced landslides. *Journal of Geophysical Research: Earth Surface, 123*, 1835–1859. https://doi.org/10.1029/2017JF004494.

Nowicki, M. A., Wald, D. J., Hamburger, M. W., Hearne, M., & Thompson, E. M. (2014). Development of a globally applicable model for near real-time prediction of seismically induced landslides. Engineering Geology, *173*, 54–65. https://doi.org/10.1016/j.enggeo.2014.02.002.

Obermeier, S. F., Pond, E. C., Olson, S. M., Green, R. A., Stark, T. D., & Mitchell, J. K. (2001). *Paleoliquefaction studies in continental settings: Geologic and geotechnical factors in interpretations and back-analysis* (Open-File Report No. 01–29). U.S. Geological Survey.

Oglesby, D. D., Archuleta, R. J., & Nielsen, S. B. (2000). The three-dimensional dynamics of dipping faults. *Bulletin of the Seismological Society of America, 90*, 616–628. https://doi.org/10.1785/0119990113.

Olsen, M. J., Ashford, S. A., Mahlingam, R., Sharifi-Mood, M., & Gillins, D. T. (2015). Impacts of potential seismic landslides on lifeline corridors. *Oregon Department of Transportation, Research Section.*

Oommen, T., Baise, L. G., & Vogel, R. M. (2011). Sampling bias and class imbalance in maximum-likelihood logistic regression. *Mathematical Geosciences, 43*, 99–120. https://doi.org/10.1007/s11004-010-9311-8.

Pagani, M., Monelli, D., Weatherill, G., Danciu, L., Crowley, H., Silva, V., Henshaw, P., Butler, L., Nastasi, M., Panzeri, L., Simionato, M., & Vigano, D. (2014). Openquake engine: An open hazard (and Risk) software for the global earthquake model. *Seismological Research Letters, 85*, 692–702. https://doi.org/10.1785/0220130087.

Paolucci, R. (2002). Amplification of earthquake ground motion by steep topographic irregularities. *Earthquake Engineering and Structural Dynamics, 31*, 1831–1853. https://doi.org/10.1002/eqe.192.

Papadopoulos, G. A., & Lefkopoulos, G. (1993). Magnitude-distance relations for liquefaction in soil from earthquakes. *83*(3), 925–938. https://doi.org/10.1785/BSSA0830030925

Peng, W.-F., Wang, C.-L., Chen, S.-T., & Lee, S.-T. (2009). Incorporating the effects of topographic amplification and sliding areas in the modeling of earthquake-induced landslide hazards, using the cumulative displacement method. *Computers & Geosciences, 35*, 946–966. https://doi.org/10.1016/j.cageo.2008.09.007 Modelling and Simulation of Dangerous Phenomena for Hazard Mapping.

Petley, D., Dunning, S., Rosser, N., & Kausar, A. B. (2010). Incipient landslides in the Jhelum Valley, Pakistan following the 8th October 2005 earthquake. In *Proceedings of the SAARC workshop on landslide risk management in South Asia* (pp. 107–116), Thimphu, Bhutan.

Plafker, G., Ericksen, G. E., & Fernández Concha, J. (1971). Geological aspects of the May 31, 1970, Perú earthquake* *Bulletin of the Seismological Society of America, 61*, 543–578. https://doi.org/10.1785/BSSA0610030543.

Qin, Y., Zhang, D., Zheng, W., Chen, X., Sun, X., & Gong, Z. (2026). Earthquake-triggered landslide susceptibility modeling based on fault geometry. Engineering Geology*, 362*, 108539. https://doi.org/10.1016/j.enggeo.2025.108539.

Quitoriano, V., & Wald, D. J. (2020). USGS "Did You Feel It?"—Science and lessons from 20 years of citizen science-based macroseismology. *8*. https://doi.org/10.3389/feart.2020.00120

Rai, M., Rodriguez-Marek, A., & Chiou, B. S. (2017). Empirical terrain-based topographic modification factors for use in ground motion prediction. *Earthquake Spectra, 33*, 157–177. https://doi.org/10.1193/071015EQS111M.

Rashidian, V., & Baise, L. G. (2020). Regional efficacy of a global geospatial liquefaction model. Engineering Geology*, 272*, 105644. https://doi.org/10.1016/j.enggeo.2020.105644.

Rathje, E. M., Abrahamson, N. A., & Bray, J. D. (1998). Simplified frequency content estimates of earthquake ground motions. *Journal of Geotechnical and Geoenvironmental Engineering, 124*, 150–159. https://doi.org/10.1061/(ASCE)1090-0241(1998)124:2(150).

Rezaeian, S., Bozorgnia, Y., Idriss, I. M., Abrahamson, N., Campbell, K., & Silva, W. (2014). Damping scaling factors for elastic response spectra for shallow crustal earthquakes in active tectonic regions: "Average" horizontal component. *Earthquake Spectra, 30*, 939–963. https://doi.org/10.1193/100512EQS298M.

Robinson, T. R., Rosser, N. J., Densmore, A. L., Williams, J. G., Kincey, M. E., Benjamin, J., & Bell, H. J. A. (2017). Rapid post-earthquake modelling of coseismic landslide magnitude and distribution for emergency response decision support. *Natural Hazards and Earth System Sciences Discussions*, 1–29. https://doi.org/10.5194/nhess-2017-83.

Rodríguez, C. E., Bommer, J. J., & Chandler, R. J. (1999). Earthquake-induced landslides: 1980–1997. *Soil Dynamics and Earthquake Engineering, 18*, 325–346. https://doi.org/10.1016/S0267-7261(99)00012-3.

Rollins, K. M., Roy, J., Athanasopoulos-Zekkos, A., Zekkos, D., Amoroso, S., & Cao, Z. (2021). A new dynamic cone penetration test–based procedure for liquefaction triggering assessment of gravelly soils. *Journal of Geotechnical and Geoenvironmental Engineering, 147*, 04021141. https://doi.org/10.1061/(ASCE)GT.1943-5606.0002686.

Roshan, P., & Pal, S. (2023). Structural challenges for seismic stability of buildings in hilly areas. *Environmental Science and Pollution Research, 30*, 99100–99126. https://doi.org/10.1007/s11356-022-23263-7.

Saade, A., Abou-Jaoude, G., & Wartman, J. (2016). Regional-scale co-seismic landslide assessment using limit equilibrium analysis. Engineering Geology, *204*, 53–64. https://doi.org/10.1016/j.enggeo.2016.02.004.

Sánchez-Sesma, F. J. (1990). Elementary solutions for response of a wedge-shaped medium to incident. *SH and SV Waves, 80*, 737–742. https://doi.org/10.1785/BSSA0800030737.

Sanger, M. D., Geyin, M., & Maurer, B. W. (2025). Mechanics-informed machine learning for geospatial modeling of soil liquefaction: Global and national surrogate models for simulation and near-real-time response. *Journal of Geotechnical and Geoenvironmental Engineering, 151*, 04025126. https://doi.org/10.1061/JGGEFK.GTENG-13737.

Santucci de Magistris, F., Lanzano, G., Forte, G., & Fabbrocino, G. (2014). A peak acceleration threshold for soil liquefaction: Lessons learned from the 2012 Emilia earthquake (Italy). *Natural Hazards, 74*, 1069–1094. https://doi.org/10.1007/s11069-014-1229-x.

Saygili, G., & Rathje, E. M. (2008). Empirical predictive models for earthquake-induced sliding displacements of slopes. *Journal of Geotechnical and Geoenvironmental Engineering, 134*, 790–803. https://doi.org/10.1061/(ASCE)1090-0241(2008)134:6(790).

Schmitt, R. G., Tanyas, H., Nowicki Jessee, M. A., Zhu, J., Biegel, K. M., Allstadt, K. E., Jibson, R. W., Thompson, E. M., van Westen, C. J., Sato, H. P., Wald, D. J., Godt, J. W., Gorum, T., Xu, C., Rathje, E., & Knudsen, K. (2017). An Open repository of earthquake-triggered ground-failure inventories. US Geol. Surv. *Data Release*. https://doi.org/10.5066/F7H70DB4.

Serey, A., Piñero-Feliciangeli, L., Sepúlveda, S. A., Poblete, F., Petley, D. N., & Murphy, W. (2019). Landslides induced by the 2010 Chile megathrust earthquake: A comprehensive inventory and correlations with geological and seismic factors. *Landslides, 16*, 1153–1165. https://doi.org/10.1007/s10346-019-01150-6.

Seyhan, E., Stewart, J. P., Ancheta, T. D., Darragh, R. B., & Graves, R. W. (2014). NGA-West2 site database. *Earthquake Spectra, 30*, 1007–1024. https://doi.org/10.1193/062913EQS180M.

Shearer, P. M. (2009). *Introduction to seismology*. Cambridge University Press.

Somerville, P. G., Smith, N. F., Graves, R. W., & Abrahamson, N. A. (1997). Modification of empirical strong ground motion attenuation relations to include the amplitude and duration effects of rupture directivity. *Seismological Research Letters, 68*, 199–222. https://doi.org/10.1785/gssrl.68.1.199.

Stein, S., & Wysession, M. (2003). *An introduction to seismology, earthquakes and earth structure* (1 edition). Malden, MA: Wiley-Blackwell.

Stone, I., Wirth, E. A., & Frankel, A. D. (2022). Topographic response to simulated Mw 6.5–7.0 earthquakes on the Seattle fault. *112*(3), 1436–1462. https://doi.org/10.1785/0120210269

Suárez, G., van Eck, T., Giardini, D., Ahern, T., Butler, R., & Tsuboi, S. (2008). The international federation of digital seismograph networks (FDSN): An integrated system of seismological observatories. *IEEE Systems Journal, 2*, 431–438.

Tanyaş, H., van Westen, C. J., Allstadt, K. E., Nowicki Jessee, M. A., Görüm, T., Jibson, R. W., Godt, J. W., Sato, H. P., Schmitt, R. G., Marc, O., & Hovius, N. (2017). Presentation and analysis of a worldwide database of earthquake-induced landslide inventories. *Journal of Geophysical Research: Earth Surface, 122*. https://doi.org/10.1002/2017JF004236.

Tanyaş, H., Van Westen, C. J., Persello, C., & Alvioli, M. (2019a). Rapid prediction of the magnitude scale of landslide events triggered by an earthquake. *Landslides, 16*, 661–676. https://doi.org/10.1007/s10346-019-01136-4.

Tanyaş, H., Westen, C. J.van, Allstadt, K. E., & Jibson, R. W. (2019b). Factors controlling landslide frequency–Area distributions. *Earth Surface Processes and Landforms, 44*, 900–917. https://doi.org/10.1002/esp.4543.

Tatard, L., & Grasso, J. R. (2013). Controls of earthquake faulting style on near field landslide triggering: The role of coseismic slip. *Journal of Geophysical Research: Solid Earth, 118*, 2953–2964. https://doi.org/10.1002/jgrb.50215.

Thompson, E. M., Hearne, M., Aagaard, B. T., Rekoske, J. M., Worden, C. B., Moschetti, M. P., Hunsinger, H. E., Ferragut, G. C., Parker, G. A., Smith, J. A., Smith, K. K., & Kottke, A. R. (2024). Automated, near real-time ground-motion processing at the U.S. Geological survey. *Seismological Research Letters, 96*, 538–553. https://doi.org/10.1785/0220240021.

Todorovic, L., & Silva, V. (2022). A liquefaction occurrence model for regional analysis. *161*, 107430. https://doi.org/10.1016/j.soildyn.2022.107430

Updike, R. G., Egan, J. A., Moriwaki, Y., Idriss, I. M., & Moses, T. L. (1988). A model for earthquake-induced translatory landslides in Quaternary sediments. *GSA Bull, 100*, 783–792. https://doi.org/10.1130/0016-7606(1988)100%3C0783:AMFEIT%3E2.3.CO;2.

U.S. Geological Survey Earthquake Hazards Program. (2019). *Advanced National Seismic System (ANSS) comprehensive catalog of earthquake events and products*. U.S. Geological Survey Data Release. https://doi.org/10.5066/F7MS3QZH

Valagussa, A., Marc, O., Frattini, P., & Crosta, G. B. (2019). Seismic and geological controls on earthquake-induced landslide size. *Earth and Planetary Science Letters, 506*, 268–281. https://doi.org/10.1016/j.epsl.2018.11.005.

Varnes, D. J. (1978). Slope movement types and processes. In *Landslides—Analysis and control* (pp. 11–33).

Vidale, J. E., & Helmberger, D. V. (1988). Elastic finite-difference modeling of the 1971 San Fernando. *California Earthquake, 78*, 122–141. https://doi.org/10.1785/BSSA0780010122.

Wald, D. J., & Allen, T. I. (2007). Topographic slope as a proxy for seismic site conditions and amplification. *Bulletin of the Seismological Society of America, 97*, 1379–1395. https://doi.org/10.1785/0120060267.

Wald, D. J., Worden, C. B., Thompson, E. M., & Hearne, M. (2022). ShakeMap operations, policies, and procedures. *38*(2), 756–777. https://doi.org/10.1177/87552930211030298

Wang, Y., & Rathje, E. M. (2015). Probabilistic seismic landslide hazard maps including epistemic uncertainty. Engineering Geology, *196*, 313–324. https://doi.org/10.1016/j.enggeo.2015.08.001.

Wilkinson, M. D., Dumontier, M., Aalbersberg, I. J., Appleton, G., Axton, M., Baak, A., Blomberg, N., Boiten, J.-W., da Silva Santos, L. B., Bourne, P. E., Bouwman, J., Brookes, A. J., Clark, T., Crosas, M., Dillo, I., Dumon, O., Edmunds, S., Evelo, C. T., ... Mons, B. (2016). The FAIR guiding principles for scientific data management and stewardship. *Scientific Data, 3*, 160018. https://doi.org/10.1038/sdata.2016.18.

Wills, C. J., & Silva, W. (1998). Shear-wave velocity characteristics of geologic units in California. *14*(3), 533–556. https://doi.org/10.1193/1.1586014

Wilson, R. C., & Keefer, D. K. (1985). Predicting areal limits of earthquake-induced landsliding. In J. I. Ziony (Ed.). *Evaluating earthquake Hazards in the los angeles region-an earth-science perspective* (pp. 317–345). USGS Professional Paper. U.S. Geological Survey.

Wirth, E. A., Vidale, J. E., Frankel, A. D., Pratt, T. L., Marafi, N. A., Thompson, M., & Stephenson, W. J. (2019). Source-dependent amplification of earthquake ground motions in deep sedimentary basins. *Geophysical Research Letters, 46*, 6443–6450. https://doi.org/10.1029/2019GL082474.

Worden, C. B., Thompson, E. M., Baker, J. W., Bradley, B. A., Luco, N., & Wald, D. J. (2018). Spatial and spectral interpolation of ground-motion intensity measure observations. *Bulletin of the Seismological Society of America, 108*, 866–875. https://doi.org/10.1785/0120170201.

Xu, S., Dimasaka, J., Wald, D. J., & Noh, H. Y. (2022). Seismic multi-hazard and impact estimation via causal inference from satellite imagery. *Nature Communication. 13*, 7793. https://doi.org/10.1038/s41467-022-35418-8.

Yamaguchi, A., Mori, T., Kazama, M., & Yoshida, N. (2012). Liquefaction in Tohoku district during the 2011 off the Pacific Coast of Tohoku Earthquake. *Soils Found. 52*, 811–829. https://doi.org/10.1016/j.sandf.2012.11.005 Special Issue on Geotechnical Aspects of the 2011 off the Pacific Coast of Tohoku Earthquake.

Yeh, T., Olsen, K. B., Steidl, J. H., & Haeussler, P. J. (2025). Near-surface material and topography generate anomalous high-frequency ground-motion amplification in Chugiak, Alaska. *Bulletin of the Seismological Society of America, 115*, 2793–2808. https://doi.org/10.1785/0120240283.

Yin, Y., Wang, F., & Sun, P. (2009). Landslide hazards triggered by the 2008 Wenchuan earthquake, Sichuan, China. *Landslides, 6*, 139–152. https://doi.org/10.1007/s10346-009-0148-5.

Yong, A., Hough, S. E., Iwahashi, J., & Braverman, A. (2012). A Terrain-Based Site-Conditions Map of California with Implications for the Contiguous United States. *Bulletin of the Seismological Society of America, 102*(1), 114–128. 10.1785/0120100262.

Youd, T. L., & Carter, B. L. (2005). Influence of soil softening and liquefaction on spectral acceleration. *Journal of Geotechnical and Geoenvironmental Engineering, 131*, 811–825. https://doi.org/10.1061/(ASCE)1090-0241(2005)131:7(811).

Youd, T. L., & Idriss, I. M. (2001). Liquefaction resistance of soils: Summary report from the 1996 NCEER and 1998 NCEER/NSF workshops on evaluation of liquefaction resistance of soils. *Journal of Geotechnical and Geoenvironmental Engineering, 127*, 297–313. https://doi.org/10.1061/(ASCE)1090-0241(2001)127:4(297).

Youd, T. L., & Perkins, D. M. (1978). Mapping Liquefaction-Induced Ground Failure Potential. *Journal of the Geotechnical Engineering Division, 104*(4), 433–446. 10.1061/ajgeb6.0000612.

Zhu, J., Baise, L. G., & Thompson, E. M. (2017). An Updated Geospatial Liquefaction Model For Global Application. *Bulletin of the Seismological Society of America, 107*, 1365–1385. https://doi.org/10.1785/0120160198.

Zhu, J., Daley, D., Baise, L. G., Thompson, E. M., Wald, D. J., & Knudsen, K. L. (2015). A geospatial liquefaction model for rapid response and loss estimation. *Earthquake Spectra, 31*, 1813–1837. https://doi.org/10.1193/121912EQS353M.

CHAPTER 5

Earth surface processes and geomorphology

Tim Davies
School of Earth and Environment, University of Canterbury, Christchurch, New Zealand

5.1 Introduction

The purpose of this chapter is to provide a basis for interpreting the landform geometries that result from the action of gravity and its consequent forces causing landscape deformation, to underpin the study of the impacts of earthquakes on the form of the landscape. These seismic impacts are potentially distinguishable from the geomorphic outcomes of nonseismic events because they result from sudden, violent forces acting on the land surface during earthquakes, rather than the less intense and more prolonged forces that characterise nonseismic processes.

The geometry and behaviour of Earth's surface, and of the processes that form it, are the subject of the science of geomorphology, which has a long and venerable history (e.g. Tinkler, 2020, and references therein). Geomorphology traditionally attempts to identify and systematise the shapes of features present in a landscape in terms of the processes that give rise to them. These processes involve the response of crustal solids to the forces acting on them due to gravity and to gravity-driven fluid motion; such responses are based on the well-known principles of fluid and solid mechanics. Thus the conventional basis of geomorphology is the application of mechanics to the analysis and modelling of observed landforms, in order to understand their past evolution and predict their future behaviour. As well as being of deep scientific interest, this topic is of fundamental importance to human society, which lives on Earth's changing surface and makes use of it for all the activities of society, thus being susceptible to its changes; understanding past changes and anticipating future changes are crucial for informing societal preadaptation to reduce the societal impacts of future geomorphic events.

This chapter is intended to provide an outline of what is known about the basics of geomorphology for scientists who are not closely acquainted with this area. Any comprehensive summary would require several volumes to cover the plethora of plausible, but not universally agreed, mechanisms proposed to explain how the landscape is formed and altered by agents of erosion and deposition. The approach taken herein is, by contrast, empirical, outlining what is known about the ways in which gravity and fluid flow cause solid materials to move in the landscape, and then outlining

Geomorphic Impacts of Earthquakes
DOI: https://doi.org/10.1016/B978-0-443-40375-0.00006-2

and illustrating (in a digital appendix) the relatively small variety of characteristic shapes that result from these motions. This approach is unconventional in that it does not provide explicit theoretical or model linkages between physical processes and the resulting forms, which has been the objective of much geomorphology to date. Instead, it is suggested that a conceptual basis for landscape evolution may be based on the developing field of complex dynamic systems (CDS).

The present approach is justified by thinking of geomorphic processes as a nested set of CDS of all scales; such systems are known to be characterised by the phenomenon of *emergence*, in which the operation of a set of known processes gives rise to generally consistent, but unexpected and unexplainable, behaviours and geometries at a larger scale. If this perspective is valid, for which there is increasing evidence, then the conventional search for explicit mechanical linkages between process and form might be fruitless at present. It would then follow that ever-more-detailed modelling might not be the best way to anticipate future geomorphic changes. The empirical approach, by contrast, relies only on known forces and geometries to anticipate future morphologies. Our rapidly increasing knowledge of CDS behaviour has the potential to usefully illuminate future approaches to geomorphology.

5.2 Perspectives on geomorphology; mechanics, complexity and modelling

Tectonic ground movement, earthquakes and volcanism can all cause externally driven changes to the local relief and hence the topographic gradient of landscapes. The material comprising the landscape experiences changed forces from these changes in gradient; these forces are exerted directly by gravity, and indirectly by the gravity-driven motion of fluids (water, air and ice) moving across the land surface. The ways in which materials move in response to these forces constitute a wide range of processes (e.g. gravity-driven landslides, growth of alluvial fans by sediment deposition, and creation of canyons as river channels incise by bed erosion). These processes, in turn, can trigger and interact with a plethora of local landscape responses, such as the blocking of rivers by landslides, the undercutting of slopes by rivers, or lake formation during glacier advance. At larger spatial scales, river erosion can cause local tectonic uplift to increase, and isostatic rebound during deglaciation can cause earthquakes, to mention only a few.

Further complications in landform change are caused by biological processes and human activities. The growth cycles of vegetation, and their changes due to alterations in climate, can affect the ability of driving forces to disturb the underlying terrain, as can human alterations of natural systems (e.g. water extraction for irrigation, flood control or power generation); both biotic and human influences can trigger substantial changes in landscape form and behaviour at decade to century and longer time scales.

Conventional geomorphology attempts to understand and explain these phenomena by embedding solid- and fluid-mechanics principles (Newton's laws of motion; conservation of mass, energy and momentum; and empirical rules for the effects of solid friction and fluid turbulence) in numerical models that simulate the behaviour of geomorphic systems within specific environments. However, attempts to develop accurate predictions of the future evolution of natural landscapes from fundamental mechanics are bedevilled by three factors: (1) the complications of the detail of initial landforms and geology, driving forces, and environmental constraints in natural situations; (2) the lack of fundamental understanding of the mechanics of solid friction and fluid turbulence which act to degrade potential, pressure and kinetic energies to heat; and (3) the substantial behavioural complexity present in any given local situation, resulting from the interactions of many forces acting on many (themselves interacting) system components, which, as will be explained, means that the resultant landforms cannot be related directly and unequivocally to the forces that are acting.

The first factor is increasingly able to be managed by modern digital technology (i.e., by ever more detailed, complicated and sophisticated computer modelling, albeit at increasing time and cost requirements). The second factor is commonly dealt with by the use of empirical data (such as solid friction coefficients and fluid flow resistance factors) derived from both field and laboratory data, which obviously introduces uncertainties into predictions for situations not previously encountered. However, the vast database of fluid turbulence and solid friction behaviour acquired over the past two centuries usually allows these uncertainties to be relatively minor.

The third factor, complexity, is less easily dealt with. In essence, the nonlinear interactions of multiple geomorphic processes both cause and interact with feedback-dominated evolution of land surface geometry, so that landscapes behave as CDS. In such systems the phenomenon of *emergence* dominates the progression from a known set of lower-level processes to a set of next-level-up processes. Emergence means that there is no indication in the lower-level processes that the specific higher-level processes are going to appear, nor do the higher-level outcomes appear to depend on lower-level details; thus the higher-level behaviour *emerges* from the lower level. The outcome in geomorphic terms is that a specific land configuration emerges from the driving forces, formative processes and environment, and evolves until further evolution is limited by intrinsic or extrinsic constraints. The set of behaviours that emerges from specific lower-level situations is generally consistent and comprises a small number of well-defined behaviour patterns. A simple example of emergence is the development of a vortex when the plug is removed from a bath full of water; the stationary body of water in the bath prior to plug removal contains no hint of the well-defined flow structure that will inevitably emerge, and the vortex emerges irrespective of the motion of the water in the bath as the plug is removed.

A geomorphic example of this emergence is the development of a series of river meanders from an initially straight water channel in erodible material; here, the meander characteristics (wavelength, amplitude and sinuosity) are not an explicit consequence of the characteristics of flow in the initial straight channel. Generally, as in this case, the details of how specific landforms emerge from the underlying processes are unknown, but often the process appears to involve a *tendency* (e.g., for river bed configurations, such as meanders or ripples, to evolve so as to increase the resistance they offer to flow past them; Davies, 1987) that is limited by a *constraint* (e.g., when the changing bed configuration achieves a local maximum of resistance to flow, so that further deformation reduces flow resistance again; or it encounters an area of inerodible material) that halts the evolution. The significant issue is that full knowledge of the forces acting on a specific land surface does not, in general, allow the subsequent land surface geometry to be predicted without empirical extrapolation from the past behaviour of the same or similar systems.

There is thus good reason to suspect that, because of inescapable system complexities, the bottom-up approach of ever more complicated, sophisticated, expensive and time-consuming modelling driven by conventional mechanics will not soon allow reliable prediction of the geomorphology resulting from specific perturbations (natural or anthropic) to present-day landscapes. This same outcome is becoming apparent in weather forecasting, where purely empirical methods based on artificial intelligence (AI) are already outstripping some aspects of physics-based models: '*... cutting-edge generative AI methods can capture very high-dimensional and complex distributions over rich temporal dynamics, with sufficient accuracy and reliability to support effective decision-making in crucial applications*' (Price et al., 2024).

Furthermore, it has long been realised that the variety of characteristic landform shapes that emerge from the complexities of geomorphic system dynamics is quite small. Originally, this limited range of shapes was considered to correspond to the limited number of formative processes available, and the idea of process-form relationships took hold, but this correspondence is undermined by the recognition that similar morphologies often occur when different processes and materials are involved ('equifinality'). It is suggested herein that the occurrence of a small variety of characteristic landform morphologies appears to parallel the emergence of characteristic spatial patterns from CDS behaviour. Thus what is known about CDS hints at the potential utility of a 'top-down' as a complement to a 'bottom-up' approach to the problems of understanding and prediction in geomorphology.

I first outline the fundamental mechanics of land surface change and then describe the limited set of landform shapes that results from these processes, without inferring any detailed mechanical connections between them. Subsequently, the basics of dynamic complexity concepts and the characteristic behaviours and patterns that result are outlined, together with some parallels in geomorphology, suggesting that these are a useful basis for conceptualising geomorphology.

5.3 Geomorphic systems behaviour

5.3.1 Motion of individual solids

The basic process of geomorphology that alters the surface geometry of a landscape is the motion of a unit of solid crustal material (e.g., a clay particle, sand grain or rock mass) from an initial location at rest in the landscape to a different location at rest. In order to move a mass, a net (unbalanced) force must be exerted on it, and in land-forming processes, the ultimate source of all such driving forces is gravity. Gravity itself acts to cause solid masses to move closer to Earth's centre if that motion is dynamically feasible, that is if restraining forces (reactions and friction) can be overcome. In addition, gravity acts on fluids (water, ice and lava) causing them to deform and flow downhill, and these moving fluids can in turn exert forces on solid masses and cause them to move. Gravity also causes air masses of different densities (due to temperature differences) to flow across the planet's surface, generating winds and thus waves in water, that are both able to cause grain motion. Even tectonic plate motion, which causes uplift and volcanism to generate land surface elevation changes and sometimes generates impulsive seismic forces, is the result of gravity-driven motion of the lithosphere above a fluid mantle.

Forces tending to oppose the motion of solid particles are:

i. Static reaction forces applied by contact-adjacent solid materials following Newton's third law of motion: action and reaction are equal and opposite. Here the geometry of adjacent static solids results in forces that act against the available directions of motion. The reaction forces also generate static friction forces that prevent sliding motion as long as the local coefficient of static friction is sufficient to prevent motion.

ii. Sliding contacts between a moving solid and its adjacent surfaces generate dynamic frictional forces that oppose motion; these supersede prior static forces at the same contacts. The magnitudes of these dynamic forces depend on the nature of the sliding surfaces, and they are generally proportional to the square of the sliding velocity. Dynamic friction coefficients are generally smaller than their static counterparts.

iii. Fluid motion necessarily involves distributed internal shear of the material, which is opposed by the material viscosity. Some fluids (e.g. ice) can resist a given level of applied shear and deform elastically when stationary, but when this level is exceeded, the material begins to deform plastically (i.e. flow) and viscosity takes over control of the material motion. Importantly, water cannot resist any shear whatsoever; this is why stationary water surfaces are horizontal.

When the fluid is water, its extremely low viscosity means that resistance to motion is almost always dominated by turbulence, which is characterised by the ubiquitous presence of eddies (cells of three-dimensional circular motion) at all scales greater than a minimum (of the order of mm) determined by the fluid viscosity.

5.3.2 Motion of collective solids

Where a small number of individual solid masses respond to applied forces in a given volume of space, the spatial and temporal frequencies of intersolid collisions are low and the collision-generated forces will have little overall effect on the collective motion, which can thus be approximated as the sum of the individual motions. As collision frequencies increase, the collective behaviour increasingly departs from the simple sum of individual motions, until at some concentration collisional forces become significant and can give rise to quite different behaviours (especially when individual solids are breaking). These behaviours are also affected by the density and rheology of the interstitial fluid, if present. Granular material comprising many individual solid particles can flow *en masse*, and such flows are controlled by intergranular friction forces, and also by fracture-generated impulse forces if the solid grains are breaking.

5.3.3 Sliding and flowing

When multiple solid grains (wet or dry) are moving in contact with a substrate, they can either slide or flow, and it is important to understand the difference between these modes of motion. When *sliding,* the solid grains move as a rigid body, that is without internal shear, by slippng across the substrate; all of the shear occurs at the substrate interface. When the material is *flowing,* there is no shear at the substrate interface, and all the shear takes place within the body of grains, which therefore moves as a fluid. In some cases, mixed motion may occur where sliding occurs at the substrate interface *and* shear simultaneously takes place within the body of the material; glacier motion is an example where the moving body, in this case ice, has a very high viscosity. Mixed motion is, however, most likely where the solid mass is stratified into layers of different properties.

5.3.4 Feedbacks

In fluid-driven motion of solids, the driving force is the fluid flow, and this is fundamentally dependent on the geometry of the stationary boundary of the flow. This boundary is also where the morphological landscape change is occurring due to grain motion, so there is feedback between the boundary and the flow; the erosion or deposition of solid material by the flow alters the boundary shape, and this in turn alters the flow, causing the erosion or deposition to change. Where the feedback is positive, the boundary change is enhanced by the previous boundary change, while in a negative feedback the previous change in boundary shape reduces the subsequent change in boundary shape. The former clearly leads to increasing morphological change whose eventual geometry cannot be determined, while the latter, if sufficiently prolonged, leads to morphological stability. Insofar as many landforms appear to be at least quasistable in timeframes of human interest, it would appear that negative feedbacks are as common as the positive feedbacks that generate the landforms from whatever forms preceded them. In order to explain a specific landform, then, both the processes of boundary deformation that generate the characteristic shape and the process(es) that halt (s) the deformation at the final shape and size need to be explained.

5.4 Processes

The processes that describe the ways in which solids are transported by gravity and fluid motion are set out in Table 5.1; here the processes are classified according to whether the solids are moving as individual coherent masses, or as a group of multiple individual masses that are moving but not significantly interacting, or as multiple masses that are interacting with each other, that is, they are sufficiently close to each other (with sufficient volumetric concentration) that significant forces result from collisions or prolonged contact between adjacent masses.

The role of interstitial pore fluids is conceptually important (e.g. Iverson, 2005), although difficult to quantify because pore fluid pressures vary rapidly and locally in moving bodies of solids. In unsaturated bodies of solids, fluids are held close to solid contact points and have concave air-water interfaces; fluid pressures are thus negative and act to hold the solids together. In saturated bodies of solids, fluid pressures are positive and tend to force solids apart, thus reducing intergranular contact pressures.

Note that process nomenclature is itself an obstacle to communicating and analysing information; for example, the flow of water containing high concentrations of sediment may be referred to as hyperconcentrated flow, mudflow, mudflood or debris flow

Table 5.1 Transport processes of single, multiple and interacting mass movement, in response to gravity and fluid flow.

Driver	Gravity	Fluid	Fluid	Fluid
Environment / Fluid		Water	Air	Ice
Individual mass	rockfall, boulder roll, blockslide	rolling, sliding, saltation, suspension	saltation, impact creep, suspension	supra/en/sub- glacial block
Multiple masses	rockfall/dry ravel, nuée ardente, powder-snow avalanche, turbidity current	bedload, suspended load	creep, saltation, sandstorm,	supra/en/sub- glacial debris
Interacting masses	rock avalanche, pyroclastic flow, block–and-ash flow, wet snow avalanche	traction carpet, slurry flow		supra/subglacial debris
Fluid-unsaturated interacting masses	landslide/debris avalanche, earthflow/debris flow			
Fluid-saturated interacting masses	Mudslide, subaqueous landslide			rock glacier

(Keaton, 2019). Alternatively, one might say that multiple processes fit within the categories or classes in Table 5.1, depending on setting and the characteristics and relative proportion or availability of solid and fluid phases. At the more detailed level, however, the mechanics of water–sediment flow processes are better defined as outlined above.

Each of these processes generates characteristic morphologies during (1) erosion, where surface material is set in motion at a greater rate than material is deposited at the same location; (2) deposition, where material is deposited on the surface at a greater rate than it is eroded from the surface; and (3) pure transport, where the rates of erosion and deposition are equal. Conservation of mass in solids transport is described by the Exner equation (e.g. Paola and Voller, 2005):

$$\frac{\delta\eta}{\delta t} = -\frac{\delta U}{\delta x}$$

where η is bed elevation relative to some fixed datum, t is time, U is average flow velocity and x is downstream distance.

These characteristic morphologies are also liable to be modified by events external to the system, such as fault surface uplift and base-level change. Local land uplift or downthrow can modify both landforms and the processes that generate them. *Base level* is the level at which rivers enter the sea or a lake, where the flow velocity becomes zero and sediment transport ceases; landscapes formed by river processes are thus altered if base level changes and river erosion alters at a given location. For example, if a coastal area is uplifted relative to sea level by tectonics or isostasy (e.g. due to deglaciation), river base level will fall and local river slope will increase; increased river erosion will cause this steeper reach to migrate upstream, and eventually the whole catchment may suffer reduction in elevation so that its landscape again corresponds to the new sea level. Note, however, that whether sea-level fall causes downcutting or aggradation on land depends on the relative gradients of the adjacent land-surface and sea-bed (Schumm, 1993).

The characteristic morphologies and the situations in which they are found are listed below:

5.5 Characteristic landscape shapes

While a very wide range of landscape features can result from differential erosion of crustal materials that are heterogeneous due to underlying geological structure, a much smaller range of characteristic landform shapes results from geomorphic processes acting to deform *homogeneous* crustal materials by erosion, transport and deposition of detrital materials.

The following classification considers geomorphic signatures in the form of surfaces, deposits, travel paths and lineations. These forms are listed below, together with some situations in which they occur. Illustrations of the specific types of landform are provided in a digital appendix under the same classification.

5.5.1 *Surfaces* are areally extensive landforms that can be planar, conical, exponential or stepped

5.5.1.1 Planar

Surfaces are extensive flat areas most often formed by deposition of sediment from the water that has transported it to the location.

If the planar surface is *horizontal,* then the water from which it was deposited was stationary, which in turn means that the sediment will be very fine and able to be suspended by very slight water movement or even Brownian motion. Horizontal planar land surfaces thus frequently reflect the past presence of lakes.

Gently inclined planar surfaces suggest that the flows that emplaced them were flowing, albeit slowly, and thus were probably rivers carrying fine sediment. Such rivers often meander across the valleys they occupy, forming floodplains.

Moderately inclined planar surfaces reflect the action of faster rivers carrying coarser sediment, which are likely to be braided rather than meandering, but still alter course from time to time so that the whole width of valleys is potentially active river bed (floodplain).

Steep (angle-of-repose) planar surfaces result from subaerial emplacement of multiple small rockfalls from a laterally extensive source such as a long cliff. Hillslopes in weak bedrock can also achieve angles of repose by surface weathering following oversteepening by basal erosion; as rock strength increases, the steady-state slope angle also increases. The angle of repose of cohesionless granular crustal materials ranges from c.30 degrees (for smooth, rounded clasts) to c.40 degrees (for angular clasts).

Very steep to vertical planar surfaces usually result from either fault uplift or from ongoing rapid basal erosion of a slope. Except in strong (highly cohesive) rock, these usually degrade by weathering and erosion to lower angles over time.

5.5.1.2 Conical

Surfaces are formed similarly to planar surfaces but are restricted to situations where the deposited sediment emanates and spreads from a localised source (e.g. water flow from a gorge, gulley or valley mouth, rockfall from a point source, or a volcano).

5.5.1.3 Curved

Surfaces are formed by sediment transport by rainsplash and/or gravity down a mild (less than angle-of-repose) hillslope, with or without sediment deposition at the foot of the slope. The combination of erosion upslope and deposition downslope causes double curvature in both the downslope and cross-slope directions.

5.5.1.4 Stepped

Stepped surfaces result from externally driven uplift or downthrow/downwasting affecting existing surfaces. This can result either from fault uplift or downthrow, or from base-level lowering and river incision.

5.5.2 Mass erosion-deposition features

These features result from the gravity-driven detachment of masses of material from their initial positions in the landscape and their emplacement downslope following runout across the landscape.

Scarp-slump features form when a mass of material becomes unstable and detaches from its parent slope, sliding a short distance to a new position under gravity without deforming significantly.

Hummocks: These are heaps of material – initially conical but which flatten out over time – formed by mass movements such as landslides or debris flows that travel longer distances. Hummocks are also frequently formed by glacially driven sediment accumulations (moraines).

Push ridges: These are transverse ridges formed by compression of sliding material as it slows down and halts; again, these can also be formed by glacial action.

Fans: These are sometimes formed by mass movements that slide/flow relatively long distances and spread across valley floors. These tend to have horizontal cross-profiles rather than the conical profiles of features in 5.5.1.2 above.

5.5.3 Travel paths

These are formed by fluids flowing across the landscape and forming their own paths, usually by erosion and deposition of sediment, but sometimes by incising, melting or dissolving the material that they flow through.

Straight flow paths are usually either artificial or controlled by a fault along which easier erosion of crushed material allows the river to flow. Some mass movements turn into rapid mass flows with straight paths.

Meanders are single flow paths which curve alternately in opposite directions; meander wavelengths generally scatter around about ten times the flow width.

Confluent flow paths converge and join in the downstream direction, as in drainage basins (catchments). These usually reflect erosion and incision of watercourses.

Divergent flow paths bifurcate repeatedly in the downstream direction, as in deltas and alluvial fans. These usually reflect sediment deposition and watercourse aggradation.

Braided flow paths both diverge and converge in the downstream direction.

Scour-hole, ripple and dune features form in erodible-bed channels when near-bed flow conditions become strongly nonuniform; local excess shear stress erodes a hole in the bed and the eroded sediment deposits just downstream to form the ripple or dune, which in turn alters the flow geometry and causes a further scour-hole. The characteristic bedform shape is controlled by the geometry of successive scour-holes.

Antidunes: Rapid shallow flows form sinusoidal antidunes.

5.5.4 Lineations

These are long, very narrow landscape features usually comprising ridges but occasionally depressions.

Straight: Lineations can be formed by faults, or by volcanic intrusions of resistant rock (dykes), or sometimes in groups formed by ice-sheet-driven or aeolian sediment deposition and erosion.

Arcuate: Lineations are most frequently formed by sand dunes or terminal moraines of glaciers, but can also be formed at the toes of large mass movements.

Wiggly: Lineations are most often formed by eskers – sediment deposited in subglacial conduits of retreated glaciers. Sand dunes can also form long, wiggly ridges.

Circular: Lineations are the result of bolide impacts, or of subsidence, or of ejection features such as volcanoes.

5.5.5 Combinations

Many combinations of erosional and depositional features are possible; for example, bowl-and-cone features (a slump scar immediately above a depositional cone) appear at all scales and in a wide variety of circumstances.

5.6 Complex systems

The following section discusses the behaviour of CDS and the parallels with geomorphology, to illustrate the potential for knowledge of these systems to contribute to the understanding of landscape evolution.

5.6.1 Systems

The systems concept has frequently been used as a perspective on geomorphology (e.g. Favis-Mortlock, 2013; Huggett, 2007). A system is a group of components that interact with each other to produce an outcome. Thus a geomorphic system is a unit of land surface whose components (solid and granular materials, gravity, flowing water, vegetation) interact following mechanical principles so that the shape of the land surface alters over time.

5.6.2 Simple systems

The properties and behaviour of a simple system (such as a mechanical device) can be understood or predicted from the addition or aggregation of its components; that is, the large-scale properties and behaviour of a simple system can be inferred from knowledge of the smaller-scale properties and behaviour of its parts. Changes in system variables that are directly proportional to time, to the system's current state, or to changes in the environment are called linear and typify simple systems. If the interactions of the landscape components are simple, in this sense, then the system outputs are simply related to the changes in the components.

5.6.3 Complex systems

In complex systems, by contrast, the properties of a whole system often cannot be understood or predicted from the knowledge of its components because the components interact with each other, giving rise to unexpected structures and behaviours at larger scales, a phenomenon known as 'emergence' (Section 5.8).

5.6.3.1 Complex system characteristics

Complex systems are typically nonlinear, meaning that their variables change at different rates depending on their states and their environment. Interactions between components of a complex system may self-organise to produce distinct spatial and temporal patterns of behaviour. As a system becomes more organised by this process, new interaction patterns may emerge over time, potentially leading to the production of even greater complexity. Some complex systems self-organise into a 'critical' state – a balance between randomness and regularity, such as avalanches from slowly-fed sandpiles. Such 'self-organised critical' states often show characteristics such as self-similarity and power-law distributions of pattern properties. Complex systems may have stable states in which they can return to the same state if perturbed, and unstable states in which the systems can be drastically disrupted by a small perturbation; in some cases, small environmental changes can completely change the system behaviour, giving rise to transitions that are known as bifurcations, phase transitions or tipping points. Some systems are extremely sensitive to small perturbations and unpredictable in the long run; these are called chaotic. The future state of a complex system may depend not only on its present state, but also on its history.

5.6.3.2 Categories of complex systems

In general terms, there are three categories of complex systems:

a. CDS: In these systems, the behaviours of the system components, and the interactions among them, are mechanical and unchanging with time. Abiotic geomorphic systems (i.e. those not containing living components) belong to this category; solid and fluid mechanics principles underpin CDS behaviour, although the outcomes are often not predictable from those principles. A geomorphic example is the response of the bed configuration in a river to changes in flow rate, forming ripples, dunes, meanders and braiding.

b. Complex *adaptive* systems (CAS): Here, the nature and behaviour of components and their interactions change with time because living systems are involved, and the system evolves. This is characteristic of biotic systems in which the nature and behaviour of system components change slowly by evolution in response to their changing environments. Geomorphology is generally liable to be affected relatively slowly by biosystems (vegetation and animals); an example is the way in which river behaviour is affected by catchment vegetation succession resulting from fire or changes in climate, causing changes in sediment supply and storage.

c. Complex *human* systems (CHS): Here, the behaviour of some components is affected by human actions, in which consciousness is involved. The behaviour of societal system components can change rapidly as a result of human behaviour, and geomorphology can be affected rapidly by this anthropogenic behaviour. An example is the way in which river behaviour is affected by dam construction, deforestation or gravel extraction.

Herein, CDS are emphasised as the main drivers of geomorphology, while CAS and CHS behaviours come into play when geomorphology is affected by the presence of living systems (mainly vegetation) and by human activities, respectively. However, it is assumed that biotic and human influences do not significantly alter the *range* of landform morphologies that emerge from the dynamics of solid motion, but rather affect their *speed* of development and the sequence of evolution.

5.7 Complex system behaviours

CDS exhibit the following behaviours (corresponding geomorphic examples are listed that correspond to each):

5.7.1 The response of a complex system to a perturbation is typically a set of self-reinforcing (positive-feedback-driven) processes leading to the appearance and development of characteristic spatial patterns

Examples: an initially flat sand river bed deforms into ripples, dunes and antidunes in response to a sufficient perturbation or increasing flow; an artificially straightened river reach forms initially small curves, which amplify into meanders and propagate downstream (e.g. Davies & Tinker, 1984; Hooke, 1977).

5.7.2 The geometries that characterise complex systems are scale-free (fractal)

Such that similar spatial behaviour and characteristic patterns occur across a range of scales and materials.

Example: the amplitudes and wavelengths of meanders are related to channel width irrespective of scale and physical materials (Davies & Sutherland, 1980); many of the landform shapes outlined in Section 5.5.5 occur across a wide range of scales.

5.7.3 The size distributions of pattern elements are space-filling; that is, the size distributions of elements or patterns are those that occupy all the available volume

Such distributions give rise to power-law probability density functions. The existence of such probability density functions in temporal distributions further suggests that complex system behaviour may result in 4-dimensional spacetime-filling distributions.

Examples include fluid shear turbulence (Mihelich et al., 2017; Xu & Yang, 2018) and primary (unsorted) particle-size distributions (Dufresne & Dunning, 2017), in both of which the elements (eddies and grains, respectively) are created (by shear and fragmentation, respectively) in proportions by size that allow smaller elements to fill the spaces between larger elements.

5.7.4 Complex dynamic systems self-organise so that minimum energy expenditure is required to accomplish a given quantity of work

This behaviour corresponds to the maximum entropy production principle (MEPP) that appears to be well-established for general physics (Kleidon, 2010; Martyushev & Seleznev, 2006; Martyushev, 2021; Mihelich et al., 2017; Ross et al., 2012). This is, in some sense, equivalent to a maximum efficiency principle, so that maximum mechanical work is achieved for given energy expenditure, which can be a constraint on emergent dynamics.

Examples: derivatives of MEPP have been shown to apply to some river morphology situations, for example that the geometries of both river meanders and lower-flow regime bedforms (ripples and dunes) correspond to local maxima of resistance to flow, implying that flow boundary shapes self-adjust so that the flows can transport sediment at the highest possible mass flow rate (e.g. Davies & McSaveney, 2006; Davies & Sutherland, 1980, 1983; Davies, 1987).

5.8 Emergence

5.8.1 Key concepts

A crucial feature of CDS behaviour is that of emergence. The term describes how a set of known processes operating in a known environment results in behaviour that is not expectable from this knowledge; in other words, things happen that cannot be explained by conventional reductionist physics (e.g. Kauffman, 2007, 2008). Recently, emergence has been the focus of a number of publications that emphasise the fundamental role it may play in understanding both how the planet and its biome function and how such complexity developed from the primordial accumulation of space junk that was early Earth (e.g. Azarian, 2022). It is useful to illustrate the key concepts that underlie emergence and its applicability to geomorphology. An excellent example in the context of fluvial geomorphology is the occurrence of meandering and braiding patterns from the flow of water across sand and gravel beds, because the equations that describe the motion of sediment by water flows do not appear to imply such behaviours.

5.8.2 Example: surface tension meanders

A simpler analogue example makes the point more clearly – that of surface tension meanders (Fig. 5.1). The figure shows the flow of a narrow (c.3 mm wide) laminar stream of high-viscosity water straight down a steeply-inclined Perspex sheet; an initial bend is introduced (1) and no further action is taken. After a few minutes, the initial bend has amplified, thus sharpening the downstream junction with the initial stream, which in turn starts to amplify into a further bend in the opposite direction (2). The disturbance continues to propagate downstream (3) and eventually generates a long

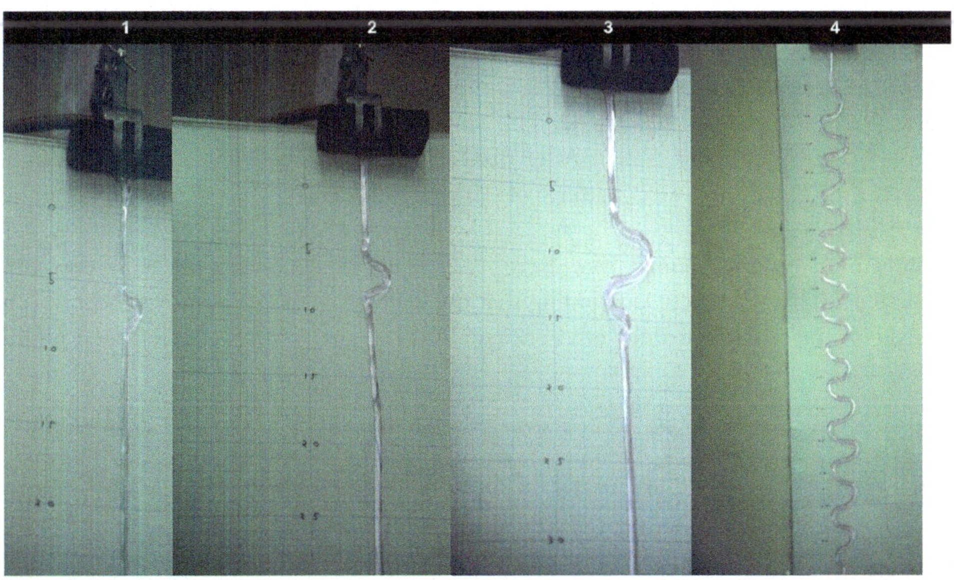

Figure 5.1 Development of meanders in the flow of high-viscosity water down a Perspex plate (Author's photo).

sequence of similar bends. The environment in which this phenomenon occurs is simple and ideal, being planar, smooth and isotropic; the flow itself is initially simple, lacking turbulence or sediment. The active process is that of the advance of the surface-tension- and angle-of-contact-controlled edge of the initial bend across the Perspex in response to fluid forces acting on the outside of the flow bends. The governing equations here are much more clearly understood than those of sediment erosion and transport, but again do not appear to imply the resultant behaviour.

5.8.3 Example: Langton's Ant

A final, even simpler, example hammers the point home – Langton's Ant (Langton, 1986). This is a 'cellular automaton' that operates within an environment consisting of a uniform two-dimensional grid of squares. The procedure identifies a square as the active square. The active square (the 'ant') can move in any of the four cardinal directions at the first time step. The 'ant' subsequently moves according to two rules:

- Arriving at a white square: turn 90 degrees clockwise, alter the colour of the square (from white to black), move forward one unit in the same direction.
- Arriving at a black square: turn 90 degrees counter-clockwise, alter the colour of the square (from black to white), move forward one unit in the same direction.

Here, both the environment and the movement rules are completely known and simple.

The behaviour, however, is anything but simple. Starting from a completely white grid, the first few hundred steps result in a set of more-or-less symmetrical patterns, as one would expect. After a few hundred moves, a large, irregular pattern of black and white squares develops (Fig. 5.2A). The ant then traces a pseudo-random path until around 10,000 steps (Fig. 5.2B). After this, a dramatic change of behaviour occurs; a recurrent diagonal 'highway' pattern of 104 steps emerges, that extends indefinitely (Fig. 5.2C,D).

All initial configurations that have been tested eventually produce the same repetitive 'highway' pattern, suggesting that the 'highway' is in some way fundamentally implicit in the combination of environment and rules, but it has not yet been proven

(A) (B) (C) (D)

Figure 5.2 Behaviour of Langton's Ant. (A) 479 steps, (B) 6179 steps, (C) 10459 steps, (D) 11464 steps.

that this is true for all such initial configurations. Here we have the consistent emergence of extremely well-defined behaviour (morphology) from an apparently simple, fully-defined situation – but we have a complete lack of any understanding of how that behaviour is implicit in the rules and environment governing the Ant's motion.

These examples illustrate the difficulty that the possibility of emergence in geomorphology poses for the development of an understanding of the morphology of landforms from knowledge of the materials and forces involved.

5.9 Complex systems geomorphology

If geomorphic processes do indeed behave as CDS, it would be expected that the resultant landforms will be scale-free, space-filling and energetically (or entropically) optimal. As noted above, there is some evidence that this is the case. This in turn suggests that it would, in principle, not be possible to predict the future landform geometry purely from knowledge of the formative forces operating on a known starting situation (geometry plus materials).

5.9.1 Attractors, characteristic morphologies and constraints

The response of a complex system to stimuli involves the emergence of higher-level dynamics and processes from lower-level dynamics and processes, so that unexpected behaviour emerges from simple rules. In complex systems terms, the tendency for the emergent behaviours to be consistent is ascribed to the existence of attractors in the behavioural phase space. In the geomorphic context, the consistent geometric patterns (morphologies) that characterise this emergent behaviour result from positive-feedback-driven development processes being limited by the action of internal or external constraints (negative feedbacks). The constraints on emergent dynamics arise internally from the principles of mechanics (including conservation of mass, energy and momentum; gravity; Newtonian mechanics; and open-system (nonlinear) thermodynamics), as well as externally from the nature of system boundaries (such as inerodible stream boundaries) and of Earth surface space (e.g. cutoffs limiting meander sinuosity). Thus while the emergence may be unexplainable, the constraints are in some cases explainable (e.g. maximisation of entropy production implying maximum friction factor and maximum sediment transport; space filling implying power-law size distributions).

5.9.2 Maximum entropy production

The apparent operation of the MEPP in the behaviour of CDS is particularly interesting in the geomorphic context. The possibility that observed landscape behaviours correspond to some energetic extremal principle was studied several decades ago in the context of rivers and generated both criticism and enthusiasm, but has been of less interest more recently (e.g. Davies & McSaveney, 2006; Davies & Sutherland, 1983; Davies, 1987; Davy & Davies, 1979; Griffiths, 1984; and references therein). Nevertheless, a recent review (Tranmer et al., 2020) concluded that the concept has validity. Given that much of geomorphology results from the transport of granular material across Earth's surface by fluid motion, there is attraction in the concept that, since the surface topography caused by these transport processes controls the transport rates, the topography will tend towards that which maximises the transport rate for the given fluid energy available. 'If it were done ..., then twere well it were done quickly: ...' (Shakespeare, 1607, cited in Shakespeare, 1992). While supported only for river behaviour to date in the geomorphic context, it is tempting to suggest that the principle might well apply to other geomorphic processes that are associated with characteristic landforms, such as aeolian transport, debris flows and some mass movements. It is noteworthy that in recent years, the concept that adaptive systems interact with their surroundings in such a manner as to co-evolve towards maximisation of entropy production rate is being applied increasingly widely and successfully (e.g. Azarian, 2022). If this can be shown to apply to other geomorphic processes it would provide a satisfactory – or at least plausible – underpinning for quantification of geomorphic change in general.

5.9.3 Equifinality

A further indicator that tendencies and constraints determine the morphology of landforms is the phenomenon of 'equifinality', which in geomorphology means that distinctly different processes can lead to similar landforms (e.g. Nicholas & Quine, 2010). If landform shape indeed results from the action of a constraint limiting the progress of a tendency, then the morphology may be independent of the nature of the process and materials on which the tendency and constraint operate. For example, Fig. 5.3 shows the result of mass movements in very different circumstances and materials: in the upper image the Allen II rock avalanche in Alaska (McSaveney, 1978) has left a digitate deposit of rock debris on the surface of a snow-covered glacier, while in the lower image a wet-snow avalanche has left a strikingly similar deposit on a snow-covered alluvial fan (Seligman & Douglas, 1962). Given the extreme differences in material properties of the avalanche materials (dry rock debris and wet snow) in the two situations, this similarity of deposit morphology is startling.

Another example of equifinality is the consistent geometry of meandering streams; as shown by Davies and Sutherland (1980), there is a well-defined relationship (Fig. 5.4) between meander wavelength λ and channel breadth B:

$$\lambda \approx 10B$$

Figure 5.3 Equifinality: spectacularly elongated, digitate deposits (arrowed) in widely different materials. Upper: supraglacial rock avalanche deposit of Allen II rock avalanche, Alaska (McSaveney, 1978). Lower: wet-snow avalanche deposit onto snow-covered pasture (Seligman & Douglas, 1962).

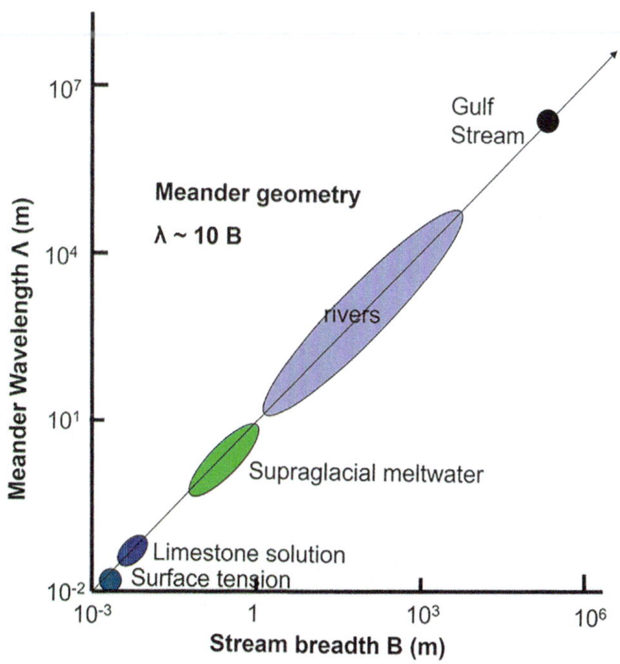

Figure 5.4 Equifinality – consistent meander geometry in water flows across scales and environments. *Data envelopes from Davies, T.R.H., & Sutherland, A.J. (1980). Resistance to flow past deformable boundaries. Earth Surface Processes, 5, 175–179.*

This relationship remains consistent across a 10^8-fold range of linear scales and a variety of environments in which flowing water is deforming to develop meanders, including those where the boundary to the flow comprises surface tension, limestone, ice, sediment and cold water.

5.10 Contribution of seismicity to geomorphology

This volume explores the ways in which earthquakes contribute to geomorphology, and the foregoing perspective is intended to encourage exploration of potentially novel aspects of this contribution. In a similar sense, this final section sets out an overview of the ways in which earthquakes can affect the driving forces and material properties that affect the motion of solids on Earth's surface.

5.10.1 Radiated energy

The seismic energy radiated by earthquakes is manifest in crustal materials as three-dimensional oscillatory seismic forces, which can be large in magnitude (comparable with static gravity forces – earthquakes can cause ground accelerations exceeding those of gravity), with frequencies in the Hz–kHz range and with durations of up to several minutes. These forces are thus capable of substantially altering the properties of Earth's surface materials, such as shear strength and density, and causing ground instability, leading to large-scale ground motions and failures while the seismic forces are in operation.

5.10.2 Ground shaking

The response to seismic forces can cause fractures to initiate and propagate in crustal materials, resulting in alteration of material strength and permeability. These can influence the propensity for a hillslope to fail (Brain et al., 2017), and the ground shaking can increase the runout of a resultant landslide (Brain et al., 2015). Severe weakening of rock adjacent to active faults is well known and, in combination with seismic forces, causes ground failures to occur preferentially close to faults (Massey et al., 2013). The amplification of ground shaking as a result of near-surface stratigraphy and surface topography (Havenith et al., 2003) can also cause ground failures to occur far from faults, such as along mountain ridge crests. In saturated fine-grained soils, coseismic liquefaction may cause considerable deformation in flat topography. At the scale of mountain edifices, resonant shaking of the mountain can cause large, deep-seated failure surfaces to initiate and propagate, and if any of these intersect with the surface they can cause edifice-scale instability and large coseismic landslides (Bazgard et al., 2011; Davies & Moretti, 2022).

5.10.3 Surface uplift and downthrow

Land can be uplifted or downthrown by earthquakes, with vertical elevation changes of up to c.5–10 m in a single event. These changes may be very localised, resulting in fault scarps, or distributed over larger areas; however, their geomorphological significance is limited except for fluvial contexts, where the resulting alteration of water flow gradients can cause dramatic geomorphic changes (e.g. McEwan et al., 2025; Yanites et al., 2010). Over a sequence of earthquakes on century and greater timescales, however, cumulative elevation changes can affect slope morphologies sufficiently to affect slope stability.

5.10.4 Horizontal offset

Lateral displacement across a fault can be up to c.10s of m in a single earthquake, and offsets are commonly seen where faults cross linear landscape features such as stream channels. Again, excepting the fluvial context, single-event offsets cause only minor geomorphological consequences, but over a sequence of events the effects can become correspondingly more significant, for example leading to river avulsion into a new course.

5.10.5 Postseismic geomorphic cascade

Often equally as geomorphically significant as the immediate consequence of an earthquake on geomorphology are the longer-term and more extensive consequences that result from the initial impacts cascading through the intricately linked network of geomorphic processes that underpin landscape evolution (Fig. 5.5; Burt & Allison, 2010; Davies & Stahl, 2022; Fan et al., 2019; Kincey et al., 2021; Marc et al., 2019). The time scales of this process cascade vary dramatically; for example, while a coseismically emplaced landslide dam in one valley may fail within a few days (e.g. Hancox et al., 2005), another nearby dam may remain in place, impounding water and sediment, for years (e.g. Massey et al., 2013) or even millennia (e.g. Augustinus et al., 2013; Wang et al., 2015). Thus it is quite conceivable that the consequence cascade of one earthquake may still be in progress when the next earthquake happens, so that no interseismic 'equilibrium' stage occurs during which no seismic influences are present.

5.10.6 Discussion

Reference to Section 5.5.5 shows that there are no characteristic landform types that occur only as the result of earthquakes. However, straight lineations – particularly single, long straight lineations – occur more frequently as fault traces than from any other cause (which can include, for example, volcanic dykes) and, particularly when they define geological boundaries, indicate the likely past (and thus future) occurrence

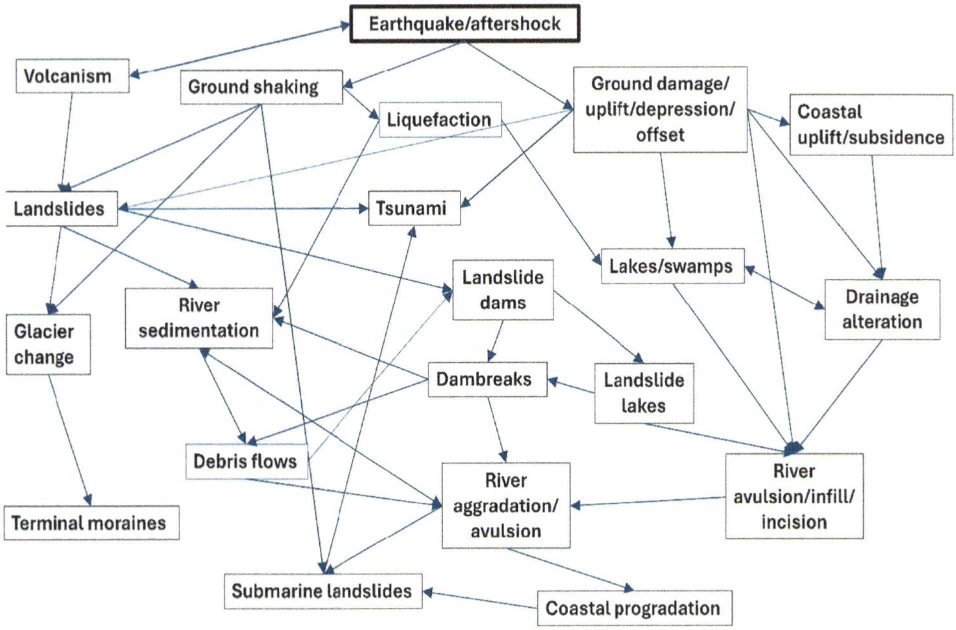

Figure 5.5 Cascade of geomorphic processes triggered by an earthquake. *Adapted from Davies, T., & Stahl, T. (2022). Geomorphic impacts of large earthquakes. In Shroder, J.J.F. (Ed.), Treatise on geomorphology vol. 9 Elsevier, Academic Press pp. 652–674. https://doi.org/10.1016/B978-0-12-818234-5.00013-4 ISBN: 9780128182345.*

of earthquakes. Thus the least equivocal indicators of earthquake activity are long, straight(ish) lineations (e.g. Fig. 5.6).

More generally, the effect of earthquakes is to alter – perhaps drastically – the time of occurrence, rate, and magnitude of the surface processes that give rise to particular landforms, by briefly altering either or both of the driving and restraining forces that affect surface deformation. Usually, seismically active regions will experience more frequent geomorphological events than inactive ones, mainly because they are subject to more rapid tectonic activity (including, but not limited to, rock uplift) and hence more rapid increase in land surface gradients, and because earthquakes can be high-intensity landscape disturbances. However, regional seismic activity can change over time; Scandinavia is an example of a region that was seismic during rapid deglaciation following the last ice age (c.10,000–20,000 years ago), and there are many lineations (fault traces) and large landslide remnants that reflect that activity (Eldholm & Bungum, 2021). Today, by contrast, with the crust having rebounded from the depression caused by a regional ice sheet, seismicity is relatively low – lower than might be expected from the geomorphic indicators (see Chapter 2) – and so the geomorphic indicators of

Figure 5.6 Long straight lineation in a mountain landscape indicates a large-scale earthquake fault (shown by the arrows): Alpine Fault, South Island, New Zealand. Copyright: GNS Science.

earthquakes can over-represent the future likelihood of earthquakes. Recent work in Labrador (Syvitski et al., 2025) suggests a similar situation there.

5.11 Conclusions

5.11.1 Emergence

The evident complexity of process interactions in geomorphology suggests that it may well be realistic to admit the possibility that, as with CDS, characteristic morphologies are emergent, and thus that the search for specific mechanical process-form relationships may not be fruitful until emergence in general is better understood.

5.11.2 Modelling

A further implication is that physically-based geomorphic modelling – e.g. of future river behaviour in response to anticipated alterations in water and sediment inputs – may not realistically represent emergence and so may yield incorrect predictions. In any case, such modelling can only yield characteristic outputs, such as ranges of possible patterns of behaviour, rather than specific behaviour predictions. This means also that progress in handling large data sets and more detailed environments (e.g. higher-resolution digital elevation models and particle-size distributions) is unlikely to yield more accurate model outputs. Use of Monte Carlo strategies to investigate the potential ranges and probabilities of behaviours will be limited by running times until further advances in computing technology occur.

5.11.3 Characteristic morphologies

The shape of the landscape, as formed by geomorphic processes, comprises a limited variety of characteristic features. These are distributed in the landscape so as to be scale-free, space-filling and to maximise entropy production (i.e. energy dissipation) rate.

5.11.4 Application

If geomorphic systems are indeed CDS, as evidence from fluvial system behaviour suggests, then the best perspective from which nonspecialists can utilise geomorphic knowledge is an empirical one – that is, to become acquainted with the limited range of characteristic geomorphic shapes found in the landscape and the ranges of circumstances in which they can form. In this way, the possible ranges of past and future landscape geomorphic activity can be realistically assessed without involvement in uncertain cause-effect issues.

5.12 Supplementary material

Supplementary material related to this chapter can be found on online at doi: https://doi.org/10.1016/B978-0-443-40375-0.00006-2.

References

Augustinus, P. M., Sweeney, C., Brideau, M., & Fink, D. (2013). Lochnagar landslide-dam, central otago, new Zealand: Geomechanics and timing of the formative event (s). *AGU Fall Meeting Abstracts, 2013*, NH21A–NH21497.

Azarian, B. (2022). The romance of reality: How the universe organizes itself to create life, consciousness, and cosmic complexity. *Benbella books*, 321.

Bazgard, A., Davies, T. R. H., & Cundall, P. A. (2011). Dynamic analysis of mountain response using PFC3D and FLAC3D. In *Continuum and distinct element modelling in geomechanics. Minneapolis, MN, USA*: Itasca, 1–7.

Brain, M. J., Rosser, N. J., Sutton, J., Snelling, K., Tunstall, N., & Petley, D. N. (2015). The effects of normal and shear stress wave phasing on coseismic landslide displacement. *Journal of Geophysical Research: Earth Surface, 120*(6), 1009–1022.

Brain, M. J., Rosser, N. J., & Tunstall, N. (2017). The control of earthquake sequences on hillslope stability. *Geophysical Research Letters, 44*(2), 865–872.

Burt, T. P., & Allison, R. J. (2010). Sediment cascades in the environment: An integrated approach. In *Sediment cascades: An integrated approach*. Wiley, 471.

Davies, T., & Stahl, T. (2022). Geomorphic impacts of large earthquakes. In J. J. F. Shroder (Vol. Ed.), *Treatise on geomorphology: vol. 9*, (pp. 652–674). Elsevier, Academic Press. https://doi.org/10.1016/B978-0-12-818234-5.00013-4 ISBN: 9780128182345.

Davies, T. R. H., & McSaveney, M. J. (2006). Geomorphic constraints on the management of bedload-dominated rivers. *Journal of Hydrology (New Zealand), 45*(2), 111–130.

Davies, T. R. H., & Moretti, D. (2022). Geomorphic precursors of large landslides: Seismic preconditioning and slope-top benches. In *Landslide hazards, risks, and disasters*. Elsevier, 641–666.

Davies, T. R. H. (1987). Channel boundary shape—Evolution and equilibrium. In K. Richards (Vol. Ed.), Rivers—Environment, process and form, *IBG Spec. Publication. 18*, (pp. 228–248). London: Inst. of Br. Geogr.

Davies, T. R. H., & Sutherland, A. J. (1980). Resistance to flow past deformable boundaries. *Earth Surface Processes, 5*, 175–179.

Davies, T. R. H., & Sutherland, A. J. (1983). Extremal hypotheses for river behaviour. *Water Resources Research, 19*(1), 141–148.

Davies, T. R. H., & Tinker, C. C. (1984). Fundamental characteristics of stream meanders. *Bulletin of the Geological Society of America. 95*, 505–512.

Davy, B. W., & Davies, T. R. H. (1979). Entropy concepts in fluvial geomorphology: A reevaluation. *Water Resources Research, 15*(1), 103–106.

Dufresne, A., & Dunning, S. A. (2017). Process dependence of grain size distributions in rock avalanche deposits. *Landslides, 14*(5), 1555–1563.

Eldholm, O., & Bungum, H. (2021). Potential secondary events caused by early Holocene paleoearthquakes in Fennoscandia – A climate-related review. *Norwegian Journal of Geology, 101*, 202108. https://doi.org/10.17850/njg101-2-3.

Fan, X., Scaringi, G., Korup, O., West, A. J., van Westen, C. J., Tanyas, H., Hovius, N., Hales, T. C., Jibson, R. W., Allstadt, K. E., & Zhang, L. (2019). Earthquake-induced chains of geologic hazards: Patterns, mechanisms, and impacts. *Reviews of Geophysics, 57*.

Favis-Mortlock, D. (2013). 1.14 Systems and complexity in geomorphology. In J. F. Shroder (Ed.). *Treatise on geomorphology* (pp. 257–270).

Griffiths, G. A. (1984). Extremal hypotheses for river regime: An illusion of progress. *Water Resources Research, 20*(1), 113–118.

Hancox, G. T., McSaveney, M. J., Manville, V. R., & Davies, T. R. (2005). The october 1999 Mt Adams rock avalanche and subsequent landslide dam-break flood and effects in Poerua river, Westland, New Zealand. *New Zealand Journal of Geology and Geophysics, 48*(4), 683–705.

Havenith, H.-B., Vanini, M., Jongmans, D., & Faccioli, E. (2003). Initiation of earthquake-induced slope failure: Influence of topographical and other site specific amplification effects. *Journal of Seismology, 7*, 397–412.

Hooke, J. M. (1977). The distribution and nature of changes in river channel patterns: The example of devon. In K. J. Gregory (Ed.). *River channel changes* (pp. 266–280). New York: Wiley.

Huggett, R. (2007). A history of the systems approach in geomorphology. *Géomorphologie: Relief, processus, environnement, 13*(2), 145–158.

Iverson, R. M. (2005). Regulation of landslide motion by dilatancy and pore pressure feedback. *Journal of Geophysical Research: Earth Surface, 110*(F2).

Kauffman, S. A. (2007). Beyond reductionism: Reinventing the sacred. *Zygon® 42*(4), 903–914.

Kauffman, S. A. (2008). *Reinventing the sacred: A new view of science, reason, and religion*. New York: Basic Books, 320.

Keaton, J. R. (2019). Review of contemporary terminology for damaging surficial processes: Stream flow, hyperconcentrated sediment flow, debris flow, mud flow, mud flood, mudslide. *Association of Environmental and Engineering Geologists special publication 28*.

Kincey, M. E., Rosser, N. J., Robinson, T. R., Densmore, A. L., Shrestha, R., Pujara, D. S., Oven, K. J., Williams, J. G., & Swirad, Z. M. (2021). Evolution of coseismic and post-seismic landsliding after the 2015 Mw 7.8 Gorkha earthquake, Nepal. *Journal of Geophysical Research: Earth Surface, 126*(3) p. e2020JF005803.

Kleidon, A. (2010). A basic introduction to the thermodynamics of the Earth system far from equilibrium and maximum entropy production. *Philosophical Transactions of the Royal Society B: Biological Sciences, 365*(1545), 1303–1315.

Langton, C. G. (1986). Studying artificial life with cellular automata. *Physica D: Nonlinear Phenomena, 22*(1-3), 120–149.

Marc, O., Behling, R., Andermann, C., Turowski, J. M., Illien, L., Roessner, S., & Hovius, N. (2019). Long-term erosion of the Nepal Himalayas by bedrock landsliding: The role of monsoons, earthquakes and giant landslides. *Earth Surface Dynamics,* 7(1), 107–128.

Martyushev, L. M. (2021). Maximum entropy production principle: History and current status. *Physics-Uspekhi, 64*(6), 558.

Martyushev, L. M., & Seleznev, V. D. (2006). Maximum entropy production principle in physics, chemistry and biology. *Physics Reports, 426*(1), 1–45.

Massey, C. I., McSaveney, M. J., & Davies, T. R. H. (2013). Evolution of an overflow channel across the Young River landslide dam, New Zealand. In *Landslide science and practice: Volume 6: Risk assessment, management and mitigation.* Berlin, Heidelberg: Springer Berlin Heidelberg, 43–49.

McEwan, E., Stahl, T., Langridge, R., Davies, T. R. H., Howell, A., & Wilson, M. (2025). Seismic hazard and shifting channels: Exploring coseismic river response. *Earth-Science Reviews,* 105042.

McSaveney, M. J. (1978). Sherman glacier rock avalanche, Alaska, USA. In *Developments in geotechnical engineering, Vol. 14,* Elsevier, 197–258.

Mihelich, M., Faranda, D., Paillard, D., & Dubrulle, B. (2017). Is turbulence a state of maximum energy dissipation? *Entropy, 19*(4), 154.

Nicholas, A. P., & Quine, T. A. (2010). Quantitative assessment of landform equifinality and palaeoenvironmental reconstruction using geomorphic models. *Geomorphology, 121*(3-4), 167–183.

Paola, C., & Voller, V. R. (2005). A generalized Exner equation for sediment mass balance. *Journal of Geophysical Research: Earth Surface, 110*(F4).

Price, I., Sanchez-Gonzalez, A., Alet, F., Andersson, T. R., El-Kadi, A., Masters, D., Ewalds, T., Stott, J., Mohamed, S., Battaglia, P., & Lam, R. (2024). Probabilistic weather forecasting with machine learning. *Nature, 2024,* 1–7.

Ross, J., Corlan, A. D., & Müller, S. C. (2012). Proposed principles of maximum local entropy production. *The Journal of Physical Chemistry B, 116*(27), 7858–7865.

Schumm, S. A. (1993). River response to baselevel change: Implications for sequence stratigraphy. *The Journal of Geology, 101*(2), 279–294.

Seligman, G., & Douglas, C. K. M. (1962). *Snow structure and ski fields* Brussels: Jos. Adam.

Shakespeare, W. (1992). *Macbeth (folger shakespeare library). Folger Shakespeare Library.*

Syvitski, J., Normandeau, A., & Lajeunesse, P. (2025). Major Holocene glacio-isostatically-induced earthquakes triggered mass-transport deposits and seabed displacements in Lake Melville. *Marine Geology,* 107482.

Tinkler, K. J. (2020). *A short history of geomorphology.* Routledge.

Tranmer, A. W., Caamano, D., & Goodwin, P. (2020). Evaluation of extremal hypotheses in an undeveloped alluvial river. *Progress in Physical Geography: Earth and Environment, 44*(4), 514–533.

Xu, H. H., & Yang, X. I. A. (2018). Fractality and the law of the wall. *Physical Review E, 97*(5), 053110.

Wang, G., Huang, R., & Kamai, T. (2015). *The internal structure and stability of some large landslide dams induced by the 2008 Wenchuan (Mw7. 9) Earthquake, China. In Engineering Geology for Society and Territory-Volume 2: Landslide Processes,* 1163–1169.

Yanites, B. J., Tucker, G. E., Mueller, K. J., & Chen, Y. G. (2010). How rivers react to large earthquakes: Evidence from central Taiwan. *Geology, 38*(7), 639–642.

CHAPTER 6

Fault trace processes

Marie-Luce Chevalier
Institute of Geology, Chinese Academy of Geological Sciences, Beijing, P.R. China

6.1 Introduction

The Earth's surface is composed of large, rigid blocks known as tectonic plates, which move relative to one another like pieces of a vast puzzle (Fig. 6.1). These movements generate large-scale active faults that accommodate the resulting deformation. While many active faults are located along plate boundaries, significant faulting also occurs within the plates because of intraplate deformation driven by regional compression, extension or shear (e.g. Tapponnier et al., 2001; Zoback, 1992). These fault systems are the primary sources of earthquakes, releasing accumulated energy through sudden rupture events that may propagate all the way to the Earth's surface. When this displacement reaches the surface, it creates what is known as a fault trace, the geomorphic expression of rupture visible at or near the ground surface, linking deep-seated tectonic activity with observable surface deformation (e.g. Yeats et al., 1997). Understanding how fault traces form, propagate and evolve is essential to the broader study of tectonic geomorphology and seismic hazard assessment (e.g. Wesnousky, 2008).

The three main types of faults (Fig. 6.2) are:

- Strike-slip faults, which involve horizontal motion, where two blocks slide past one another laterally along the fault plane. These faults are typically steep and accommodate shear stress. Well-known examples are the San Andreas Fault in California and the Alpine Fault in New Zealand. Other examples discussed throughout this chapter are taken from the large-scale strike-slip faults on the Tibetan Plateau (e.g. Kunlun, Karakorum, Xianshuihe, Altyn Tagh).
- Normal faults, where the crust experiences extension. In these faults, the hanging wall moves downward relative to the footwall, creating features like fault scarps and grabens. They are common in regions of continental rifting and extensional tectonics, such as in the Basin and Range region of the western United States, the Rhine graben in western Europe, or southern Tibet.
- Reverse faults (or thrusts) form in compressional settings, where the hanging wall moves upward relative to the footwall. These faults often generate prominent scarps and folds and are typical of convergent plate boundaries and mountain belts such as the Himalayas.

Geomorphic Impacts of Earthquakes
DOI: https://doi.org/10.1016/B978-0-443-40375-0.00012-8

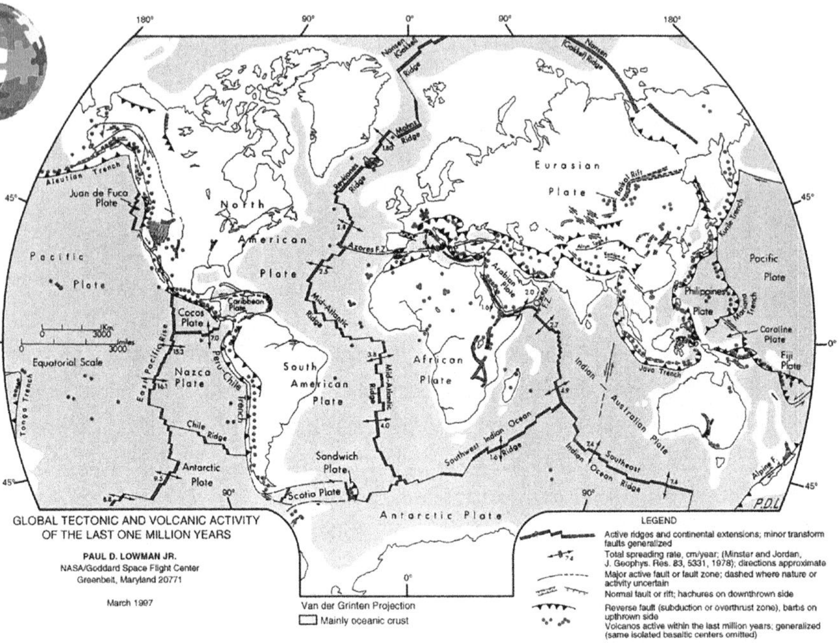

Figure 6.1 Distribution of tectonic activity over the past one million years, highlighting active mid-ocean ridges, continental extension, transform and strike-slip faults, spreading rates, continental rifts, subduction zones and thrusts. (Paul D, Lowman Jr., Code 921, NASA/Goddard Space Flight Centre, Greenbelt, MD 20771). Note: Map lines delineate study areas and do not necessarily depict accepted national boundaries.

During major earthquakes, slip on a fault may reach all the way to the surface, producing coseismic surface ruptures (sections 6.2 and 6.4). The nature of these surface ruptures depends on factors like rock type, fault geometry, as well as the depth and magnitude of the event (e.g. Wesnousky, 2008). These expressions, collectively referred to as fault trace processes, show how the ground deforms during earthquakes and reveal how faults move and how surface materials respond to stress.

Fault trace processes also operate as long-term agents of landscape development. In the context of geomorphology – the scientific study of landforms, their processes, history and the patterns they form on the Earth's surface – repeated surface ruptures over geological time can offset, laterally or vertically, landforms such as rivers, river fill or strath terraces and the risers between them, alluvial fans or glacial moraines; they can also form shutter ridges, dam drainages to create sag ponds or generate cumulative fault scarps (section 6.3; e.g. Burbank & Anderson, 2011; Wallace, 1977). These features shape the evolution of tectonically active landscapes and provide a visible history of seismic recurrence. Fault traces thus serve as both short-term records of individual earthquakes and long-term archives of tectonic deformation.

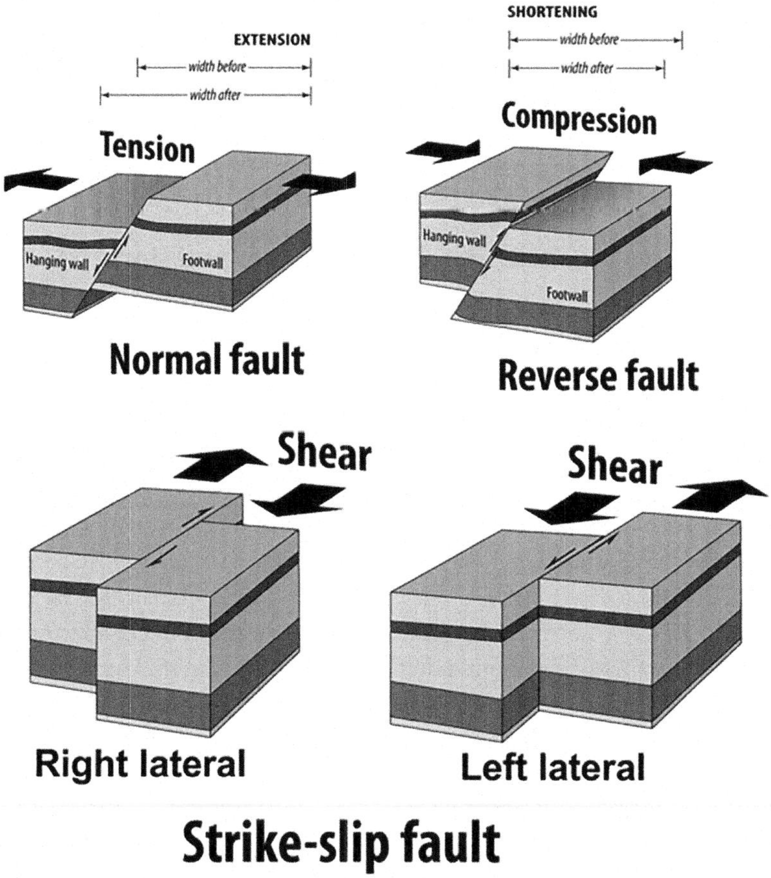

Figure 6.2 Types of faults. Top: Dip-slip faults, including normal faults formed under extensional (divergent) tectonic settings and reverse or thrust faults resulting from compressional (convergent) forces. Bottom: Strike-slip faults, right-lateral (dextral) or left-lateral (sinistral), accommodating shear. *Image credit: Physical Geology, 3rd Adapted Edition by Karla Panchuk used under a CC BY 4.0 international license.*

Recent advances in remote sensing – particularly high-resolution light detection and ranging (LiDAR), interferometric synthetic aperture radar (InSAR), and uncrewed aerial vehicle (UAV)-based photogrammetry, but also high-resolution satellite image (WorldView and Pléiades) correlation – have revolutionised the study of fault traces (e.g. Antoine et al., 2021; Lemrabet et al., 2023; Li et al., 2025; Massonnet et al., 1993; Oskin and Arrowsmith, 2012; Pan et al., 2022; Ren et al., 2022b; Zielke et al., 2010). These technologies enable precise mapping of surface ruptures and offset landforms, often revealing subtle or distributed deformation patterns that were previously undetectable (section 6.4.1). As a result, our understanding of the complexity and spatial variability of surface faulting has deepened significantly in recent years.

This chapter explores the nature and expression of fault trace processes, from the mechanisms that produce surface ruptures to their geomorphic signatures, landscape alteration and implications for seismic hazard analysis. By examining these features, we gain not only a clearer picture of the geomorphic consequences of earthquakes but also insight into the dynamic processes that continually reshape tectonically active regions.

6.2 Fault rupture mechanisms and surface expressions

Surface ruptures mark the final and most visible stage of earthquake processes, where displacement originating at seismogenic depths propagates upward through the brittle crust. This propagation is governed by dynamic rupture mechanisms and the mechanical properties of the overlying materials (e.g. Avouac, 2003; Scholz, 2002; Sibson, 1983). While ruptures on continental faults typically initiate at 5–15 km depth, their ability to reach the surface depends on energy release, slip distribution, hypocenter depth and attenuation through overlying materials. Large-magnitude earthquakes (generally $M_w \geq 6.7$, e.g. Somerville, 2003) or moderate events (M_w 4.6 to 6) at shallow depths (≤ 4 km) (e.g. Bonilla, 1988; Di et al., 2023) often produce coseismic ruptures. These manifest as steep scarps (free faces), horizontal or vertical offsets, extensional cracks, compressional push-ups or mole tracks or distributed deformation zones (Fig. 6.3A; section 6.4). Their morphology is controlled by lithology, near-surface material properties and structural complexities such as stepovers or bends (Fig. 6.3A, B) (e.g. King & Nabelek, 1985; Klinger, 2010; Li et al., 2016; Oskin and Arrowsmith, 2012; Wesnousky, 2006; Zinke et al., 2019). The resulting offset landforms, sag ponds and shutter ridges (e.g. Bai et al., 2021; McCalpin, 2009) preserve surface deformation, providing a detailed record of strain accommodation during earthquakes. These features offer critical insights for seismic hazard assessment and serve as direct archives of fault-related deformation (Cunningham and Mann, 2007). Alternatively, ruptures may terminate beneath the surface as blind faults, generating subtle distributed deformation without clear scarps (e.g. Li et al., 2025; Shaw & Shearer, 1999) – a configuration that complicates hazard assessment (e.g. Milliner et al., 2015).

Dynamic rupture models and field observations reveal complex relationships between fault behaviour and surface expression. While slip typically peaks at depth and diminishes toward the surface, shallow unconsolidated materials can locally amplify surface displacement (e.g. Dunham et al., 2011; Oglesby et al., 2000). Fault segmentation and stepovers govern rupture geometry and displacement distribution (e.g. Biasi & Wesnousky, 2016; Lettis et al., 2002; Oglesby, 2005; Wesnousky, 2006). Segment boundaries – marked by stepovers, bends or cross-faults – act as barriers or sites of complex deformation. Releasing stepovers promote extension (e.g. forming pull-apart basins), while restraining bends generate localised uplift (e.g. forming pressure ridges). Recent earthquakes demonstrate how ruptures navigate these complexities: some rupture

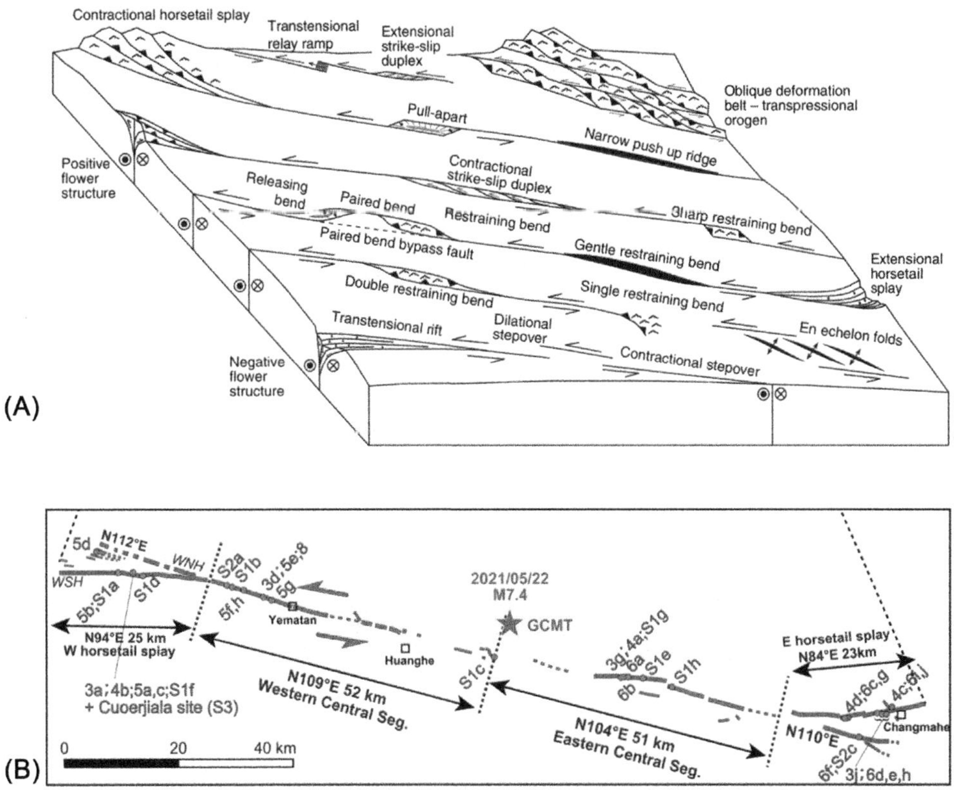

Figure 6.3 (A) Block diagram illustrating fault segments and stepovers, including releasing and restraining bends, where rupture may jump or terminate. (B) Structural complexity (bifurcation at both ends) of the 2021 M_w 7.4 Maduo earthquake surface rupture on the Tibetan Plateau. *(A) From Cunningham, W.D., & Mann, P. (2007). Tectonics of strike-slip restraining and releasing bends. London: Geological Society, 290, 1–12 https://doi.org/10.1144/SP290.1 Special Publications. (B) From Pan, J., Li, H., Chevalier, M.L., Tapponnier, P., Bai, M., Li, C., & Chen, P. (2022). Coseismic rupture of the 2021, M_w7. 4 Maduo earthquake (northern Tibet): Short-cutting of the Kunlun fault big bend. Earth and Planetary Science Letters, 594, 117703.*

across multiple geometric barriers (e.g. the 2023 M_w 7.8 Türkiye earthquake rupturing 300 km; Barbot et al., 2023; Ren et al., 2024), while rupture in others is arrested at segment boundaries. The 2019 Ridgecrest earthquake sequence in California, for example, illustrates both behaviours: the M_w 6.4 foreshock rupture arrested at a fault intersection, while the M_w 7.1 mainshock propagated through minor structural barriers (e.g. Chen et al., 2020; Shelly, 2020). Segmented rupture was also observed in the 2021 M_w 7.4 Maduo (Tibetan Plateau) earthquake (e.g. Pan et al., 2022; Ren et al., 2022a) and as cascading multifault failure in the 2016 M_w 7.8 Kaikōura (New Zealand) earthquake (Hamling et al., 2017). Such events underscore the need to integrate segmentation models with dynamic rupture processes for robust hazard assessments.

Several key factors influence whether and how a rupture is expressed at the surface:

- Depth of rupture initiation: Deeper hypocentres reduce the likelihood of surface rupture, while shallow events (e.g. <10 km depth) have a higher probability of reaching the surface.
- Lithology and rock strength: Strong, brittle bedrock tends to localise rupture more clearly, producing steep scarps (Fig. 6.4A, B). In contrast, Quaternary deposits or highly weathered rocks can absorb strain and promote distributed deformation (Fig. 6.4C, D).
- Soil cover and surficial materials: Thick alluvial or colluvial deposits can mask or diffuse fault rupture. In these settings, the surface rupture may be expressed as subtle folding rather than discrete breaks (e.g. Li et al., 2025).
- Preexisting structures: Earthquakes may reactivate older, now inactive structures or zones of weakness, thus influencing the style and geometry of surface rupture by being more diffuse (e.g. Armijo et al., 1986).

Surface ruptures can be categorised into primary and secondary faulting based on their kinematic and structural relationships to the main rupture. This distinction is critical for seismic hazard assessment, surface rupture characterisation, and paleoseismological interpretations (e.g. Choi et al., 2018; Li et al., 2016; Yuan et al., 2022).

- Primary faulting occurs along the master fault that accommodates most of the coseismic slip (Fig. 6.5A, B). These ruptures are generally well-defined and often exhibit the largest surface offsets. The geometry of primary ruptures tends to be relatively linear or gently curvilinear, and they typically align closely with deeper seismogenic fault planes as constrained by seismological and geodetic data.
- Secondary faulting encompasses a wide range of structures that accommodate distributed deformation in the broader fault damage zone (Fig. 6.5C). These include antithetic and synthetic faults (which dip in the opposite or same direction as the main fault, respectively), fissures, en echelon cracks, pressure ridges and mole tracks. Secondary ruptures may occur tens to hundreds of metres away from the principal trace, often reflecting localised variations in stress or structural inheritance. Although they may exhibit smaller displacements, these features can be broadly distributed.

Distinguishing between primary and secondary ruptures is particularly important in paleoseismology, where misinterpretation of secondary features as principal ruptures can lead to overestimation or mislocation of past surface ruptures and seismic moment (e.g. McCalpin, 2009). Moreover, when documenting surface rupture length (SRL) for seismic hazard models or empirical magnitude–SRL regressions (e.g. Wells & Coppersmith, 1994), inclusion or exclusion of secondary ruptures can significantly affect the inferred earthquake size (e.g. DuRoss et al., 2020; Hamling et al., 2017). As such, clear criteria for rupture classification are essential for consistent geological and engineering analyses.

Figure 6.4 Comparative photographs of normal fault traces in bedrock and in Quaternary deposits. (A) Bedrock fault surface in central Italy. (B) Bedrock fault trace in southern Tibet. (C–E) Aerial and field photographs of fault traces cutting alluvial fans in the northern Xainza rift, south-central Tibet. Red arrows show individual fault traces. Double white arrows and numbers in E indicate fault scarp heights. *(A) From Palumbo et al. (2004). (B) Photo courtesy Chunrui Li. (C–E) From Chevalier et al. (2025).*

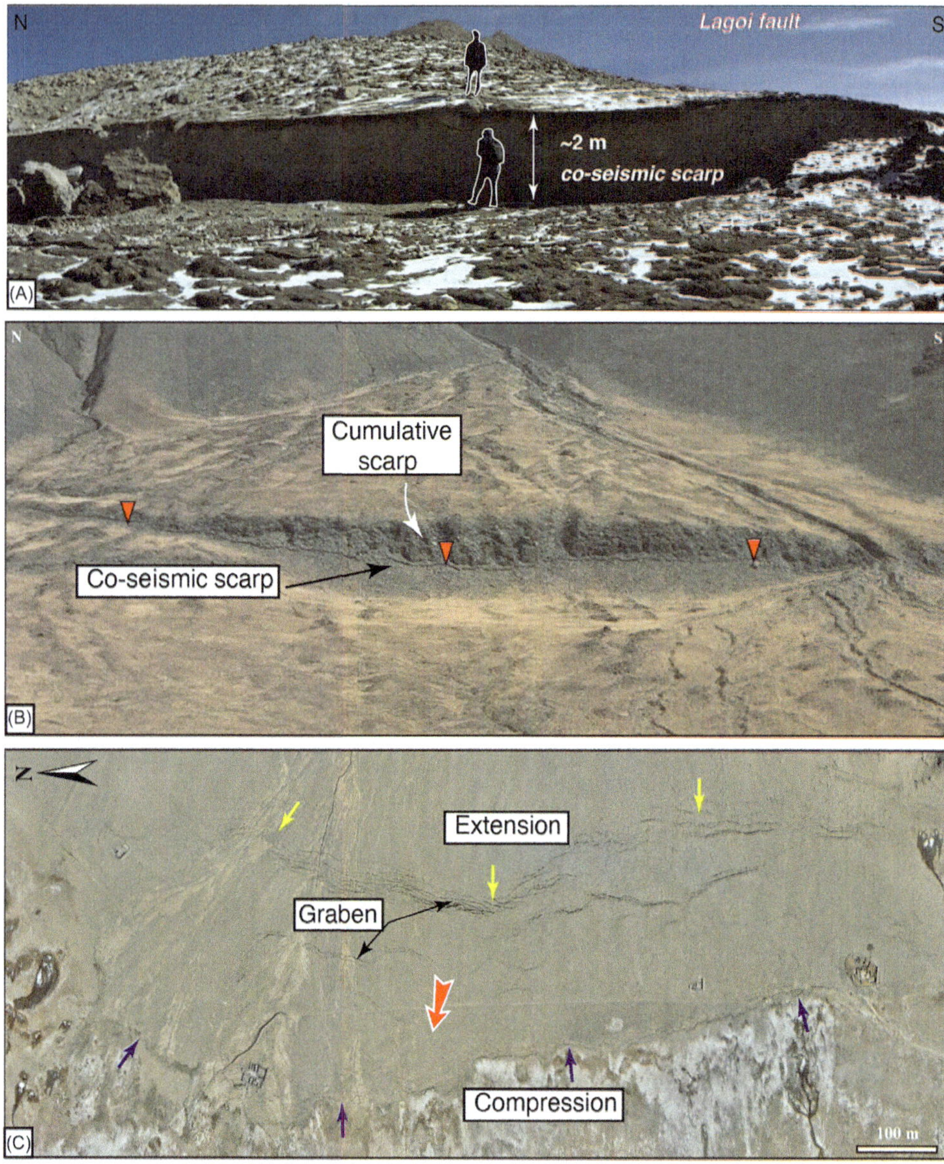

Figure 6.5 Field photographs of the 2025 M_w 7.1 Tingri earthquake, southern Tibet (Liu et al., 2025; Wu et al., 2025) showing primary (A, B) versus secondary (C) fault traces. (A) Two meters-high coseismic scarp (free face) along the Lagoi fault segment. (B) The free face along the Dingmuco master fault is shown by red triangles at the base of a cumulative scarp, whereas secondary cracks (yellow arrows) and mole tracks (blue arrows) farther downstream are shown in (C). Note the coseismic scarp superimposed on the cumulative scarp, contributing to the increased vertical offset of the fan.

6.3 Geomorphic features of fault traces

Fault traces manifest at the Earth's surface through diverse geomorphic features that record both short-term seismic events and long-term landscape evolution. These features – including fault scarps, linear valleys, offset streams, shutter ridges, and sag ponds – encode information about the style, magnitude and frequency of fault movement over time (e.g. Wallace, 1977). Over geological time scales, repeated surface ruptures systematically reshape landscapes by offsetting rivers and terrace risers, damming drainages to form sag ponds or lakes, and building cumulative scarps and triangular facets (Fig. 6.6). Such features serve dual roles: as immediate indicators of individual earthquakes and as long-term archives of tectonic deformation (Bull, 2008; Burbank & Anderson, 2011). In addition to natural features, earthquakes can also induce direct modifications to anthropogenic landscapes (Fig. 6.7). Roads, railways, fences, bridges and buildings may be laterally or vertically displaced by surface faulting. These human-made structures, when disrupted, not only provide recent evidence of rupture but can also influence postseismic geomorphic processes by diverting drainage, altering sediment pathways or creating new depressions. Such anthropogenic impacts – often more visually striking – complement natural geomorphic markers and are increasingly integrated into palaeoseismic and hazard assessments.

These geomorphic signatures are necessary for identifying active faults, reconstructing palaeoseismic histories and assessing seismic hazards (e.g. Klinger et al., 2011). Moreover, they illustrate how fault trace processes operate as fundamental agents of landscape development in tectonically active regions, creating visible records that bridge time scales from individual seismic events to million-year tectonic evolution. The geomorphic expression of faulting is influenced by a combination of tectonic parameters – including slip rate, fault geometry, rupture length, and earthquake recurrence interval – as well as surface conditions such as lithology, vegetation cover, and topographic setting. Additionally, climatic factors strongly modulate the preservation or erosion of fault-related landforms by governing rates of sedimentation, weathering and runoff (e.g. Whipple, 2004). In arid environments or in regions where high slip rates are not counteracted by rapid erosion or sedimentation, scarps and offsets may remain preserved for centuries (e.g. Chevalier et al., 2012; Li et al., 2021). In contrast, rapid erosion may obscure or degrade fault geomorphology shortly after formation in humid or high-relief settings (e.g. the 2008 M_w 7.9 Wenchuan earthquake). As such, deciphering fault traces requires a multidisciplinary understanding of tectonic processes and surface dynamics.

Linear to curvilinear geomorphic features are among the most characteristic and diagnostic expressions of fault traces in the landscape. They are generally extremely clear in aerial photographs, satellite images (Fig. 6.8B), or high-resolution LiDAR data (Fig. 6.8C, D), as they cut across topographic and geomorphic features. These include

~30 m
~6 m
~3 m
(A)
SE
NW
13±1 m
(B)
(C)
~44 m
~44 m
(D)
(E)
(F)
~1 m
(G)
SW
NE
T4
31 m
T3
sag pond
T3
T2'
T2
22-25 m
T2
T1
T0
(H)

fault scarps, linear valleys, aligned ridgelines or slope breaks, vegetation lineaments, and spring alignments – each providing visual and spatial clues to the location and nature of underlying fault structures (e.g. Wallace, 1977). The morphology of fault scarps can provide insights into the kinematics of past faulting events, including recurrence intervals and displacement magnitudes (e.g. Benedetti et al., 2013; Iezzi & Roberts, 2021). Fracturing often increases subsurface permeability, facilitating the upward migration of groundwater. As a result, springs, seeps and associated vegetation often align along fault traces (Fig. 6.8A, B), forming linear oases that may be diagnostic in otherwise dry or sparsely vegetated settings (e.g. Caine et al., 1996; Faulkner et al., 2010). These indirect surface expressions are particularly valuable in regions where recent rupture is obscured by sediment cover, vegetation or anthropogenic modification.

One of the most conclusive geomorphic expressions of active faulting is the measurable offset of landscape features that serve as tectonic markers. These markers (Fig. 6.6) – such as stream channels, gullies, terrace risers, alluvial fan edges, glacial moraines, levees, etc. – can be displaced laterally or vertically, depending on the style of faulting. Such displacements provide direct evidence of cumulative fault slip and are critical for reconstructing the kinematic behaviour of faults over time (e.g. Sieh, 1978). Strike-slip faults typically produce horizontal (lateral) displacements of linear geomorphic features, including anthropogenic structures, which can be preserved across multiple seismic cycles. Quantifying the magnitude of these lateral offsets and dating the displaced geomorphic surfaces enables the calculation of long-term slip rates (see Chapter 12; e.g. Bai et al., 2018; Chevalier et al., 2005; Li et al., 2021, 2022; McGill & Sieh, 1991; Mériaux et al., 2004; Sieh & Jahns, 1984; Van der Woerd et al., 2002) and informs probabilistic models of earthquake recurrence intervals (e.g. Li et al., 2021; Sieh, 1978; Wu et al., 2025). In contrast, normal faults are primarily associated with

Figure 6.6 Photographs illustrating coseismic and cumulative offsets, both horizontal and vertical, along strike-slip and normal faults in Tibet (highlighted by red arrows). (A) Coseismic surface rupture from the 2001 M_w 7.9 Kokoxili (or Kunlun) earthquake, horizontally cutting across alluvial terraces and their risers. (B) Cumulative horizontal offset of levees in the Chinese Pamir. (C) Deep coseismic crack on steep mountain slopes created during the 2008 M_w 7.1 Yutian earthquake (western Tibet), with extension direction shown by red arrows. (D) Cumulative vertical offset across moraines in the Yangbajing Graben, southern Tibet. (E) Repetitive vertical offsets forming triangular facets and contributing to mountain building in northern Tibet. (F) Large-scale, cumulative horizontal offset of a mountain range in the Tarim Basin, northwest China. (G) Coseismic surface rupture from the 2014 M_w 6.9 Yutian earthquake, northwestern Tibet, horizontally offsetting vehicle tracks. (H) Cumulative horizontal offsets along the Gozha Co Fault, western Tibet, with fan edges highlighted by snow. *(A) Modified from Li et al. (2005). (B) Modified from Chevalier et al. (2015). (C) Photo courtesy Haibing Li. (D) Modified from Chevalier et al. (2020). (G) Modified from Li et al. (2016). (H) Modified from Chevalier et al. (2017).*

(A)
(B)
(C)
(D)
(E)
(F)
(G)
(H)
(I)

vertical displacements, leading to the progressive rock uplift of the fault footwall relative to the hanging wall. This often results in prominent fault scarps ranging from subtle slope breaks to towering escarpments and triangular facets, particularly in arid regions where erosion is minimal, as well as vertically offset terraces, fans or moraines (e.g. Chevalier et al., 2015, 2020, 2025; Li et al., 2023a; Wang et al., 2020; Wu et al., 2025). Triangular facets are planar to slightly concave, triangular-shaped landforms that commonly develop along the base of fault scarps or mountain fronts in tectonically active regions. They form through the interaction of fault slip and erosional processes, typically where normal faulting uplifts resistant bedrock or Quaternary deposits, creating steep, fault-controlled slopes. Subsequent erosion by surface processes, such as rockfall, fluvial or glacial incision, truncates spurs and leaves behind sharp, straight-edged facets (Fig. 6.6D, E). These features are often aligned along linear mountain fronts and are indicative of relatively recent and repeated fault movement. Triangular facets are recognisable by their geometric shape, alignment along fault traces and abrupt slope changes, and they are useful indicators of active normal faulting. Reverse or thrust faults, on the other hand, uplift and tilt the hanging wall, producing ground surface folding that can be traced in deformed terraces, hillslopes and soil horizons (e.g. Hetzel et al., 2004; Ritz et al., 2003; Wu et al., 2019). These geomorphic signatures not only help in identifying active fault traces but are also integral to paleoseismological studies aimed at constraining past seismic events and landscape evolution in tectonically active regions.

Many active faults, however, do not exhibit purely vertical or purely horizontal motion but rather oblique slip, generating compound landforms like scarps with associated pressure ridges or ponds (Fig. 6.9A, C) that require integrated structural, geomorphic and sometimes geophysical approaches to decipher fault geometry and slip history. Sag ponds form at extensional zones (releasing bends, stepovers or pull-apart basins) along strike-slip or normal faults, where subsidence creates depressions (sometimes filled with water) that preserve fine-grained sediments and organic material ideal for paleoseismic studies (e.g. Sylvester, 1988) (Fig. 6.9C). Because of their capacity to trap and preserve stratigraphic records, sag ponds are valuable sites for trenching as they often record fault motion and provide datable organic material for earthquake

Figure 6.7 Photographs illustrating coseismic damage to anthropogenic infrastructure. (A) Right-lateral offset in an agricultural field during the 1979 M_w 6.4 Imperial Valley earthquake, California. (B) Displaced railroad tracks during the 2010 M_w 7.1 Canterbury earthquake in New Zealand. (C, G) Bridge collapse caused by the 2021 M_w 7.4 Maduo earthquake, northern Tibet. (D) Uplift due to thrust faulting during the 2008 M_w 7.9 Wenchuan earthquake, eastern Tibet. (E) Push-up across the highway following the 2021 M_w 7.4 Maduo earthquake. (F) Road offset by a mountain slope collapse triggered during the 2022 M_s 6.8 Luding earthquake, eastern Tibet. (H, I) Damage to a railroad bridge and tunnel during the 2022 M_w 6.9 Menyuan earthquake, northern Tibet. *(D) Modified from Li et al. (2008). (F) Photo courtesy Haibing Li. (H, I) Photos courtesy Jiawei Pan.*

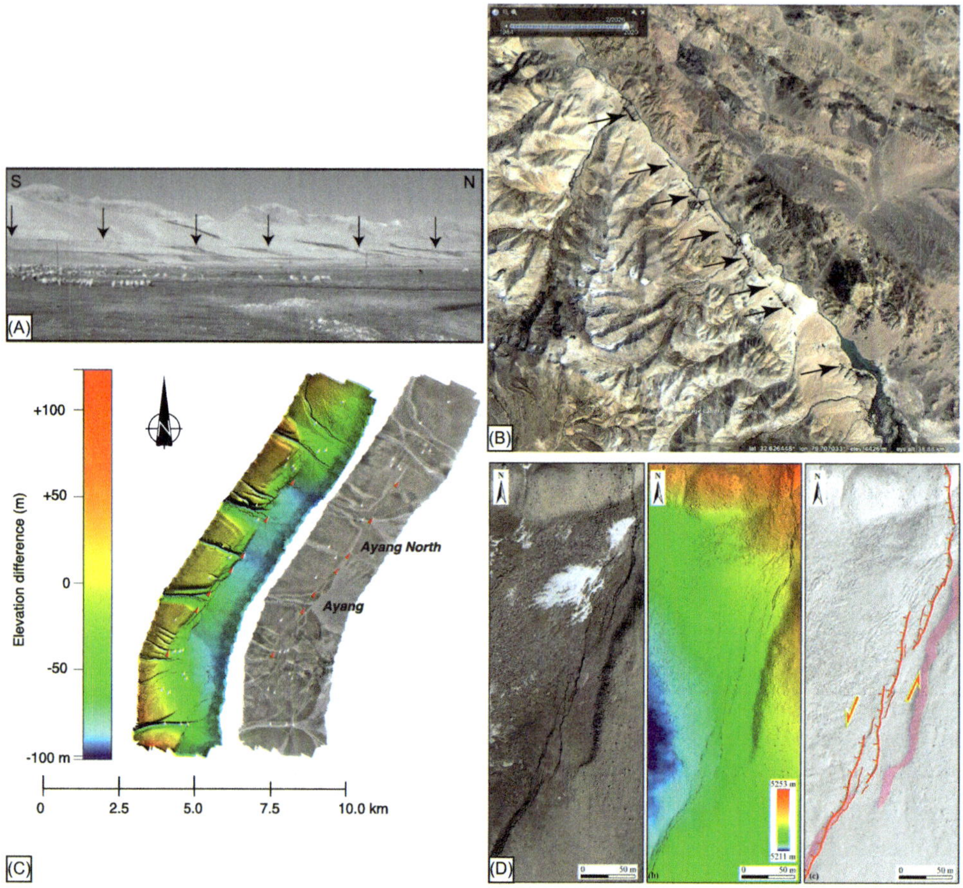

Figure 6.8 Examples of tectonic landforms and surface features in Tibet from field photos and high-resolution imagery. (A) Field photograph and (B) Google Earth satellite image showing a linear alignment of springs (indicated by black arrows and dark patches) along the Karakorum Fault, western Tibet. (C) High-resolution digital elevation model (DEM) derived from uncrewed aerial vehicle (UAV) photographs showing multiple linear normal faults (highlighted by red and white triangles) along the northern Xainza Rift, southern Tibet. (D) UAV photo and UAV-derived DEM of curvilinear surface rupture in the 2025 M_w 7.1 Tingri earthquake in southern Tibet. *(B) From Chevalier et al. (2016). (C) Google Earth Pro: Image Landsat / Copernicus, 1984. (D) From Liu et al. (2025).*

chronologies (e.g. Sieh, 1978). In compressional settings, pressure ridges (or mole tracks) (Fig. 6.9B, D, G, H, J) develop as uplifted folds of deformed sediment at restraining bends or fault overlaps, which record cumulative tectonic deformation. Shutter ridges arise from strike-slip displacement that blocks drainages, diverting streams and preserving measurable slip histories (e.g. McGill & Sieh, 1991) (Fig. 6.9K).

Figure 6.9 Field photographs of sag ponds, extensional cracks, pressure ridges and shutter ridges across the Tibetan Plateau. (A) Pond dammed by the sinistral Cuopu Fault, eastern Tibet. (B, G, H) Push-ups and mole tracks formed by the 2022 M_w 6.9 Menyuan earthquake, northeastern Tibet. (C) Pull-apart basin occupied by a lake along the Xianshuihe Fault, eastern Tibet. (D) Mole tracks formed by the 2021 M_w 7.4 Maduo earthquake. (E, F) Extensional cracks (~ 6m deep in E) generated by the 2022 M_w 6.9 Menyuan earthquake. (I) Extensional crack from the 2021 M_w 7.4 Maduo earthquake. (J) Push-up as secondary deformation produced by the 2025 M_w 7.1 Tingri earthquake, southern Tibet. (K) Shutter ridges along the Karakax fault in northwestern Tibet. *(A) From Chevalier and Replumaz (2019). (C) From Bai et al. (2018). (D) From Pan et al. (2022). (J) From Wu et al. (2025). (K) From Li et al. (2021).*

6.4 Surface deformation and coseismic offsets

Earthquake surface deformation encompasses a suite of geomorphic phenomena resulting directly from fault rupture, including fracturing, offsets and secondary ground failures such as liquefaction. These features are essential for interpreting the mechanics of earthquake rupture and its geomorphic consequences. This section addresses methods for mapping precise surface deformation, highlights the geomorphic signatures produced during coseismic events and discusses secondary processes that may modify or obscure primary rupture features. Accurate mapping of surface ruptures is indeed critical for understanding fault kinematics, assessing seismic hazard and identifying zones of coseismic deformation. Traditional field mapping methods remain fundamental for assessing rupture characteristics, particularly where vegetation or infrastructure limits remote sensing capabilities.

6.4.1 Mapping surface rupture using field surveys, LiDAR, InSAR, and photogrammetry

Modern geospatial tools such as airborne and terrestrial LiDAR, InSAR and drone-based structure-from-motion (SfM) photogrammetry have dramatically expanded our ability to document surface deformation at high spatial resolutions and over large areas (e.g. Ji et al., 2024; Lemrabet et al., 2023). These technologies produce topographic models capable of revealing detailed fault geometry, deformation patterns and rupture characteristics that were previously difficult or impossible to observe (e.g. Glennie et al., 2013; Oskin & Arrowsmith, 2012; Ren & Zhang, 2019; Ren et al., 2022b) (left panel in Fig. 6.10). For example, InSAR has proven particularly valuable in mapping blind ruptures or distributed deformation not expressed as discrete surface breaks (e.g. Li et al., 2025) (right panel in Fig. 6.10), but it is less effective in vegetated or steep terrain due to signal decorrelation. SfM photogrammetry offers a cost-effective, flexible alternative for generating 3D digital elevation models (DEMs) from overlapping photographs acquired via drones and is particularly suited for postearthquake surveys of rupture traces (Fig. 6.8D and left panel in 6.10), enabling rapid documentation in remote areas (e.g. Li et al., 2023b; Liu et al., 2025; Pan et al., 2022; Wu et al., 2025).

6.4.2 Coseismic landform development or modification

Coseismic rupture commonly results in the formation or modification of landforms such as lateral offsets, fault scarps, mole tracks, pressure ridges, cracks and grabens, which are influenced by fault segmentation, local topography and sedimentary context. In regions with loose Quaternary sediments, the expression of rupture may be diffuse and poorly preserved, complicating efforts to reconstruct the full displacement field. High-resolution datasets from UAV-DEMs, InSAR and field surveys have been instrumental

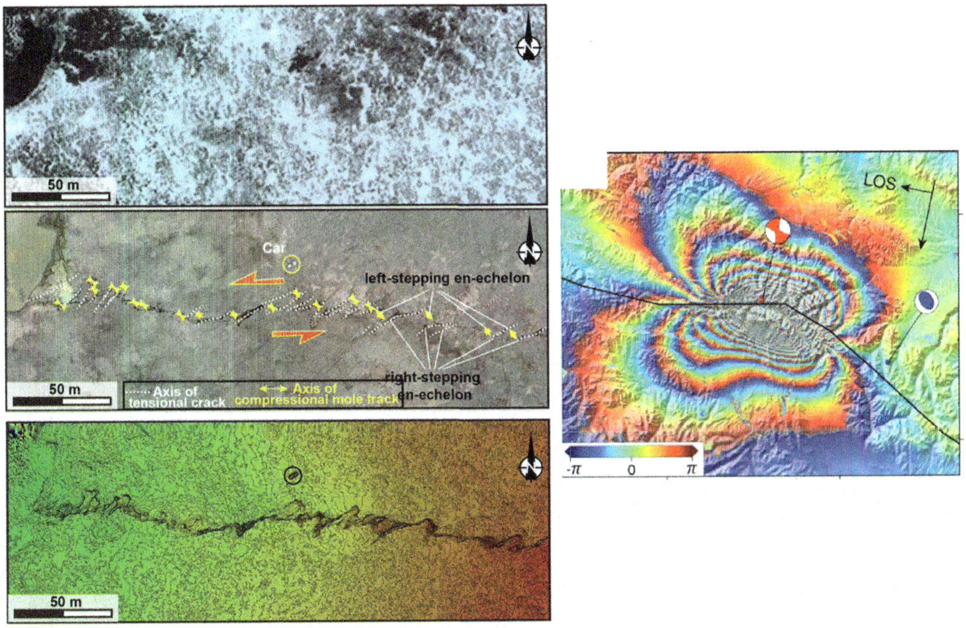

Figure 6.10 Left: Comparison of pre- (Bing satellite image, top) and post- (orthorectified Uncrewed Aerial Vehicle (UAV) photograph, middle, and UAV-derived digital elevation model, bottom) earthquake imagery from the 2021 M_W 7.4 Maduo earthquake in northeastern Tibet (Pan et al., 2022), highlighting prominent coseismic surface ruptures with cracks and mole tracks. Right: Coseismic interferogram of the M_W 6.9 Menyuan earthquake. The epicentre is plotted as a red beachball (Luo & Wang, 2022).

in documenting these features where preservation is limited or deformation is distributed.

Lateral offsets along active faults are displacements of geomorphic features – such as stream channels or gullies, terrace risers or alluvial fan edges (Fig. 6.6A, H), levees (Fig. 6.6B), glacial moraines, entire mountain ranges (Fig. 6.6F) or anthropogenic structures (Figs. 6.6G and 6.7A, B) – caused by horizontal (strike-slip) motion along a fault plane at depth. Repeated earthquakes can produce cumulative offsets over time (Figs. 6.6A, B, F, H and 6.9A, K), which can be preserved in stable landscapes with low erosion and sedimentation rates. Dating the surfaces on either side of the fault and measuring the horizontal offset of the feature yields horizontal slip rates (e.g. Chevalier et al., 2005, 2012; Mériaux et al., 2004; Van der Woerd et al., 2002). However, these features are subject to degradation by natural processes such as fluvial erosion, landslides and vegetation growth, which can obscure or destroy geomorphic evidence of fault motion.

Fault scarps (vertical offsets) are perhaps the most striking coseismic landforms, forming where vertical or oblique slip propagates to the surface (Figs. 6.4, 6.5 and 6.11A, B, E, F). Scarp morphology varies with fault geometry and surficial materials, often evolving into long-term tectonic landforms through cumulative displacement (Figs. 6.4, 6.5B, 6.6D, E, and 6.8C; e.g. Chevalier et al., 2015; Wallace, 1977). Fresh scarps may exhibit sharp profiles that degrade rapidly due to weathering, erosion and human activity (e.g. Arrowsmith et al., 1998; Avouac, 1993; Wallace, 1977) (Fig. 6.12).

Ground cracking and tensile fractures (and grabens) are common along and adjacent to the main rupture, especially in areas of extensional or compressional stepovers (Figs. 6.5C, 6.6A, C, 6.9E, F, I and right panel in Fig. 6.10). These features may be tens to hundreds of metres wide and are often distributed across a fault zone (e.g. Milliner et al., 2015; Yuan et al., 2022).

Compressional coseismic deformation commonly manifests as mole tracks, push-ups, and localised folding along or near the main fault trace (Figs. 6.7D, E, 6.9B, D, G, H, J and 6.11E, F). These features typically form in transpressional segments, restraining bends, or compressional stepovers, where the rupture front encounters geometric or structural barriers. Mole tracks appear as linear ridges of uplifted ground, while push-ups form low-amplitude pressure ridges or small-scale anticlines composed of surficial materials displaced by fault-parallel shortening. Push-ups in frozen ground or permafrost, such as on the Tibetan Plateau, are particularly spectacular (Fig. 6.9B, G, H, J). These structures reveal how slip and deformation were distributed within the uppermost crust during faulting. Since they can develop several 100 m away from the main rupture, they often make it challenging to distinguish the primary fault trace in the field (Fig. 6.5C).

Liquefaction represents a secondary but significant coseismic phenomenon that commonly accompanies strong ground shaking, particularly in water-saturated unconsolidated sediments (see Chapter 8). It occurs when cyclic loading from seismic waves increases pore-water pressure, reducing effective stress and causing the soil to temporarily behave like a fluid. This process can lead to the formation of sand blows (Fig. 6.9C, D), often affecting broad areas (e.g. Huang & Yu, 2013; Ishihara, 1993; Pan et al., 2022; Youd & Perkins, 1978). Though not directly caused by fault slip, liquefaction profoundly alters the geomorphic and hazard expression of earthquakes. Its surface manifestations may obscure or mimic tectonic deformation, complicating the identification of primary fault rupture in the field. While often classified as secondary effects of an earthquake, liquefaction features are integral to understanding the full spectrum of earthquake-induced surface processes, particularly in deltaic, alluvial and lacustrine settings where their geomorphic impact may dominate.

Figure 6.11 Field photos of vertical coseismic surface rupture and sand blows produced by recent large earthquakes on the Tibetan Plateau. (A) Thrust fault scarp (free face) produced by the 2008 M_w 7.9 Wenchuan earthquake. (B) 90 cm-high normal fault scarp (free face) from the 2025 M_w 7.1 Tingri earthquake. (C, D) Sand blows resulting from the 2021 M_w 7.4 Maduo earthquake. (E, F) Coseismic thrust scarp from the 2008 M_w 7.9 Wenchuan earthquake. *(A) Photo courtesy Haibing Li. (B) Modified from Liu et al. (2025).*

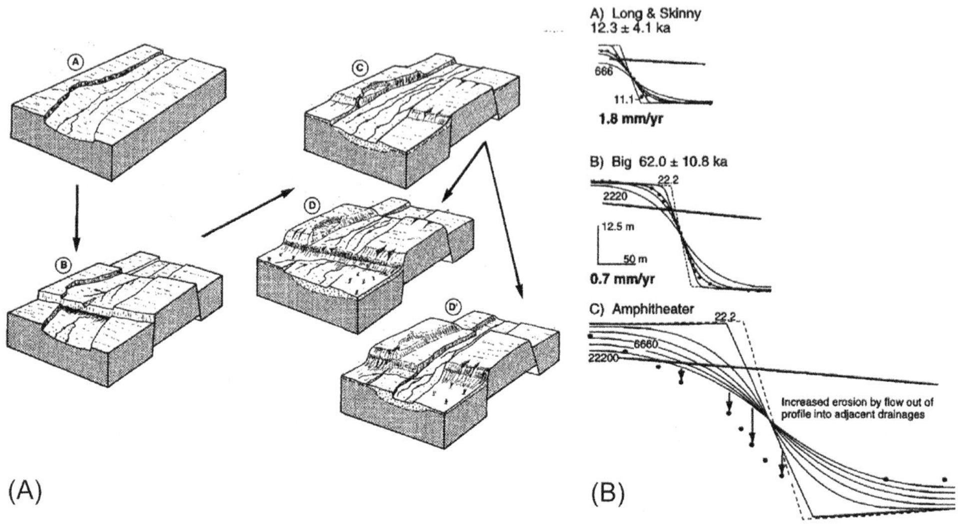

Figure 6.12 Fault scarp formation (A, Avouac & Peltzer, 1993) and degradation over time (B, Arrowsmith et al., 1998). Diffusion models such as those shown in (B) are often used to reconstruct the evolution of a scarp and assess its potential age, given assumptions about its initial geometry. See Chapter 12 for more details.

6.5 Postseismic and interseismic modification of fault traces

While the geomorphic expressions of surface-rupturing earthquakes are initially well-defined, these features are transient in many landscapes and subject to significant modification in the postseismic and interseismic periods. The long-term preservation of fault traces is governed by a complex interplay of tectonic activity, geomorphic regime and environmental setting. For example, in regions with ongoing tectonic activity and low erosion rates, multiple generations of scarps and offset landforms may be preserved, allowing for the construction of long palaeoseismic records (e.g. Li et al., 2021; Wu et al., 2025). Conversely, in dynamic alluvial settings, evidence of past ruptures, even large-magnitude ones, may be erased rapidly (e.g. Chevalier et al., 2025). Climatic variability over Quaternary time scales also affects preservation – glacial-interglacial cycles may alternately expose and bury fault features, challenging the correlation of geomorphic markers with specific seismic events (e.g. Wang et al., 2020). Understanding those processes that degrade, obscure or preserve fault traces is critical for reconstructing past surface ruptures and interpreting the paleoseismic record.

Following a surface rupture, newly formed fault scarps (Figs. 6.5A and 6.11A, B) are exposed to the elements and undergo degradation through mechanical and chemical weathering processes (Figs. 6.4E and 6.5B). The rate of scarp degradation depends on lithology, climate, slope aspect and the magnitude of initial vertical displacement. In arid environments, scarps may be preserved for thousands of years (e.g. Chevalier et al., 2020, 2025; Li et al., 2021; Wu et al., 2025), whereas in humid high-relief settings, they may be eroded beyond recognition within a few years or decades. Diffusion-based models of scarp evolution have been used to estimate the age of faulting events by comparing modelled profiles with present-day morphology (Fig. 6.12) (e.g. Arrowsmith et al., 1998; Avouac, 1993; Nash, 1980; Wallace, 1977).

Depositional processes may rapidly obscure or bury fault-related features. In fluvial environments, postseismic aggradation can mask channel offsets or linear traces. Similarly, colluvial and alluvial processes may fill in sag ponds and depressions, making identification of older ruptures difficult without subsurface investigations. In addition, anthropogenic modification of the land surface by, for example, agriculture or construction, can contribute to sediment redistribution and the obliteration of geomorphic fault indicators. In densely populated regions, this can dramatically complicate the interpretation of earthquake history and the evaluation of seismic hazard. Biological activity also plays a subtle but persistent role in degrading fault traces. Root growth, animal burrowing and soil mixing processes (collectively termed bioturbation) gradually homogenise the soil profile and can disrupt subtle stratigraphic or geomorphic evidence of past ruptures.

6.6 Fault trace mapping

Accurate identification and mapping of fault traces are foundational to earthquake geomorphology, paleoseismology and seismic hazard assessment. Fault traces often manifest subtly in the landscape and require integration of geomorphic, structural and remote sensing observations for reliable interpretation. High-resolution imagery, field mapping and geophysical surveys are often required to confirm tectonic origins. Misidentifying nontectonic lineaments as active faults can lead to significant errors in seismic hazard assessments, underscoring the importance of multidisciplinary verification.

A major interpretive challenge in tectonic geomorphology lies in distinguishing true tectonic lineaments – features indicative of active faulting – from those generated by nontectonic processes. Surface expressions such as linear drainage patterns, vegetation anomalies or anthropogenic structures (e.g. roads, fences) may align with the landscape in ways that mimic fault traces, leading to potential misinterpretation (Fig. 6.13A, B). This difficulty is compounded in regions where human activity or erosional processes have modified the landscape. To differentiate genuine tectonic features, geomorphologists apply a suite of diagnostic criteria, including: (1) persistence of linearity across diverse landforms and scales, (2) alignment with the orientation of regional stress fields and known fault systems, (3) direct geomorphic or stratigraphic evidence of recent displacement (e.g. offset channels, scarps or disturbed sediments) and (4) spatial association with broader fault networks or seismicity patterns.

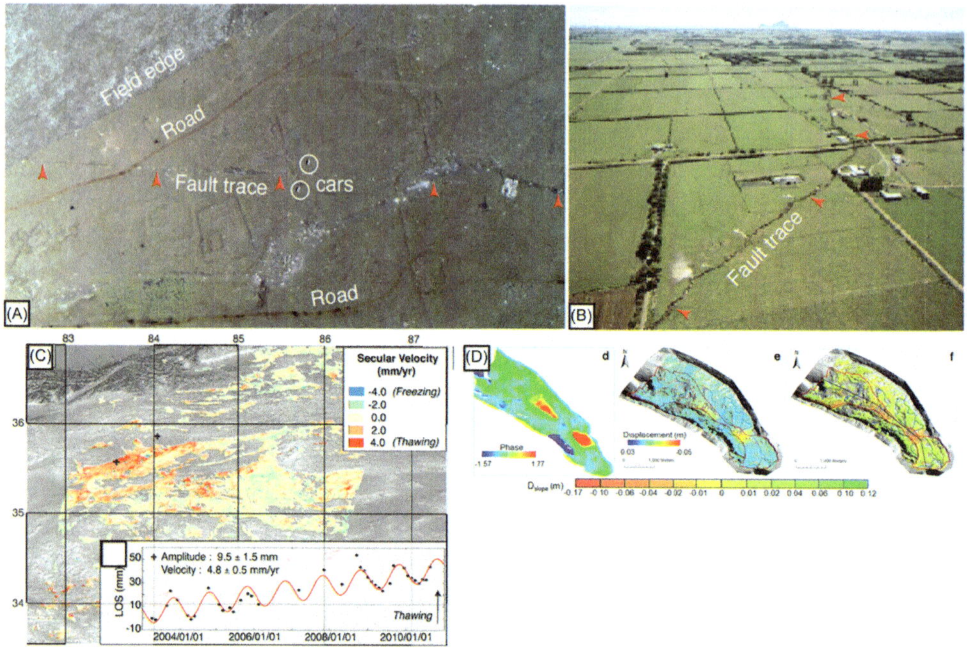

Figure 6.13 (A) Aerial photograph of the 1948 M_S 7.3 Litang earthquake in eastern Tibet, showing surface rupture (red arrows) amongst roads and former housing foundations (polygons), as well as field edges. (B) Surface rupture (red arrows) of the 1987 M_W 6.5 Edgecumbe earthquake amongst pastures in New Zealand. (C) Mapping of permafrost thawing in northwestern Tibet and (D) landslide monitoring in northeastern Tibet using interferometric synthetic aperture radar data. *(C) From Daout et al. (2017). (D) From Zhang et al. (2018).*

Effective fault trace mapping relies on the integration of multiscale observations and multidisciplinary datasets to validate tectonic origins and enhance interpretive precision. Multitemporal and multisensor techniques – such as repeat aerial imagery, time-series InSAR and correlated seismicity data – are valuable for detecting active ground deformation (right panel in Fig. 6.10) and identifying associated geomorphic expressions, whether directly linked to fault activity or not. These datasets can reveal subtle deformation patterns, but they also frequently capture nontectonic processes such as permafrost degradation (Fig. 6.13C) (Daout et al., 2017), landslides (Fig. 6.13D) (Zhang et al., 2018) or anthropogenically-induced subsidence (e.g. caused by pumping water or oil). As such, careful contextual interpretation is required to distinguish tectonic deformation from other sources of surface change.

Geomorphic features reflect recent surface processes, while structural geology provides insights into long-term fault architecture and kinematics. By combining remote sensing, field mapping, structural measurements and stratigraphic analysis, one can better resolve complex fault zones with segmented behaviour or multiple rupture pathways.

A critical component of this integrative approach is geographic information system (GIS)-based analysis, which enables the systematic compilation, visualisation and interpretation of diverse datasets essential for fault trace identification. By overlaying geomorphic indicators (e.g. linear scarps, offset drainages, sag ponds), fault databases (e.g. existing tectonic maps, paleoseismic trench logs) and topographic data (e.g. hillshade models, slope maps, curvature analyses), researchers can enhance the reliability of fault mapping (e.g. Bemis et al., 2014). GIS tools facilitate the detection of subtle tectonic signatures by (1) spatial correlation of fault-related landforms with known tectonic structures; (2) multicriteria assessment using weighted evidence layers (e.g. lineament density, deformation patterns); and (3) change detection through time-series analysis (e.g. InSAR, LiDAR differencing; Bürgmann et al., 2000).

These methods are particularly essential in tectonically complex regions (e.g. distributed fault zones) or data-sparse areas, where field access is limited (high mountains, politically unstable countries). By integrating field observations, remote sensing and geodetic data within a GIS framework, researchers can produce more accurate and dynamic fault maps, crucial for earthquake preparedness and infrastructure planning.

6.7 Fault traces in palaeoseismology and seismic hazard assessment

Fault traces record past surface-rupturing earthquakes, providing insights into slip behaviour, recurrence intervals and seismic hazards. Palaeoseismology – the study of prehistoric earthquakes – relies on fault-related geomorphic features and trenching across the fault to identify past ruptures. Excavations perpendicular to faults expose cross-sections (Fig. 6.14) of displaced stratigraphy, upward fault terminations, event horizons (e.g. colluvial wedges, fissure fills) and liquefaction features (e.g. McCalpin, 2009; Schwartz & Coppersmith, 1984). Radiocarbon or optically stimulated luminescence (OSL) dating of these units allows estimation of earthquake timing and recurrence. Successful trenching requires careful site selection – typically where sedimentation rates are moderate and deformation is preserved in stratified deposits. When the age of the offset surface is known (e.g. via cosmogenic nuclide dating such as ^{10}Be or ^{26}Al, or OSL) and the amount of offset is measurable, a slip rate can be calculated (e.g. Ryerson et al., 2006).

Palaeoseismic data can also reveal whether rupture behaviour is periodic, clustered or random. Recurrence intervals, derived from dated faulting events, contribute to probabilistic seismic hazard models. Where possible, combining palaeoseismic records with historical and instrumental catalogues provides a more complete temporal perspective on fault behaviour. Data from fault trace analysis feed directly into seismic

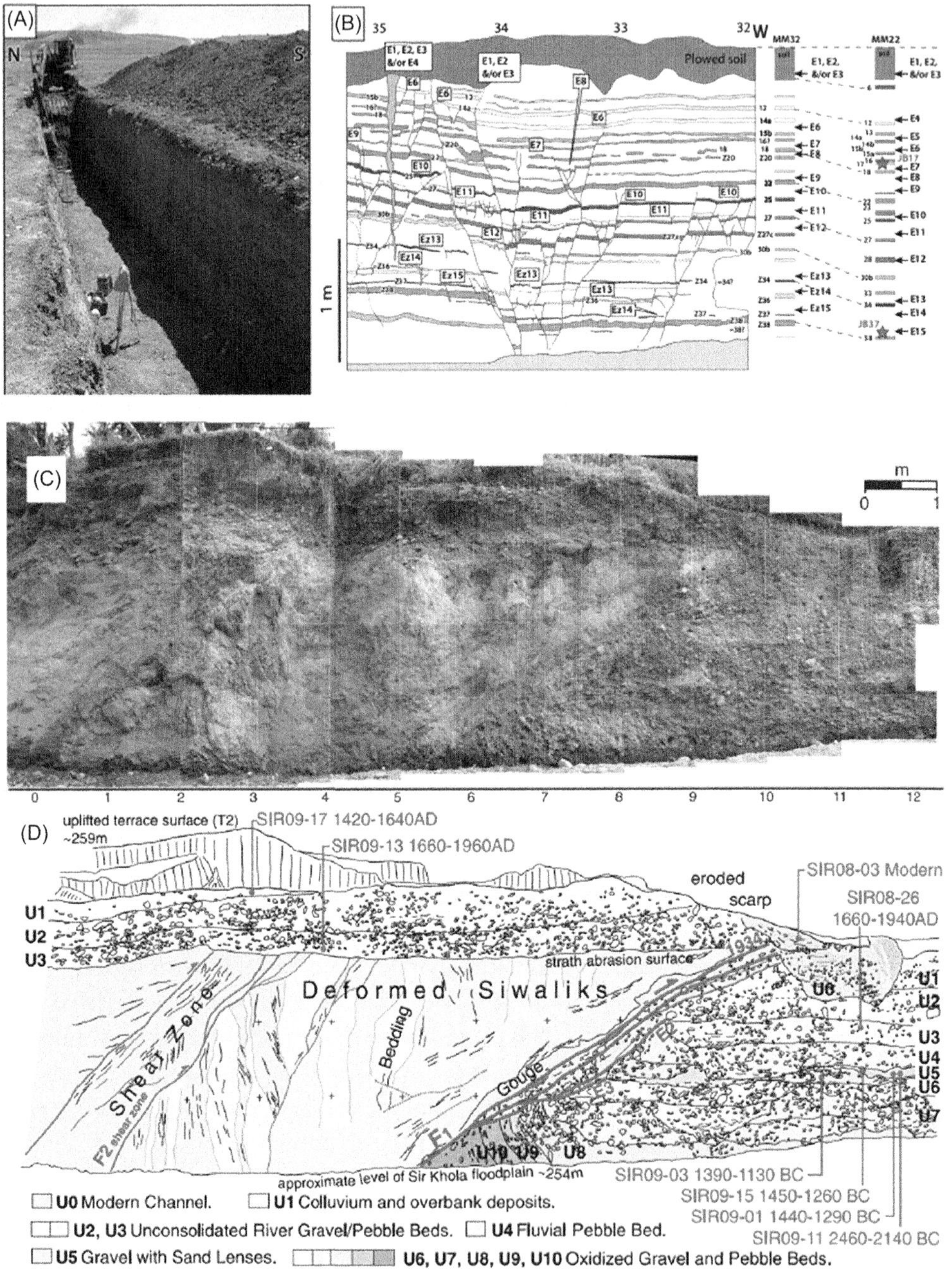

Figure 6.14 (A) Excavated and (B) annotated trench log across the active Yammouneh Fault, Lebanon. (C) Uninterpreted and (D) interpreted river cut across the Himalayan Main Frontal thrust. *(B) From Le Béon et al. (2018). (D) From Bollinger et al. (2014); Sapkota et al. (2013).*

hazard assessments at regional and national scales. Parameters such as slip rate, fault length, rupture segmentation and recurrence interval are used to estimate earthquake probabilities (e.g. Wells & Coppersmith, 1994) and ground shaking intensities. These models guide building codes, insurance underwriting and emergency planning.

6.8 Emerging approaches

The study of fault traces is being transformed by advances in remote sensing, numerical modelling, data integration and community science, enabling new approaches to detect and analyse surface deformation – even where evidence is subtle or obscured. Cutting-edge tools like UAV-based SfM photogrammetry (e.g. Johnson et al., 2014) and terrestrial laser scanning (e.g. Telling et al., 2017) now provide sub-metre topographic resolution almost anywhere in the world, revealing previously undetectable fault features. Combining optical, thermal and multispectral imagery with these datasets offers unprecedented detail.

Machine learning techniques, particularly convolutional neural networks, are revolutionising fault mapping by automatically identifying scarps and offset landforms in large topographic datasets (Han et al., 2023). While these methods enable rapid regional-scale analysis – particularly valuable in inaccessible areas – they require careful training using expert-curated datasets.

Citizen science initiatives are creating novel data streams for fault studies. The United States Geological Survey's 'Did You Feel It?' system (e.g. Atkinson & Wald, 2007) and social media (e.g. Zahra et al., 2020) have proven effective for documenting coseismic effects. When properly verified (Minson et al., 2015), such crowdsourced data can capture ephemeral ground ruptures that might otherwise go unrecorded, particularly in urban environments. These interdisciplinary approaches are advancing both fundamental understanding of fault systems and practical hazard assessment.

6.9 Conclusions

Fault traces are the primary interface between tectonic processes and the Earth's surface. Their study offers a direct view into the mechanics of earthquake rupture, the formation and evolution of landforms and the geological record of past seismic activity. From dramatic scarps to subtle lineaments, fault traces encode valuable information about the behaviour of faults over time scales ranging from seconds to millennia.

In this chapter, I have outlined how fault traces form through surface rupture, how they are expressed in the morphology of the Earth's surface and how they evolve through coseismic and postseismic processes. I have discussed the tools used to map and measure them, from field observations and remote sensing to numerical models and machine learning. Crucially, fault traces are not only scientific records but also practical guides – palaeoseismology and seismic hazard assessments depend on their accurate interpretation.

Looking forward, the convergence of high-resolution data, real-time sensing and interdisciplinary methodologies is transforming the study of fault traces. Yet many challenges remain: understanding distributed deformation, improving global fault mapping and distinguishing tectonic signals from nontectonic noise.

As landscapes continue to be shaped by active faults – and as human settlements increasingly intersect these zones – fault trace studies will remain central to both geomorphology and earthquake hazard science. Their importance lies not only in explaining the past, but in anticipating the future.

References

Antoine, S. L., Klinger, Y., Delorme, A., Wang, K., Bürgmann, R., & Gold, R. D. (2021). Diffuse deformation and surface faulting distribution from submetric image correlation along the 2019 Ridgecrest, California, ruptures. *Bulletin of the Seismological Society of America, 111*(5), 2275–2302.

Armijo, R., Tapponnier, P., Mercier, J. L., & Han, T. L. (1986). Quaternary extension in southern Tibet: Field observations and tectonic implications. *Journal of Geophysical Research, 91*(B14), 13803–13872.

Arrowsmith, J. R., Rhodes, D. D., & Pollard, D. D. (1998). Morphologic dating of scarps formed by repeated slip events along the San Andreas Fault, Carrizo Plain, California. *Journal of Geophysical Research, 103*(B5), 10141–10160.

Atkinson, G. M., & Wald, D. J. (2007). Did You Feel It?" intensity data: A surprisingly good measure of earthquake ground motion. *Seismological Research Letters, 78*(3), 362–368.

Avouac, J. P. (1993). Analysis of scarp profiles: Evaluation of errors in morphologic dating. *Journal of Geophysical Research, 98*(B4), 6745–6754.

Avouac, J. P. (2003). Mountain building, erosion, and the seismic cycle in the Nepal Himalaya. *Advances in Geophysics, 46*, 1–80. https://doi.org/10.1016/S0065-2687(03)46001-9.

Avouac, J. P., & Peltzer, G. (1993). Active tectonics in southern Xinjiang, China: Analysis of terrace riser and normal fault scarp degradation along the Hotan-Qira fault system. *Journal of Geophysical Research: Solid Earth, 98*(B12), 21773–21807.

Bai, M., Chevalier, M. L., Leloup, P. H., Li, H., Pan, J., Replumaz, A., & Zhang, J. (2021). Spatial slip rate distribution along the SE Xianshuihe fault, eastern Tibet, and earthquake hazard assessment. *Tectonics, 40*(11), e2021TC006985.

Bai, M., Chevalier, M. L., Pan, J., Replumaz, A., Leloup, P. H., Métois, M., & Li, H. (2018). Southeastward increase of the late Quaternary slip-rate of the Xianshuihe fault, eastern Tibet. Geodynamic and seismic hazard implications. *Earth and Planetary Science Letters, 485*, 19–31.

Barbot, S., Luo, H., Wang, T., Hamiel, Y., Piatibratova, O., Javed, M. T., & Gurbuz, G. (2023). Slip distribution of the February 6, 2023 M_w7.8 and M_w7.6, Kahramanmaraş, Turkey earthquake sequence in the East Anatolian fault zone. *Seismica, 2*(3).

Bemis, S. P., Micklethwaite, S., Turner, D., James, M. R., Akciz, S., Thiele, S. T., & Bangash, H. A. (2014). Ground-based and UAV-Based photogrammetry: A multi-scale, high-resolution mapping tool for structural geology and paleoseismology. *Journal of Structural Geology, 69*, 163–178.

Benedetti, L., Manighetti, I., Gaudemer, Y., Finkel, R., Malavieille, J., Pou, K., Arnold, M., Aumaıtre, G., Bourles, D., & Keddadouche, K. (2013). Earthquake synchrony and clustering on Fucino faults (Central Italy) as revealed from in situ ^{36}Cl exposure dating. *Journal of Geophysical Research, 118*. https://doi.org/10.1002/jgrb.50299.

Biasi, G. P., & Wesnousky, S. G. (2016). Steps and gaps in ground ruptures: Empirical bounds on rupture propagation. *Bulletin of the Seismological Society of America, 106*(3), 1110–1124. https://doi.org/10.1785/0120150175.

Bollinger, L., Sapkota, S. N., Tapponnier, P., Klinger, Y., Rizza, M., Van der Woerd, J., Tiwari, D. R., Pandey, R., Bitri, A., & Bes de Berc, S. (2014). Estimating the return times of great Himalayan earthquakes in eastern Nepal: Evidence from the Patu and Bardibas strands of the Main Frontal Thrust. *Journal of Geophysical Research: Solid Earth, 119*, 7123–7163. https://doi.org/10.1002/2014JB010970.

Bonilla, M. G. (1988). Minimum earthquake magnitude associated with coseismic surface faulting. *Bulletin of the Association of Engineering Geologists, 25*(1), 17–29.

Bull, W. B. (2008). *Tectonic geomorphology of mountains: A new approach to paleoseismology*. John Wiley & Sons.

Burbank, D. W., & Anderson, R. S. (2011). *Tectonic geomorphology* (2nd ed.). Wiley-Blackwell ISBN: 978-1-444-33887-4.

Bürgmann, R., Rosen, P. A., & Fielding, E. J. (2000). Synthetic aperture radar interferometry to measure Earth's surface topography and its deformation. *Annual Review of Earth and Planetary Sciences, 28*(1), 169–209.

Caine, J. S., Evans, J. P., & Forster, C. B. (1996). Fault zone architecture and permeability structure. *Geology, 24*(11), 1025–1028.

Chen, K., Avouac, J. P., Aati, S., Milliner, C., Zheng, F., & Shi, C. (2020). Cascading and pulse-like ruptures during the 2019 Ridgecrest earthquakes in the Eastern California Shear Zone. *Nature Communications, 11*(1), 22.

Chevalier, M. L., Fang, Z., Chen, S., Pan, J., Li, H., Wu, K., & Ding, L. (2025). Extension rate variations across the South Tibetan rifts–New data from the northern Xainza-Dinggye Graben. *Tectonophysics*, 230824.

Chevalier, M. L., Pan, J., Li, H., Liu, D., & Wang, M. (2015). Quantification of both normal and right-lateral late Quaternary activity along the Kongur Shan extensional system, Chinese Pamir. *Terra Nova, 27*(5), 379–391.

Chevalier, M. L., Pan, J., Li, H., Sun, Z., Liu, D., Pei, J., & Wu, C. (2017). First tectonic-geomorphology study along the Longmu–Gozha Co fault system, Western Tibet. *Gondwana Research, 41*, 411–424.

Chevalier, M. L., & Replumaz, A. (2019). Deciphering old moraine age distributions in SE Tibet showing bimodal climatic signal for glaciations: Marine Isotope Stages 2 and 6. *Earth and Planetary Science Letters, 507*, 105–118.

Chevalier, M. L., Ryerson, F. J., Tapponnier, P., Finkel, R. C., Van Der Woerd, J., Li, H., & Qing, L. (2005). Slip-rate measurements on the Karakorum fault may imply secular variations in fault motion. *Science (New York, N.Y.), 307*(5708), 411–414.

Chevalier, M. L., Tapponnier, P., Van Der Woerd, J., Leloup, P. H., Wang, S., Pan, J., & Li, H. (2020). Late Quaternary extension rates across the northern half of the Yadong-Gulu rift: Implication for east-west extension in southern Tibet. *Journal of Geophysical Research, 125*(7), e2019JB019106.

Chevalier, M. L., Tapponnier, P., Van der Woerd, J., Ryerson, F. J., Finkel, R. C., & Li, H. (2012). Spatially constant slip rate along the southern segment of the Karakorum fault since 200 ka. *Tectonophysics, 530*, 152–179.

Chevalier, M. L., Van der Woerd, J., Tapponnier, P., Li, H., Ryerson, F. J., & Finkel, R. C. (2016). Late Quaternary slip-rate along the central Bangong-Chaxikang segment of the Karakorum fault, western Tibet. *Geological Society of America Bulletin, 128*(1-2), 284–314.

Choi, J.-H., Klinger, Y., Ferry, M., Ritz, J. F., Kurtz, R., Rizza, M., Bollinger, L., Davaasambuu, B., Tsend-Ayush, N., & Demberel, S. (2018). Geologic inheritance and earthquake rupture processes: The 1905 M ≥ 8 Tsetserleg-Bulnay strike-slip earthquake sequence, Mongolia. *Journal of Geophysical Research, 123*, 1925–1953. https://doi.org/10.1002/2017JB013962.

Cunningham, W. D., & Mann, P. (2007). *Tectonics of strike-slip restraining and releasing bends. London: Geological Society, 290*, 1–12. https://doi.org/10.1144/SP290.1 Special Publications.

Daout, S., Doin, M. P., Peltzer, G., Socquet, A., & Lasserre, C. (2017). Large-scale InSAR monitoring of permafrost freeze-thaw cycles on the Tibetan Plateau. *Geophysical Research Letters, 44*(2), 901–909.

Di, N., Li, C., Li, T., Hu, W., Chen, Z., Zhang, Y., Lü, L., Chen, J., & Shan, X. (2023). The 2021 M_w5.2 Baicheng Earthquake: Implications for the hazards of extremely shallow earthquakes. *Seismological Society of America, 94*(4), 1775–1790.

Dunham, E. M., Belanger, D., Cong, L., & Kozdon, J. E. (2011). Earthquake ruptures with strongly rate-weakening friction and off-fault plasticity, Part 1: Planar faults. *Bulletin of the Seismological Society of America, 101*(5), 2296–2307.

DuRoss, C. B., Gold, R. D., Dawson, T. E., Scharer, K. M., Kendrick, K. J., Akciz, S. O., & Zinke, R. (2020). Surface displacement distributions for the July 2019 Ridgecrest, California, earthquake ruptures. *Bulletin of the Seismological Society of America, 110*(4), 1400–1418.

Faulkner, D. R., Jackson, C. A. L., Lunn, R. J., Schlische, R. W., Shipton, Z. K., Wibberley, C. A. J., & Withjack, M. O. (2010). A review of recent developments concerning the structure, mechanics and fluid flow properties of fault zones. *Journal of Structural Geology, 32*(11), 1557–1575.

Glennie, C. L., Carter, W. E., Shrestha, R. L., & Dietrich, W. E. (2013). Geodetic imaging with airborne LiDAR: The Earth's surface revealed. *Reports on Progress in Physics, 76*(8), 086801.

Hamling, I. J., Hreinsdóttir, S., Clark, K., Elliott, J., Liang, C., Fielding, E., & Stirling, M. (2017). Complex multifault rupture during the 2016 M w 7.8 Kaikōura earthquake, New Zealand. *Science (New York, N.Y.), 356*(6334), eaam7194.

Han, W., Zhang, X., Wang, Y., Wang, L., Huang, X., Li, J., ... Wang, Y. (2023). A survey of machine learning and deep learning in remote sensing of geological environment: Challenges, advances.

Hetzel, R., Tao, M., Stokes, S., Niedermann, S., Ivy-Ochs, S., Gao, B., & Kubik, P. W. (2004). Late Pleistocene/Holocene slip rate of the Zhangye thrust (Qilian Shan, China) and implications for the active growth of the northeastern Tibetan Plateau. *Tectonics, 23*(6).

Huang, Y., & Yu, M. (2013). Review of soil liquefaction characteristics during major earthquakes of the twenty-first century. *Natural Hazards, 65*, 2375–2384.

Iezzi, F., Roberts, G., Faure Walker, J., Papanikolaou, I., Ganas, A., Deligiannakis, G., & Gheorghiu, D. (2021). Temporal and spatial earthquake clustering revealed through comparison of millennial strain-rates from ^{36}Cl cosmogenic exposure dating and decadal GPS strain-rate. *Scientific Reports, 11*(1), 23320. https://doi.org/10.1038/s41598-021-02131-3.

Ishihara, K. (1993). Liquefaction and flow failure during earthquakes. *Geotechnique, 43*(3), 351–451.

Ji, H., Ren, Z., Zhu, X., Bai, M., Bao, G., Liu, J., et al. (2024). Slip distribution along the Chenghai fault from airborne LiDAR and tectonic implications for the 1515 Yongsheng earthquake, China. *Tectonics, 43*, e2024TC008285. https://doi.org/10.1029/2024TC008285.

Johnson, K., Nissen, E., Saripalli, S., Arrowsmith, J. R., McGarey, P., Scharer, K., & Blisniuk, K. (2014). Rapid mapping of ultrafine fault zone topography with. *Geosphere, 10*(5), 969–986.

King, G., & Nabelek, J. (1985). Role of fault bends in the initiation and termination of earthquake rupture. *Science (New York, N.Y.), 228*(4702), 984–987. https://doi.org/10.1126/science.228.4702.984.

Klinger, Y. (2010). Relation between continental strike-slip earthquake segmentation and thickness of the crust. *Journal of Geophysical Research, 115*(B07306). https://doi.org/10.1029/2009JB006550.

Klinger, Y., Etchebes, M., Tapponnier, P., & Narteau, C. (2011). Characteristic slip for five great earthquakes along the Fuyun fault in China. *Nature Geoscience, 4*(6), 389–392.

Le Béon, M., Tseng, Y. C., Klinger, Y., Elias, A., Kunz, A., Sursock, A., & Jomaa, R. (2018). High-resolution stratigraphy and multiple luminescence dating techniques to reveal the paleoseismic history of the central Dead Sea fault (Yammouneh fault, Lebanon). *Tectonophysics, 738*, 1–15.

Lemrabet, L., Doin, M. P., Lasserre, C., & Durand, P. (2023). Referencing of continental-scale InSAR-derived velocity fields: Case study of the eastern Tibetan Plateau. *Journal of Geophysical Research, 128*, e2022JB026251. https://doi.org/10.1029/2022JB026251.

Lettis, W., Bachhuber, J., Witter, R., Brankman, C., Randolph, C. E., Barka, A., & Kaya, A. (2002). Influence of releasing step-overs on surface fault rupture and fault segmentation: Examples from the 17 August 1999 Izmit earthquake on the North Anatolian fault, Turkey. *Bulletin of the Seismological Society of America, 92*(1), 19–42.

Li, H., Chevalier, M. L., Tapponnier, P., Pan, J., van Der Woerd, J., Mériaux, A. S., & Xu, X. (2021). Block tectonics across Western Tibet and multi-millennial recurrence of great earthquakes on the Karakax Fault. *Journal of Geophysical Research, 126*(12), e2021JB022033.

Li, H., Pan, J., Chevalier, M.-L., Liu, D., Wang, S., Luo, H., Zhang, L., Fang, L., Wang, T., Liu, F., Chen, S., He, X., Zhu, X., Zhang, Y., Wu, Q., & Li, C. (2025). Aftershock-induced surface ruptures overshadow the 2024 Mw 7.0 Wushi mainshock, China. *Geology, 54*(2), 105–110. https://doi.org/10.1130/g54078.1.

Li, H., Pan, J., Lin, A., Sun, Z., Liu, D., Zhang, J., & Gong, Z. (2016). Coseismic surface ruptures associated with the 2014 M w 6.9 Yutian earthquake on the Altyn Tagh fault, Tibetan plateau. *Bulletin of the Seismological Society of America, 106*(2), 595–608.

Li, H., Wang, Z., Fu, X., Hou, L., Si, J., Qiu, Z., Li, N., & Wu, F. (2008). The surface rupture zone distribution of the Wenchuan earthquake (Ms8.0) happenned on may 12th, 2008. *Geology in China, 35*(5) In Chinese.

Li, H., Van der Woerd, J., Tapponnier, P., Klinger, Y., Xuexiang, Q., Jingsui, Y., & Yintang, Z. (2005). Slip rate on the Kunlun fault at Hongshui Gou, and recurrence time of great events comparable to the 14/11/2001, Mw~ 7.9 Kokoxili earthquake. *Earth and Planetary Science Letters, 237*(1-2), 285–299.

Li, K., Tapponnier, P., Xu, X., Ren, J., Wang, S., & Zhao, J. (2022). Holocene slip rate along the beng co fault and dextral strike-slip extrusion of central eastern tibet. *Tectonics, 41*(8), e2022TC007230.

Li, K., Xu, X., Tapponnier, P., Kang, W., & Wang, S. (2023a). Lake level changes and holocene extension rate across the ns-trending peiku co-gyirong faults, southern tibet. *Tectonics, 42*(5), e2022TC007644.

Li, K., Tapponnier, P., Xu, X., & Kang, W. (2023b). The 2022, Ms 6.9 Menyuan earthquake: Surface rupture, Paleozoic suture re-activation, slip-rate and seismic gap along the Haiyuan fault system, NE Tibet. *Earth and Planetary Science Letters, 622*, 118412.

Liu, F., Pan, J., Li, H., Yang, S., Wu, K., Su, Q., Zhang, S., Chevalier, M. L., & Lu, H. (2025). Coseismic surface rupture of the 2025 Mw7.1 Tingri earthquake and potential seismic risk in southern Tibetan Plateau. *Acta Geologica Sinica, 99*(3), 685–703 In Chinese.

Luo, H., & Wang, T. (2022). Strain partitioning on the western Haiyuan fault system revealed by the adjacent 2016 Mw5.9 and 2022 Mw6.7 Menyuan earthquakes. *Geophysical Research Letters, 49*, e2022GL099348. https://doi.org/10.1029/2022GL099348.

Massonnet, D., Rossi, M., Carmona, C., Adragna, F., Peltzer, G., Feigl, K., & Rabaute, T. (1993). The displacement field of the Landers earthquake mapped by radar interferometry. *Nature, 364*(6433), 138–142.

McCalpin, J. P. (2009). *Paleoseismology* (2nd ed.). Academic Press, 1–613 ISBN: 978-0-12-373576-8, ISSN: 0074-6142.

McGill, S. F., & Sieh, K. (1991). Surficial offsets on the central and eastern Garlock fault associated with prehistoric earthquakes. *Journal of Geophysical Research, 96*(B13), 21597–21621.

Mériaux, A. S., Ryerson, F. J., Tapponnier, P., Van der Woerd, J., Finkel, R. C., Xu, X., & Caffee, M. W. (2004). Rapid slip along the central Altyn Tagh Fault: Morphochronologic evidence from Cherchen He and Sulamu Tagh. *Journal of Geophysical Research: Solid Earth, 109*(B6).

Milliner, C. W. D., Dolan, J. F., Hollingsworth, J., Leprince, S., Ayoub, F., & Sammis, C. G. (2015). Quantifying near-field and off-fault deformation patterns of the 1992 Mw 7.3 Landers earthquake. *Geochemistry, Geophysics, Geosystems, 16*, 1577–1598. https://doi.org/10.1002/2014GC005693.

Minson, S. E., Brooks, B. A., Glennie, C. L., Murray, J. R., Langbein, J. O., Owen, S. E., & Hauser, D. L. (2015). Crowdsourced earthquake early warning. *Science Advances, 1*(3), e1500036. https://doi.org/10.1126/sciadv.1500036.

Nash, D. B. (1980). Morphologic dating of degraded normal fault scarps. *The Journal of Geology, 88*(3), 353–360. https://doi.org/10.1086/628513.

Oglesby, D. D. (2005). The dynamics of strike-slip step-overs with linking dip-slip faults. *Bulletin of the Seismological Society of America, 95*(5), 1604–1622. https://doi.org/10.1785/0120050058.

Oglesby, D. D., Archuleta, R. J., & Nielsen, S. B. (2000). The three-dimensional dynamics of dipping faults. *Bulletin of the Seismological Society of America, 90*(3), 616–628.

Oskin, M. E., Arrowsmith, J. R., Corona, A. H., Elliott, A. J., Fletcher, J. M., & Fielding, E. J. (2012). Near-field deformation from the El Mayor–Cucapah earthquake revealed by differential LIDAR. *Science, 335*(6069), 702–705.

Palumbo, L., Benedetti, L., Bourlès, D., Cinque, A., & Finkel, R. (2004). Slip history of the Magnola fault (Apennines, Central Italy) from ^{36}Cl surface exposure dating: Evidence for strong earthquakes over the Holocene. *Earth and Planetary Science Letters, 225*(1-2), 163–176.

Pan, J., Li, H., Chevalier, M. L., Tapponnier, P., Bai, M., Li, C., & Chen, P. (2022). Coseismic rupture of the 2021, M_w7. 4 Maduo earthquake (northern Tibet): Short-cutting of the Kunlun fault big bend. *Earth and Planetary Science Letters, 594*, 117703.

Ren, C., Wang, Z., Taymaz, T., Hu, N., Luo, H., Zhao, Z., & Ding, H. (2024). Supershear triggering and cascading fault ruptures of the 2023 Kahramanmaraş, Türkiye, earthquake doublet. *Science, 383*(6680), 305–311.

Ren, J., Xu, X., Zhang, G., Wang, Q., Zhang, Z., Gai, H., & Kang, W. (2022a). Coseismic surface ruptures, slip distribution, and 3D seismogenic fault for the 2021 M_w7.3 Maduo earthquake, central Tibetan Plateau, and its tectonic implications. *Tectonophysics, 827*, 229275.

Ren, Z., & Zhang, Z. (2019). Structural analysis of the 1997 M_w 7.5 Manyi earthquake and the kinematics of the Manyi fault, central Tibetan Plateau. *Journal of Asian Earth Sciences, 179*, 149–164.

Ren, Z., Zielke, O., Chevalier, M. L., Nissen, E., & Zhang, H. (2022b). Resolving Quaternary tectonic activity with high-resolution data in space and time. *Lithosphere, 2022*(1), 9823275.

Ritz, J. F., Bourles, D., Brown, E. T., Carretier, S., Chery, J., Enhtuvshin, B., & Yiou, F. (2003). Late Pleistocene to Holocene slip rates for the Gurvan Bulag thrust fault (Gobi-Altay, Mongolia) estimated with ^{10}Be dates. *Journal of Geophysical Research, 108*(B3).

Ryerson, F. J., Tapponnier, P., Finkel, R. C., Mériaux, A. S., Van der Woerd, J., Lasserre, C., & King, G. C. (2006). Applications of morphochronology to the active tectonics of Tibet. Geological Society of America Special Paper 415.

Sapkota, S. N., Bollinger, L., Klinger, Y., Tapponnier, P., Gaudemer, Y., & Tiwari, D. (2013). Primary surface ruptures of the great Himalayan earthquakes in 1934 and 1255. *Nature Geoscience, 6*(1), 71–76.

Scholz, C. H. (2002). *The mechanics of earthquakes and faulting* (2nd ed.). *Cambridge University Press.*

Schwartz, D. P., & Coppersmith, K. J. (1984). Fault behavior and characteristic earthquakes: Examples from the Wasatch and San Andreas fault zones. *Journal of Geophysical Research: Solid Earth, 89*(B7), 5681–5698.

Shaw, J. H., & Shearer, P. M. (1999). An elusive blind-thrust fault beneath metropolitan Los Angeles. *Science (New York, N.Y.), 283*(5407), 1516–1518. https://doi.org/10.1126/science.283.5407.1516.

Shelly, D. (2020). A high-resolution seismic catalog for the initial 2019 Ridgecrest earthquake sequence: Foreshocks, aftershocks, and faulting complexity. *Seismological Research Letters 2020, 91*(4), 1971–1978. https://doi.org/10.1785/0220190309.

Sibson, R. H. (1983). Continental fault structure and the shallow earthquake source. *Journal of the Geological Society, 140*, 741–767. https://doi.org/10.1144/gsjgs.140.5.0741.

Sieh, K. (1978). Slip along the San Andreas fault associated with the great 1857 earthquake. *Bulletin of the Seismological Society of America, 68*(5), 1421–1448.

Sieh, K. E., & Jahns, R. H. (1984). Holocene activity of the San Andreas fault at Wallace creek, California. *Geological Society of America Bulletin, 95*(8), 883–896.

Somerville, P. G. (2003). Magnitude scaling of the near fault rupture directivity pulse. *Physics of the Earth and Planetary Interiors, 137*(1-4), 201–212.

Sylvester, A. G. (1988). Strike-slip faults. *Geological Society of America Bulletin, 100*(11), 1666–1703.

Tapponnier, P., Xu, Z. Q., Roger, F., Meyer, B., Arnaud, N., Wittlinger, G., & Yang, J. S. (2001). Oblique stepwise rise and growth of the Tibet Plateau. *Science (New York, N.Y.), 294*, 1671–1677. https://doi.org/10.1126/science.105978.

Telling, J., Lyda, A., Hartzell, P., & Glennie, C. (2017). Review of Earth science research using terrestrial laser scanning. *Earth-Science Reviews, 169*, 35–68.

Van der Woerd, J. V., Tapponnier, P., Ryerson, F. J., Meriaux, A. S., Meyer, B., Gaudemer, Y., & Zhiqin, X. (2002). Uniform postglacial slip-rate along the central 600 km of the Kunlun Fault (Tibet), from ^{26}Al, ^{10}Be, and ^{14}C dating of riser offsets, and climatic origin of the regional morphology. *Geophysical Journal International, 148*(3), 356–388.

Wallace, R. E. (1977). Profiles and ages of young fault scarps, north-central Nevada. *Geological Society of America Bulletin, 88*(9), 1267–1281.

Wang, S., Chevalier, M. L., Pan, J., Bai, M., Li, K., Li, H., & Wang, G. (2020). Quantification of the late Quaternary throw rates along the Yadong rift, southern Tibet. *Tectonophysics, 790*, 228545.

Wells, D. L., & Coppersmith, K. J. (1994). New empirical relationships among magnitude, rupture length, rupture width, rupture area, and surface displacement. *Bulletin of the Seismological Society of America, 84*(4), 974–1002.

Wesnousky, S. G. (2006). Predicting the endpoints of earthquake ruptures. *Nature, 444*(7117), 358–360. https://doi.org/10.1038/nature05275.

Wesnousky, S. G. (2008). Displacement and geometrical characteristics of earthquake surface ruptures: Issues and implications for seismic hazard analysis and the process of earthquake rupture. *Bulletin of the Seismological Society of America, 98*(4), 1609–1632. https://doi.org/10.1785/0120070111.

Whipple, K. X. (2004). Bedrock rivers and the geomorphology of active orogens. *Annual Review of Earth and Planetary Sciences, 32*, 151–185. https://doi.org/10.1146/annurev.earth.32.101802.120356.

Wu, C., Zheng, W., Zhang, P., Zhang, Z., Jia, Q., Yu, J., & Chen, J. (2019). Oblique thrust of the Maidan fault and late Quaternary tectonic deformation in the southwestern Tian Shan, Northwestern China. *Tectonics, 38*(8), 2625–2645.

Wu, K., Chevalier, M. L., Pan, J., Liu, F., Yang, S., Zhang, S., & Li, H. (2025). Recurrence of large earthquakes along the southern Xainza-Dinggye rift and comparison with the 7 January 2025, M_w7. 1 Tingri earthquake, southern Tibet. *Tectonophysics*, 230827.

Yeats, R. S., Sieh, K., & Allen, C. R. (1997). *The geology of earthquakes*. Oxford University Press.

Youd, T. L., & Perkins, D. M. (1978). Mapping liquefaction-induced ground failure potential. *Journal of the Geotechnical Engineering Division, 104*(4), 433–446.

Yuan, Z., Li, T., Su, P., Sun, H., Ha, G., Guo, P., & Thompson Jobe, J. (2022). Large surface-rupture gaps and low surface fault slip of the 2021 M_w7.4 Maduo earthquake along a low-activity strike-slip fault, Tibetan Plateau. *Geophysical Research Letters, 49*(6), e2021GL096874.

Zahra, K., Imran, M., & Ostermann, F. O. (2020). Automatic identification of eyewitness messages on Twitter during disasters. *Information Processing & Management, 57*(1) 102107.

Zhang, Y., Meng, X., Jordan, C., Novellino, A., Dijkstra, T., & Chen, G. (2018). Investigating slow-moving landslides in the Zhouqu region of China using InSAR time series. *Landslides, 15*(7), 1299–1315.

Zielke, O., Arrowsmith, J. R., Ludwig, L. G., & Akçiz, S. O. (2010). Slip in the 1857 and earlier large earthquakes along the Carrizo Plain, San Andreas fault. *Science (New York, N.Y.), 327*, 1119–1122. https://doi.org/10.1126/science.1182781.

Zinke, R., Hollingsworth, J., Dolan, J. F., & Van Dissen, R. (2019). Three-dimensional surface deformation in the 2016 M_w7.8 Kaikōura, New Zealand, earthquake from optical image correlation: Implications for strain localization and long-term evolution of the Pacific-Australian plate boundary. *Geochemistry, Geophysics, Geosystems, 20*, 1609–1628. https://doi.org/10.1029/2018GC007951.

Zoback, M. L. (1992). First- and second-order patterns of stress in the lithosphere: The World Stress Map Project. *Journal of Geophysical Research, 97*(B8) 11,703–711,728.

CHAPTER 7

Coseismic landslides and rock avalanches

Gonghui Wang
Research Center for Landslide Risk Cognition and Reduction, DPRI, Kyoto University, Uji, Kyoto, Japan

7.1 Introduction

Coseismic landslides (including falls, topples, slides, spreads, flows and other types) are mobilised when ground shaking reduces the stability of geo-materials on a slope, and they involve a sudden and massive release of gravitational potential energy. In most cases, these failures are not only destructive hazards, accounting for a significant fraction of earthquake-related casualties and economic losses (Budimir et al., 2014; Keefer, 1984; Rodríguez et al., 1999), but also transformative geomorphic processes that reshape mountain landscapes, redistribute sediment on a regional scale and set in motion long-term environmental changes (Fan et al., 2019; Hewitt, 1998; Hovius et al., 2011; Korup, 2005).

The importance of coseismic landslides extends beyond their immediate hazards. Their geomorphic records, such as rock-avalanche deposits, disrupted colluvium or landslide-dammed lakes, are markers that persist long after the triggering earthquake (Harp et al., 1992; Hewitt, 1999; Hovius et al., 2011; Korup, 2005; Owen et al., 2008). These features form an archive of seismic events in the landscape and exert long-lasting influences on hydrology, sediment fluxes and ecosystems (Dai et al., 2005; Li et al., 2017; Marc et al., 2016; Dai et al., 2011; Korup & Montgomery, 2008). Systematic documentation of coseismic landslides has grown rapidly with advances in remote sensing and digital mapping (Gorum et al., 2011; Kawamura et al., 2019; Keefer, 1984; Kieffer et al., 2006; Meunier et al., 2008; Ohno et al., 2024; Rodríguez et al., 1999; Tanyaş et al., 2017; Xu et al., 2016; Yagi et al., 2007), providing new insights into their scale and variability. Historical and recent earthquakes illustrate this destructive potential, such as the 2008 M_w 7.9 Wenchuan earthquake in China (see Chapter 14), the 2016 M_w 7.8 Kaikōura earthquake in New Zealand, the 2018 M_w 6.6 Hokkaido Eastern Iburi earthquake in Japan and the 2023 Turkey–Syria earthquakes, among others; each of these triggered huge numbers of landslides, dramatically altering terrain, river systems and sediment cascades (e.g. Gorum et al., 2023; Huang & Fan, 2013; Kawamura et al., 2019; Keefer, 1984; Massey et al., 2018; Roback et al., 2018; Yin et al., 2009).

Geomorphic Impacts of Earthquakes
DOI: https://doi.org/10.1016/B978-0-443-40375-0.00007-4

From the perspective of attempting to quantify hazard and risk, coseismic landslides pose extraordinary challenges. Unlike rainfall-induced landslides, they can occur suddenly and simultaneously over thousands of square kilometres, generating cascading hazards such as landslide dams, outburst floods, debris flows and even landslide-triggered tsunamis (Chigira et al., 2010; Dohmen et al., 2025; Fan et al., 2019; Tang et al., 2011, 2016; Wang, Huang, et al., 2014; Xu et al., 2022, 2022). Their synchronous, large-scale nature magnifies their impact on infrastructure, economies, resilience and disaster response. At the same time, from a geomorphic perspective, they are fundamental drivers of erosion and mountain-belt evolution, often supplying sediment volumes comparable to centuries of aseismic processes (Hovius et al., 1997; Korup et al., 2007; Li et al., 2015).

This chapter synthesises current knowledge of coseismic landslides with special focus on slides and rock/debris avalanches, emphasising both their significance as geomorphic archives of tectonic activity and their implications for hazard assessment. It begins with an introduction to different types of coseismic landslides and a discussion of the geomorphic characteristics of source areas and deposits, followed by a brief introduction to the physical bases and triggering mechanisms that underlie the initiation of landslides during earthquakes. Finally, case examples from some major recent earthquakes introduce new results that increase our understanding of the initiation and movement of specific slides and rock/debris avalanches.

7.2 Types of earthquake-triggered landslides

As summarised by Varnes (1978) and later refined by Cruden and Varnes (1996), landslides can be classified into five fundamental types according to their kinematics, namely fall, topple, slide, spread and flow. Hungr et al. (2014) further updated this classification by adding slope deformation as a sixth type and distinguishing 32 subtypes based on both movement style and the material (rock or soil) involved. Based on data from 40 historical worldwide earthquakes, Keefer (1984, 2002) classified coseismic landslides into fourteen types belonging to three fundamental categories based on material type, movement mechanism, degree of internal disruption and geologic setting. The first category, '*disrupted slides and falls*', includes rock/soil falls, disrupted slides and rock and soil avalanches, characterised by highly fragmented material moving rapidly by falling, bouncing, rolling or sliding from steep slopes; these failures are usually shallow (<3 m), except for large rock avalanches (with volumes of more than c.1 × 10^6 m^3) that may travel great distances. The second category, '*coherent slides*', comprises translational and rotational block slides and slow earth flows, where relatively intact blocks move with limited internal deformation; these failures occur on moderately steep slopes, move more slowly, travel generally less than 100 m and are typically thicker (>3 m). The third category, '*lateral spreads and flows*', consists of failures dominated by fluid-like motion that initiate on gentle slopes and can travel long distances. Keefer (1984, 2002) also pointed out that in terms of frequency, disrupted slides and falls overwhelmingly dominate, accounting for ~86% of documented cases from 40 earthquakes, compared with ~8% coherent slides and ~6% lateral spreads and flows. Similar tendencies have been reported by Rodríguez et al. (1999), based on analysis of landslides triggered by earthquakes between 1980 and 1997.

It is worth noting that although Keefer's classification primarily focuses on the spatial distribution of landslides on hillslopes during earthquakes, which may be associated with different levels of mobility of the displaced materials, it generally aligns with the categories defined by Varnes (1978) and Cruden and Varnes (1996). It is also noted that the classification of debris avalanche in Hungr et al. (2014) is basically the same as that of soil avalanche in Keefer (1984), and hence the term 'debris avalanche' will be used in the present work. As this chapter mainly concentrates on rockslides and rock/debris avalanches, the next section presents a concise overview of their key characteristics. Note also that landslides in the 'slope deformation' category defined by Hungr et al. (2014) are not discussed here, as their direct geomorphic and environmental impacts are typically limited, although it is reasonable to infer that earthquake-induced slope deformation may help in preconditioning hillslopes for subsequent rainfall-induced failures after major earthquakes.

7.2.1 Slides

As stated by Varnes (1978), a slide is a downslope movement of soil or rock mass occurring dominantly on the surface of rupture or on relatively thin zones of intense shear strain. Slides can be classified according to their modes of sliding, including rotational slides, translational (planar) slides, compound slides, rock wedge slides, irregular rockslides and gravel-, sand-, or debris-slides (Cruden & Varnes, 1996; Hungr et al., 2014).

According to the thickness of displaced materials, slides can be classified as shallow or deep-seated (Chigira, 2012; Keefer, 1984). Shallow slides are among the most common coseismic landslides. They typically have initial failure depths of less than several metres in most cases (Keefer, 1984) and can involve surficial regolith, colluvium and weathered rock. Their failure surfaces typically coincide with the interface between surficial deposits and bedrock or with preexisting weathered horizons. Although the volume of shallow slides is limited (in most cases ranging from ~10^2 to 10^5 m^3), their abundance and widespread distribution during earthquakes can result in extensive cumulative damage to landscapes, infrastructure and communities. This is exemplified by the slides occurring in pyroclastic fall deposits triggered by the 2018 M_w 6.6 Hokkaido Eastern Iburi earthquake (Fig. 7.1), during which more than 6000 shallow slides occurred (Kawamura et al., 2019).

Figure 7.1 Oblique view of shallow slides induced by the 2018 M_w 6.6 Hokkaido Eastern Iburi Earthquake in the area of Takaoka, Hokkaido. *Photo by G. Wang on October 21, 2018.*

Deep-seated slides are large-scale slope failures characterised by sliding surfaces that can extend up to several hundred metres below the ground surface. Unlike shallow slides, they mobilise enormous volumes of weathered and intact bedrock, colluvium and regolith. Their displacement mechanisms are governed by deep-seated structural discontinuities, lithological contrasts and long-term weathering processes that predispose slopes to instability. Such movements often develop over extensive areas, covering from several hectares to many square kilometres.

7.2.2 Rock/debris avalanches

According to Hungr et al. (2014), the term 'rock avalanche' refers to those landslides that involve 'extremely rapid, massive, flow-like motion of fragmented rock from a large rock slide or rock fall', while 'debris avalanche' refers to those with 'very rapid to extremely rapid shallow flow of partially or fully saturated debris on a steep slope, without confinement in an established channel'. On the other hand, Keefer (1984) pointed out that the movement of rock avalanches triggered by earthquakes is 'complex, involving sliding and/or flow, as a stream of rock fragments' with extremely rapid velocity, while the movement of soil avalanches triggered by earthquakes involves 'translational sliding with subsidiary flow' with very rapid to extremely rapid velocity.

Rock/debris avalanches are characterised by very large volumes (commonly exceeding ~10^6 to 10^8 m^3), high velocities and exceptionally long runouts, making them one of the most geomorphically significant and hazardous coseismic slope phenomena (Hungr & Evans, 2004; Keefer, 1984). For example, during the 2008 M_w 7.9 Wenchuan earthquake in China, a giant rock avalanche with a volume of about

1.2 billion m^3 occurred in the Daguangbao area of Sichuan Province (Fig. 7.2), representing one of the largest coseismic landslides in recent centuries (Cui et al., 2021; Huang & Fan, 2013; Yin et al., 2009). The displaced mass travelled about 4.5 km and formed a landslide dam approximately 500 m high. Another catastrophic case is the Donghekou debris avalanche (Fig. 7.3), which was also triggered by the 2008 Wenchuan earthquake but was located about 200 km from the epicentre. About 20 million m^3 of material failed, burying four villages, damming a river and travelling ~2.5 km, resulting in more than 780 fatalities (Wang et al., 2014, 2021). These landslides are described in more detail later.

7.3 Topographical and geological features of source areas

7.3.1 Geomorphic features

Coseismic landslides develop under a broad spectrum of conditions, reflecting the combined influence of terrain morphology, lithology and hydrological conditions. As noted by Keefer, 2002, slopes and geological settings prone to earthquake-induced landslides vary greatly, from overhanging slopes formed in well-indurated bedrock to nearly level surfaces underlain by unconsolidated sediments, with minimum slope angles for different types of failures ranging from 0.3 to 40 degrees.

Based on a range of studies conducted in Japan (Abe & Hayashi, 2011; Chigira & Yagi, 2006; Tanaka, 1998), Higaki et al. (2012) summarised the topographic characteristics of coseismic landslides and emphasised that certain geomorphic settings

Figure 7.2 Daguangbao rock avalanche with a volume of about 1.2 billion m^3 triggered by the 2008 Wenchuan earthquake, China. Width and length of failure are approximately 2 and 4 km, respectively. *Photo courtesy of Runqiu Huang, Chengdu University of Technology.*

Figure 7.3 Debris avalanche in the Donghekou area triggered by the 2008 Wenchuan earthquake, China. *Photo by G. Wang on July 7, 2008.*

are systematically more susceptible to failure during earthquakes. They classified these settings into two principal categories: (1) regions of generally high risk, determined primarily by regional geology and landform type, and (2) specific hazardous locations within these high-risk regions, where slope failure is governed by detailed topographic and hydrological conditions.

For regions of generally high risk, the principal features include:

(1) areas in the vicinity of active fault systems, where strong shaking and fault-related fracturing are concentrated;
(2) volcanic terrain composed of pyroclastic and volcaniclastic materials, which are typically weak, highly fractured and prone to amplification of seismic waves;
(3) unconsolidated sandy deposits in alluvial and coastal plains (e.g. sand dunes and beach ridges), where low cohesion and high pore-pressure sensitivity enhance failure under shaking;
(4) layered Tertiary–Quaternary sedimentary or volcanic rocks with slope-parallel bedding, which facilitate planar or wedge failures during earthquakes;
(5) terrains underlain by deeply weathered plutonic rocks, which commonly host thick, weakened regolith susceptible to large-scale slope failures;
(6) artificial fill and reclaimed land, which often contain unconsolidated materials prone to deformation or sliding under seismic loading; and
(7) accretionary prism and related hard-rock complexes, where pervasive shearing fabrics and brittle structures can localise coseismic failures.

For the specific hazardous locations, their principal features include:

(1) areas close to major faults;
(2) geomorphic settings that are prone to collapse or complete detachment-type slides, such as sea cliffs, river terraces, areas below knickpoints (convex breaks in slope) and other steep or convex-up slopes;
(3) valley walls where groundwater flow tends to converge toward the surface;
(4) hillslopes with exposed bedrock;
(5) hillslopes consisting of old landslide deposits, especially those whose toes have been eroded by rivers;
(6) hillslopes where ground motion during earthquakes is likely to be amplified due to slope geometry (the topographic effect); and
(7) hillslopes with a prominent flat bench near the summit.

7.3.2 Geological features

Keefer (2002) concluded that the materials most susceptible to earthquake-induced landslides were (1) weakly cemented, weathered, sheared, intensely fractured or closely jointed rocks, (2) better-indurated rocks having prominent discontinuities, (3) sandy residual or colluvial soils, (4) saturated volcanic soils containing sensitive clay, (5) loess, (6) cemented soils, (7) granular deltaic sediments, (8) granular flood-plain alluvium and (9) uncompacted, or poorly compacted, granular artificial fill. Thus, in contrast, intact and well-indurated bedrock with few discontinuities, dry and dense granular soils and well-compacted engineered fills are generally less susceptible to coseismic landsliding, except under exceptionally strong ground shaking or unfavourable slope geometry.

Chigira (2012) further summarised the geological features of earthquake-induced landslides and stated that they range from shallow surface failures and rockfalls involving weathered mantles or colluvium to deep-seated movements of bedrock masses. Shallow surface slides are generally governed by the properties of the surficial deposits and by topographic amplification of seismic motion. In contrast, deep-seated landslides reflect the structural fabric of the underlying rock mass (which may have been weakened by previous earthquakes) and its hydrogeological conditions, as well as mountain-scale topographic resonance, in which the entire rock mass (often including the ridge crest) responds to and amplifies ground shaking at frequencies close to its natural modes of vibration (Bazgard et al., 2011). In this sense, the distinction between shallow and deep-seated failures is not strictly a matter of depth but rather of the dominant geological and physical processes involved.

Besides the basic geological features of the landslide sites, seismic fault proximity also plays a controlling role in the occurrence of coseismic landslides. Landslide density or abundance correlates more closely with distance from the seismogenic fault than from the epicentre (Keefer, 1984; Rodriguez et al., 1999; Massey et al., 2018; Tang et al., 2011). Reverse-fault earthquakes typically produce far more landslides on the hanging wall side than on the footwall. Recent studies on the landslides triggered by the 2005 M_w 7.6 Kashmir (Sato et al., 2007) and by the 2008 M_w 7.9 Wenchuan earthquakes (Dai et al., 2011; Huang & Fan, 2013; Tang et al., 2011, 2016) have further supported this relationship. Directionality of seismic radiation also exerts some control, with dense landsliding often occurring perpendicular to the fault strike.

Among the most characteristic Japanese examples of coseismic landslides are those involving weathered pyroclastic materials. Pyroclastic fall deposits (such as air-fall ash and pumice layers) frequently undergo rapid strength loss during seismic shaking, particularly when they have high water contents, transforming slides into highly mobile flows. For example, the 2016 M_w 7.0 Kumamoto earthquake induced large numbers of flow-type slides on the slopes of pyroclastic fall deposits (Doi et al., 2019; Xu et al., 2017), while the 2018 M_w 6.6 Hokkaido Eastern Iburi earthquake triggered several thousand flow-type slides on hillslopes underlain by pyroclastic fall deposits (Fig. 7.1; Kawamura et al., 2019; Wang et al., 2019). Field investigation in both earthquakes revealed that the soil layers near the sliding surface after the earthquake had water contents up to 200% (Wang et al., 2019), and the slip surfaces typically formed within halloysite-bearing layers with high water content (Chigira & Suzuki, 2016; Wang et al., 2019).

In sedimentary rock terrains, structural weakness exerts a primary control. Bedding-parallel landslides, such as the Horonai landslide during the 2018 Hokkaido Eastern Iburi earthquake, originate along shale or bedding-plane faults preconditioned by flexural slip or prior movement. Folded sequences can also be destabilised, as in the giant Hattian Bala landslide triggered by the 2005 Kashmir earthquake in Pakistan, where synclinal geometry and undercutting of the toe promoted collapse (Dunning et al., 2007). Chigira et al. (2010) also reported that in carbonate rocks, dissolution along bedding planes reduces contact area and shear strength and thus predisposes slopes to failure, a process responsible for numerous prehistoric and historical collapses in limestone and marl. Even mudstones, once considered relatively resistant, may weaken through chemical weathering: oxidation of pyrite produces acidic groundwater that dissolves calcite and generates a fragile 'dissolution front', a possible reason inferred for many landslides triggered by the 2004 M_w 6.6 Mid-Niigata earthquake in Japan.

Large rock masses in mountains commonly undergo slow gravitational deformation prior to failure, perhaps exacerbated by prior seismicity. These mountain-mass gravitational deformations (referred to as *sackung* or rock-mass creep) can provide a crucial precursor to coseismic collapse. Depending on the dip of planar structures or discontinuities relative to the slope, deformation can be manifest as buckling, sliding or toppling, any of which can rapidly evolve into catastrophic movement during strong shaking. Recognising uphill-facing scarps and linear depressions therefore offers valuable clues to latent or potential future instability.

Special geological architectures further complicate slope response. Caprock sequences of hard rock overlying weak strata, buttressed slopes where strong toe rock supports softer material upslope and deposits of old landslides whose toes have been eroded all show high seismic susceptibility. Likewise, valley-fill and artificial-fill deposits – often water-saturated – readily fluidise under shaking, as demonstrated by numerous urban disasters in Japan from the 1978 M_s 7.7 Miyagi-ken-oki (Miura et al., 1998) and the 1995 M_w 6.9 Kobe earthquakes (Kamai et al., 2000).

Finally, surface slides and rockfalls, though individually small, can occur in vast numbers during strong earthquakes. They concentrate along ridges, cliffs and terrace edges where topographic convexity amplifies ground motion. Failures typically involve competent blocks or thin weathered mantles detached from intact bedrock, and their distribution closely follows zones of high peak ground acceleration (PGA).

In summary, the geological features controlling coseismic landslides encompass the structural fabric, lithological weakness, hydrochemical alteration and preexisting gravitational (and likely seismic) deformation of rock masses. Understanding how these factors interact with seismic forcing provides the foundation for interpreting past landslide distributions and for forecasting the style and mobility of future earthquake-induced slope failures.

7.4 Physical basis and triggering mechanisms

The initiation of coseismic landslides is fundamentally controlled by strong ground shaking. Seismic waves impose transient accelerations on topographic edifices that may temporarily exceed the resisting forces provided by material strength and edifice geometry, leading to detachment and motion of a volume of material. A commonly used indicator is PGA or the ratio of PGA to gravitational acceleration; this indicates the maximum force exerted on the slope by shaking, with higher values strongly correlated with increased landslide occurrence (Harp & Jibson, 1996; Keefer, 1984). In addition to PGA, the duration, phasing (Brain et al., 2015) and frequency content of shaking play critical roles. Long-period motions can initiate or destabilise deep-seated landslides and result in larger accumulated displacements, which may further result in reduction of shear strength along the sliding surface if the materials within the shear zone or along the sliding surface exhibit strain-weakening behaviour (Cui et al., 2021; Sassa et al., 2004). Conversely, high-frequency vibrations are especially effective in triggering landslides at small scales (Bazgard et al., 2011; Meunier et al., 2007).

The minimum earthquake magnitude causing landsliding has been stated as M_w ~4.5 (Ghahramani et al., 2011). However, according to Wistuba et al. (2018), earthquakes of magnitude M_w <5 can trigger activity of preexisting landslides and are responsible for 50% of landslide events on adjacent slopes, and the impact of M_w <5 earthquakes is increased by cooccurrence of precipitation; thus even small to moderate earthquakes can be an underestimated landslide trigger in areas of low seismic activity. Most widespread landsliding (mainly disrupted slides or falls of rock and soil) is caused by shallow earthquakes (with epicentral depths of <45 km) of magnitudes ranging from M_w 6.2 to 8.2. The minimum MM (Modified Mercalli) shaking intensity threshold for landsliding is MM VI, while the most common intensities for significant landsliding are MM VII and VIII. Very large landslides occur mainly at MM IX and X (Hancox et al., 1997).

Local topographic amplification significantly influences the occurrence of coseismic landslides. Ridges, spurs and convex slope crests often experience enhanced shaking compared to adjacent valleys or flat ground (Geli et al., 1988; Meunier et al., 2008): for example, Buech et al. (2010) measured shaking intensities on a bedrock ridge-top that were twelve times more intense than corresponding intensities on nearby flat ground. This effect explains the concentration of rockfalls and small-scale failures at ridge tops observed during numerous earthquakes, such as the 1994 M_w 6.7 Northridge earthquake in California (Harp & Jibson, 1996), the 2015 M_w 7.8 Gorkha earthquake in Nepal (Kargel et al., 2016; see Chapter 15) and the 2024 M_w 7.5 Noto Peninsula earthquake in Japan (Ohno et al., 2024). Subsurface conditions also modulate shaking intensity, with soft sedimentary layers and weathered bedrock amplifying seismic energy and lowering slope stability thresholds (Asano et al., 2006; Ma et al., 2019).

Lithology and structural fabric strongly govern slope sensitivity to seismic shaking. Weak materials such as tephra, weathered tuff, loess or colluvium are particularly prone to shallow failures under moderate shaking (Keefer, 1984; Rodríguez et al., 1999). Rock masses with abundant discontinuities (joints, faults, bedding planes) are highly susceptible to coseismic rockfalls and toppling (Keefer, 2002). On the other hand, structural orientation is also critical and slopes with discontinuities dipping out of the slope face or aligned with the direction of seismic wave propagation are more likely to fail. Likewise, slopes with preexisting landslide scars or long-term creep deformation may undergo rapid reactivation under seismic loading (Hovius et al., 2011; Wang et al., 2023; Chigira et al., 2010; Sassa et al., 2005). Bazgard et al. (2011) showed that whole-of-edifice resonance with seismic shaking can initiate deep-seated failures corresponding with the source-area scars of large coseismic bedrock landslides.

Groundwater conditions exert a major influence on coseismic slope stability. Elevated pore-water pressures reduce effective stress and shear strength, enhancing susceptibility to failure during shaking (Seed & Idriss, 1971). Saturated conditions can shift otherwise stable slopes to a critical state, particularly in fine-grained soils or loose sands susceptible to liquefaction (Wang, Huang, et al., 2014; Zhang & Wang, 2007; see Chapter 8). Further increase in pore-water pressure within the landslide material during displacement can elevate the postfailure landslide mobility (Sassa et al., 2004). Note, however, that in the upper parts of slopes where coseismic instabilities are most likely, high pore-water pressures are relatively unlikely compared to the lower parts of slopes where rainfall-induced landslides are more commonly initiated (Densmore & Hovius, 2000).

In addition to shaking, direct fault rupture and coseismic surface deformation can trigger landslides. Displacements along fault scarps often generate over-steepened slopes that collapse immediately or fail during subsequent aftershocks. Examples include abundant landslides along rupture zones during the 1999 M_w 7.7 Chi-Chi earthquake in Taiwan (see Chapter 13), the 2008 Wenchuan earthquake (see Chapter 14) and the 2016 Kaikōura earthquake, which illustrate the combined effects of seismic shaking and tectonic displacement (Chigira et al., 2003; Xu et al., 2014). In reality, coseismic landslides rarely result from a single factor; instead, they reflect the complex interplay of shaking, site conditions, lithology and hydrogeology. For instance, in the 2016 Kaikōura earthquake, large rock avalanches were concentrated in steep, fractured greywacke slopes near the rupture trace, where topographic amplification and high groundwater further contributed to instability (Massey et al., 2018). Recognising such cascading triggers is essential for hazard assessment and for interpreting the geomorphic record of past earthquakes.

In the following subsections, some representative case studies of landslides triggered by recent earthquakes are presented. These examples aim to offer new insights into the prediction of coseismic landslides and to better understand the geomorphic, hydrological, geological and geotechnical factors that govern their occurrences and postfailure movements. Further information is provided in the case study Chapters 13–15.

7.4.1 Coseismic landslides in fills

Coseismic landslides occurring in artificial fill constitute a distinctive class of earthquake-triggered failures, as their mechanical behaviour is strongly controlled by hydraulic conditions established during and after seismic shaking. The Tsukidate landslide (Fig. 7.4), triggered by the 2003 M_w 7.0 Miyagi earthquake in Japan, provides a clear illustration of this mechanism. Based on detailed field investigations conducted by Wang and Sassa (2009), the failure was shown to have developed on a gentle slope (13.5 degrees) that had been artificially infilled several decades earlier to bury a former gully. The fill consisted predominantly of loose, saturated pyroclastic deposits, characterised by a shallow groundwater table and limited internal drainage capacity. Field surveys revealed that the displaced mass was rapidly mobilised and travelled approximately 130 m onto an adjacent rice paddy, displaying morphology and mobility typical of a flow slide. To examine the underlying mechanical processes, Wang and Sassa (2009) conducted laboratory ring-shear tests and complementary in situ investigations, which demonstrated that the fill was poorly compacted and retained a high water content. These conditions favoured rapid pore-water pressure buildup during shaking and only limited dissipation afterward. The ring-shear test results

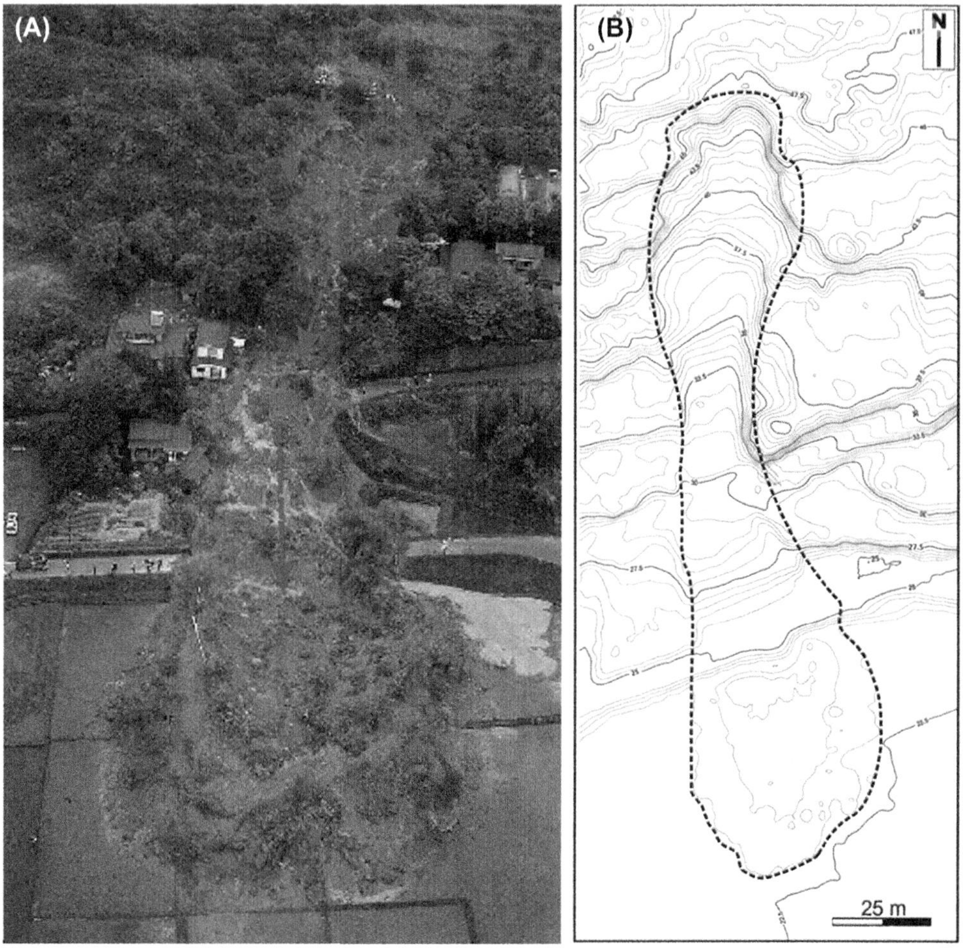

Figure 7.4 (A) Oblique side view of the Tsukidate landslide induced by the 2003 M_w 7.0 Miyagi earthquake in Japan. (B) Contour map of the Tsukidate landslide area obtained by airborne laser scanner. The contour interval is 50 cm, and the dashed line outlines the landslide area. Note: Map lines delineate study areas and do not necessarily depict accepted national boundaries. *(A, B) Photo courtesy of Yomiuri-Shimbun, (courtesy Aero Asahi Corporation, Japan).*

confirmed that seismic loading could generate high excess pore-water pressures, induce undrained shear failure and sustain postseismic movement through persistent pore-pressure elevation.

The behaviour observed at Tsukidate highlights the crucial role of groundwater control in the seismic performance of engineered fill slopes. As reported by Wang and Sassa (2009), when drainage is limited, excess pore-water pressure generated during the earthquake can be maintained long after the shaking has ceased if the saturated soil layer above the sliding surface is sufficiently thick (Fig. 7.5a). This delayed dissipation reduces

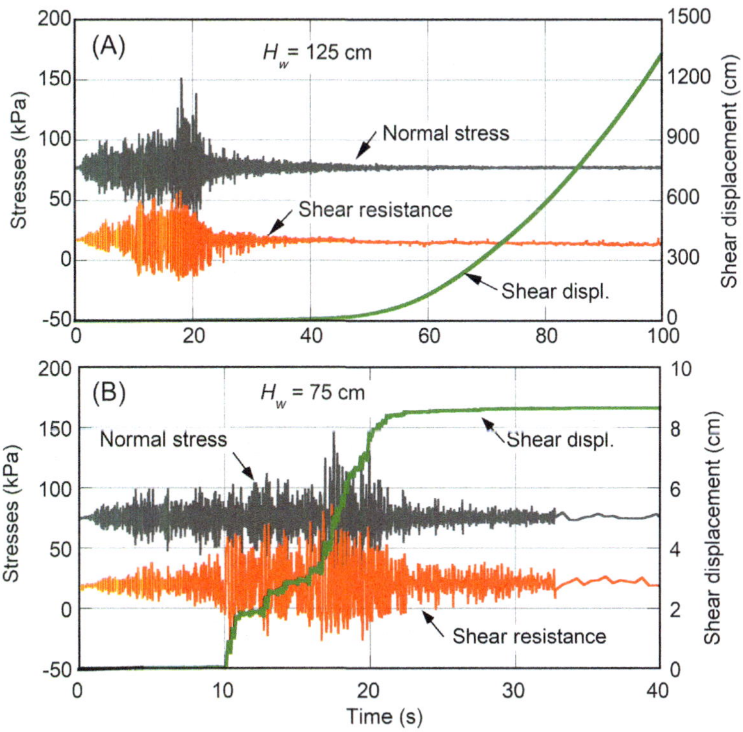

Figure 7.5 Seismic-simulating tests on water-saturated samples under partially drained conditions with groundwater tables above the sliding surface being about 125 cm (A) (void ratio e = 1.11) and 75 cm (B) (e = 1.10), respectively. In both figures, the red and grey traces show the seismic normal and shear stresses, respectively, while the green trace presents the resultant shear displacement. *These results are modified from Wang, G., & Sassa, K. (2009). Seismic loading impacts on excess pore-water pressure maintain landslide triggered flowslides. Earth Surf. Process. Landforms, 34, 232–241.*

effective stress, weakens the soil skeleton and enables continued movement. Conversely, when the drainage path is short (Fig. 7.5b) or when the groundwater table is lowered, pore-pressure dissipation is rapid, reducing postseismic mobility even if an initial failure is triggered. These findings emphasise that effective drainage, such as horizontal drains, subdrains or lowering of the groundwater table, is one of the most important measures for mitigating the risk of flowslide-type failures in artificial fill slopes subjected to strong seismic loading.

7.4.2 Coseismic landslides in loess

Major earthquakes in loess regions have repeatedly produced extensive and highly destructive landslides. For example, the 1920 M_w 8.5 Haiyuan earthquake in China triggered more than 30,000 collapses and flows, causing about 100,000 fatalities (Close

& McCormick, 1922; Zhang & Wang, 2007). Similar concentrations of landslides were reported during the 1927 M_w 7.7 Gulang earthquake, also in China (Zhong et al., 2022). In addition, widespread flow slides occurred in loess deposits in Soviet Tajikistan during the 1989 M 5.5 Gissar earthquake (Ishihara et al., 1990). More recently, the 2023 M_w 5.9 Jishishan earthquake in China produced a large flow slide on a loess terrace, destroying two villages and resulting in 13 fatalities (Wang et al., 2025; Zhuang et al., 2025). Studies of loess landslides triggered by recent earthquakes have revealed that the flow-like landsliding phenomena resulted from liquefaction of saturated loess layers due to preearthquake irrigation (Ishihara et al., 1990; Wang et al., 2025; Zhuang et al., 2025).

On the other hand, the loess landslides triggered by the 1920 M_w 7.9 Haiyuan earthquake in China were historically regarded as 'dry loess flows' driven by collapse of metastable loess structure and fluidisation by pore-air compression (Ter-Stepanian, 1998; Varnes, 1978). This interpretation stemmed from the region's semiarid climate, high-porosity loess and early field observations. However, later systematic investigations of representative failure sites, combined with advanced laboratory testing, have since challenged this view and demonstrated that liquefaction of water-saturated loess layers played the key role in generating rapid sliding and long runout (Wang, Huang, et al., 2014; Zhang & Wang, 2007).

Detailed surveys of the Dangjiacha landslide (Fig. 7.6a), one of the largest Haiyuan earthquake-triggered events, revealed that the failure initiated on a gentle (~20 degrees) concave slope and displaced more than $2 \times 10^7\,m^3$ of material by over ~3.2 km, damming a valley and forming a lake. Field evidence observed eight decades after the earthquake reveals persistent groundwater, shallow wells (Fig. 7.6B) and standing water within the source area (Fig. 7.6C), indicating that the loess layers adjacent to the sliding surface were likely located below the groundwater table prior to failure. Although Fig. 7.6C shows that the current groundwater table lies below the sliding surface, this lowering is likely a consequence of postfailure topographic changes rather than a reflection of preearthquake conditions. This observation contradicts the assumption of a purely dry loess environment and suggests favourable hydrogeological conditions for undrained loading and pore-pressure rise during shaking. Wang, Zhang, et al. (2014) conducted laboratory triaxial and ring-shear tests on undisturbed and remoulded loess taken from the source area of this landslide and confirmed a strong susceptibility to flow liquefaction when saturated, with dramatic strength reduction to extremely low steady-state shear resistance values. In contrast, air-dry specimens showed only minor pore-air pressure development and no significant strength loss, indicating that air entrapment alone could not generate high mobility.

Complementary regional investigations of 39 large loess landslides triggered by the Haiyuan earthquake further show that these failures preferentially occurred on gentle slopes (<15 degrees) with concave morphologies and exhibited markedly higher

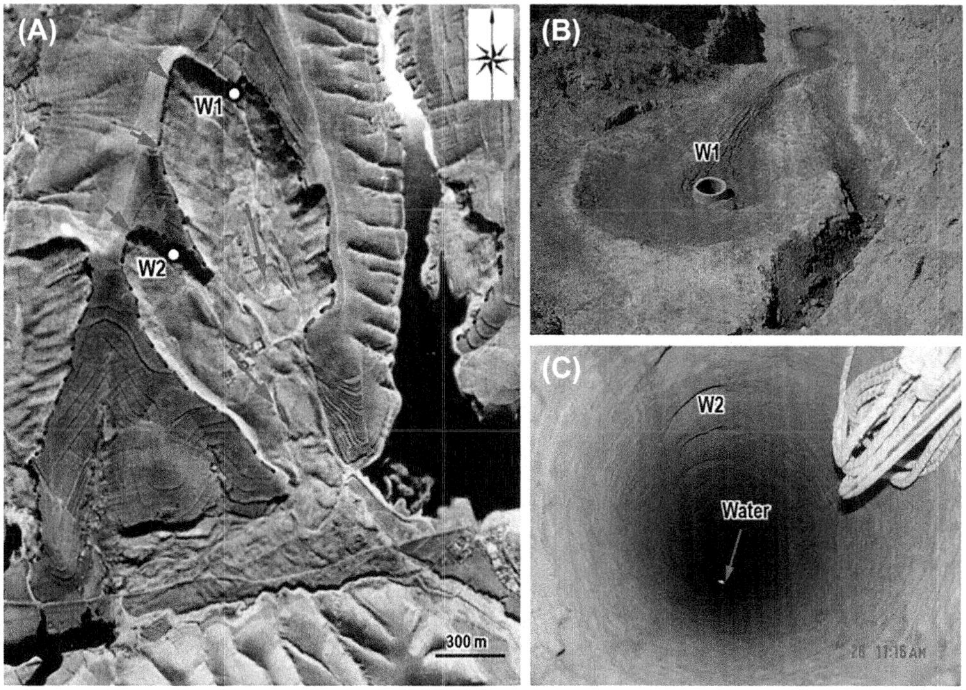

Figure 7.6 (A) Aerial photograph of the Dangjiacha landslide in Xiji, Ningxia Province, China, triggered by the 1920 Haiyuan earthquake. W1 and W2 refer to wells shown in panels (B) and (C). (B) and (C) Wells (photos on 26 March 2003) located in the eastern (W1) and western (W2) slide blocks, respectively. The white point in (C) is a reflection showing the water surface in the well. Well in (B) is for irrigation, and well in (C) is for local residents' drinking water.

mobility – characterised by very low apparent friction coefficients (Fig. 7.7) – than loess landslides induced by rainfall or other landslides reported worldwide, indicating liquefaction-related landsliding behaviour. Together, these findings provide evidence refuting the traditional 'dry flow' hypothesis and establish that seismic liquefaction of water-saturated loess is a plausible mechanism driving the high mobility of the Haiyuan earthquake-triggered landslides. They underscore the critical role of groundwater in loess slope stability and highlight the importance of recognising hydro-mechanical processes in seismic hazard assessment across loess terrains.

7.4.3 Reactivation of old landslides with rapid movement

Reactivation of preexisting landslide masses during earthquakes has often been considered less hazardous than first-time failures, because landslide deposits typically contain preweakened shear zones and are thought to deform mainly through slow or limited displacement. However, recent earthquake events have demonstrated that ancient landslide bodies can

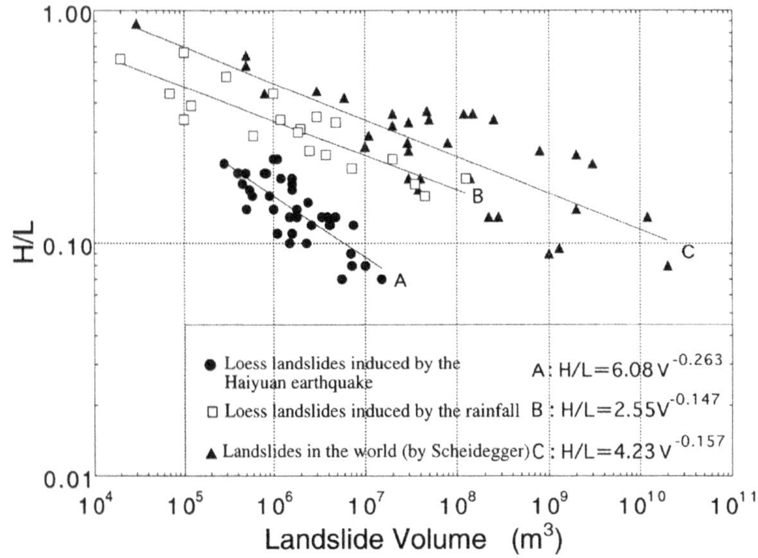

Figure 7.7 Relationship of landslide volume to the equivalent coefficient of friction, defined as landslide height to length or H/L ratio, for loess landslides triggered by the 1920 Haiyuan earthquake (black circles), loess landslides triggered by rainfall (white squares) and other types of landslides worldwide (black triangles). Solid lines show least-squares power-law fits to each dataset. *Modified from Zhang, D.X., & Wang, G. (2007). Study of the 1920 Haiyuan earthquake-induced landslides in loess (China). Engineering Geology, 94, 76–88.*

undergo sudden acceleration and travel rapidly over long distances when critical hydro-mechanical conditions are met. In particular, the presence of saturated granular layers within palaeo-landslide deposits can promote undrained weakening or liquefaction during seismic shaking, reducing shear resistance dramatically and enabling high-mobility movement. These cases challenge the conventional assumption that reactivated landslides are intrinsically slow and instead highlight the possibility that dormant landslide terrain in seismic regions may conceal the potential for catastrophic, fast-moving reactivation.

For example, the 2004 M_w 6.6 Mid-Niigata earthquake provided clear evidence that reactivated landslides, even those originating within ancient landslide masses, can undergo rapid and catastrophic coseismic movement. The Higashi-Takezawa landslide (Fig. 7.8), one of the largest failures triggered during the earthquake, reactivated an older landslide body, accelerated suddenly and travelled approximately 100 m down-slope, striking the opposite bank of the Imokawa River and destroying infrastructure including an elementary school (Sassa et al., 2005). Airborne LiDAR data (Fig. 7.8B) revealed an old landslide above the 2004 head scarp, confirming that the sliding mass belonged to an ancient landslide deposit. Despite the gentle slope angle (≈14–15 degrees), the landslide mobilised at high speed, with an apparent mobilised friction angle of only ~7.5 degrees, indicating extremely low shear resistance during motion.

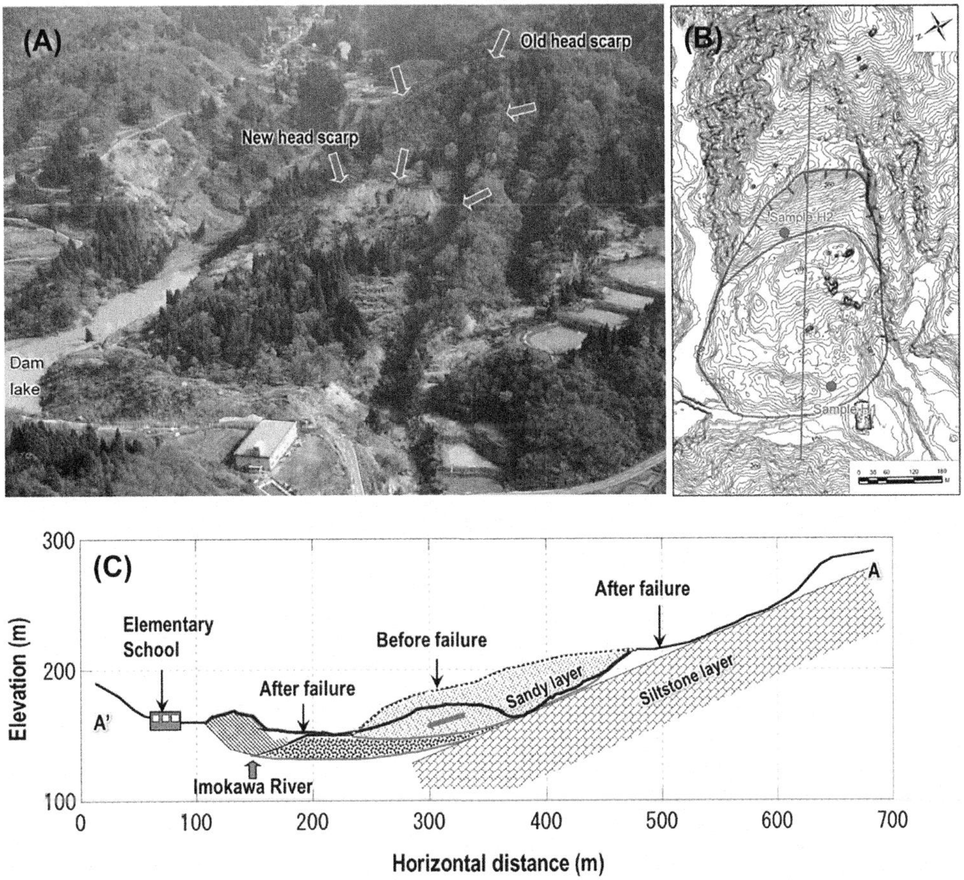

Figure 7.8 Higashi-Takezawa landslide triggered by the 2004 Mid-Niigata earthquake, Japan. (A) Oblique aerial view of the landslide (photo courtesy of Kyoji Sassa). Arrows indicate the headscarp of the 2004 landslide as well as an older headscarp located upslope of the 2004 event. (B) Postearthquake topographic contour map (contour interval: 2m) obtained by airborne laser scanning (courtesy of Aero Asahi Corporation, Japan). Line A–A′ indicates the location of the cross section shown in (C). The upper curve delineates the headscarp, whereas the lower closed curve outlines the deposition area of the displaced landslide mass. (C) Cross section along A–A′. The dashed line represents the prefailure slope topography, while the solid line labelled 'after failure' shows the postlandslide topography. The sliding surface is clearly located within the sandy layer. Map lines delineate study areas and do not necessarily depict accepted national boundaries.

As reported in Sassa et al. (2005), the sliding surface intersected a saturated sand layer overlying a stiff silt bed (Fig. 7.9a), with groundwater confined above the less-permeable silt horizon. Ring-shear tests conducted on samples from the sliding zone reproduced sliding-surface liquefaction under cyclic and real seismic loading, with the

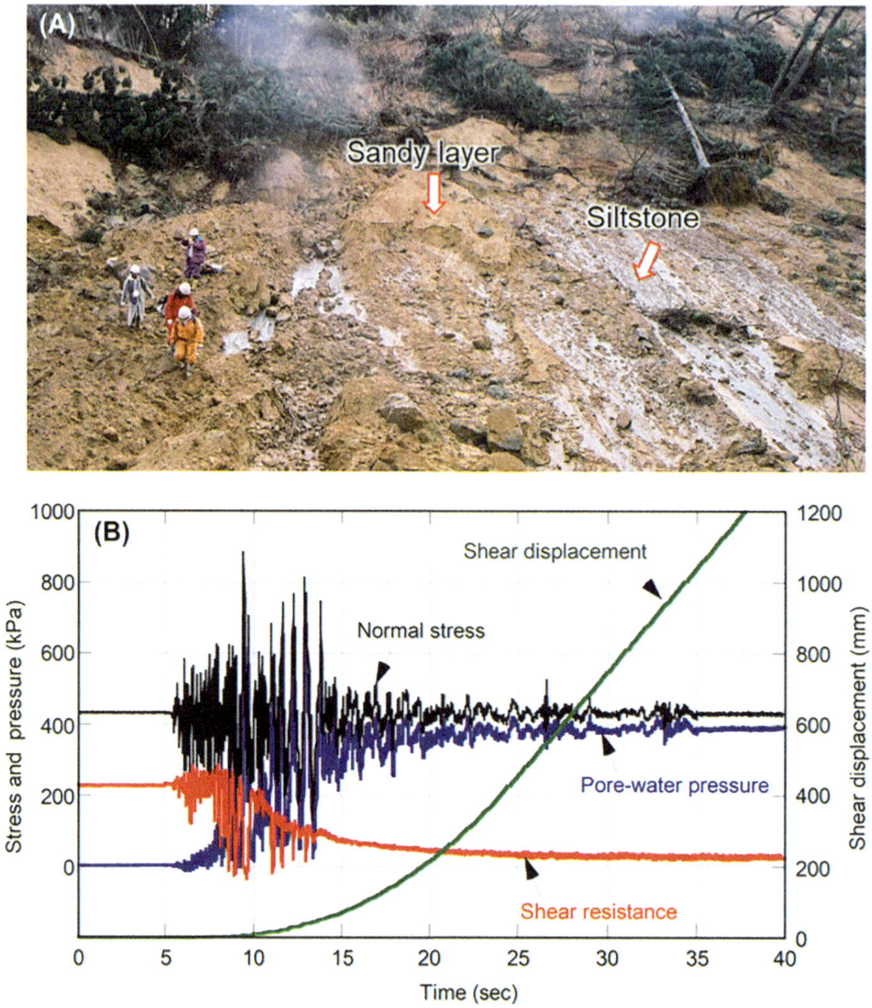

Figure 7.9 Sliding surface and experimental examination results for the Higashi-Takezawa landslide. (A) Photograph of the sliding surface with exposures of siltstone overlaid by sandy soil. (B) Undrained ring shear test-based geotechnical simulation of landsliding triggered by the earthquake by using sample taken from the sandy layer shown in (A). The black and red traces show the seismic normal and shear stresses, respectively, while the green line shows the resultant shear displacement. Landsliding continued after the cessation of seismic motion with the shear strength being greatly lowered. *From Sassa, K., Fukuoka, H., Wang, F., & Wang, G. (2005). Dynamic properties of earthquake-induced large-scale rapid landslides within past landslide masses. Landslides, 2(2), 125–134.*

sand displaying dramatic loss of strength (Fig. 7.9B). The results confirmed that seismic shaking triggered liquefaction within the sand layer of the paleo-landslide mass, rather than within the underlying siltstone, enabling rapid acceleration and long runout despite the gentle slope and preexisting failure surface. This mechanism illustrates that old landslide masses, especially those containing saturated sand layers, may transition to rapid motion under strong ground shaking. The Higashi-Takzawa case again demonstrates that reactivation of ancient landslide deposits under seismic loading can evolve into high-mobility failures, emphasising the need to reassess hazard assumptions for dormant landslide zones in seismic regions.

A further compelling example of rapid coseismic reactivation of an old landslide is the Aratozawa landslide triggered by the 2008 M_W 6.9 Iwate–Miyagi Nairiku earthquake (Fig. 7.10a). This giant slope failure mobilised approximately 67 million m^3 of material from a prehistoric landslide deposit (Fig. 7.10B) and travelled about 300 m along a sliding surface dipping only 0–2 degrees, before colliding with the opposite

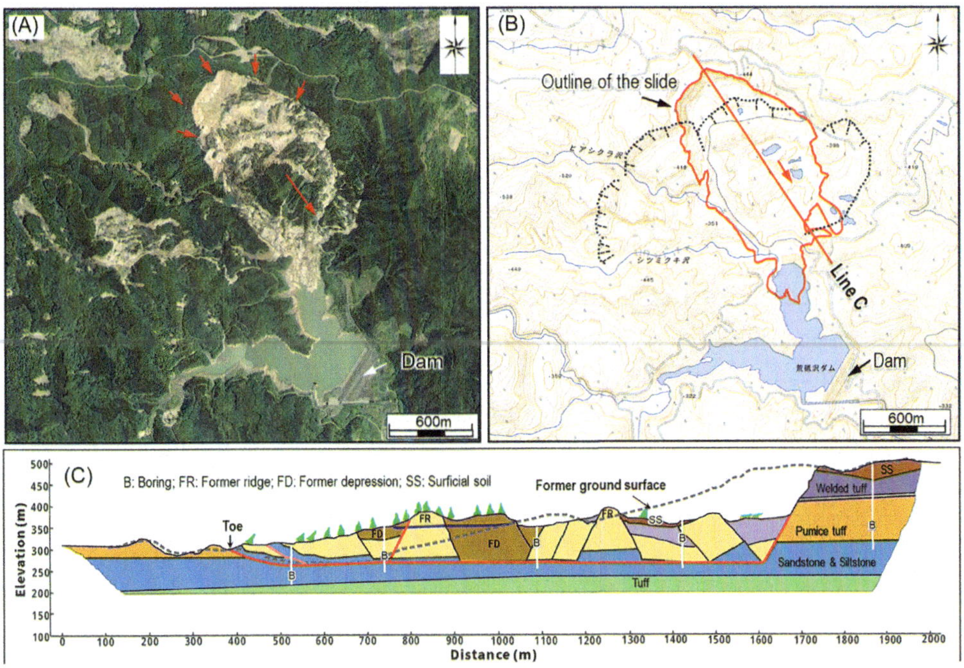

Figure 7.10 Side-by-side comparison of aerial photography and topographic data for the Aratozawa landslide triggered by the 2008 Iwate–Miyagi Nairiku earthquake. (A) Aerial photograph and (B) topographic map (after GSI Maps; retrieved 9 August 2025, from https://maps.gsi.go.jp/). The arrows and red line indicate the extent of the 2008 landslide. In (A), shorter arrows mark the headscarp, whereas the longer arrow indicates the sliding direction. In (B), dotted lines delineate scarps of different blocks of older landslides in the Aratozawa area; the red closed curve outlines the landslide area, and the straight line indicates the location of the cross section shown in (C). (C) Cross section along Line C. The dotted line represents the former ground surface of the original landslide deposit, while the red lines indicate the sliding surfaces. Note: Map lines delineate study areas and do not necessarily depict accepted national boundaries.

valley slope and partially entering the upstream reservoir of the Aratozawa Dam (Fig. 7.10C) (Miyagi et al., 2011; Moriya et al., 2010; Yamashina et al., 2009). Despite its gentle basal inclination, the landslide accelerated suddenly and displaced *en masse*, with most of the mass preserving internal integrity except near the base and surface, indicating predominantly translational failure behaviour (Wang et al., 2023). Drilling and geophysical investigations confirmed that the sliding surface developed within a saturated sandy layer in interbedded sandstone–siltstone bedrock, and ring-shear experiments demonstrated that this layer was highly susceptible to liquefaction under seismic loading, leading to dramatic pore-pressure generation and a marked loss of shear resistance. Newmark analyses incorporating measured seismic waves further indicated that both a high groundwater table and considerable landslide thickness (~100 m) were required to achieve the observed long runout and postseismic motion. These findings emphasise that reactivation of ancient landslide masses under strong shaking can produce unexpectedly rapid and long-travelling failures when hydrogeological conditions favour undrained weakening and when the displaced mass is sufficiently large to sustain low effective stresses. As with the Higashi-Takezawa event, the Aratozawa case overturns the traditional assumption that reactivated landslides tend to move slowly and instead demonstrates that dormant landslide deposits can transition to catastrophic, high-mobility failures under appropriate seismic and groundwater conditions.

7.4.4 Occurrence of giant rock avalanches and geomorphic precursors

The Daguangbao landslide (Figs. 7.2 and 7.11), the largest landslide triggered by the 2008 Wenchuan earthquake, occurred within a complex thrust-faulted mountain setting in the central Longmenshan Belt of Sichuan Province, China. The hillslope that failed lies near the intersection of the Yingxiu–Beichuan fault and the Jiudingshan thrust fault, where repeated tectonic uplift and faulting have produced steep, high-relief terrain and pervasive structural discontinuities. The landslide involved failure of a massive block within Sinian-age dolomite and dolomitic rocks, dipping gently downslope and underlain by a major bedding-parallel fault zone (Figs. 7.11B, C). Gouge, mylonite, breccia and damaged rock along the fault zone form a mechanically weak horizon with markedly reduced strength and elevated permeability. Numerous slickensides on the sliding surface indicate long-term shear localisation (Fig. 7.11B). It is inferred that, prior to failure, groundwater circulation along this fault created persistent saturation, an important factor in seismic weakening and liquefaction of the breccia layer during strong shaking (Cui et al., 2021).

Topographically, the Daguangbao source area formed a prominent horseshoe-shaped amphitheatre bounded by steep head- and lateral scarps. The sliding surface corresponded to a broad, planar bedding surface dipping approximately 16 degrees along the movement direction, with an overall elevation drop of about 870 m (from the

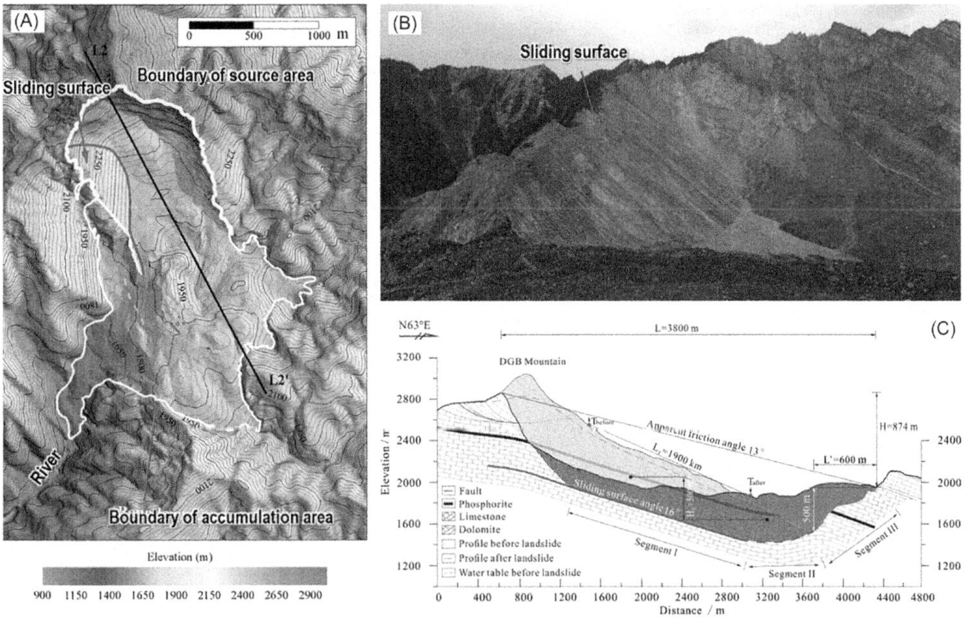

Figure 7.11 Daguangbao landslide triggered by the 2008 Wenchuan earthquake, Sichuan Province, China. (A) Digital elevation map of the postlandslide area. (B) View of the source area from uppermost part of the landslide deposit. The smooth surface shows the area outlined by red curve shown in (A). (C) Longitudinal section along L2–L2′ in (A). Note: Map lines delineate study areas and do not necessarily depict accepted national boundaries. *Panels (A) and (C) are modified from Cui, S., Pei, X., Jiang, Y., Wang, G., Fan, X., Yang, Q., & Huang, R. (2021). Liquefaction within a bedding fault: Understanding the initiation and movement of the Daguangbao landslide triggered by the 2008 Wenchuan Earthquake (Ms=8.0). Engineering Geology, 295106455.*

distal toe of the landslide deposits to the headscarp of the sliding surface) over a travel distance of nearly 4 km (Fig. 7.11C). Prior to failure, subtle but diagnostic geomorphic precursors were present (Fig. 7.12; Chigira et al., 2010). Remote-sensing and field observations identified linear depressions and tension features at the future headscarp, aligned with bedding strike, indicating progressive gravitational deformation and incipient detachment along the weak bedding horizon (Chigira et al., 2010). These features suggest that the slope was in a metastable state prior to the earthquake and did not fail under static conditions, but became unstable only when subjected to strong seismic shaking. Past earthquakes may have contributed to the progressive development or intermittent reactivation of these structures (Davies & Moretti, 2022). Similar fractures along the northern margin (Fig. 7.12A, B) further constrained a wedge-shaped block between the bedding fault and lateral discontinuities, consistent with later field interpretations that failure was ultimately initiated along a saturated bedding fault surface during the main shock.

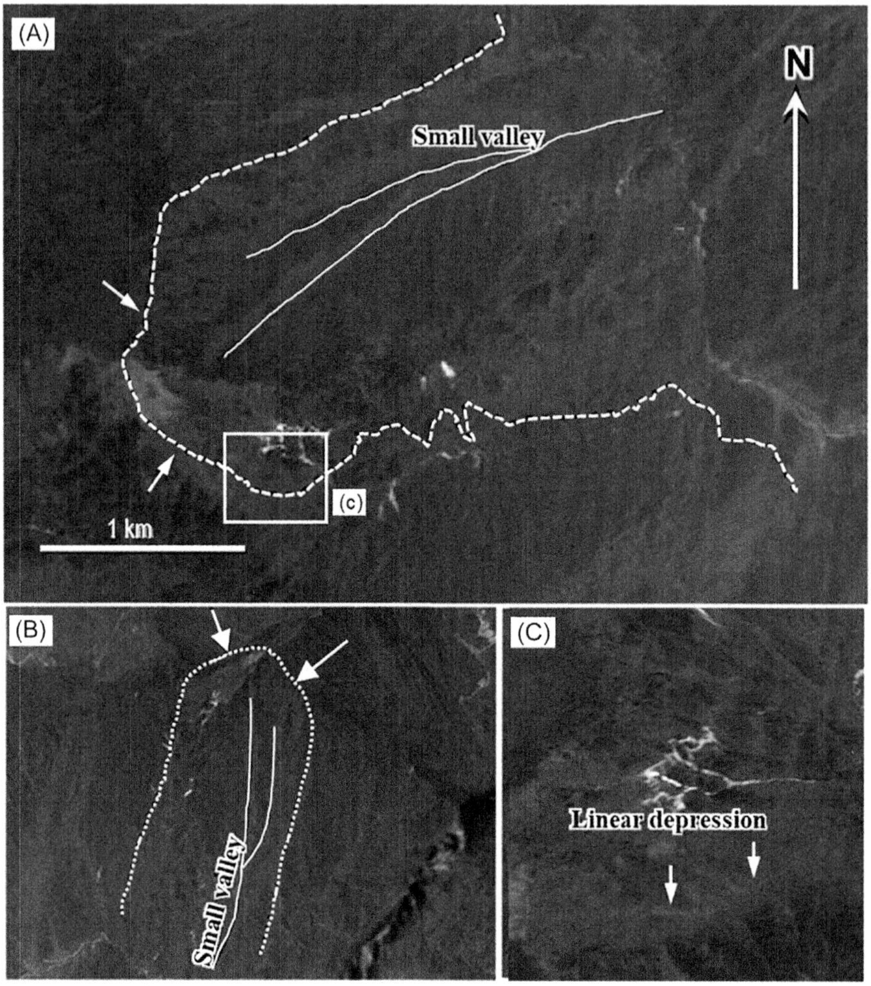

Figure 7.12 Satellite images of the Daguangbao landslide area before the 2008 Wenchuan earthquake. (A) Outline of the Daguangbao landslide and linear depressions (arrowed) at the top of the landslide area before the slide (Google Earth). (B) Oblique view of the linear depressions and (C) closeup view of one of the linear depressions in the source area. *Note that (A) and (B) are modified after Chigira, M., Wu, X.Y., Inokuchi, T., & Wang, G. (2010). Landslides induced by the 2008 Wenchuan earthquake, Sichuan, China. Geomorphology, 118(2010), 225–238.*

These precursor features, though subtle, presaged slope instability in a setting where lithological anisotropy, tectonic weakening and groundwater concentration combined to reduce shear resistance. It is inferred that the long-duration strong shaking and topographic amplification of the 2008 earthquake caused rapidly elevating pore pressure within the breccia zone and triggered runaway sliding (Cui et al., 2021). Recognition of such linear depressions and bedding-parallel loosened zones in steep slopes is therefore critical for identifying latent giant-landslide hazards in active mountain belts.

Recent research has also highlighted the importance of identifying geomorphic precursors that may signal the potential for future giant landslides. In seismically active mountain belts, conventional precursors such as increasing numbers of small rockfalls or accelerating deformation are often absent, as large earthquakes usually occur without clear foreshocks and can trigger catastrophic slope failure suddenly. A recent study by Davies and Moretti (2022) pointed out that slope-top benches, relatively flat surfaces developed near the crest of steep slopes, can serve as diagnostic precursors to large landslides in such environments. These benches may form through cumulative coseismic deformation, where repeated strong shaking produces episodic slumping and progressive extension of deep-seated failure surfaces, gradually reducing slope stability until a through-going rupture develops and catastrophic failure occurs (Davies & Moretti, 2022). Field examples from New Zealand, Japan and the United States demonstrate that slope-top benches frequently occur adjacent to prehistoric or modern large landslide scars. Experimental physical modelling and numerical simulations further show that topographic amplification and resonant response during earthquakes concentrate deformation in upper slope zones, promoting deep-seated cracking and slump formation. In several documented cases reported by Davies and Moretti (2022), these benches exhibit subtle slope-parallel ridges that record early-stage deformation of an incipient failure mass. This work emphasises that slope-top benches in tectonically active regions, particularly near major faults, should be considered potential indicators of future catastrophic landslides. Their recognition, especially when combined with geological conditions conducive to deep-seated failure, highlights sites that warrant detailed geotechnical assessment and monitoring. As such, slope-top benches can provide a valuable geomorphic tool for early identification of giant landslide hazards in mountainous seismic regions.

7.4.5 Movement of debris avalanches

Understanding the movement characteristics of earthquake-induced debris avalanches is essential for clarifying the mechanisms responsible for their exceptional mobility, capacity for material entrainment, deposit morphology and extent and severe downstream impacts. Such events can rapidly transform from localised slides into long-runout mass flows that reshape valley systems and trigger further secondary hazards. As a representative case from the 2008 Wenchuan earthquake, the Donghekou debris avalanche (Figs. 7.3, 7.13) provides a valuable opportunity to examine these processes in detail, from initiation and rapid acceleration to entrainment and final deposition (Fig. 7.14).

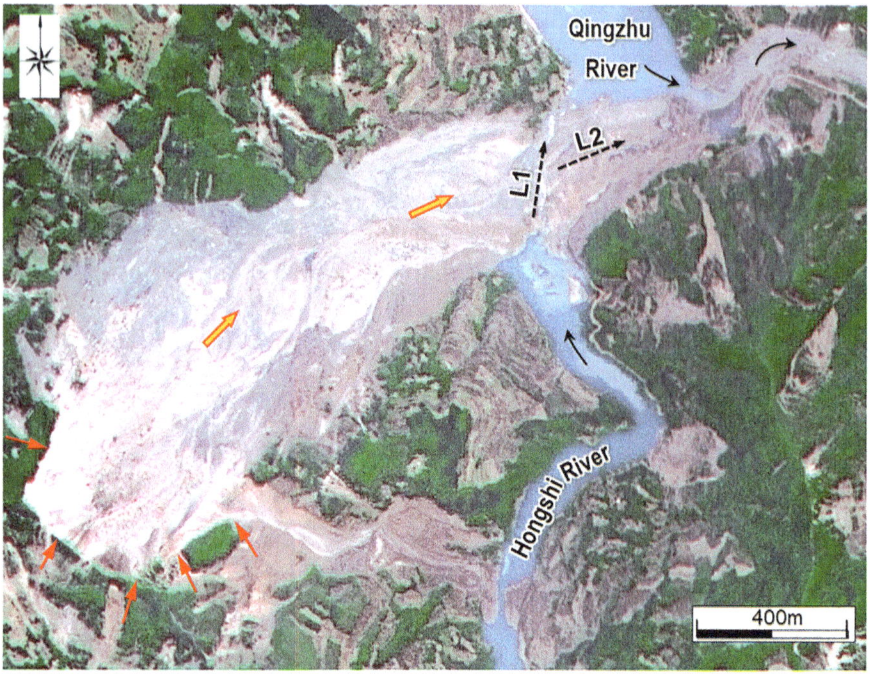

Figure 7.13 Oblique aerial view of the Donghekou landslide triggered by the 2008 Wenchuan earthquake. The landslide deposit dammed both the Hongshi and Qingzhu rivers. Red arrows delineate the main scarp, while yellow arrows indicate the approximate movement directions. Black dashed arrows (L1 and L2) mark the locations and orientations of the geophysical survey lines (shear-wave velocity and electrical resistivity tomography) shown in Fig. 7.14. Note: Map lines delineate study areas and do not necessarily depict accepted national boundaries.

The Donghekou landslide is located in Qingchuan County, Sichuan, China. The failure originated on a steep mountain slope composed mainly of unstable Sinian carbonate and phyllite units adjacent to the Hongkan Fault, while the runout path was underlain by colluvial deposits (Wang et al., Huang, et al., 2014, 2021). Approximately $1 \times 10^7\ m^3$ of material detached with relief of nearly 500 m and rapidly transformed from a rockslide into a high-velocity debris avalanche. The mass travelled about 2 km downslope, burying four villages and destroying nearly all structures along the valley, resulting in about 780 casualties. The landslide also dammed both the Qingzhu and Hongshi rivers (Fig. 7.13), forming a temporary lake which subsequently overtopped and partially drained. Eyewitness accounts and field observations suggest that, during the earthquake, the colluvial deposits started to move first, before the arrival of the displaced landslide debris from the main source area.

The transport and depositional behaviour of the Donghekou debris avalanche exemplifies the complexity of flow processes in large earthquake-triggered mass

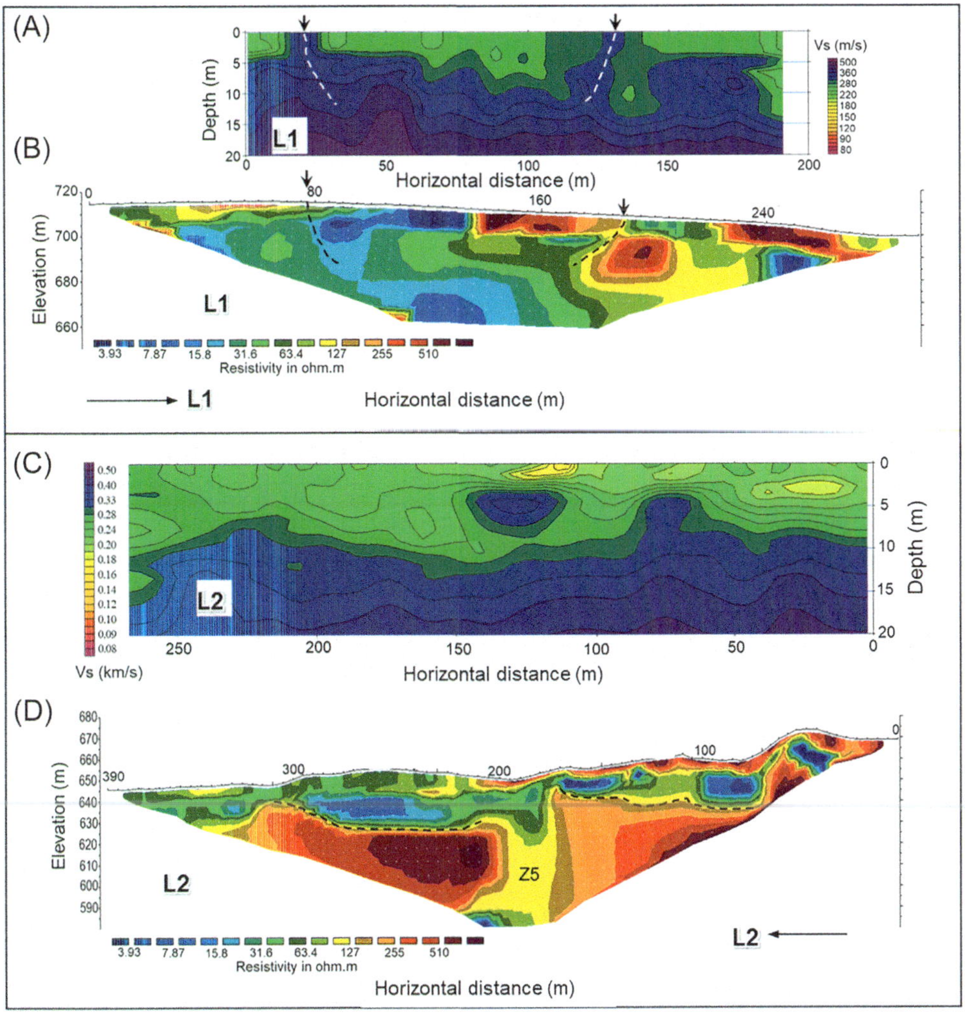

Figure 7.14 Shear-wave velocity (Vs) profiles (A, B) and electrical resistivity tomography profiles (C, D) along survey lines L1 and L2, respectively. Dashed lines in panels (A) and (B) indicate possible boundaries of channelised sliding, while the vertical arrows mark the corresponding locations. Dashed lines in panel (D) denote a possible shear surface parallel to the sliding direction. Horizontal bars in panels (B) and (D) show the resistivity legends (unit: Ω·m). *The figures are modified from Wang, G., Zhang, D., Furuya, G., & Yang, J. (2014). Pore-pressure generation and fluidization in a loess landslide triggered by the 1920 Haiyuan earthquake, China: A case study. Engineering Geology, 174, 36–45; Zhang, et al. (2014, 2021).*

movements. During runout, the sliding mass entrained loose valley-fill materials and displayed variable velocities among different flow sectors, forming surging fronts and mud waves within saturated channel sediments. The deposit exhibits pronounced lateral spreading and contains coarse, block-rich debris at the surface and front, underlain by finer, slurry-like materials toward the interior and basal zones (Wang, Huang, et al., 2014). Geophysical investigations, including multichannel surface-wave analysis and electrical resistivity imaging, reveal a heterogeneous internal architecture characterised by a shallow, low-velocity, disaggregated carapace and denser interior layers overlying intact bedrock at depths of about 20–26 m (Fig. 7.14; Wang et al., Huang, et al., 2014, 2021).

This stratigraphy, as with many rock avalanches, results from the structure of the preevent rock mass and from the damage to the mass during detachment and motion of the landslide. In this case the initial sliding mass probably disaggregated into joint-controlled fragments in the early stages, while the extreme stresses and strain rates at the slide base caused fragmentation of intact blocks at the basal sliding surface. As motion accelerated, intact rock masses attempting to shear within the body of the landslide started to crush and fragment (Davies, 2018), recycling elastic strain energy as high-frequency body waves that caused further fragmentation and increased mobility. Close to the surface of the moving mass, overburden pressure was insufficient to confine blocks so they could shear past each other, forming the coarser, less dense carapace. These results show the complex depositional facies associated with the Donghekou event, providing valuable insights into runout mechanisms and postfailure behaviour of large coseismic debris avalanches.

7.5 Conclusions

Coseismic landslides are one of the most destructive and geomorphically significant consequences of major earthquakes. They occur across a wide spectrum of geological settings and slope morphologies, from steep bedrock cliffs to gentle loess and colluvial slopes and can mobilise enormous volumes of material with rapid deformation and long runout. Studies of recent catastrophic events have greatly advanced our understanding of their initiation and mobility mechanisms, revealing that strong seismic shaking interacts with edifice shape, lithological conditions and groundwater regime in complex ways to produce high-mobility failures and long-term geomorphic evolution. Evidence synthesised in this chapter demonstrates that (1) saturated loess can liquefy and generate highly mobile, long-runout landslides; (2) artificial fills may undergo delayed and progressive weakening due to excess pore-water pressure accumulation; (3) ancient landslide deposits containing saturated granular layers may reactivate catastrophically during earthquakes; and (4) subtle geomorphic precursors, such as linear depressions and slope-top benches, can indicate the presence of slopes susceptible to catastrophic collapse during later earthquakes. These findings expand traditional concepts that once linked high mobility primarily to steep topography or first-time rock failures, underscoring the need to consider hydro-mechanical coupling and preexisting slope deformation as critical controls on, and indicators of, coseismic slope behaviour.

Despite major advances, key knowledge gaps remain. Future research should prioritise:

- Quantifying hydro-mechanical thresholds for liquefaction and strain-weakening in natural slopes, including loess, pyroclastic soils and buried colluvium.
- Monitoring precursory deformation of large, potentially unstable hillslopes using high-resolution InSAR and LiDAR after minor earthquakes, along with continuous GPS, to detect stages of deformation prior to rupture.
- Improving runout models that incorporate entrainment, internal fragmentation and dynamic pore-pressure evolution to better predict downstream impacts.
- Integrating coseismic landslide hazard into seismic risk planning, including inventory-based susceptibility mapping, early-warning criteria for reservoir and infrastructure corridors, and protocols for postearthquake reconnaissance and stabilisation.

A comprehensive understanding of coseismic landslides therefore requires bridging geomorphology, geotechnical testing, seismology and numerical modelling. As observation and analytical capacities deepen and expand, the combination of empirical field evidence with physical and data-driven models will be crucial for developing predictive tools capable of anticipating where and under what conditions catastrophic landslides will occur in future large earthquakes.

References

Abe, S., & Hayashi, K. (2011). Morphologic features of landslides induced by recent large-scale earthquakes and their geomorphologic and geologic features. *Journal of Japan Landslide Society, 48*(1), 52–61.

Asano, S., Ochiai, H., Kuroiwa, S., & Okada, Y. (2006). Topographic effects of seismic motion in mountainous areas and their influence on slope failures. *Journal of the Japan Landslide Society, 42*(6), 1–10.

Bazgard, A., Davies, T., and Cundall, P. (2011) Dynamic analysis of mountain response using PFC3D and FLAC3D. *Continuum and distinct element modelling in geomechanics.* Itasca, Minneapolis, MN, USA, pp. 1-7.

Brain, M. J., Rosser, N. J., Sutton, J., Snelling, K., Tunstall, N., & Petley, D. N. (2015). The effects of normal and shear stress wave phasing on coseismic landslide displacement. *Journal of Geophysical Research: Earth Surface, 120*, 1009–1022. https://doi.org/10.1002/2014JF003417.

Budimir, M. E. A., Atkinson, P. M., & Lewis, H. G. (2014). Earthquake-and-landslide events are associated with more fatalities than earthquakes alone. *Natural Hazards, 72*, 895–914. https://doi.org/10.1007/s11069-014-1044-4.

Buech, F., Davies, T. R., & Pettinga, J. R. (2010). The Little Red Hill seismic experimental study: Topographic effects on ground motion at a bedrock-dominated mountain edifice. *Bulletin of the Seismological Society of America, 100*(5A), 2219–2229.

Chigira, M. (2012). Earthquake-induced landslides and geology. In *Earthquake-induced landslides.* the Japan Landslide Society, 103–110 published by.

Chigira, M., & Yagi, H. (2006). Geological and geomorphological characteristics of landslides triggered by the 2004 Niigata Chuetsu earthquake in Japan. *Engineering Geology, 82*(4), 202–221.

Chigira, M., Wang, W. N., Furuya, T., & Kamai, T. (2003). Geological causes and geomorphological precursors of the Tsaoling landslide triggered by the 1999 Chi-Chi earthquake, Taiwan. *Engineering Geology, 68*, 259–273.

Chigira, M., Wu, X. Y., Inokuchi, T., & Wang, G. (2010). Landslides induced by the 2008 Wenchuan earthquake, Sichuan, China. *Geomorphology, 118*(2010), 225–238.

Chigira, M., and Suzuki, T. (2016) Prediction of earthquake-induced landslide of pyroclastic fall deposits. In Landslides and Engineered Slopes: Experience, Theory and Practice (edited by: Aversa, S., Cascini, L., Picarelli, L., Scavia, C.). Proceedings of the 12th international symposium on landslides (Napoli, Italy, 12-19 June 2016), 93-100.

Close, U., & McCormick, E. (1922). Where the mountains walked. *The National Geographic Magazine, 41*(5), 445–464.

Cruden, D.M., and Varnes, D.J. (1996) Landslide types and processes. In: Turner AK, Schuster RL (eds) *Landslides Investigation and Mitigation*. Transportation research board, US National Research Council. Special Report 247, Washington, DC, Chapter 3, pp. 36–75.

Cui, S., Pei, X., Jiang, Y., Wang, G., Fan, X., Yang, Q., & Huang, R. (2021). Liquefaction within a bedding fault: Understanding the initiation and movement of the Daguangbao landslide triggered by the 2008 Wenchuan Earthquake (Ms=8.0). *Engineering Geology, 295*, 106455.

Dai, F. C., Xu, C., Yao, X., Xu, L., Tu, X., & Gong, Q. M. (2011). Spatial distribution of landslides triggered by the 2008 Wenchuan earthquake. *China. Geomorphology, 133*, 152–167.

Dai, F. C., Lee, C. F., Deng, J. H., & Tham, L. G. (2005). The 1786 earthquake-triggered landslide dam and subsequent dam-break flood on the Dadu River, southwestern China. *Geomorphology, 65*, 205–221.

Davies, T. R. H. (2018). Rock Avalanches: Processes, Significance and Hazards. In *Oxford research encyclopedia of natural hazard science*. Oxford University Press.

Davies, T., & Moretti, D. (2022). Geomorphic precursors of large landslides: seismic preconditioning and slope-top benches. In *Landslide hazards, risks, and disasters*. Elsevier, 641–666.

Densmore, A. L., & Hovius, N. (2000). Topographic fingerprints of bedrock landslides. *Geology, 28*(4), 371–374.

Dohmen, K., Blum, P., Braun, A., & Fernandez-Steeger, T. M. (2025). Landslide-triggered tsunamis – A review. *Natural Hazards*. https://doi.org/10.1007/s11069-025-07551-2 published online: 09 October 2025.

Doi, I., Kamai, T., Azuma, R., & Wang, G. (2019). A landslide induced by the 2016 Kumamoto Earthquake adjacent to tectonic displacement - Generation mechanism and long-term monitoring. *Engineering Geology, 248*, 80–88.

Dunning, S. A., Mitchell, W. A., Rosser, N. J., & Petley, D. N. (2007). The Hattian Bala rock avalanche and associated landslides triggered by the Kashmir Earthquake of 8 October 2005. *Engineering Geology, 93*(3–4), 130–144.

Fan, X., Scaringi, G., Korup, O., West, A. J., van Westen, C. J., Tanyas, H., et al. (2019). Earthquake-induced chains of geologic hazards: Patterns, mechanisms, and impacts. *Reviews of Geophysics, 57*, 421–503. https://doi.org/10.1029/2018RG000626.

Geli, L., Bard, P. Y., & Jullien, B. (1988). The effect of topography on earthquake ground motion: a review and new case studies. *Bulletin of the Seismological Society of America, 78*, 42–63.

Ghahramani, M., Evans, S.G. and Natural Disaster Systems Research Group (2011) The occurrence of earthquake-triggered landslides. In AGU fall meeting abstracts (2011, pp. NH31A-1529).

Gorum, T., Fan, X. M., van Westen, C. J., Huang, R. Q., Xu, Q., Tang, C., & Wang, G. (2011). Distribution pattern of earthquake-induced landslides triggered by the 12 May 2008 Wenchuan earthquake. *Geomorphology, 133*(3-4), 152–167.

Gorum, T., Tanyas, H., Karabacak, F., Yılmaz, A., Girgin, S., Allstadt, K., Suzen, M. L., & Burgi, P. (2023). Preliminary documentation of coseismic ground failure triggered by the February 6, 2023 Türkiye earthquake sequence. *Engineering Geology, 327*, 107315.

Hancox, G. T., Dellow, G. D., & Perrin, N. D. (1997). *Earthquake-induced landsliding in New Zealand and implications for MM intensity and seismic hazard assessment*. Institute of Geological & Nuclear Sciences.

Harp, E.L., and Jibson, R.W. (1996) Landslides triggered by earthquakes. In: Turner, A.K., & Schuster, R. L. (eds.), Landslides: Investigation and mitigation. Transportation Research Board Special Report 247, Washington, D.C., pp. 277–280.

Harp, E. L., & Keefer, D. K. (1992). Remote sensing of earthquake-induced landslides. *U.S. Geological Survey Bulletin, 1939*, 35.

Hewitt, K. (1998). Catastrophic landslides and their effects on the Upper Indus streams, Karakoram Himalaya, northern Pakistan. *Geomorphology, 26*(1-3), 47–80.

Hewitt, K. (1999). Quaternary moraines vs catastrophic rock avalanches in the Karakoram Himalaya, northern Pakistan. *Quaternary Research, 51*(3), 220–237.

Higaki, D., Yagi, H., Kamiyama, J., & Sasaki, A. (2012). Topographic characteristics of earthquake-induced landslides. In *Earthquake-induced landslides*. Japan Landslide Society, 99–103.

Hovius, N., Meunier, P., Lin, C. W., Chen, H., Chen, Y. G., Dadson, S., Horng, M. J., & Lines, M. (2011). Prolonged seismically induced erosion and the mass balance of a large earthquake. *Earth and Planetary Science Letters, 304*, 347–355.

Hovius, N., Stark, C. P., & Allen, P. A. (1997). Sediment flux from a mountain belt derived by landslide mapping. *Geology, 25*, 231–234.

Huang, R., & Fan, X. (2013). The landslide story. *Nature Geoscience, 6*, 325–326.

Hungr, O., & Evans, S. G. (2004). Entrainment of debris in rock avalanches: An analysis of long runout mechanisms. *GSA Bulletin, 116*, 1240–1252.

Hungr, O., Leroueil, S., & Picarelli, L. (2014). The Varnes classification of landslide types, an update. *Landslides, 11*, 167–194. https://doi.org/10.1007/s10346-013-0436-y.

Ishihara, K., Okusa, S., Oyagi, N., & Ischuk, A. (1990). Liquefaction-induced flow slide in the collapsible loess deposit in Soviet Tajik. *Soils and Foundations, 30*(4), 73–89.

Kamai, T., Kobayashi, Y., Jinbo, C., & Shuzui, H. (2000). Earthquake risk assessments of fill-slope instability in urban residential areas in Japan. *Proceedings 8th ISL, Landslides*, 801–806.

Kargel, J. S., Leonard, G. J., Shugar, D. H., et al. (2016). Geomorphic and geologic controls of geohazards induced by Nepal's 2015 Gorkha earthquake. *Science (New York, N.Y.), 351*(6269), aac8353.

Kawamura, S., Kawajiri, S., Hirose, W., & Watanabe, T. (2019). Landslides over a wide area in the 2018 Hokkaido Eastern Iburi earthquake. *Soils and Foundations, 59*(6), 2376–2395.

Keefer, D. K. (1984). Landslides caused by earthquakes. *Geological Society of America Bulletin, 95*, 406–421.

Keefer, D. K. (2002). Investigating landslides caused by earthquakes – A historical review. *Surveys in Geophysics, 23*, 473–510.

Kieffer, D. S., Jibson, R., Rathje, E. M., & Kelson, K. (2006). Landslides triggered by the 2004 Niigata Ken Chuetsu, Japan, earthquake. *Earthquake Spectra, 22*(S1), 47–73.

Korup, O. (2005). Large landslides and their effect on sediment flux in South Westland, New Zealand. *Earth Surface Processes and Landforms, 30*, 305–323.

Korup, O., & Montgomery, D. R. (2008). Tibetan plateau river incision inhibited by glacial stabilization of the Tsangpo gorge. *Nature, 455*, 786–790. https://doi.org/10.1038/nature07322.

Korup, O., Densmore, A. L., & Schlunegger, F. (2007). The role of landslides in mountain range evolution. *Geomorphology, 88*, 77–99.

Li, G., West, A. J., Densmore, A. L., Jin, Z., Parker, R. N., Hilton, R. G., Lai, Z., & Zhang, F. (2017). Seismic mountain building: Landslides associated with the 2008 Wenchuan earthquake. *Earth Surface Processes and Landforms, 42*, 2367–2382.

Li, Y., Zhou, R., Zhao, G., Li, H., Su, D., Ding, H., Yan, Z., Yan, L., Yun, K., & Ma, C. (2015). Uplift and erosion driven by Wenchuan earthquake and their effects on geomorphic growth of Longmen Mountains: A case study of Hongchun gully in Yingxiu, China. *Journal of Chengdu University of Technology (Science & Technology Edition), 42*(1), 5–17.

Ma, N., Wang, G., Doi, I., Kamai, T., & Chigira, M. (2019). Amplification of seismic response of a large deep-seated landslide in Tokushima, Japan. *Engineering Geology, 249*, 218–234.

Marc, O., Hovius, N., Meunier, P., Gorum, T., & Uchida, T. (2016). A seismologically consistent expression for the total area and volume of earthquake-triggered landsliding. *Journal of Geophysical Research: Earth Surface, 121*, 640–663.

Massey, C., Townsend, D., Rathje, E., Allstadt, K. E., Lukovic, B., Kaneko, Y., Bradley, B., Wartman, J., Jibson, R. W., Petley, D. N., Horspool, N., Hamling, I., Carey, J., Cox, S., Davidson, J., Dellow, S., Godt, J. W., Holden, C., Jones, K., ... Villeneuve, M. (2018). Landslides triggered by the 14 November 2016 M_w 7.8 Kaikōura Earthquake, New Zealand. *Bulletin of the Seismological Society of America, 108*(3B), 1630–1648. https://doi.org/10.1785/0120170305.

Meunier, P., Hovius, N., & Haines, A. J. (2007). Regional patterns of earthquake-triggered landslides and their relation to ground motion. *Geophysical Research Letters, 34*, L20408.

Meunier, P., Hovius, N., & Haines, A. J. (2008). Topographic site effects and the location of earthquake-induced landslides. *Earth and Planetary Science Letters, 275*, 221–232.

Miura, K., Yoshida, N., Nishimura, M., and Wakamatsu, K. (1998) Stability analysis of the fill embankment damaged by recent two major earthquakes in Hokkaido, Japan. In Proceedings of the 1998 geotechnical earthquake engineering and soil dynamics specialty conference, Seattle, Washington, 3-6 August. ASCE Geo-Institute Geotechnical Special Publication No. 75, 2, pp. 926-937.

Miyagi, T., Yamashita, S., Esaka, F., & Abe, S. (2011). Massive landslide triggered by 2008 Iwate–Miyagi inland earthquake in the Aratozawa Dam area, Tohoku, Japan. *Landslides, 8*, 99–108.

Moriya, H., Abe, S., Ogita, S., & Higaki, D. (2010). Structure of the large-scale landslide at the upstream area of Aratozawa Dam induced by the Iwate–Miyagi Nairiku Earthquake in 2008. *Journal of the Japan Landslide Society, 47*(2), 77–83.

Ohno, H., Tsutsumi, D., Furuya, G., Takiguchi, S., Ikeda, M., Miyagi, A., Miike, T., & Sawa, Y. (2024). Sediment related disasters induced by the Noto Peninsula Earthquake in January 2024. *Journal of Japan Society of Erosion Control Engineering, 77*(2), 27–34.

Owen, L. A., Kamp, U., Khattak, G. A., Harp, E. L., Keefer, D. K., & Bauer, M. A. (2008). Landslides triggered by the October 8, 2005 Kashmir earthquake. *Geomorphology, 94*, 1–9.

Roback, K., Clark, M. K., West, A. J., Zekkos, D., Li, G., Gallen, S. F., Godt, J. W., & Puskas, C. M. (2018). The size, distribution, and mobility of landslides caused by the 2015 Gorkha earthquake, Nepal. *Nature Geoscience, 11*, 521–526.

Rodríguez, C. E., Bommer, J. J., & Chandler, R. J. (1999). Earthquake-induced landslides: 1980–1997. *Soil Dynamics and Earthquake Engineering, 18*, 325–346.

Sassa, K., Fukuoka, H., Wang, G., & Ishikawa, N. (2004). Undrained dynamic-loading ring-shear apparatus and its application to landslide dynamics. *Landslides, 1*(1), 9–17.

Sassa, K., Fukuoka, H., Wang, F., & Wang, G. (2005). Dynamic properties of earthquake-induced large-scale rapid landslides within past landslide masses. *Landslides, 2*(2), 125–134.

Sato, H. P., Hasegawa, H., Fujiwara, S., Tobita, M., Koarai, M., Une, H., & Iwahashi, J. (2007). Interpretation of landslide distribution triggered by the 2005 Northern Pakistan earthquake using SPOT 5 imagery. *Landslides, 4*, 113–122.

Seed, H. B., & Idriss, I. M. (1971). Simplified procedure for evaluating soil liquefaction potential. *Journal of Soil Mechanics and Foundations Div., ASCE, 97*(SM9), 1249–1273.

Tanaka, K. (1998). Site prediction of earthquake-induced landslides. *Journal of Japan Landslide Society, 34*(4), 435–438.

Tang, C., Zhu, J., Qi, X., & Ding, J. (2011). Landslides induced by the Wenchuan earthquake and the subsequent strong rainfall event: A case study in the Beichuan area of China. *Engineering Geology, 122*, 22–33.

Tang, C., van Asch, T. W. J., Chang, M., Chen, G. Q., & Zhao, X. H. (2016). Coseismic landslides triggered by the 2008 Wenchuan earthquake and long-term hazards. *Engineering Geology, 212*, 66–79.

Tanyaş, H., van Westen, C. J., Allstadt, K. E., et al. (2017). Presentation and analysis of a worldwide database of earthquake-induced landslide inventories. *Journal of Geophysical Research: Earth Surface, 122*(10), 1991–2015.

Ter-Stepanian, G. (1998) Suspension force induced landslides. Proceedings of 8th international congress international association for engineering geology and the environment, Vancouver, Canada, 3. Balkema, Rotterdam, pp. 1905–1912.

Varnes, D.J. (1978) Slope movement types and processes. In: Schuster, R.L. and Krizek, R.J., Eds., Landslides, Analysis and Control, Transportation Research Board, Special report no. 176, National Academy of Sciences, pp. 11-33.

Wang, G., & Sassa, K. (2009). Seismic loading impacts on excess pore-water pressure maintain landslide triggered flowslides. *Earth Surface Processes and Landforms, 34*, 232–241.

Wang, G., Huang, R., Lourenço, S. D. N., & Kamai, T. (2014). A large landslide triggered by the 2008 Wenchuan (M8.0) earthquake in Donghekou area: Phenomena and mechanisms. *Engineering Geology, 182*, 148–157. https://doi.org/10.1016/j.enggeo.2014.07.013.

Wang, G., Zhang, D., Furuya, G., & Yang, J. (2014). Pore-pressure generation and fluidization in a loess landslide triggered by the 1920 Haiyuan earthquake, China: A case study. *Engineering Geology, 174*, 36–45.

Wang, G., Furuya, G., Watanabe, N., Doi, I., & Ma, N. (2019). Features of landslides triggered by the 2018 Hokkaido Eastern Iburi Earthquake. *Disaster Prevention Research Institute Annuals (Kyoto University), 62*(A), 48–56.

Wang, G., Zhang, F. Y., Furuya, G., Hayashi, K., Hu, W., McSaveney, M., & Huang, R. (2021). The debris avalanche in Donghekou area triggered by the 2008 Wenchuan (M8.0) earthquake: Features and possible transportation mechanisms. *Engineering Geology, 280*, 105922. https://doi.org/10.1016/j.enggeo.2020.105922.

Wang, G., Marui, H., and Liu, F. (2023) On the initiation and movement mechanisms of a massive coseismic landslide trigged on an extremely gentle slope: a case study. 6th World Landslide Forum.

Wang, R., Wang, S., Li, D., Kang, X., & Xin, P. (2025). Elevated pore-water pressure regulating dynamic liquefaction of a flow-like landslide in loess. *Journal of Geophysical Research: Earth Surface, 130* e2025JF008318.

Wistuba, M., Malik, I., Krzemień, K., Gorczyca, E., Sobucki, M., Wrońska-Wałach, D., & Gawior, D. (2018). Can low-magnitude earthquakes act as a triggering factor for landslide activity? Examples from the Western Carpathian Mts, Poland. *Catena, 171*, 359–375.

Xu, C., Xu, X., Shen, L., Yao, Q., Tan, X., Kang, W., Ma, S., Wu, X., Cai, J., Gao, M., & Li, K. (2016). Optimized volume models of earthquake-triggered landslides. *Scientific Reports, 6*, 29797.

Xu, C., Ma, S., Tan, Z., Xie, C., Toda, S., & Huang, X. (2017). Landslides triggered by the 2016 Mj 7.3 Kumamoto, Japan, earthquake. *Landslides, 15*, 551–564.

Xu, C., Xu, X., Yao, X., & Dai, F. C. (2014). Inventories of landslides triggered by the 2008 Wenchuan earthquake and statistical analysis of their spatial distribution. *Landslides, 11*, 441–461.

Xu, Q., Li, W. L., Tang, C., Fan, X., Luo, Y. H., Zhu, K., Xu, C., Zhuang, Y., & Towhata, I. (2022). Report on landslides triggered by the 2008 Wenchuan earthquake. In Towhata, (Ed.). *Coseismic landslides*. Springer.

Yagi, H., Yamasaki, T., & Atsumi, M. (2007). GIS analysis on geomorphological features and soil mechanical implication of landslides caused by 2004 Niigata Chuetsu earthquake. *Journal of the Japan Landslide Society, 43*(5), 294–306.

Yamashina, S., Yamazaki, T., Hashimoto, J., Kasai, S., Agatsuma, T., & Shibuya, K. (2009). Aratozawa landslide triggered by the Iwate-Miyagi Nairiku Earthquake in 2008. *Journal of the Japan Landslide Society, 45*(5), 42–47.

Yin, Y., Wang, F., & Sun, P. (2009). Landslide hazards triggered by the 2008 Wenchuan earthquake, Sichuan, China. *Landslides, 6*(2), 139–152.

Zhang, D. X., & Wang, G. (2007). Study of the 1920 Haiyuan earthquake-induced landslides in loess (China). *Engineering Geology, 94*, 76–88.

Zhong, X., Xu, X., Chen, W., Liang, Y., & Sun, Q. (2022). Characteristics of loess landslides triggered by the 1927 *Mw*8.0 earthquake that occurred in Gulang County, Gansu Province, China. *Frontiers in Environmental Science, 10*, 973262. https://doi.org/10.3389/fenvs.2022.973262.

Zhuang, J. Q., Ma, P. H., Chang, L., Ma, Z. L., Zhan, J. W., Zhu, Y., Kong, J. X., Chen, S. Y., & Peng, J. B. (2025). Mudflows due to liquefaction triggered by earthquakes in terrains with gentle lower slopes: A case study. *Landslides, 22*, 95–108. https://doi.org/10.1007/s10346-024-02383-w.

CHAPTER 8

Geomorphic consequences of seismic liquefaction and lateral spreading

Mark Quigley
School of Geography, Earth and Atmospheric Sciences, University of Melbourne, Parkville, VIC, Australia

8.1 What is seismic liquefaction?

Seismic liquefaction (i.e. *earthquake-induced liquefaction, cyclic liquefaction*) is a geohydrologic process whereby saturated, unconsolidated sediments transiently lose strength and stiffness during earthquake ground shaking and undergo a transformation from solid-like toward fluid-like behaviour (Seed & Idriss, 1971). As seismic waves pass through saturated granular sediments, dynamic shear stresses disrupt interparticle contact forces (Fig. 8.1). Overburden load is transferred from the soil skeleton to interstitial pore water, increasing pore-fluid pressure (PFP), reducing effective stress and causing a loss of shear strength (Wang et al., 2020). Under such conditions, sedimentary masses may undergo vertical fluidisation (i.e. flow behaviour) and/or lateral transport (i.e. lateral spreading), even on gentle slopes (Elfilali et al., 2024). Seismic liquefaction may commence within seconds to minutes of first initiation of seismic ground motions (Kramer et al., 2016). Excess pore water pressures may persist in liquefied sediments for minutes to >1 hour following the initiating event (e.g. Quigley et al., 2013, 2016).

There is substantive engineering and seismological literature on the mechanical behaviour and risk mitigation aspects of liquefaction (Seed & Idriss, 1982; Kramer, 1996; Youd et al., 2001; National Academies, 2021). However, the geomorphic consequences of liquefaction (i.e. how liquefaction can affect landscape forms and processes) and vice versa (i.e. how geomorphic factors may affect liquefaction susceptibility and characteristics) are less studied and are the foci of this chapter.

It is important to note that liquefaction can also be caused by a variety of other nonseismic processes that affect stresses and PFPs in sedimentary media (e.g. groundwater fluctuations, meteorite impacts, permafrost thawing, rapid sediment loading, tidal shear, tsunamis, water flows, water waves). Some of these processes induce '*flow liquefaction*' via static monotonic loading. This chapter does not discuss nonseismic or flow liquefaction herein but instead refers readers to Owen and Moretti (2011a) and references therein for information on these processes.

Geomorphic Impacts of Earthquakes
DOI: https://doi.org/10.1016/B978-0-443-40375-0.00003-7

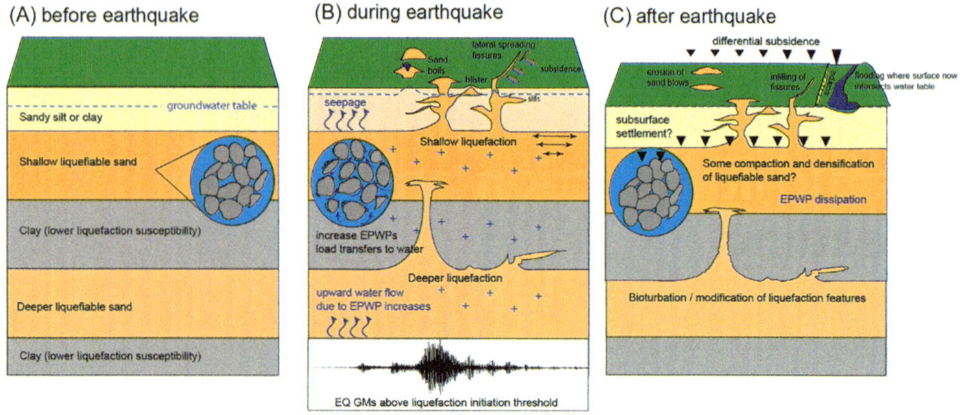

Figure 8.1 Schematic representation of the liquefaction process with some geological and geomorphic aspects. (A) Before the earthquake, sediments below the groundwater table (saturated) contain sand deposits susceptible to liquefaction and interlayered less susceptible sediments. Overburden pressures are supported by the soil skeleton (contact forces between sedimentary grains). (B) During the earthquake, transient shear stresses increase due to passing seismic waves, which reduce grain boundary contact forces. Effective pore water pressures (*EPWPs*) increase and upward flow transfers overburden load to interstitial pore water. Water fountaining and eruption of liquefied material in the form of sand boils and lateral spreading fissures may occur at the surface. Note that lateral spreading is more likely in kinematically favourable locations such as the vicinity of free faces such as riverbanks or terrace risers. Sills and boils may occur in the subsurface. There may be multiple layers of liquefaction. The process of liquefaction influences ground motions. Seepage may occur and raise water tables. (C) After the earthquake, EPWPs dissipate. Some compaction may occur and sediments may increase in density. The vertical position of subsurface layers may subside. Changes in the land elevation might cause certain locations to intersect the groundwater table and form permanent or ephemeral water bodies. Scarps may form. Erosion of sand blows and infilling of surface fissures may occur. Evidence of subsurface liquefaction persists but biochemical processes such as bioturbation may influence the texture and appearance of subsurface features. See Quigley et al. (2016) for additional information.

This chapter aims to synthesise current knowledge on the geomorphic signatures of earthquake liquefaction, summarise case studies from recent and historic seismic events, and explore the broader implications for landscape evolution and hazard assessment. By integrating field observations, remote sensing data, and theoretical models, this chapter highlights the role of liquefaction as a significant agent of geomorphic change in earthquake-prone environments.

8.2 Under what conditions does seismic liquefaction occur?

8.2.1 Geohydrologic conditions

Seismic liquefaction commonly occurs in fluvial environments (e.g. Güven et al., 2023; Obermeier, 1996a, 1996b), lacustrine environments (e.g. Boncio et al., 2020), tidal estuaries, interdune depressions and swamps (e.g. Bucci et al., 2017, 2018; Quigley et al., 2016). Seismic liquefaction is also common in engineered environments with vulnerable site conditions such as reclaimed land (e.g. Holzer, 1998; Kazaoka et al., 2015; Yasuda & Ishikawa, 2012).

Seismic liquefaction requires hydrogeologic conditions that enable PFPs to rise during earthquake ground shaking, including shallow water tables (e.g. within 0–3 m of surface), undrained and poorly drained conditions (so that PFPs can increase without dissipation), and unconsolidated, permeable coarse-grained silt, silty sand and/or fine-grained sand (Andrews and Martin, 2000; Obermeier, 1996a, 1996b; Owen, 2003). The presence of artesian aquifers (Cox et al., 2021) and interspersed layers of impermeable sediments (Owen and Moretti, 2011b) may influence the depth distributions of liquefaction and enhance the potential for PFPs to reach liquefaction-triggering thresholds under imposed seismic shaking. Hydraulic gradients, sediment porosity and permeability, sediment sorting and resistance to deformation exert major controls on liquefaction susceptibility (Brandes and Winsemann, 2013; Lowe, 1976; Maltman and Bolton, 2003; Collinson et al., 2019).

Sediments with elevated clay fractions, which enhance interparticle cohesion and reduce PFP generation under cyclic loading, are typically regarded as having lower liquefaction potential (Boulanger & Idriss, 2004; Gratchev & Sassa, 2004; Liu et al., 2025). Coarse-grained sediments such as gravels may possess higher hydraulic conductivity and enhanced permeability that can inhibit the development of PFPs sufficient to induce liquefaction. Gravel liquefaction is typically restricted to extreme loading scenarios, such as high-magnitude seismic shaking, or to cases where particle entrainment occurs within zones of adjacent liquefied sand (e.g. Postma, 1983; Wang and Manga, 2021).

The presence of impermeable surface caps (natural or engineered) can impede surface manifestations of liquefaction despite its occurrence at depth (Ishihara, 1985; Liu and Dobry, 2000; Sonmez et al., 2008). Engineered structures that penetrate surface caps can provide liquefaction pathways and influence the spatial distributions of liquefaction surface ejecta (Borella et al., 2020). The potential for liquefaction decreases with increasing overburden pressure. Consequently, liquefaction tends to be restricted to depths of c.1–10 m (Crespellani et al., 1988; Obermeier, 1996a, 1996b). Deeper liquefaction may occur in highly susceptible sediments subjected to more intense seismic loading (e.g. Yuan et al., 2009; Bhattacharya et al., 2011), although the direct geomorphic ramifications of deep liquefaction may be too subtle to easily recognise.

Directly observed seismic liquefaction phenomena are primarily restricted to sediments of Modern-Holocene to Late Pleistocene age (Youd and Hoose, 1977; Youd and Perkins, 1978; Martin and Clough, 1994). As sediments age, associated changes can include physical and chemical restructuring of grain geometries and boundaries, compaction-induced densification, formation of new mineral phases, chemical binding and biological transformations. As a result, sediment ageing tends to increase sediment stiffness and strength, reduce porosity and permeability, and therefore reduce liquefaction susceptibility (Schmertmann, 1991). Qualitative classifications of soil liquefaction susceptibility as a function of deposit age, derived from postearthquake field observations (Youd and Perkins, 1978), suggest that:

1. Sediments <500 years old typically exhibit high to very high liquefaction susceptibility
2. Holocene sediments (<10,000 years old) generally possess moderate to high susceptibility
3. Pleistocene deposits (10,000–1.8 million years old) are generally characterised by low to very low susceptibility
4. Pre-Pleistocene sediments display a very low susceptibility to liquefaction

The 'age' of a sedimentary unit from a liquefaction perspective may be reset or partially reset by the triggering of liquefaction, even if large strains do not develop (Andrus et al., 2009; Hayati and Andrus, 2009; Maurer et al., 2014), suggesting that unconsolidated sediments deposited before the Holocene could retain high liquefaction potential if liquefied in the Holocene.

In situ tests, laboratory tests, empirical and analytical approaches and geophysical methods are commonly used to evaluate the liquefaction susceptibility of sediments (Table 8.1). Detailed descriptions of these methods are beyond the scope of this review; we refer interested readers to the extensive National Academies (2021) report and references therein for relevant reviews.

8.2.2 Seismological conditions

Liquefaction is triggered in susceptible sediments when the applied seismic loading exceeds the liquefaction resistance of the sediment (Seed & Idriss, 1971). The evaluation of liquefaction triggering therefore requires measures of earthquake loading and liquefaction resistance. Measures of liquefaction resistance can be estimated through some of the approaches listed in Table 8.1.

The intensity of earthquake loading is commonly described by two parameters: the peak cyclic shear stress and the earthquake moment magnitude (M_w). Peak cyclic shear stress is derived from an estimate of peak ground acceleration (PGA) in the earthquake; typically this is computed as the geometric mean of the horizontal components of recorded ground accelerations. Because the number of effective liquefaction-triggering cycles is anticipated to scale with the earthquake duration, which in turn scales with M_w, it is common practice to use a magnitude scaling factor to scale the anticipated cyclic shear stresses or PGAs to a standardised M_w (typically M_w 7.5; see Quigley et al., 2013 for an example).

Table 8.1 Summary of in situ testing, laboratory testing, empirical-analytical approaches and geophysical methods of investigating physical-hydrological-compositional characteristics of sediments to evaluate liquefaction susceptibility in diverse environments.

Method	Category	Principle	Key Data Required	Advantages	Limitations	Typical Applications	Best For
Standard Penetration Test (SPT)	In-Situ Testing Method	Drive split-barrel sampler; measure blow count (N-value)	Corrected N-values, fines content, unit weight	Widely used, large empirical database	Disturbed samples, energy variability, less reliable in gravels	General liquefaction screening, site classification	Fluvial, coastal, engineered
Cone Penetration Test (CPT)	In-Situ Testing Method	Measure tip resistance (q_t) & sleeve friction (f_s) during penetration	q_t, f_s, pore pressure (optional), soil classification	Continuous profile, repeatable, no borehole needed	Cannot sample soil, corrections for overburden stress needed	High-resolution liquefaction mapping	Fluvial, coastal, engineered
Shear Wave Velocity (Vs) Profiling	In-Situ Testing Method	Measure seismic wave travel times	Vs profile, unit weight, stress history	Non-destructive, works through pavements	Requires correlation with CRR, less sensitive to loose sands	Screening for liquefaction susceptibility	Fluvial, lacustrine, coastal
Becker Penetration Test (BPT)	In-Situ Testing Method	Drive casing with high-energy hammer	Blow counts, energy calibration	Suitable for coarse gravelly soils	Limited empirical database, noisy operation	Liquefaction assessment in gravels	Fluvial, engineered
Dilatometer / Pressuremeter	In-Situ Testing Method	Measure deformation under applied pressure	Modulus, strength parameters	Direct stiffness measurement, minimal disturbance	Expensive, specialized equipment	Supplementary liquefaction resistance evaluation	Engineered
Cyclic Triaxial Test	Laboratory Testing Method	Apply cyclic axial loads in lab	Stress-strain curves, cyclic resistance	Controlled stress path, well established	Requires high-quality undisturbed samples	Research, detailed site evaluation	Fluvial, lacustrine, coastal
Cyclic Simple Shear Test	Laboratory Testing Method	Apply cyclic shear under constant volume	Shear strain, cyclic resistance	Simulates earthquake shear mode	Sample disturbance affects results	Engineering design, advanced studies	Fluvial, coastal, engineered
Cyclic Torsional Shear Test	Laboratory Testing Method	Apply cyclic torque to specimen	Shear modulus, damping, cyclic resistance	Tests anisotropy and complex loading	Requires special apparatus	Advanced research	Engineered
Resonant Column Test	Laboratory Testing Method	Apply low-strain cyclic loading	Shear modulus (G), damping ratio	Captures small-strain behavior	Not directly a liquefaction test	Dynamic soil property characterization	Lacustrine, coastal
Seed & Idriss Simplified Procedure	Empirical & Analytical Approach	Compare CSR to CRR	Earthquake magnitude, PGA, field resistance	Standard in guidelines (NCEER, NCEI)	Based on empirical correlations	Regulatory compliance, hazard analysis	All environments
Magnitude Scaling & Correction Factors	Empirical & Analytical Approach	Adjust CRR for magnitude, stress, and fines	Scaling factors, soil gradation	Improves accuracy of empirical methods	Dependent on database quality	Refinement of simplified procedure	All environments
Probabilistic Liquefaction Hazard Analysis (PLHA)	Empirical & Analytical Approach	Integrate seismic hazard curves with site data	Ground motion hazard, geotechnical parameters	Accounts for uncertainty	Requires detailed hazard analysis	Urban planning, infrastructure risk	All environments
MASW	Geophysical / Remote Method	Analyze dispersion of surface waves	Vs profile, layering	Non-invasive, cost-effective	Lower resolution for deep layers	Regional liquefaction screening	Fluvial, lacustrine, coastal
Seismic Refraction/ Reflection	Geophysical / Remote Method	Measure refracted/reflected seismic waves	Velocity profiles	Maps stratigraphy	Expensive, requires seismic sources	Subsurface characterization	Lacustrine, coastal
Electrical Resistivity Tomography (ERT)	Geophysical / Remote Method	Map resistivity variations	Resistivity profiles, moisture data	Identifies saturated loose zones	Affected by salinity, interpretation needed	Supplementary investigation	Lacustrine, coastal

A global review of surface observations of seismic liquefaction in moderate (M_w 4–6) earthquakes suggests that seismic liquefaction can be triggered in earthquakes as small as M_w 4.5 in extremely susceptible soil deposits (Green & Bommer, 2019). For sedimentary media suitable for building structures, the minimum moment magnitude for liquefaction triggering is about M_w 5 (Green & Bommer, 2019). The minimum PGA for triggering liquefaction based on surface evidence is 0.08–0.1 g (Quigley et al., 2016; Santucci de Magistris et al., 2014), although PFP increases and liquefaction occurrence in extremely susceptible sediments may occur at lower PGAs (c.≥0.05 g) particularly in larger M_w earthquakes with longer ground shaking durations (Yasuda et al., 2004). Lateral spreading requires broader connectivity within the liquefiable layer compared with isolated sand blows and thus is likely to require larger M_w and PGA thresholds for initiation (Quigley et al., 2016). Quigley et al. (2016) proposed a minimum M_w of 5.3 and a minimum PGA of c.0.2 g (equivalent to a M_w 7.5; weighted PGA of 0.16 g) for lateral spreading initiation in Christchurch during the 2010–11 Canterbury earthquake sequence. A variety of other deterministic and probabilistic approaches exist for estimating the ground shaking parameters associated with liquefaction and lateral spreading initiation (e.g. stress-based methods, performance-based methods, experimental methods; for a review, see National Academies, 2021).

It is important to acknowledge that ground motions may be amplified in the near surface, that PFP increases, including liquefaction initiation, modify ground motions, and that ground motions may often be inferred at a liquefaction site based on ground motion models. Thus it may be challenging to precisely determine liquefaction-initiating 'threshold' ground shaking intensities in subsurface sediments. Further sources of uncertainty include inferences that liquefaction did not occur based on an absence of surface evidence, and uncertainties in characterising the depth and thickness of the liquefied layer (National Academies, 2021).

8.3 Geologic and geomorphic features produced by seismic liquefaction

Surface evidence for seismic liquefaction is commonly manifested as sand boils and ground fissures (Fig. 8.1). Sand boils (also known as sand volcanoes) form when the hydraulic gradients generated by excess pore pressure drive upward water flow with enough force and volume to carry subsurface liquefied sediment to the surface. Typically, liquefiable sediments are covered by a nonliquefiable layer of varying thickness that enables PFP buildup in the underlying layers, so the pore water and liquefied sediment usually escape through existing or newly formed cracks and openings in this crust. If the overlying nonliquefied layer is too thick (i.e. the liquefied zone is too deep beneath the surface) or the sediment 'cap' is structurally intact and stiff enough to prohibit rupture pathways for liquefied sediment to ascend, then sand boils may not appear at the surface, even when liquefaction and sediment mobility occur at depth. In some cases, liquefied sediment may remain in the subsurface but may create topographic bulges or 'blisters' on the surface.

The subsurface geological 'plumbing system' that feeds sand boils or 'blows' commonly comprises subvertical dykes sourced from the liquefied layer that intersect the surface in the vent zone of the sand blows. Sand blows commonly form in linear arrays, suggesting corresponding subsurface arrays of feeder dykes. Extensive studies of feeder dyke geometries in earthquake sequences indicate repeated dyke reactivations in successive liquefaction events, with localised preservation of penultimate and older dyke remnants (e.g. Quigley et al., 2013, 2016). This implies that many liquefaction events may be accompanied by tensional forces orthogonal to average dyke orientations that further facilitate cracking of the sediment cap to enable ascent of liquefied material (Fig. 8.2). If the magnitude of tension exceeds the upward sedimentation rate of sand blows, sediment-filled fissures may develop instead. These features are commonly associated with liquefaction-induced lateral spreading.

Lateral spreading fissures and sand blow arrays are commonly oriented perpendicular to the spreading direction of the liquefied sediment mass and the local downslope vector

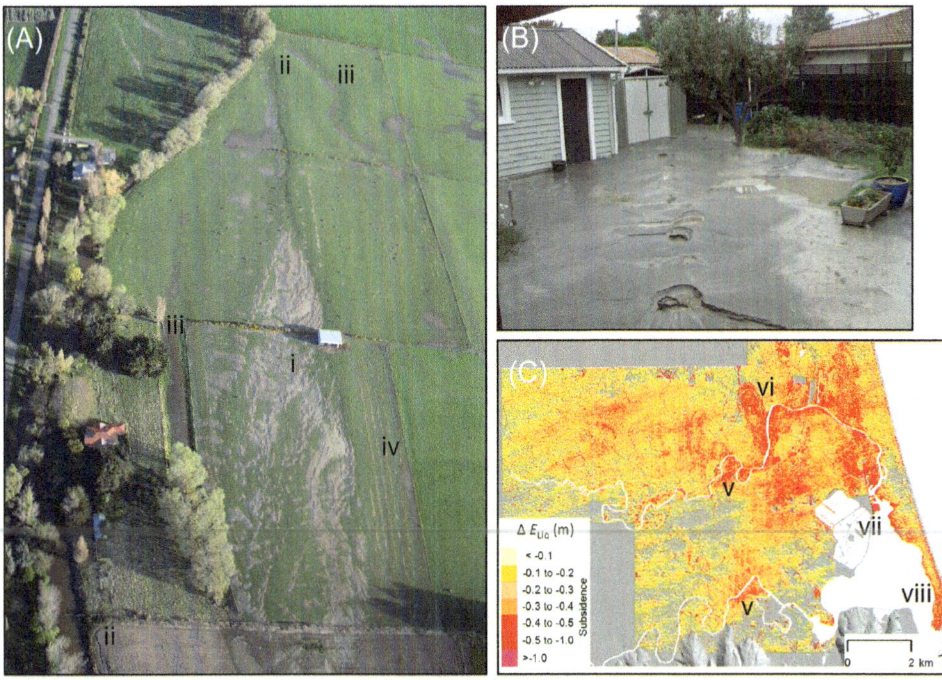

Figure 8.2 (A) Liquefaction effects in the Canterbury region of New Zealand's South Island after the 2010 M_W 7.1 Darfield earthquake. Image taken from aircraft centred at lat. 43.620°S, long. 172.5681°E. Effects include (i) linear arrays of sand blows along point bars of Holocene alluvial deposits and associated subsidence, (ii) lateral spreading cracks on terrace risers and edges of active channels, (iii) groundwater and liquefied sediment expulsion into abandoned / ephemeral alluvial channels and associated relief reduction, and (iv) linear arrays of sand blows through crop lines, showing anthropogenic influence on sand blow surface expressions. (B) Sand blows produced in eastern Christchurch showing linear array of eruption vents with compound stratigraphy reflecting multiple liquefaction events and sedimentary sorting during the liquefaction process. Photo location lat. 43.528025°S, long. 172.663921°E. (C) Differential light detection and ranging model showing liquefaction-related vertical ground movements (Δ E_{Liq}) in Christchurch through the 2010–11 Canterbury earthquake sequence. Effects include (v) enhanced liquefaction-induced subsidence in inner meanders of active streams, (vi) enhanced subsidence in paleochannels and inner meanders, (vii) enhanced subsidence in tidal estuaries and associated late Holocene sediments, and (viii) enhanced subsidence in tidal spits. Centre of image lat. 43.520916°S, long. 172.659588°E. Map lines delineate study areas and do not necessarily depict accepted national boundaries.

(Fig. 8.2). Spreading directions are driven primarily by gravitational forces that cause horizontal displacement towards free faces such as riverbanks or terrace risers. Lateral displacements produce tensile and shear cracking at the ground surface, often manifesting as long, open fissures parallel to the free face (Bartlett & Youd, 1995; Idriss & Boulanger, 2008). The magnitude and extent of lateral spreading cracks are influenced by parameters such as the thickness of liquefiable layers, slope gradient, distance to the free face and ground motion intensity (Zhang et al., 2004). Patterns of lateral spreading in geomorphic settings with youthful sedimentary sequences fringed by free faces (e.g. the inner meander bends of alluvial rivers; Fig. 8.2) typically display fissure arrays subparallel to the stream channel margins with increasing fissure width and intensity closer to the channel edge. Empirical and numerical models developed over the past decades have been used to predict lateral displacement, providing essential input for geotechnical seismic hazard assessment and mitigation design (Cubrinovski & Robinson, 2016a; Youd et al., 2002). Field observations from events such as the 2010–11 Canterbury earthquakes in New Zealand and the 1999 M_w 7.6 Kocaeli earthquake in Turkey have underscored the critical role of lateral spreading in infrastructure damage (Bray & Travasarou, 2007; Cubrinovski et al., 2011a, 2011b). Because lateral spreading is gravitationally driven, it tends to reduce surface elevations and relief in the affected areas.

Liquefaction-induced landslides represent a key geomorphic process linking seismic shaking to rapid and often catastrophic reorganisation of surface materials and landforms. The 1964 M_w 7.6 Niigata and 2004 M_w 6.6 Niigata-Chuetsu earthquakes in Japan, for instance, triggered extensive lateral spreading and flow failures in river terraces and reclaimed lowlands, altering floodplain topography and redirecting surface drainage (Youd, 1973; Towhata et al., 2004). During the 2018 M_w 7.5 Sulawesi earthquake in Indonesia, widespread liquefaction and lateral spreading transformed entire suburban landscapes in the Palu Valley, causing horizontal displacements exceeding 5 m and substantially modifying local relief and sedimentary architecture (Bradley et al., 2019; Mason et al., 2019). These examples demonstrate that liquefaction-induced landslides not only damage infrastructure but also operate as significant geomorphic agents, redistributing sediment, infilling depressions and creating new topographic lows that influence postseismic drainage, sedimentation, and landscape evolution (Sassa and Fukuoka, 2006; Cubrinovski & Robinson, 2016a). Other geologic and geomorphic features produced by liquefaction include formation of diapirs, sills, and sinkholes (e.g. see Bastin et al., 2016).

8.4 Environmental effects of seismic liquefaction: topography, geomorphology, hydrology

Topographic effects of liquefaction can include both immediate and delayed changes in vertical and lateral positions of the land surface (Figs. 8.1 and 8.2). Subsidence (i.e. vertical ground displacements or settlement) may be induced through several mechanisms including liquefaction-induced densification of subsurface layers (Idriss & Boulanger, 2008; Seed & Idriss, 1982; Tokimatsu & Seed, 1987), sediment transport (e.g. lateral spreading, movement of sediment from subsurface to surface ejecta; Bartlett & Youd, 1995; Cubrinovski et al., 2011a, 2011b; Wang and Wong, 2020), and surface loading effects (e.g. buildings; Bray & Sancio, 2006; Ishihara, 1993). Localised uplift can also occur (e.g. lateral translation of higher topography into areas of lower topography, sedimentation associated with sand blows and other deposits; Obermeier, 1996a, 1996b; Rodríguez-Pascua et al., 2000). Lateral displacements may result from lateral spreading and other mass movements (Youd et al., 2002; Zhang et al., 2004). Liquefaction, although transient in duration, can alter the dynamics of geomorphic processes and cause significant transformations of the Earth's surface (Cubrinovski & Robinson, 2016a; Obermeier, 1996a, 1996b; Quigley et al., 2013).

Liquefaction-induced ground subsidence can accentuate the vertical separation between geomorphic features such as (1) terrace risers, (2) fault scarps, and (3) channel margins. The formation and amplification of localised depressions increase accommodation space for surface water ponding. The associated lowering of the ground surface relative to the water table reduces the soil infiltration capacity and enhances groundwater ejection, thereby promoting transient or sustained flooding (Hughes et al., 2015). As a result, saturated and liquefaction-prone sediments amplify both the spatial extent and duration of inundation, contributing to the development of temporary lakes in fine-grained alluvial settings. In contrast, gravelly substrates generally exhibit greater resistance to liquefaction because their coarse particle size and higher permeability inhibit the buildup of excess pore water pressure during seismic shaking. However, liquefaction can still occur in coarse-grained deposits when gravel layers are overlain by impermeable strata or when the fine-to-coarse material ratio is sufficiently high to impede drainage (Do et al., 2017; Evans & Zhou, 1995).

Liquefaction-induced lateral spreading can decrease the width and depth of alluvial channels as liquefied sediment in channel-proximate deposits flows downslope into the thalweg. Decreased channel cross-sectional areas result in reduced flow capacity and increased flood hazard, thereby changing the fluvial dynamics and geomorphology of the affected area. Associated liquefaction-induced subsidence of proximate floodplains also increases flood hazard. These impacts can have cascading effects on the geomorphic development of affected landscapes (e.g. floral-faunal changes and associated changes in landscape stability, changes in the geometry and distribution of sedimentary deposits; see Quigley et al., 2016).

Inactive or ephemeral channels in alluvial landscapes may also provide conduits for more effective expulsion of groundwater and liquefied sediments due to their lower elevations, lower cohesivity, and potential for thinner or absent soil caps relative to surrounding sediments (Fig. 8.1). In some cases, geomorphic 'smoothing' (i.e. micro-relief reduction) due to sedimentation into alluvial channel depressions may occur.

In coastal environments, liquefaction-induced ground deformation can significantly modify hydrological and geomorphic conditions by altering the tidal range of estuarine rivers, local relative sea level, groundwater table elevation, and the spatial extent of tidal flooding (Delano et al., 2023; Hughes et al., 2015). Such deformation has been widely documented following major seismic events, including New Zealand's 1931 M_w 7.7 Napier earthquake (Delano et al., 2023) and the 2010–11 Canterbury earthquake sequence (Hughes et al., 2015; Quigley et al., 2013). These ground-level changes can substantially influence coastal flood hazards, modify groundwater dynamics, and reshape patterns of surface inundation in low-lying nearshore areas (Delano et al., 2023). Moreover, coseismic coastal subsidence can exacerbate the effects of tsunami waves (generated by either the same earthquake or subsequent ones) that may propagate inland along river channels, causing abrupt stage rises and extensive overbank flooding in areas inland (Tanaka et al., 2020; Delano et al., 2023). Large, in-channel dip-slip fault displacements may induce fluvial surges; for example, fault rupture along the Reelfoot fault in the 1812 M_w c.7.5 New Madrid earthquake produced approximately 4–6 m of vertical throw (Johnston & Schweig, 1996; Van Arsdale, 2000), generating a rapid upstream stage increase and widespread overbank flooding in the Mississippi River (USGS, 2019).

Changes to the spatial distributions of floral and faunal assemblages can occur in response to liquefaction-induced subsidence in coastal environments and proximal to river channels (Quigley et al., 2016). These changes may in turn have localised geomorphic consequences (e.g. liquefaction-induced tree death may destabilise landforms and render them more susceptible to erosion). Other cascading geomorphic effects can include liquefaction-induced changes to soil profiles and chemistry, which may influence floral-faunal distributions and erosional susceptibilities.

8.5 Case studies from recent and historic seismic events

Table 8.2 presents a list of selected historical (observed and documented) and prehistorical ('paleo' events inferred from geological evidence) liquefaction events. Readers are encouraged to consult the associated references for more detailed descriptions of these events. This chapter summarises only two of these case studies:

8.5.1 2010–2011 Canterbury earthquake sequence (Christchurch, New Zealand)

The 2010–2011 Canterbury earthquake sequence, particularly the 22 February 2011 M_w 6.2 and 13 June 2011 M_w 6.0 Christchurch earthquakes, caused widespread liquefaction across Christchurch, New Zealand. Liquefaction in Christchurch led to several notable geomorphic changes:

Table 8.2 Summary of some historic and prehistoric (paleo) seismic liquefaction events.

Date	Location/region	Mw (if available)	Liquefaction evidence summary	Primary citation(s)
~2350 BCE	New Madrid seismic zone, USA	7.5–8.0	Sand blows, dykes, sand volcanoes; radiocarbon-dated Holocene paleoliquefaction	Obermeier et al. (2001), Tuttle et al. (2002)
~1400 BCE; ~929 BCE; ~302 BCE; ~1000 BCE; 1750–2200 BCE; 3550 BCE; 5000 BCE; 5300–6300 BCE	Charleston seismic zone, South Carolina, USA	5.5–7.5	Burled sand blows and vent structures; multiple Holocene paleoliquefaction layers	Obermeier (1989), Talwani and Cox (1985), Talwani and Schaeffer (2001)
~500 BCE	Coastal Oregon and Washington (Cascadia subduction zone)	-	Liquefaction and subsidence in estuarine muds and sand layers	Atwater (1987), Kelsey et al. (2005)
300 ± 200 BCE	New Madrid seismic zone, USA	-	Prehistoric sand blows and vent complexes	Tuttle et al. (2002)
600–900 BCE	Charleston seismic zone, USA	-	Multiple liquefaction horizons identified in trenching studies	Obermeier et al. (2001)
900–1100 BCE	New Madrid seismic zone, USA	-	Regional sand blow horizon (NMSZ Event 1)	Holbrook et al. (2006), Tuttle et al. (2002)
1020 ± 100 BCE	Cascadia subduction zone, USA/Canada	-	Liquefaction and coastal subsidence in tidal marsh sequences	Atwater and Hemphill-Haley (1997)
1300–1500 BCE	Wairarapa, New Zealand	-	Paleoliquefaction in river terraces, sand ejection features	Berryman et al. (2011)
995–1395 BCE	Christchurch, New Zealand	-	Buried sand blows and vent structures	Villamor et al. (2016)

(*Continued*)

Table 8.2 (Continued)

Date	Location/region	Mw (if available)	Liquefaction evidence summary	Primary citation(s)
1415−1910 BCE	Christchurch, New Zealand	-	Paleoliquefaction dykes	Bastin et al. (2015)
1450−1600	New Madrid seismic zone, USA	-	Liquefaction deposits underlying 1811−12 layers	Holbrook et al. (2006), Tuttle et al. (2002)
1660−1905 BCE	Christchurch, New Zealand	-	Paleoliquefaction dykes	Bastin et al. (2015)
1693-01-11	Catania, Sicily, Italy	~7.4	Liquefaction, sand venting, coastal subsidence	Gutscher et al. (2006), Scicchitano et al. (2022)
1700-01-26	Cascadia subduction zone, USA/Canada	8.7–9.2	Liquefaction, buried soils, and drowned forests	Atwater and Hemphill-Haley (1997), Atwater et al. (2014), Kelsey et al. (2005), Satake et al. (1996)
1755-11-01	Lisbon, Portugal	8.5–9.0	Historical liquefaction, sand ejection, and settlement	Moreira (1985), Yasuda and Tohno (1988)
1811−1812	New Madrid, USA	7.5–7.8	Massive sand blows, lateral spreads, ground subsidence	Fuller (1912), Tuttle et al. (2002)
1822-11-19	Valparaiso, Chile	8.0–8.5	Liquefaction and ground failure in coastal plains	Udías et al. (2012)
1848-10-16	Marlborough, New Zealand	7.1	Sand boils, ground failure	Fairless and Berrill (1984)
1855-01-23	Wairarapa, New Zealand	8.2	Sand boils, water ejection, lateral spreading	Grapes and Downes (1997)
1868-08-13	Arica, Peru−Chile	8.5–9.3	Sand ejection and coastal ground failure	Comte and Pardo (1991)

1886-08-31	Charleston, South Carolina, USA	6.8–7.3	Widespread sand blows and lateral spreading	Dutton (1889), Obermeier (1989)
1888-08-31	Canterbury, New Zealand	7.1	Sand boils, ground failure	Fairless and Berrill (1984)
1891-10-28	Nobi, Japan	7.5	Sand boils and liquefaction in alluvial plains	Yasuda and Tohno (1988)
1894-03-18	Shonai, Japan	7	Liquefaction and sand venting in river deltas	Seed and Idriss (1967)
1901-11-15	Cheviot, New Zealand	7.0	Sand boils, ground failure	Fairless and Berrill (1984)
1906-04-18	San Francisco, California, USA	7.9	Liquefaction in waterfront and filled areas	Seed and Idriss (1967), USGS (1988a, 1988b)
1913-02-03	Westport, New Zealand	6.8	Sand boils, subsidence	Fairless and Berrill (1984)
1923-09-01	Great Kanto, Japan	7.9	Liquefaction and ground failure in reclaimed land	Japan Meteorological Agency (1924), Yasuda and Tohno (1988)
1929-03-09	Arthur's Pass	6.9	Liquefaction in alluvial areas	Fairless and Berrill (1984)
1929-06-17	Murchison	7.8	Liquefaction in alluvial areas	Fairless and Berrill (1984)
1931-02-03	Hawke's Bay	7.9	Coastal liquefaction	Fairless and Berrill (1984)
1964-03-27	Prince William Sound, Alaska, USA	9.2	Widespread coastal liquefaction and ground settlement	USGS Prof. Paper 542-G (1969)
1964-06-16	Niigata, Japan	7.5	Severe liquefaction, building tilting, sand blows	Kishida (1966), Yasuda and Tohno (1988)
1971-02-09	San Fernando, California, USA	6.6	Liquefaction in alluvial areas and lateral spreading	Seed et al. (1973)
1979-10-15	Imperial Valley, California, USA	6.5	Sand both and lateral spreading along canals	Holzer et al. (1989)

(*Continued*)

Table 8.2 (Continued)

Date	Location/region	Mw (if available)	Liquefaction evidence summary	Primary citation(s)
1983-05-02	Coalinga, California, USA	6.7	Localised liquefaction in valley deposits	Youd and Bennett (1983)
1987-03-02	Edgecumbe, New Zealand	6.5	Liquefaction in alluvial areas and lateral spreading	Franks (1988)
1989-10-17	Loma Prieta, California, USA	6.9	Liquefaction in bay mud (Marina District)	Harder and Seed (1986), USGS (1990)
1994-01-17	Northridge, California, USA	6.7	Localised liquefaction in valley fills	Holzer et al. (1999)
1995-01-17	Kobe (Hyögoken-Nanbu), Japan	6.9	Liquefaction in reclaimed port areas	Ishihara (1997)
1999-08-17	Kocaeli, Turkey	7.5	Lateral spreading and sand ejecta	Cetin et al. (2000)
1999-09-21	Chi-Chi, Taiwan	7.6	Liquefaction in river and coastal plains	Chu et al. (2004)
2001-01-26	Bhuj (Gujarat), India	7.6	Liquefaction and ground failure in alluvial plains	Rastogi (2004a, 2004b)
2004-12-26	Sumatra-Andaman, Indian Ocean	9.1	Liquefaction in coastal sediments	Baba et al. (2005)
2008-05-12	Wenchuan, China	7.9	Liquefaction in river valleys and basins	Wang et al., 2009
2010-02-27	Maule, Chile	8.8	Liquefaction and settlement in coastal areas	Motta et al. (2010)
2010-09-04	Canterbury (Darfield), New Zealand	7.1	Liquefaction near rivers and fills	Quigley et al. (2016)
2011-02-22	Christchurch, New Zealand	6.2	Severe liquefaction across Christchurch	Quigley et al. (2016)
2011-03-11	Tohoku, Japan	9	Widespread liquefaction in coastal plains	Yasuda et al. (2012)

2011-06-13	Christchurch, New Zealand	6.1	Severe liquefaction across Christchurch	Quigley et al. (2016)
2013-07-21	Cook Strait, New Zealand	6.5	Liquefaction in alluvial areas	Van Dissen et al. (2013)
2013-08-16	Lake Grassmere, New Zealand	6.6	Liquefaction in alluvial areas	Van Dissen et al. (2013)
2014-08-24	South Napa, California, USA	6	Localised liquefaction near rivers	Bray et al. (2014)
2016-04-16	Kumamoto, Japan	7	Liquefaction and sand ejecta in plains	Towhata et al. (2016)
2016-11-13	Kaikōura, New Zealand	7.8	Localised liquefaction and lateral spreading	Stirling et al. (2017)
2018-09-28	Sulawesi (Palu), Indonesia	7.5	Catastrophic flow failures due to liquefaction	Mason et al. (2020)
2019-07-04	Ridgecrest, California, USA	7.1	Localised sand ejecta and settlement	Brandenberg et al. (2020)
2020-01-07	Puerto Rico sequence	6.4	Liquefaction in coastal areas	USGS (2020)
2023-02-06	Kahramanmaraş, Türkiye–Syria	7.8	Widespread liquefaction and lateral spreading	AFAD (2023), Bilgin et al. (2023)

Surface fissuring and lateral spreading: Strong ground shaking induced lateral spreading, causing ground fissures and displacements proximal to rivers and other water bodies (Fig. 8.2), severely affecting roads and infrastructure.

Sand boils and ejecta: Pressurised groundwater expelled fine-grained materials, forming sand boils and ejecta deposits, which were particularly evident in residential areas (Fig. 8.2).

Ground settlement: Uneven settlement of the ground surface occurred, leading to tilting of buildings and infrastructure, especially in areas with loose, water-saturated soils.

Alteration of drainage patterns: Liquefaction-induced ground deformation disrupted natural and artificial drainage systems, leading to localised flooding and changes in surface water flow.

Floodplain subsidence: Shaking resulted in floodplain subsidence exceeding 0.5–1 m along tidal stretches of the city's two main urban rivers, notably increasing the spatial extent and severity of inundation hazards posed by floods, storm surges, and sea-level rise (Fig. 8.2).

River channel modifications: Liquefaction-induced lateral spreading and sedimentation led to river channel narrowing and shallowing, further elevating flood risks (Fig. 8.3).

Tectonic contributions: Surface uplift and downthrow associated with tectonic displacements on local blind faults reduced or enhanced surface elevations and contributed to geomorphic change (see Quigley et al., 2016).

Using pre- and post-earthquake light detection and ranging data, Hughes et al. (2015) identified extensive subsidence of 0.5–1 m across low-lying floodplains in Christchurch, primarily driven by liquefaction and lateral spreading of saturated alluvial sediments. These processes also produced pronounced narrowing and shallowing of river channels, reducing their flow capacity and altering sediment dynamics. Tectonic displacements above blind faults further modified local gradients, compounding the hydrological impacts. The resulting subsidence increased exposure to flooding from high tides, storm surges and extreme rainfall events, while reduced channel gradients led to upstream water ponding and slower drainage. The spatial overlap between zones of liquefaction and measured subsidence confirmed a strong coupling between seismic and geomorphic processes. The authors emphasised that even moderate-magnitude earthquakes (M_W 6–7) can generate substantial and enduring topographic change, challenging the assumption that only large subduction events reshape floodplains. Human modifications such as stopbanks, drainage works, and the artificial confinement of rivers influenced surface and subsurface hydrology and the distribution and geometry of liquefaction features (Borella et al., 2020). Liquefaction-induced subsidence (with some contribution from tectonic subsidence) lowered the city's terrain, increased tidal influence in estuarine reaches, and reduced the hydraulic efficiency of its waterways, thereby elevating long-term flood risk. Hughes et al. (2015) concluded that effective hazard management in coastal urban settings must account for the dynamic interdependence between seismic activity, ground deformation, and fluvial processes.

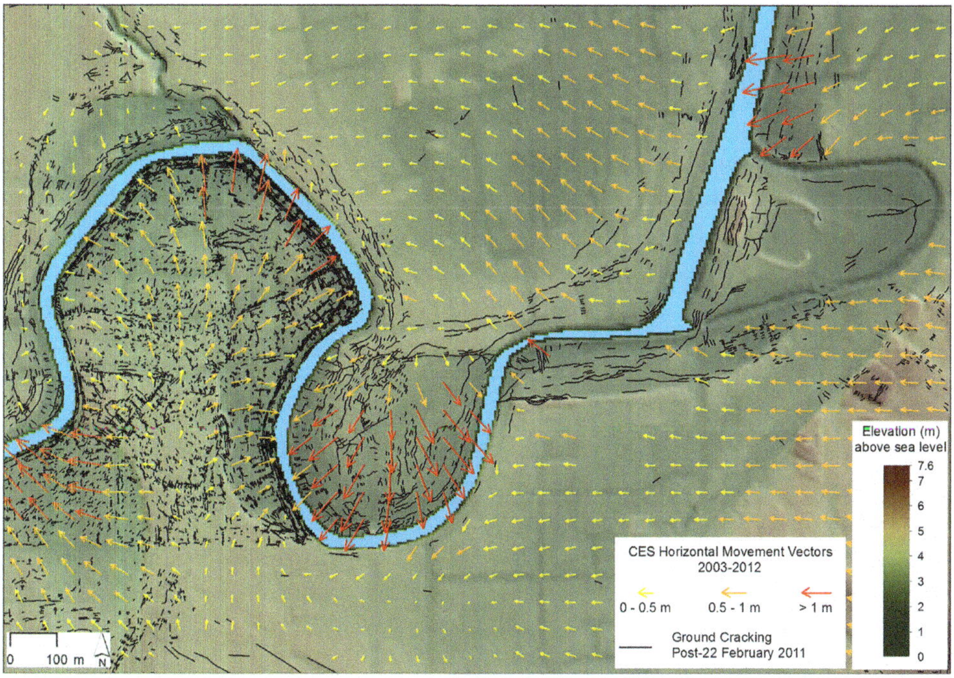

Figure 8.3 Cumulative horizontal ground displacement vectors along the Avon River, South Island, New Zealand, showing effects of lateral spreading during the 2010–11 Canterbury earthquake sequence. Image centred on lat. 43.517510°S, long. 172.680117°E. Displacement measurements and ground cracks mapping were derived from field inspections and surveys, and aerial surveys using light detection and ranging, aerial photography and satellite images (Cubrinovski & Robinson, 2016b). Displacements in inner meanders trend towards the most proximal free faces along the channel margins. Lateral spreading cracks tend to orient perpendicular to lateral spreading vectors and parallel to channel margins. Largest lateral spreading cracks and largest displacements tend to occur in topographically lowest surfaces with youngest sediments most proximal to channels. Further details are available in Quigley et al. (2016). Map lines delineate study areas and do not necessarily depict accepted national boundaries.

Quigley and Duffy (2020) describe in detail how this earthquake sequence exacerbated flood risks by altering hydrological systems and landforms. Earthquake-induced liquefaction processes are generally categorised as: (1) those that directly changed the volume and flow of surface and subsurface waters, and (2) those that caused permanent or transient physical changes to natural and engineered environments – such as elevation, topographic relief and permeability – that subsequently affect hydrological regimes. Quigley and Duffy (2020) highlight how earthquakes can lead to significant land surface and waterway deformations, which in turn increase flood hazards. These deformations can alter the flows, volumes, and distributions of surface and subsurface waters, thereby affecting both natural and engineered environments.

An exceptional aspect of the Canterbury earthquake sequence was the occurrence of at least 11 liquefaction-triggering earthquakes over approximately 16 months (Quigley et al., 2013), of which at least 7 generated substantial elevation changes that affected local geomorphology (e.g. Quigley et al., 2016).

A series of geological investigations was conducted in areas of liquefaction during the 2010–11 earthquake sequence to determine whether evidence for earlier liquefaction events was present in Holocene sediments (e.g. Bastin et al., 2015, 2016; Villamor et al., 2016). Dating of associated sedimentary sequences provided evidence for preceding liquefaction events with similar attributes (e.g. feeder dyke widths) to the 2010–11 features. The spatial correspondence between paleoliquefaction sites and areas that liquefied during the 2010–11 Canterbury earthquake sequence suggests long-term persistence of liquefaction susceptibility related to inherited geomorphic and sedimentologic controls (Green et al., 2014; Quigley et al., 2016). Maurer et al. (2015) considered potential seismic sources for triggering the paleoliquefaction events using probabilistic magnitude-bound relations. The preferred 'historical' source earthquakes include the 1717 M_w >8 plate boundary earthquake emanating from the Alpine Fault >120 km west of Christchurch, and the 1869 M_w <5 earthquake local to Christchurch, highlighting the potential of diverse earthquakes (by size and site distance) to generate liquefaction (Maurer et al., 2015).

8.5.2 Liquefaction in the New Madrid seismic zone (USA)

Seismic liquefaction in the New Madrid seismic zone, which last had a major earthquake sequence in 1811–12, has produced a suite of distinctive geomorphic effects across floodplain, terrace and escarpment settings, including large sand blows and sand dykes, lateral-spread deposits and associated open fissures, localised subsidence and ground settlement, river-channel disturbance (bank collapse, avulsion and temporary ponding), and ridge-scale coseismic mass movements (sackungen or ridgetop spreading) along topographic contrasts (Delano et al., 2018; Gold et al., 2019; Obermeier et al., 2005; Obermeier, 1996a, 1996b; Tuttle et al., 2019; USGS, 1988a, 1988b).

Sand blows and injected sand dykes exposed in Mississippi Embayment cutbanks and archaeological sites record repeated high-intensity shaking and preserve multievent stratigraphy that is widely used for paleoearthquake reconstruction; these paleoliquefaction features are commonly tens of centimetres to metres across and may be preserved for $>10^3$ years in fine overbank deposits. Lateral spreading along stream banks and terrace edges has produced hummocky back-swamp topography, displaced levee and terrace segments, and generated long, narrow zones of vented sand that can extend many kilometres from inferred source faults, collectively altering local drainage patterns. Field and geophysical surveys show that liquefaction effects during the 1811–12 sequence covered an area on the order of 10,000 km^2 and were induced (in the strongest shocks) at distances of >200 km from epicentral locations, indicating very large source magnitudes and long-range propagation of seismic shaking in the low-attenuation Mississippi Embayment sediments. The spatial distribution and preservation of these geomorphic markers are strongly conditioned by local stratigraphy; thick, fine-grained overbank deposits with interbedded sands are most susceptible to liquefaction and may best preserve geological evidence if liquefied sedimentary features such as dykes intrude into lithologically distinct units such as clays, gravels or swamp deposits. The geomorphic setting (cutbanks, low terraces and abandoned channels concentrate evidence for liquefaction) and postdepositional processes (alluviation, pedogenesis may obscure evidence for liquefaction) significantly influence whether features remain visible or are rapidly buried. This, in turn, governs how reliably paleoliquefaction records can be used for seismic hazard characterisation in the New Madrid seismic zone and other analogous settings globally.

8.6 Implications for landscape evolution and hazard assessment

Seismic liquefaction can induce landscape changes through the movement of liquefied sediments in the subsurface and at the surface. There is a tendency for reduction of small-scale topographic relief in liquefied areas, as material tends to flow into topographic depressions such as river channels via lateral spreading. However, relief may also be increased in liquefied areas; for example, liquefied alluvial terrace sequences might subside relative to surrounding older and higher terrace sequences, resulting in enhanced relief. Through modifications to alluvial channel geometries and floodplain elevations, liquefaction can enhance the tendency for channel network re-arrangements, avulsion events and reoccupation of ephemeral or abandoned channels. Erosion may be enhanced on topographic slopes; for example, lateral-spreading fissures may enhance infiltration of surface waters and increase erosion. Flood hazards and associated geomorphic impacts may also be enhanced due to floodplain subsidence. Changes in coastal elevations may enhance the influence of tidal fluctuations and marine events such as coastal surges and tsunamis on the geomorphic evolution of coastal environments. In areas of liquefaction hazard, the geomorphic consequences of future liquefaction events may be considered in the context of land-use planning decisions aimed at reducing disaster risks including fluvial, pluvial and coastal flooding (e.g. Hughes et al., 2015; Quigley et al., 2019a, 2019b, 2020).

8.7 Conclusions

1. Seismic liquefaction alters the physical and hydrological properties of saturated, unconsolidated sediments through rapid increases in PFP and reductions in effective

stress. These processes can commence within seconds of the onset of ground shaking and may persist beyond the cessation of seismic motions, enabling substantial sediment mobility and ground deformation.

2. Surface and subsurface geomorphic expressions of seismic liquefaction reflect interactions between topography, sedimentology, hydrology and seismic loading. Liquefaction commonly generates sand blows, feeder dykes, ground fissures, lateral spreading, subsidence, uplift and associated changes to micro- and meso-scale landforms. The morphology, distribution and preservation of these features are strongly conditioned by local stratigraphy, sediment permeability, water-table position, and the presence of natural or engineered surface caps.
3. Liquefaction produces significant and often enduring modifications to topography and drainage networks. Subsidence, lateral spreading, and channel narrowing reduce flood capacity and increase inundation hazards, while elevation changes in coastal and estuarine environments alter tidal reach, groundwater levels and susceptibility to marine flooding. These geomorphic impacts can amplify hazard exposure long after the triggering earthquake, with implications for both natural landscapes and built environments.
4. Case studies demonstrate the geomorphic scale and persistence of liquefaction impacts. The Canterbury earthquake sequence illustrates how repeated liquefaction events can accumulate substantial subsidence, reorganise drainage systems and exacerbate flood hazards in low-lying alluvial settings. Paleoliquefaction fields in the New Madrid seismic zone reveal how liquefaction features may persist for millennia in fine-grained overbank sediments, providing key insights into prehistoric seismicity and landscape evolution.
5. Liquefaction influences landscape evolution by redistributing sediment and modulating relief. Mass movement of liquefied sediments smoothes or enhances local relief depending on geomorphic context, while modifications to channel geometry promote avulsion, reoccupation of abandoned channels and long-term reorganisation of fluvial systems. In coastal settings, coseismic subsidence can shift the balance between marine and fluvial processes, shaping floodplain trajectories over decades to centuries.
6. Understanding liquefaction–geomorphology interactions is critical for seismic hazard and land-use planning. Geomorphic expressions of liquefaction provide valuable evidence for paleoearthquake reconstruction, inform probabilistic and deterministic assessments of liquefaction susceptibility, and highlight the need to incorporate earthquake-induced ground deformation into urban planning and critical-infrastructure design. Effective hazard mitigation requires a holistic approach that recognises the dynamic coupling between seismic processes, sedimentary architecture, hydrological systems and surface geomorphology.

References

AFAD (Disaster and Emergency Management Authority) (2023). *2023 Kahramanmaraş earthquakes: Field damage and liquefaction observations*. AFAD Earthquake Department.

Andrews, D. C. A., & Martin, G. R. (2000). Criteria for liquefaction of silty soils. In *Proceedings of the 12th World Conference on Earthquake Engineering* (Paper No. 0312). Auckland, New Zealand.

Andrus, R. D., Hayati, H., & Mohanan, N. P. (2009). Correcting liquefaction resistance of aged sands using measured-to-estimated velocity ratio. *Journal of Geotechnical and Geoenvironmental Engineering, 135*(6), 735–744.

Atwater, B. F. (1987). Evidence for great Holocene earthquakes along the outer coast of Washington State. *Science (New York, N.Y.), 236*(4804), 942–944. https://doi.org/10.1126/science.236.4804.942.

Atwater, B. F., & Hemphill-Haley, E. (1997). *Recurrence intervals for great earthquakes of the past 3,500 years at northeastern Willapa Bay, Washington*. United States Geological Survey Professional Paper 1576. https://doi.org/10.3133/pp1576

Atwater, B. F., Tuttle, M. P., Wells, R. E., Yamaguchi, D. K., & Hemphill-Haley, E. (2014). *The orphan tsunami of 1700—Japanese clues to a parent earthquake in North America*. United States Geological Survey Professional Paper 1707.

Baba, T., Satake, K., & Kaneda, Y. (2005). Simulation of the Indian Ocean tsunami of December 26, 2004. *Earth, Planets and Space, 57*(3), 167–172. https://doi.org/10.1186/BF03351809.

Bartlett, S. F., & Youd, T. L. (1995). Empirical prediction of liquefaction-induced lateral spread. *Journal of Geotechnical Engineering, 121*(4), 316–329. https://doi.org/10.1061/(ASCE)0733-9410(1995)121:4(316).

Bastin, S. H., Quigley, M. C., & Bassett, K. N. (2015). Paleoliquefaction in Christchurch, New Zealand. *Geological Society of America Bulletin, 127*(9–10), 1348–1365.

Bastin, S. H., Bassett, K., Quigley, M. C., Maurer, B., Green, R. A., Bradley, B., & Jacobson, D. (2016). Late Holocene liquefaction at sites of contemporary liquefaction during the 2010–2011 Canterbury earthquake sequence, New Zealand. *Bulletin of the Seismological Society of America, 106*(3), 881–903.

Berryman, K. R., Villamor, P., & Little, T. A. (2011). Surface rupture and paleoseismic history of the Wairarapa Fault, New Zealand. *New Zealand Journal of Geology and Geophysics, 54*(3), 231–252. https://doi.org/10.1080/00288306.2011.563088.

Bhattacharya, S., Hyodo, M., Goda, K., Tazoh, T., & Taylor, C. A. (2011). Liquefaction of soil in the Tokyo Bay area from the 2011 Tohoku (Japan) earthquake. *Soil Dynamics and Earthquake Engineering, 31*(11), 1618–1628.

Bilgin, H., Ural, D. N., & Sezen, H. (2023). Preliminary geotechnical observations of the 2023 Kahramanmaraş earthquakes, Türkiye. *Soil Dynamics and Earthquake Engineering, 170*, 108047. https://doi.org/10.1016/j.soildyn.2023.108047.

Boncio, P., Amoroso, S., Galadini, F., Galderisi, A., Iezzi, G., & Liberi, F. (2020). Earthquake-induced liquefaction features in a late Quaternary fine-grained lacustrine succession (Fucino Lake, Italy): Implications for microzonation studies. *Engineering Geology, 272*, 105621.

Borella, J., Quigley, M., Riley, M., Trutner, S., Jol, H., Borella, M., Hampton, S., & Gravley, D. (2020). Influence of anthropogenic landscape modifications and infrastructure on the geological characteristics of liquefaction. *Anthropocene, 29*, 100235.

Boulanger, R. W., & Idriss, I. M. (2004). Analytical procedures for evaluating the cyclic failure potential of low-plasticity silts and clays. *Proceedings, U.S. Army Corps of Engineers*.

Bradley, B. A., Razafindrakoto, H. N. T., & Polak, V. (2019). Ground motion and seismic source aspects of the 28 September 2018 Mw 7.5 Sulawesi earthquake. *Earthquake Spectra, 35*(1), 1–15. https://doi.org/10.1193/101118EQS217M.

Brandenberg, S. J., Stewart, J. P., & Hudson, M. B. (2020). Geotechnical aspects of the 2019 Ridgecrest earthquake sequence. *Earthquake Spectra, 36*(4), 1751–1774. https://doi.org/10.1177/8755293020944161.

Brandes, C., & Winsemann, J. (2013). Soft-sediment deformation structures in NW Germany caused by Late Pleistocene seismicity. *International Journal of Earth Sciences, 102*(8), 2255–2274.

Bray, J. D., Boulanger, R. W., & Abrahamson, N. A. (2014). *Engineering geologic effects of the 2014 South Napa earthquake*. United States Geological Survey Open-File Report 2014–1249. https://doi.org/10.3133/ofr20141249.

Bray, J. D., & Sancio, R. B. (2006). Assessment of liquefaction-induced ground failure at specific sites. *Earthquake Spectra, 22*(2), 31–51. https://doi.org/10.1193/1.2191556.

Bray, J. D., & Travasarou, T. (2007). Simplified procedure for estimating earthquake-induced deviatoric slope displacements. *Journal of Geotechnical and Geoenvironmental Engineering, 133*(4), 381–392. https://doi.org/10.1061/(ASCE)1090-0241(2007)133:4(381).

Bucci, M. G., Almond, P. C., Villamor, P., Ries, W., Smith, C. M. S., & Tuttle, M. P. (2017). When the earth blisters: Exploring recurrent liquefaction features in the coastal system of Christchurch, New Zealand. *Terra Nova, 29*(3), 162–172.

Bucci, M. G., Almond, P. C., Villamor, P., Tuttle, M. P., Stringer, M., Smith, C. M. S., & Ries, W. (2018). Controls on patterns of liquefaction in a coastal dune environment, Christchurch, New Zealand. *Sedimentary Geology, 377*, 17–33.

Cetin, K. O., Seed, R. B., & Der Kiureghian, A. (2000). Liquefaction case histories and CPT-based probabilistic assessment of liquefaction triggering. *Journal of Geotechnical and Geoenvironmental Engineering, 126*(12), 1106–1115.

Chu, D. B., Chen, C. C., & Lee, C. T. (2004). Liquefaction hazard mapping in the Chi-Chi earthquake area, Taiwan. *Engineering Geology, 71*(1–2), 31–47. https://doi.org/10.1016/S0013-7952(03)00121-2.

Comte, D., & Pardo, M. (1991). Reappraisal of great historical earthquakes in the northern Chile and southern Peru seismic gaps. *Natural Hazards, 4*(1), 23–44. https://doi.org/10.1007/BF00126557.

Cox, S. C., van Ballegooy, S., Rutter, H. K., Harte, D. S., Holden, C., Gulley, A. K., ... Manga, M. (2021). Can artesian groundwater and earthquake-induced aquifer leakage exacerbate the manifestation of liquefaction? *Engineering Geology, 281*, 105982.

Cubrinovski, M., et al. (2011a). Geotechnical aspects of the 2010 Darfield (Canterbury) earthquake. *Bulletin of the New Zealand Society for Earthquake Engineering, 44*(4), 205–226.

Cubrinovski, M., Bray, J. D., Taylor, M., Giorgini, S., Bradley, B., Wotherspoon, L., & Zupan, J. (2011b). Soil liquefaction effects in the central business district during the February 2011 Christchurch earthquake. *Seismological Research Letters, 82*(6), 893–904. https://doi.org/10.1785/gssrl.82.6.893.

Cubrinovski, M., & Robinson, K. (2016a). Lateral spreading and its impacts in urban areas in the 2010–2011 Christchurch earthquakes. *New Zealand Journal of Geology and Geophysics, 59*(4), 450–469. https://doi.org/10.1080/00288306.2016.1180591.

Cubrinovski, M., & Robinson, K. (2016b). Lateral spreading: Evidence and interpretation from the 2010–2011 Christchurch earthquakes. *Soil Dynamics and Earthquake Engineering, 91*, 187–201.

Collinson, J. D., Mountney, N. P., & Thompson, D. B. (2019). *Sedimentary structures* (4th ed.). Dunedin Academic Press.

Crespellani, T., Nardi, R., & Simoncini, C. (1988). Evaluation of liquefaction potential in sandy deposits. In *Proceedings of the International Symposium on Penetration Testing (ISOPT-1)* (Vol. 2, pp. 883–888). Orlando, FL.

Delano, J. E., et al. (2018). Coseismic sackungen in the New Madrid Seismic Zone, USA. *Geophysical Research Letters.*

Delano, J. E., Howell, A., Clark, K. J., & Stahl, T. A. (2023). Upper plate faults may contribute to the paleoseismic subsidence record along the central Hikurangi subduction zone, Aotearoa New Zealand. *Geochemistry, Geophysics, Geosystems, 24*(10) e2023GC011060.

Do, T. Q., Zimmaro, P., & Stewart, J. P. (2017). Liquefaction of gravelly soil: Case history from the 2011 Christchurch earthquake. *Soil Dynamics and Earthquake Engineering, 100*, 237–251. https://doi.org/10.1016/j.soildyn.2017.04.014.

Dutton, C.E. (1889). The Charleston earthquake of August 31, 1886. *U.S. Geological Survey Ninth Annual Report, 1887–1888.*

Elfilali, O., Berkat, B., & Akhssas, A. (2024). Assessing liquefaction potential in alluvial plains through spatiotemporal analysis using Liquefaction Probability Index. *Civil Engineering Journal, 10*(6), 2007–2018.

Evans, S. G., & Zhou, Y. (1995). The 1988 Tien Shan earthquakes, China—Liquefaction and landslides. *Canadian Geotechnical Journal, 32*(4), 620–636. https://doi.org/10.1139/t95-062.

Fairless, G., & Berrill, J. (1984). Liquefaction during historic earthquakes in New Zealand. *Bulletin of the New Zealand Society for Earthquake Engineering, 17*, 280–291.

Franks, C. A. (1988). Engineering geological aspects of the Edgecomb, New Zealand earthquake of 2 March 1987. *Quarterly Journal of Engineering Geology, 21*, 337–345.

Fuller, M. L. (1912). *The New Madrid earthquake*. U.S. Geological Survey Bulletin 494.

Gold, R. D., et al. (2019). Four major Holocene earthquakes on the Reelfoot fault, Mississippi Embayment, USA. *Journal of Geophysical Research: Solid Earth*. https://doi.org/10.1029/2018JB016806.

Grapes, R., & Downes, G. (1997). The 1855 Wairarapa, New Zealand, earthquake: Analysis of historical data. *Bulletin of the New Zealand National Society for Earthquake Engineering, 30*(4), 271–368. https://doi.org/10.5459/bnzsee.30.4.271-368.

Gratchev, I. B., & Sassa, K. (2004). The mechanism of liquefaction of clayey soils: Influence of clay content and mineralogy. In The influence of clay mineralogy and ground water chemistry on the mechanism of landslides *(ICL Project M124)*. Springer.

Green, R. A., & Bommer, J. J. (2019). What is the smallest earthquake magnitude that needs to be considered in assessing liquefaction hazard? *Earthquake Spectra, 35*(3), 1441–1464. https://doi.org/10.1193/032218EQS064M.

Green, R. A., Cubrinovski, M., Cox, B. R., Wood, C. M., Wotherspoon, L. M., Bradley, B. A., & Maurer, B. W. (2014). Select liquefaction case histories from the 2010–2011 Canterbury earthquake sequence. *Earthquake Spectra, 30*(1), 131–153.

Gutscher, M.-A., Roger, J., Baptista, M. A., Miranda, J. M., & Tinti, S. (2006). The source of the 1693 Catania earthquake and tsunami (southern Italy): New evidence from tsunami modeling. *Geophysical Research Letters, 33*(11), L11310. https://doi.org/10.1029/2006GL026065.

Güven, C., Wolf, L. W., Tuttle, M. P., & Rogers, S. R. (2023). The influence of sedimentary architecture on the formation of earthquake-induced liquefaction features: A case study in the Mississippi River floodplain, New Madrid seismic zone, USA. *Engineering Geology, 313*, 106946.

Harder, L. F., & Seed, H. B. (1986). Determination of penetration resistance for evaluation of liquefaction potential. *Journal of Geotechnical Engineering, 112*(4), 367–387.

Hayati, H., & Andrus, R. D. (2009). Updated liquefaction resistance correction factors for aged sands. *Journal of Geotechnical and Geoenvironmental Engineering, 135*(11), 1683–1692.

Holbrook, J., Schumm, S. A., & Tuttle, M. P. (2006). Holocene liquefaction in the New Madrid Seismic Zone: Implications for paleoseismicity and landscape evolution. *Geological Society of America Special Paper, 425*, 49–64. https://doi.org/10.1130/2006.2425(04).

Holzer, T. L. (Ed.). (1998). *The Loma Prieta, California, earthquake of October 17, 1989—Marina District*. U.S. Geological Survey Professional Paper 1551–F. U.S. Geological Survey.

Holzer, T. L., Bennett, M. J., Ponti, D. J., & Tinsley, J. C. (1999). Liquefaction and postliquefaction deformation of sand, Northridge earthquake, California. *Journal of Geotechnical and Geoenvironmental Engineering, 125*(6), 500–509.

Holzer, T. L., Youd, T. L., & Hanks, T. C. (1989). Dynamics of liquefaction during the 1979 Imperial Valley earthquake. *Bulletin of the Seismological Society of America, 79*(1), 81–103.

Hughes, M. W., Quigley, M. C., van Ballegooy, S., Deam, B. L., Bradley, B. A., Hart, D. E., & Measures, R. (2015). The sinking city: Earthquakes increase flood hazard in Christchurch, New Zealand. *GSA Today: a Publication of the Geological Society of America, 25*(3–4), 4–10. https://doi.org/10.1130/GSATG221A.1.

Idriss, I. M., & Boulanger, R. W. (2008). *Soil liquefaction during earthquakes*. Earthquake Engineering Research Institute.

Ishihara, K. (1985). Stability of natural deposits during earthquakes. In *Proceedings of the Eleventh International Conference on Soil Mechanics & Foundation Engineering* (pp. 321–376).

Ishihara, K. (1993). Liquefaction and flow failure during earthquakes. *Geotechnique, 43*(3), 351–415. https://doi.org/10.1680/geot.1993.43.3.351.

Ishihara, K. (1997). *Geotechnical aspects of the 1995 Kobe earthquake*. A. A. Balkema.

Japan Meteorological Agency (JMA) (1924). *The great Kanto earthquake of 1923: Seismological report*. Imperial Government Printing Office.

Johnston, A. C., & Schweig, E. S. (1996). The enigma of the New Madrid earthquakes of 1811–1812. *Annual Review of Earth and Planetary Sciences, 24*, 339–384.

Kramer, S. L. (1996). *Geotechnical earthquake engineering*. Upper Saddle River, NJ: Prentice Hall.

Kramer, S. L., Sideras, S. S., & Greenfield, M. W. (2016). Empirical evaluation of pore-pressure dissipation and liquefaction duration following earthquakes. *Earthquake Spectra, 32*, 1733–1753.

Kazaoka, O., Kameyama, S., Shigeno, K., Suzuki, Y., Morisaki, M., Kagawa, A., ... Furuno, K. (2015). Subsidence by liquefaction–fluidization in man-made strata around Tokyo Bay, Japan: From geological survey on damaged parts of the 2011 off the Pacific coast of Tohoku Earthquake. *Proceedings of the International Association of Hydrological Sciences, 372*, 221–226.

Kelsey, H. M., Witter, R. C., & Hemphill-Haley, E. (2005). Tsunami history of an Oregon coastal lake reveals a 4600 yr record of great earthquakes on the Cascadia subduction zone. *Geological Society of America Bulletin, 117*(7–8), 1009–1032. https://doi.org/10.1130/B25452.1.

Kishida, H. (1966). Damage to reinforced concrete buildings in Niigata City during the earthquake of June 16, 1964. *Soils and Foundations, 6*(1), 71–88.

Liu, A. H., & Dobry, R. (2000). Seismic response of shallow foundations on liquefiable sand. *Journal of Geotechnical and Geoenvironmental Engineering, 126*(6), 557–567.

Liu, Y., Wang, P., Quigley, M. C., et al. (2025). Liquefaction susceptibility of clay-bearing sediments: New insights from recent earthquakes. *Engineering Geology*.

Lowe, D. R. (1976). Subaqueous liquefied and fluidized sediment flows and their deposits. *Sedimentology, 23*(3), 285–308.

Maltman, A., & Bolton, A. (2003). How sediments become mobilized and deformed. *Sedimentary Geology, 159*(1–2), 3–18.

Martin, G. R., & Clough, G. W. (1994). Seismic parameters from liquefaction evidence. *Journal of Geotechnical Engineering, 120*(8), 1345–1361.

Mason, H. B., Hossain, M. A., & Pradel, D. (2019). Liquefaction and lateral spreading observed from the 2018 Palu earthquake in Sulawesi, Indonesia. *Seismological Research Letters, 90*(5), 1840–1853. https://doi.org/10.1785/0220190089.

Mason, H. B., Pradel, D., & Wartman, J. (2020). Liquefaction-induced ground failures during the 2018 Palu earthquake, Indonesia. *Geotechnical Earthquake Engineering and Soil Dynamics VII: Geotechnical Special Publication, 331*, 409–418. https://doi.org/10.1061/9780784482872.040.

Maurer, B. W., Green, R. A., Cubrinovski, M., & Bradley, B. A. (2014). Evaluation of the liquefaction potential index for assessing liquefaction hazard in Christchurch, New Zealand. *Journal of Geotechnical and Geoenvironmental Engineering, 140*(7), 04014032.

Maurer, B. W., Green, R. A., Quigley, M. C., & Bastin, S. (2015). Development of magnitude-bound relations for paleoliquefaction analyses: New Zealand case study. *Engineering Geology, 197*, 253–266.

Moreira, V. S. (1985). The 1755 Lisbon earthquake: A review of historical sources and modern interpretations. *Tectonophysics, 118*(1–2), 55–63. https://doi.org/10.1016/0040-1951(85)90213-0.

Motta, E., Verdugo, R., & Bray, J. D. (2010). Liquefaction and ground failure during the 2010 Maule earthquake in Chile. *Proceedings of the 9th U.S. National and 10th Canadian Conference on Earthquake Engineering*, Toronto, Canada.

National Academies of Sciences, Engineering, and Medicine (2021). *State of the art and practice in the assessment of earthquake-induced soil liquefaction and its consequences*. The National Academies Press. https://doi.org/10.17226/23474.

Obermeier, S. F. (1989). The Charleston, South Carolina earthquake of 1886—Liquefaction evidence for recurrent strong earthquakes. *Annals of the New York Academy of Sciences, 558*(1), 196–214. https://doi.org/10.1111/j.1749-6632.1989.tb22448.x.

Obermeier, S. F. (1996a). Field occurrences of liquefaction-induced features: A primer for engineering geologic analysis of paleoseismic shaking. *Engineering Geology, 76*(3–4), 65–100.

Obermeier, S. F. (1996b). Using liquefaction-induced features for paleoseismic analysis. In J. P. McCalpin (Ed.). *Paleoseismology* (pp. 331–396). Academic Press.

Obermeier, S. F., Olson, S. M., & Green, R. A. (2001). Paleoliquefaction studies in continental settings: Geologic and geotechnical factors in interpretations and back-analysis. *U.S. Geological Survey Open-File Report 01–29*.

Obermeier, S. F., Olson, S. M., & Green, R. A. (2005). Field occurrences of liquefaction-induced features: A primer for engineering geologic analysis of paleoseismic shaking. *Engineering Geology, 76*(3–4), 209–234. https://doi.org/10.1016/j.enggeo.2004.07.009.

Owen, G. (2003). Load structures: Gravity-driven sediment mobilization in the shallow subsurface. In M. Van Rensbergen, R. R. Hillis, A. J. Maltman, & C. K. Morley (Vol. Eds.), *Subsurface sediment mobilization: 216*, (pp. 21–34). London: Geological Society Geological Society Special Publication.

Owen, G., & Moretti, M. (2011a). Identifying triggers for liquefaction-induced soft-sediment deformation in sands. *Sedimentary Geology, 235*(3–4), 141–147.

Owen, G., & Moretti, M. (2011b). Towards a model of soft-sediment deformation in the shallow subsurface. *Geological Society, London, Special Publications, 352*(1), 37–59.

Postma, G. (1983). Water-escape structures in the context of a depositional model of a mass-flow-dominated conglomeratic fan delta. *Sedimentology, 30*(1), 91–103.

Quigley, M., & Duffy, B. (2020). Effects of earthquakes on flood hazards: A case study from Christchurch, New Zealand. *Geosciences, 10*(3), 114.

Quigley, M. C., Bastin, S., & Bradley, B. A. (2013). Recurrent liquefaction in Christchurch, New Zealand, during the Canterbury earthquake sequence. *Geology, 41*(4), 419–422.

Quigley, M. C., Bennetts, L. G., Durance, P., Kuhnert, P. M., Lindsay, M. D., Pembleton, K. G., ... White, C. J. (2019a). The provision and utility of science and uncertainty to decision-makers: Earth science case studies. *Environment Systems and Decisions, 39*(3), 307–348.

Quigley, M. C., Bennetts, L. G., Durance, P., Kuhnert, P. M., Lindsay, M. D., Pembleton, K. G., ... White, C. J. (2019b). The provision and utility of earth science to decision-makers: Synthesis and key findings. *Environment Systems and Decisions, 39*(3), 349–367.

Quigley, M. C., Hughes, M. W., Bradley, B. A., van Ballegooy, S., Reid, C., Morgenroth, J., ... Pettinga, J. R. (2016). The 2010–2011 Canterbury earthquake sequence: Environmental effects, seismic triggering thresholds and geologic legacy. *Tectonophysics, 672*, 228–274.

Quigley, M. C., Saunders, W., Massey, C., Van Dissen, R., Villamor, P., Jack, H., & Litchfield, N. (2020). The utility of earth science information in post-earthquake land-use decision-making: The 2010–2011 Canterbury earthquake sequence in Aotearoa New Zealand. *Natural Hazards and Earth System Sciences, 20*, 3361–3385. https://doi.org/10.5194/nhess-20-3361-2020.

Rastogi, B. K. (2004a). Damage and liquefaction in the 2001 Bhuj earthquake. *Seismological Research Letters, 75*(3), 350–362.

Rastogi, B. K. (2004b). The Bhuj earthquake of 26 January 2001. *ISET Journal of Earthquake Technology, 37*(4), 1–14.

Rodríguez-Pascua, M. A., Calvo, J. P., De Vicente, G., & Gómez-Gras, D. (2000). Soft-sediment deformation structures interpreted as seismites in lacustrine sediments of the Prebetic Zone, SE Spain, and their potential use as indicators of earthquake magnitudes. *Sedimentary Geology, 135*(1–4), 117–135. https://doi.org/10.1016/S0037-0738(00)00067-2.

Santucci de Magistris, F., Lanzano, G., Forte, G., & Fabbrocino, G. (2014). A peak acceleration threshold for soil liquefaction: Lessons learned from the 2012 Emilia earthquake (Italy). *Natural Hazards, 74*(2), 1069–1094.

Sassa, K., & Fukuoka, H. (2006). Monitoring and prediction of earthquake-induced landslides. *Landslides, 3*(1), 1–4.

Satake, K., Shimazaki, K., Tsuji, Y., & Ueda, K. (1996). Time and size of a giant earthquake in Cascadia inferred from Japanese tsunami records of January 1700. *Nature, 379*(6562), 246–249. https://doi.org/10.1038/379246a0.

Schmertmann, J. H. (1991). The mechanical aging of soils. *Journal of Geotechnical Engineering, 117*(9), 1288–1330.

Scicchitano, G., Antonioli, F., & Monaco, C. (2022). The enigmatic 1693 AD tsunami in the eastern Mediterranean: Geological and historical evidence. *Scientific Reports, 12*(1), 13577. https://doi.org/10.1038/s41598-022-17688-3.

Seed, H. B., & Idriss, I. M. (1967). Analysis of soil liquefaction: Niigata earthquake case history. *Journal of Soil Mechanics and Foundations Division, ASCE, 93*(SM3), 83–108.

Seed, H. B., & Idriss, I. M. (1971). Simplified procedure for evaluating soil liquefaction potential. *Journal of Soil Mechanics & Foundations Division, 97*(9), 1249–1273.

Seed, H. B., & Idriss, I. M. (1982). *Ground motions and soil liquefaction during earthquakes.* Earthquake Engineering Research Institute.

Seed, H. B., Youd, T. L., & Idriss, I. M. (1973). Earthquake-induced liquefaction in San Fernando, California. *Journal of the Soil Mechanics and Foundations Division, ASCE, 99*(SM8), 593–606.

Sonmez, H., Papathanassiou, G., & Gokceoglu, C. (2008). Assessing the effect of non-liquefiable cap layer thickness on surface manifestation of liquefaction-induced failures. *Natural Hazards and Earth System Sciences, 8*(4), 727–740.

Stirling, M. W., et al. (2017). The Mw 7.8 Kaikōura earthquake: Surface rupture and seismic hazard implications. *Bulletin of the New Zealand Society for Earthquake Engineering, 50*(2), 73–84.

Talwani, P., & Cox, J. (1985). Paleoseismic evidence for recurrent earthquakes near Charleston, South Carolina. *Science (New York, N.Y.), 229*(4712), 379–381. https://doi.org/10.1126/science.229.4712.379.

Talwani, P., & Schaeffer, W. T. (2001). Recurrence rates of large earthquakes in the South Carolina Coastal Plain based on paleoliquefaction data. *Journal of Geophysical Research: Solid Earth, 106*(B4), 6621–6642.

Tanaka, H., Kokusho, T., & Towhata, I. (2020). Tsunami and liquefaction interactions in coastal lowlands following major earthquakes. *Soil Dynamics and Earthquake Engineering, 132*, 106063.

Tokimatsu, K., & Seed, H. B. (1987). Evaluation of settlements in sands due to earthquake shaking. *Journal of Geotechnical Engineering, 113*(8), 861–878.

Tuttle, M. P., Schweig, E. S., III, Lafferty, R. H., Collier, J., Wolf, L. W., & Haynes, M. L. (2019). Paleoliquefaction studies and the evaluation of seismic hazard. *Geosciences, 9*(7), 311. https://doi.org/10.3390/geosciences9070311.

Tuttle, M. P., Schweig, E. S., III, Sims, J. D., Lafferty, R. H., Wolf, L. W., & Haynes, M. L. (2002). The earthquake potential of the New Madrid Seismic Zone. *Bulletin of the Seismological Society of America, 92*(6), 2080–2089. https://doi.org/10.1785/0120010227.

Towhata, I., Yasuda, S., & Tokida, K. (2004). Liquefaction-induced ground damage during recent Japanese earthquakes. *Soils and Foundations, 44*(4), 1–16.

Towhata, I., Yasuda, S., Kobayashi, M., & Kuwano, R. (2016). Liquefaction-induced damage associated with the 2016 Kumamoto earthquake. *Soils and Foundations, 56*(6), 1069–1084.

U. S. Geological Survey. (1969). *The Alaska earthquake, March 27, 1964—Effects on the communities of southern Alaska* (Professional Paper 542-G). Washington, DC: U.S. Government Printing Office.

U.S. Geological Survey. (1988a). *Investigations of the New Madrid, Missouri, earthquake sequence (1811–1812)* (U.S. Geological Survey Professional Paper 1236). U.S. Government Printing Office.

U.S. Geological Survey. (1988b). *The San Francisco Earthquake of 1906 and the California earthquake hazard.* U.S. Geological Survey Professional Paper 1515. U.S. Government Printing Office.

U.S. Geological Survey. (1990). *The Loma Prieta, California, Earthquake of October 17, 1989.* USGS Professional Paper 1550. U.S. Government Printing Office.

U.S. Geological Survey. (2019). *Summary of the 1811–1812 New Madrid earthquakes sequence.* U.S. Geological Survey Earthquake Hazards Program Science Summary.

U.S. Geological Survey. (2020). *Preliminary report on the 2020 Puerto Rico earthquake sequence.* U.S. Geological Survey.

Udías, A., Madariaga, R., Buforn, E., Muñoz, D., & Ros, M. (2012). The large earthquakes of the Chilean subduction zone: A historical review. *Earth-Science Reviews, 113*(1–2), 28–57. https://doi.org/10.1016/j.earscirev.2012.04.004.

Van Arsdale, R. B. (2000). Displacement history and slip rate on the Reelfoot fault of the New Madrid seismic zone. *Engineering Geology, 55*(4), 219–226.

Van Dissen, R. J., Townsend, D. B., Pettinga, J. R., Yetton, M. D., & Villamor, P. (2013). Liquefaction observations from the Cook Strait and Lake Grassmere earthquake sequence of 2013. *Bulletin of the New Zealand Society for Earthquake Engineering, 46*(4), 297–327.

Villamor, P., Almond, P., Tuttle, M. P., Giona-Bucci, M., Langridge, R. M., Clark, K., ... Vandergoes, M. (2016). Liquefaction features produced by the 2010–2011 Canterbury earthquake sequence in southwest Christchurch, New Zealand and preliminary assessment of paleoliquefaction features. *Bulletin of the Seismological Society of America, 106*(4). https://doi.org/10.1785/0120150223.

Villamor, P., Ries, W. F., Ortuño, M., Bastin, S., Quigley, M. C., & Van Dissen, R. J. (2016). Liquefaction features as evidence of prehistoric earthquakes in Christchurch, New Zealand. *New Zealand Journal of Geology and Geophysics, 59*(3), 417–430.

Wang, C. Y., & Manga, M. (2021). *Earthquake hydrology*. Springer Nature.

Wang, G., Kamai, T., Kokusho, T., et al. (2009). Liquefaction and ground deformation during the 2008 Wenchuan earthquake. *Engineering Geology, 104*(3–4), 170–188.

Wang, Y., & Wong, I. G. (2020). Liquefaction-induced settlement and ground deformation in alluvial environments. *Engineering Geology, 272*, 105620.

Wang, Y., Stewart, J. P., & Kayen, R. E. (2020). State-of-practice review of liquefaction triggering and manifestation. *Earthquake Spectra, 36*(S1), 3–26.

Yasuda, S., Harada, K., & Ishikawa, K. (2012). Liquefaction and its countermeasures observed in the 2011 Great East Japan earthquake. *Soils and Foundations, 52*(5), 793–810. https://doi.org/10.1016/j.sandf.2012.11.002.

Yasuda, S., Harada, K., Ishikawa, K., & Kanemaru, Y. (2004). Characteristics of liquefaction in Tokyo Bay area and effects on lifelines. In *Proceedings of the 13th World Conference on Earthquake Engineering*. Vancouver, Canada.

Yasuda, S., & Ishikawa, K. (2012). Characteristics of liquefaction in the Tokyo Bay area by the 2011 Great East Japan Earthquake. *Soils and Foundations, 52*(5), 793–810.

Yasuda, S., & Tohno, I. (1988). Liquefaction during historical earthquakes in Japan and abroad. *Soils and Foundations, 28*(2), 65–78. https://doi.org/10.3208/sandf1972.28.2_65.

Youd, T. L., & Hoose, S. N. (1977). Liquefaction susceptibility and geologic setting. *In Proceedings of the 6th World Conference on Earthquake Engineering, Vol. 3*, 2189–2194.

Youd, T. L. (1973). Liquefaction, flow, and associated ground failure. *U.S. Geological Survey Circular, 688*, 1–12. https://doi.org/10.3133/cir688.

Youd, T. L., & Bennett, M. J. (1983). *Liquefaction and ground failure in the 1983 Coalinga earthquake*. Earthquake Engineering Research Institute Reconnaissance Report.

Youd, T. L., Hansen, C. M., & Bartlett, S. F. (2002). Revised multilinear regression equations for prediction of lateral spread displacement. *Journal of Geotechnical and Geoenvironmental Engineering, 128*(12), 1007–1017. https://doi.org/10.1061/(ASCE)1090-0241(2002)128:12(1007).

Youd, T. L., & Perkins, D. M. (1978). Mapping liquefaction-induced ground failure potential. *Journal of the Geotechnical Engineering Division, 104*(GT4), 433–446.

Youd, T. L., Idriss, I. M., Andrus, R. D., Arango, I., Castro, G., Christian, J. T., Dobry, R., Finn, W. D. L., Harder, L. F., Hynes, M. E., Ishihara, K., Koester, J. P., Liao, S. S. C., Marcuson, W. F., III, Martin, G. R., Mitchell, J. K., Moriwaki, Y., Power, M. S., Robertson, P. K., ... Stokoe, K. H., II (2001). Liquefaction resistance of soils: Summary report from the 1996 NCEER and 1998 NCEER/NSF workshops on evaluation of liquefaction resistance of soils. *Journal of Geotechnical and Geoenvironmental Engineering, 127*(10), 817–833.

Yuan, R., Deng, Q., & Li, H. (2009). Liquefaction-induced ground deformation associated with the 2008 Wenchuan earthquake, China. *Engineering Geology, 104*(3–4), 170–179.

Zhang, G., Robertson, P. K., & Brachman, R. W. I. (2004). Estimating liquefaction-induced lateral displacements using the standard penetration test or cone penetration test. *Journal of Geotechnical and Geoenvironmental Engineering, 130*(8), 861–871. https://doi.org/10.1061/(ASCE)1090-0241(2004)130:8(861.

CHAPTER 9

Fluvial responses to fault offset and uplift, and coseismic sediment input

Sarah J. Boulton[1], Alexander C. Whittaker[2], and Tristram C. Hales[3]
[1]School of Geography, Earth and Environmental Sciences, University of Plymouth, Plymouth, United Kingdom,
[2]Department of Earth Science and Engineering, Royal School of Mines, Imperial College, London, United Kingdom,
[3]School of Earth and Environmental Sciences, Cardiff University, Cardiff, United Kingdom

9.1 Introduction

Rivers are a fundamental landscape element, setting the pace at which landscapes respond to changes in base-level driven by extrinsic (i.e. tectonic, eustatic, climatic changes) or intrinsic (i.e. drainage reorganisation) factors, modulated by processes that control the river response rate (Fig. 9.1). Numerous processes can shape terrestrial topography over time in areas of active faulting and which may limit relief, including, but not limited to, landsliding and mass wasting (e.g. Huang et al., 2017; Roda-Boluda et al., 2018), debris flow activity (D'Arcy et al., 2017; Stock & Dietrich, 2003), and chemical dissolution (e.g. Ott et al., 2019); it is river incision, however, that typically plays a key role in nonglaciated areas (Watkins et al., 2020; Whittaker et al., 2007a). This is because rivers: (1) are ubiquitous on Earth (Gawthorpe & Leeder, 2000; Kent et al., 2016); (2) are clearly sensitive to tectonics in terms of their channel gradient and width (Kirby & Whipple, 2012; Whittaker, 2012); (3) set the boundary conditions for hillslopes (Densmore et al., 2004; Roda-Boluda et al., 2018; Whipple & Tucker, 2002), and (iv) may even influence fault behaviour and size via erosional unloading (e.g. Andrés-Martínez et al., 2019; Fernández-Blanco et al., 2020; Olive et al., 2014). Rivers also set the local base-level for watersheds and hillslopes, thus not only transporting sediment from source to sink but also influencing the amount and grade of sediment delivered to the system (e.g. Harvey et al., 2022; Pain & Bowler, 1973). Indeed, rivers not only respond to base-level changes, but they can also act as records of changing base-level with signals preserved in channel form and alluvial deposits (e.g. Whittaker, 2012).

However, not all rivers will respond in the same way, and the mode of response is dependent on the processes by which the river adjusts. River systems can be broadly grouped into three types, depending on the nature of the channel and amount of transported material (Fig. 9.1). Bedrock or detachment-limited rivers incise through and expose underlying bedrock in channels; sediment load is variable, typically coarse, and the ability of the river to respond to external forcing is dependent on the ability of the

Geomorphic Impacts of Earthquakes
DOI: https://doi.org/10.1016/B978-0-443-40375-0.00010-4

Figure 9.1 (A) Panorama of a large earthquake-triggered landslide from the 2023 Karamanmaraş earthquakes (Türkiye) and resultant dam causing significant fluvial modification, including the

(continued)

river to erode the channel. Thus these rivers typically respond over long time scales and have stable channel forms, especially in regions dominated by strong rock lithologies. By contrast, alluvial or transport-limited rivers have channel beds and banks composed of previously transported alluvial sediments. As this material is generally weak, channel forms are dynamic and can evolve rapidly over short time scales, dependent on the river's ability to transport the bedload sediment, which is generally finer than in bedrock streams. Finally, semicontrolled or intermediate rivers represent the middle ground between the end members of bedrock and alluvial rivers. Along such reaches, a mixture of bedrock and alluvium will be exposed and will exhibit variable stability depending on the underlying channel material.

Rivers will typically exhibit a 'graded' river profile when in dynamic equilibrium or steady state with the landscape (i.e. when the rate of material removed through sediment transport equals the rate of supply through erosion and uplift). Such graded rivers have a smoothly changing longitudinal profile that is steeper in the headwaters than downstream, typically evolving from bedrock to alluvial along the river course. First noted by Hack (1957), this form can be described using a power-law relationship between local channel slope, S, and the upstream drainage area, A:

$$S = k_s A^{-\Theta} \tag{9.1}$$

where Θ is known as the concavity index (Flint, 1974) and k_s is the channel steepness index.

Interestingly, various studies have shown that Θ at steady state generally falls within a narrow range (~0.4 to 0.6; e.g. Duvall et al., 2004; Kirby & Whipple, 2001; Snyder et al., 2000; Wobus et al., 2006) and it is relatively insensitive to changes or differences in rock uplift rates, climate, or lithology. By contrast, k_s does vary with uplift rate (Snyder et al., 2000), potentially making it useful for understanding the underlying tectonics. Despite numerous studies demonstrating empirical scaling relationships between the normalised steepness index

development of a lake upstream of the blockage. Note the large size of the sediment delivered to the channel by the landslide, including numerous boulders. (B) Shallow landslides from the 2015 Gorkha earthquake (Nepal) potentially delivering sediment into the deeply incised river systems. (C) Low-relief landscape incised by steep rivers with slope-break knickzones along the southern margin of the Gediz Graben (Türkiye). (D) Example transport-limited river reach with significant bedload (Gediz Graben, Türkiye). (E) Example of a bedrock river incising basalts with little bedload in this reach (New Zealand). (F) Vertical-step knickpoint/waterfall along the Sparta Fault (Greece), where the footwall is formed of resistant Mesozoic limestones. (G) Wider landscape response of the passage of a knickzone through the landscape, shown by the formation of narrow channels in deep gorges (Samaria Gorge, Crete) (Aktas et al., 2023). *All photographs copyright S.J. Boulton apart from (A) which was created from photos from the Turkish earthquakes EEFIT mission (Aktas, Y. D., So, E., & Johnson, C. (2023). The Turkiye earthquake sequence of February 2023: A longitudinal study report by EEFIT).*

(k_{sn}—where a fixed reference concavity, θ_{ref}, is defined allowing consistent comparison across systems with differing gradients; generally $\theta_{ref} = 0.45$) and erosion rates (Kirby & Whipple, 2001, 2012; Wobus et al., 2006), the ability to explicitly link k_s/k_{sn} to rates of rock uplift or slip rates on individual faults remains elusive.

As a result of this difficulty, transient river profiles (that is, rivers that do not exhibit a steady state concave-up longitudinal profile) have been increasingly the focus of research trying to infer tectonics from topography. Of particular relevance here is the fact that a tectonic perturbation, such as an increase in slip rate along a fault, causes a change in local relative base-level and consequent river profile adjustment (Whipple & Tucker, 2002; Whittaker et al., 2007a). In detachment-limited and intermediate rivers, this adjustment takes the form of a knickpoint or knickzone; transport-limited rivers may also develop localised knickpoints, but these diffuse or decay upstream and are not maintained owing to the low cohesiveness of bank sediments and sediment transport.

The knickzone is a steep channel reach upstream of the local base-level or controlling structure, and represents the part of the river that has responded to the relative fall in base-level that has initiated the transient response of the river (Wobus et al., 2006). The top of the knickzone is the knickpoint, the upstream extent of the wave of incision identified as the transition from relatively low upstream to relatively high downstream channel slopes (Boulton & Whittaker, 2009). A knickpoint formed in this way is known as a slope-break knickpoint owing to the increase in steepness index across the structure (Haviv et al., 2010). Slope-break knickpoints migrate upstream through the river system, where the retreat rate is dependent on the size of the drainage area and the rate of slip on the causative fault (Crosby & Whipple, 2006; Whittaker & Boulton, 2012). By contrast, vertical-step knickpoints, where there is no change in the river steepness above and below the localised steeper reach, are typically fixed (e.g. on a lithological contrast that may or may not be structurally significant) with limited or no upstream migration (Haviv et al., 2010; Kirby & Whipple, 2012). Waterfalls, the specific term for a vertical knickpoint, typically fall within this second class of knickpoint, forming where hard rocks overlie soft rocks (Haviv et al., 2010). Therefore some knickpoints can be used to infer information on long-term and short-term fault behaviour. Furthermore, the upstream migration of a knickzone will steepen the adjacent hillsides, resulting in gorge formation and increased rates of landsliding and sediment supply (e.g. Roda-Boluda et al., 2018). Increased sediment may, in turn, affect the ability of the river to erode, resulting in negative feedback through the evolution of the incised reaches.

In this chapter, we firstly explore the short-term response of the fluvial system to coseismic fault offset, drawing upon recent, well-documented events such as the 1999 Chi Chi, 2008 Wenchuan, and 2015 Gorkha earthquakes, which are described in more detail in Chapters 13–15, and the 2016 Kaikoura earthquake. Those events highlight recent advances in understanding the channel response to localised uplift, and the complex landscape response that results in variable amounts of sediment being delivered to the fluvial system. Subsequently, the long-term effects of active tectonics on bedrock rivers are described, demonstrating that large-scale geomorphology reflects the time-averaged effect of multiple earthquakes but that increasingly this can be used to determine the rates and evolution of active fault systems even in the absence of other data. Finally, we use this review to highlight some knowledge gaps and research questions.

9.2 Short time scale response to faulting

9.2.1 Geomorphic response to coseismic fault offset and uplift on river channels

A coseismic displacement of sufficient magnitude along an active fault has the potential to generate knickpoints or waterfalls (e.g. Yanites et al., 2010a) along a river, as well as causing significant upstream and/or downstream flooding owing to changes in channel morphology (McEwan et al., 2023; Quigley & Duffy, 2020). Yet, there are relatively few well-documented examples of the fluvial response to individual earthquakes, and linking individual earthquakes, or earthquake clusters, to knickpoints remains an outstanding scientific challenge.

This lack of documentary evidence likely arises for a couple of reasons. Firstly, the magnitude of the causative event is a key criterion. Offsets associated with lower-magnitude earthquakes (i.e. $M_w < 5$) are unlikely to be easily detectable owing to the lack of surface rupture or the generation of only small vertical offsets across the fault. This problem is likely to be a major issue with normal faults, which, although they can produce significant vertical displacements, generally only have coseismic displacements of a few decimetres unless they are part of a wider earthquake cluster (c.f. Boulton & Stewart, 2013), and also with strike-slip faults, which typically have minimal vertical offset. In addition, field (Ewiak et al., 2015) and modelling (Steer et al., 2019) studies suggest that only large-magnitude events have the potential to form knickpoints along channels as smaller events get 'censored' (not formed or not preserved), potentially as a result of aseismic slip at shallow depths (Steer et al., 2019) or owing to rapid postearthquake erosion within channels (Ewiak et al., 2015). Secondly, a number of studies have proposed that knickpoints are lacking owing to burial of active faults preventing exposure of an identifiable coseismic offset (DiBiase et al., 2015; Finnegan & Balco, 2013; Malatesta & Lamb, 2017; Rothman et al., 2023). This process, termed 'the fault-burial mechanism' by Malatesta and Lamb (2017), is where an active fault exposed at a range front is immediately buried owing to aggradation of alluvial fans, raising the catchment's relative base-level and further promoting sedimentation. Only if the fan is subsequently incised, for example, through climate change or autogenic processes, can a waterfall or knickpoint be exposed at the fault. In this case, a knickpoint will form, but it will not represent a coseismic event, nor will the height correlate with coseismic throw (Malatesta & Lamb, 2017). The fault-burial mechanism may also explain autogenic waterfall formation in the absence of external forcing (Rothman et al., 2023). The presence of knickpoints formed through this process has been proposed along rivers crossing active faults in the San Gabriel Mountains (DiBiase et al., 2015) and Death Valley (Malatesta & Lamb, 2017).

In spite of these difficulties, an increasing number of case studies from recent events demonstrate the significant impact that a single coseismic displacement can have on the fluvial geomorphology of a river. Perhaps the best-known example is from the 1999 M_w 7.6 Chi Chi (Taiwan) earthquake (see also Chapter 13). The Chi Chi earthquake generated an 80 km long surface rupture with vertical coseismic offsets of 0.5–8 m (Chen et al., 2001) along the ~ N–S striking Chelungpu thrust fault. This fault intersects a number of west-to-east draining river systems that are rapidly incising the uplifting mountain chain (Yanites et al., 2010a; Yanites et al., 2010b). Multiple alluviated bedrock rivers crossing the surface rupture experienced direct uplift, such as the Daan River with 10 m vertical uplift over ~1 km of river channel (Lee et al., 2005), the Dajia River with ≤9 m uplift, and the Zhuoshui River that was affected by ~2 m of coseismic uplift (Chen et al., 2024). Repeat surveys using aerial photographs, UAVs, and LiDAR have provided unprecedented detail of the subsequent evolution of knickpoints (Chen et al., 2024; Cook, 2017). For example, along the Daan River, Cook et al. (2013, 2014) documented the evolution of the Daan gorge formed as a result of the uplift. This river channel was uplifted by 10 m during the earthquake, generating complex deformation of the channel bed, which included a steep downstream-facing scarp at the base of the western limb of a fold, and an upstream-facing ~6.5 m high scarp ~600 m farther upstream (Cook et al., 2013). This zone of uplift was subsequently eroded to form a 20 m deep gorge. The evolution of the gorge can be broadly separated into distinct periods. Immediately after the earthquake, the uplift caused ponding upstream, which was alleviated through the construction of artificial channels to prevent further flooding. Uplift also resulted in the exposure of the bedrock, as prior to the earthquake the channel bed was completely mantled by coarse-grained sediment. During the period from 2000 to 2004, the channel was mostly stable despite most of the sediment bedload being stripped from the gorge as a result of both natural (typhoon-related heavy rainfall) and anthropogenic (gravel extraction) factors. However, erosion of the knickpoint rapidly increased after 2004 following the passage of several typhoons and the onset of coarse bedload transport, resulting in elevated rates of bedrock erosion through the gorge (Cook et al., 2013). This caused deep incision of the bedrock reach, erosion of the initial scarp, and formation of a clear knickzone (Fig. 9.2) which then experienced ~830 m of headward retreat between 2004 and 2011 (Chen et al., 2024). Chen et al. (2024) calculated the knickpoint retreat rate as 103.00–140.67 m/yr from 2004 to 2011. Google Earth imagery (Fig. 9.2) indicates that from 2006 to 2024, the river

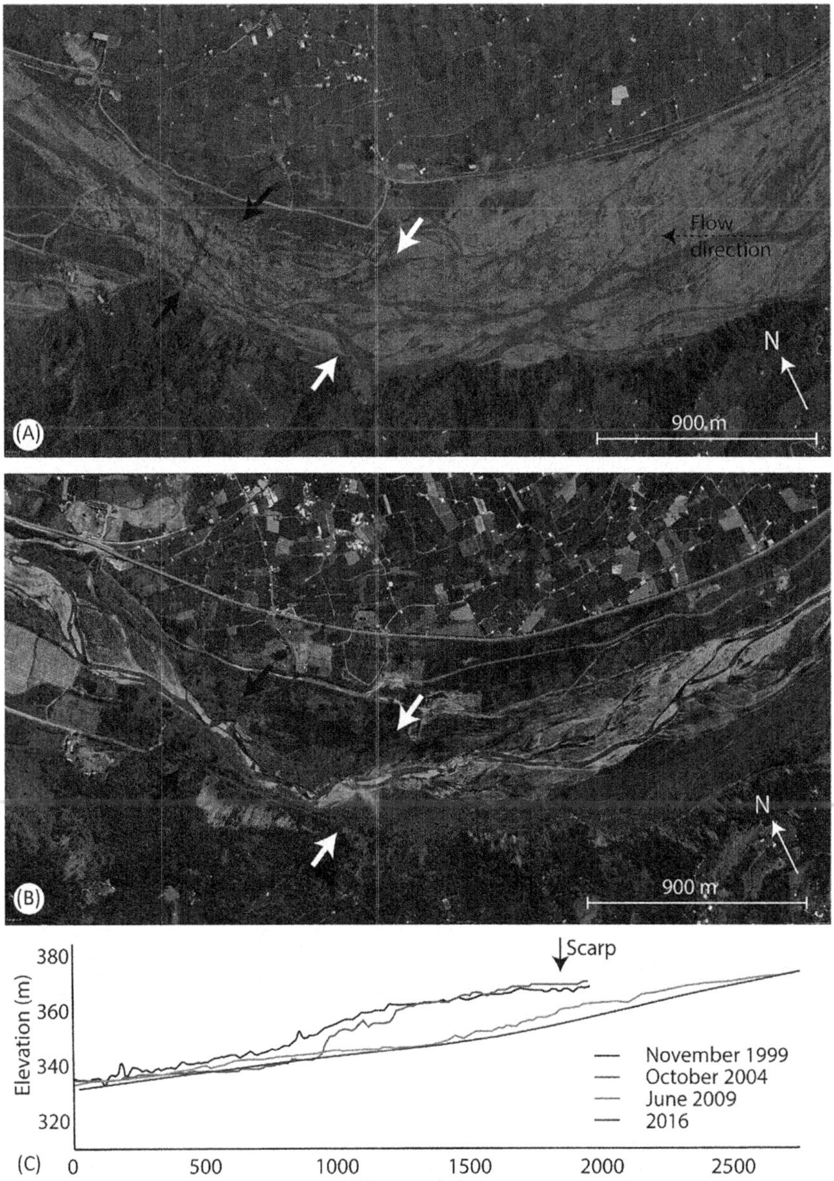

Figure 9.2 Satellite imagery from Google Earth showing the geomorphological change of the Daan Gorge, Taiwan, between (A) 31 March 2006 (©Maxar) and (B) 24 June 2024 (©Airbus). Note the narrowing of the river and the entrenchment of the gorge. Black arrows indicate the position of the anticline hinge, and white arrows indicate the position of the upstream scarp that formed in the 1999 Chi Chi earthquake from Cook et al. (2013). (C) Long profiles of the gorge from 1999 to 2016; profiles for 1999–2009 from Cook et al. (2013), derived from aerial photograph DEMs; profile for 2016 from Chen et al. (2024).

entrenched its position within the floodplain with a single channel through the main region of uplift and a narrower active floodplain upstream, whereas prior to and directly after the earthquake, the river was wider with multiple braided channels. Cook et al. (2014) estimated that it will take ~50 years for the gorge to dissipate and the uplifted topography to be removed.

A second example showing the effects of coseismic uplift on fluvial geomorphology comes from recent events in New Zealand. The complex rupture of the 2016 M_w 7.8 Kaikōura earthquake resulted in significant uplift and numerous earthquake environmental effects. In particular, slip along the 19 km long oblique sinistral reverse Papatea Fault caused displacement of the Waiau Toa/Clarence River channel in four locations in the lower reaches of the river (McEwan et al., 2023). The Papatea Fault rupture resulted in some of the highest recorded displacements in the earthquake, with a maximum throw of 9.5 ± 0.5 m (Langridge et al., 2018). Along the Waiau Toa/Clarence River throws of 7.0–9.0 m in height and sinistral offsets in the order of 4.3–4.4 m were recorded; in addition, several upstream-facing 6.5–7.5 m high fault scarps also formed (McEwan et al., 2023). These field observations are supported by a DEM of difference (DoD) indicating ~7.5–8.5 m of uplift in the footwall of the Papatea Fault (Fig. 9.3). This deformation resulted in upstream flooding and river avulsion (Fig. 9.3), termed a 'fault rupture-induced river avulsion' (FIRA) by McEwan et al. (2023), owing to uplift of the downstream reach in combination with formation of the fault scarp, forming a barrier to flow.

Along a small tributary nearby, the George Stream, Langridge et al. (2023) observed that a gully >2.5 m in depth with a migrating knickpoint had subsequently formed owing to erosion of the scarp. The main river has largely abandoned its previous channel following a period of readjustment of ~3 years (Fig. 9.4C), which included the transient formation of 'Lake Murray' upstream of a second fault barrier near the southern extent of perturbation (Fig. 9.4B). Slightly further downstream, a knickpoint formed where the surface rupture crosses the Wharekiri alluvial fan. A DoD (Fig. 9.3B) constructed from 1 m resolution airborne LiDAR data flown shortly after the earthquake (on or around 16th December 2016) shows that the knickpoint had already migrated upstream by ~400 m and had caused 4.5 m of incision downstream (Fig. 9.5A) at the location of profile A by the time of the data acquisition. Resurvey of the area undertaken between May and September 2022 indicated a further ~2000 m of headward erosion. No further deepening of the river had occurred at profile A, but upstream at profile B, where no erosion had occurred between 2012 and 2016, around 6 m of incision occurred from 2016 to 2022 (Fig. 9.5B). Recent satellite data (April 2024) clearly show that a small knickzone has formed, exposing bedrock in the reach immediately upstream of the initial rupture site (Fig. 9.4D).

Similar flooding and river avulsion have also been described after the 2010 M_w 7.1 Darfield (New Zealand) earthquake along the Hororata River. Flooding was the result

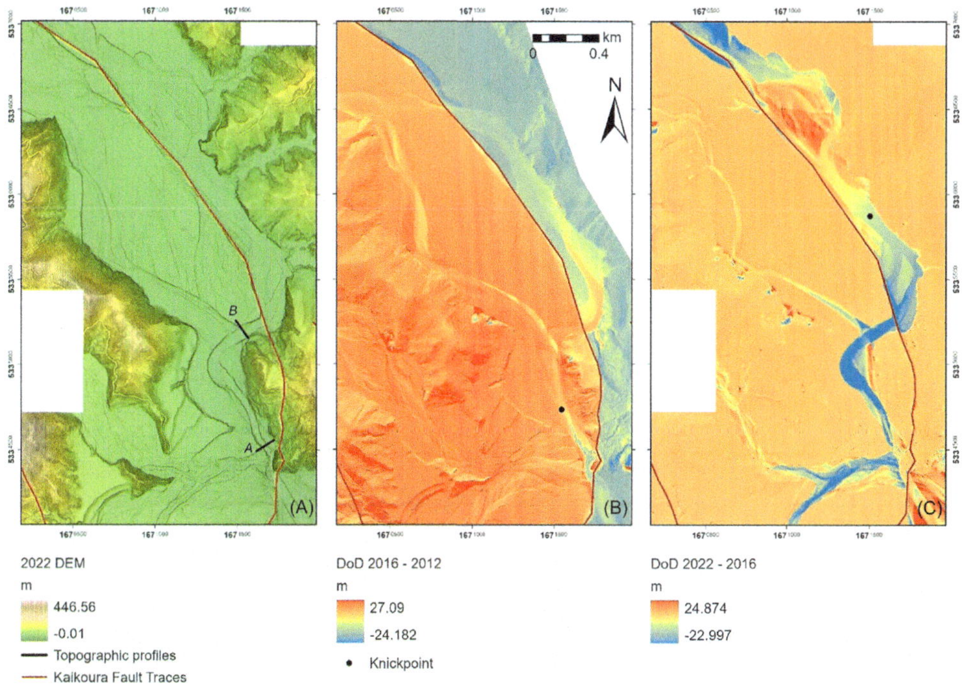

Figure 9.3 Evolution of the lower Waiau Toa/Clarence River. Flow direction is from north to south. (A) The most recent (2022) high-resolution DEM (Kaikōura and Waimakariri LiDAR 1 m DEM, 2022) and the location of the topographic profiles shown in Fig. 9.5. (B) DEM of difference (DoD) showing the change between 2012 and 2016—note the significant uplift on the west side of the fault and areas of river incision and aggradation downstream and upstream, respectively (Canterbury–Kaikōura LiDAR 1 m DEM, 2012; 2016). (C) DoD between 2016 and 2022, highlighting the significant incision along the river since 2016, especially upstream of the fault, indicative of the knickzone (Kaikōura and Waimakariri LiDAR 1 m DEM, 2022) (Langridge et al., 2016). *Data sourced from the LINZ Data Service and licensed by Toitū Te Whenua Land Information New Zealand, for re-use under CC BY 4.0. Fault trace from GNS NZAFD-AF250 database (Langridge, R., Ries, W., Litchfield, N., Villamor, P., Van Dissen, R., Barrell, D. et al. (2016). The New Zealand active faults database. New Zealand Journal of Geology and Geophysics, 59, 86–96. https://doi.org/10.1080/00288306.2015.1112818.).*

of surface deformation along the Greendale Fault, which formed a 1.6 m high fault scarp along an upstream-facing dextral-normal fault (Quigley & Duffy, 2020; Quigley et al., 2016). In this case, flooding was short-lived owing to rapid remediation of the channel by local landowners (Duffy et al., 2013). Interestingly, similar FIRA events have been documented for other recent (e.g. 1980 El Asnam earthquake; Meghraoui et al., 1988) and historical (e.g. 1812 New Madrid earthquake; Stahle et al., 1992) events where lowland rivers are intersected by surface ruptures. In these cases, it seems likely that the geomorphic imprint of these events will be short-lived (years to decades) and, barring the development of sedimentary records (e.g. Chamberlain

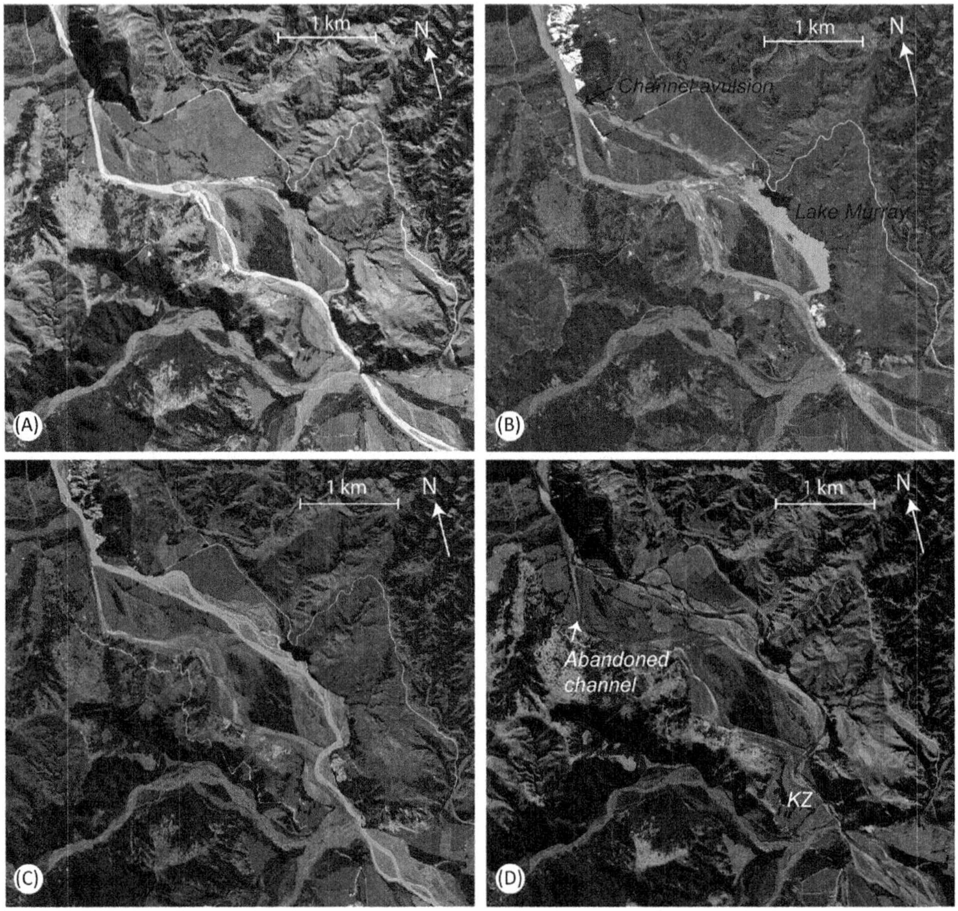

Figure 9.4 Satellite imagery (©Google Earth) showing the evolution of the lower Waiau Toa/Clarence River from (A) preearthquake on 26 March 2016, ©Maxar Technologies, and (B) shortly after the earthquake on 21 November 2016, showing the formation of Lake Murray as an example of a fault rupture-induced river avulsion, ©Maxar Technologies. (C) Three years later, on 13 November 2019, the river had largely abandoned the pre-2016 channel, and the lake had drained ©CNES/Airbus; and (D) almost eight years after the earthquake on 7 April 2024 ©Airbus. KZ=knickzone.

et al., 2024; Langridge et al., 2023; Meghraoui et al., 1988), will not be preserved longer term in the landscape.

Yet despite the theoretical issues and short-lived nature of observed events as described above, a number of recent papers have proposed knickpoints as a type of palaeoseismic record. For example, Wei et al. (2015) documented knickpoints along the 60 km long Huoshan Piedmont Fault, China, an active normal fault with a historic and palaeoseismic record of infrequent yet large (M_w~7–8) earthquakes. They used a

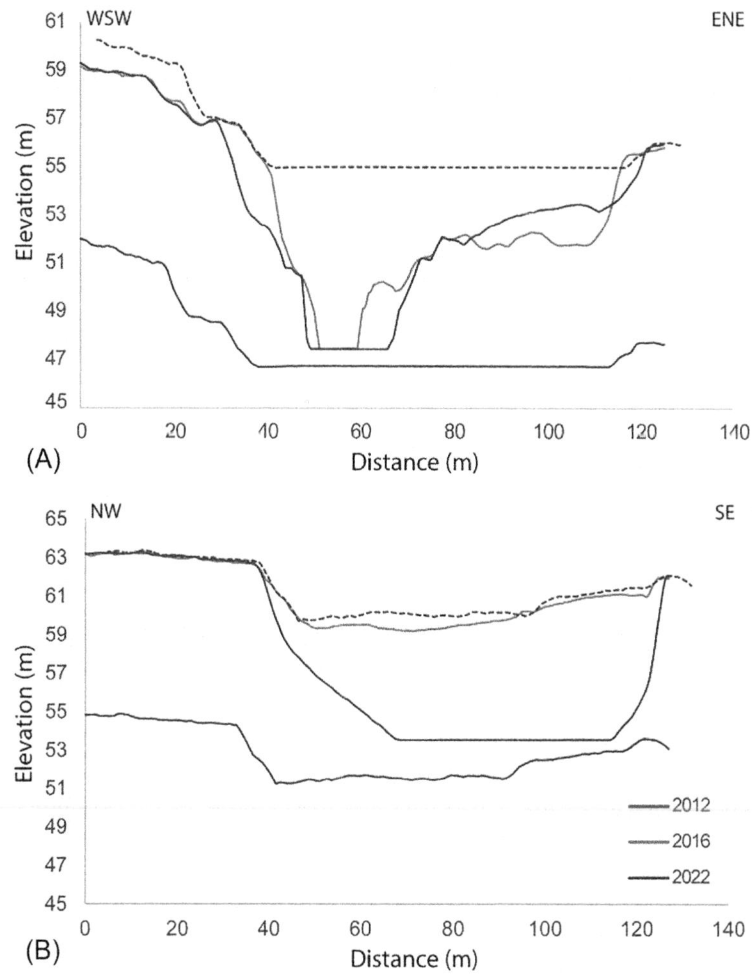

Figure 9.5 Topographic profiles across the Waiau Toa/Clarence River at the locations shown on Fig. 9.3A. Note coseismic uplift during the earthquake means the absolute elevation of the 2012 profile is lower than the subsequent profiles. The dashed line indicates the inferred relative position of the preearthquake (2012) topography to enable comparison to subsequent time slices.

high-resolution (2.5 m) DEM to identify >47 small knickpoints on the order of 2.5–20 m in height and within 500 m upstream of the causative fault in small catchments with drainage areas of <1 km^2. Using the date of the most recent earthquake on the fault and taking the position of the lowermost knickpoint along the rivers (assuming the knickpoint formed in that event), Wei et al. (2015) determined a knickpoint retreat rate for each channel. The rate was then applied to the higher knickpoints, allowing the date of knickpoint formation to be estimated. This approach indicates that three large earthquakes have taken place along the fault with estimated timings and offsets in line with previous observations from palaeoseismic studies. Similarly, Sun et al. (2016) identified six palaeoearthquakes from knickpoints along the Luoyunshan Piedmont Fault, also in China, through correlation with dated terraces and palaeoseismic trenches, and detailed differential GPS surveying of small river channels and gulleys. They identified 28 gulleys (<2 km in length) with 133 small (1–7 m high) knickpoints. In this case knickpoint retreat rates were calculated through correlation with dated terraces. These studies rely on small catchments with limited erosion and sediment deposition, and they assume constant retreat rates through time and limited influence of lithology/rock strength on knickpoint expression, which is known to be a significant control on small-scale knickpoints (i.e. Chilton & Spotila, 2022). As recognised by He & Ma (2015), individual coseismic knickpoints formed on rivers with drainage areas $>1\ km^2$ are unlikely to be preserved in the long-term owing to high erosion rates.

9.2.2 Fluvial response to coseismic sediment input

Mountain river networks are transformed by changes to sediment load entering the river network, primarily from coseismic and postseismic landslide sources (Dadson et al., 2004; Pain & Bowler, 1973; Yanites et al., 2010a). The record of changes to rivers by inputs of coseismic sediment is growing but remains limited to a small number of studies associated with major recent earthquakes. Our record of changes to the suspended sediment record of rivers immediately after earthquakes has demonstrated the role of landslide distributions in the input of sediment to the channel (Wang et al., 2015). However, the record of the rate of landslide sediment input into the fluvial network and subsequent bedload transport remains patchy (Li et al., 2025; Tunnicliffe et al., 2024; Yanites et al., 2010a).

Increased suspended sediment has been recorded after the 1923 Kanto earthquake (Koi et al., 2008), the 1970 Madang earthquake (Pain & Bowler, 1973), the 1999 Chi Chi earthquake (Dadson et al., 2004), and the 2008 Wenchuan earthquake (Wang et al., 2015). The magnitudes and spatial distributions of the increases can be simply related at a catchment scale to the distributions of peak ground accelerations (Dadson et al., 2004; Lin, 2018). For example, suspended sediment concentrations increased 3–7 times following the Wenchuan earthquake, with higher rates in areas with greater landslide concentrations and where not impacted by reservoirs or landslide dams (Wang et al., 2015). Similar increases in storm-related sediment discharges were observed after the Chi Chi earthquake (Dadson et al., 2004; Hovius et al., 2011). However, a lack of gauge data means that many other earthquakes have few observations of increased suspended sediment. However, optical satellite imagery demonstrated increased turbidity in the Neelum River associated with the 2005 Kashmir earthquake (Das et al., 2007), suggesting that there was a short-term increase in suspended sediment load associated with this event. Higher concentrations of suspended sediment most likely derive from landslide material that enters the fluvial system immediately after the earthquake and is eroded by river processes (Hovius et al., 2011). However, the presence of increased postearthquake sediment concentrations is not universal and may depend both on the magnitude of the coseismic landsliding and on subsequent rainfall distribution (Tolorza et al., 2019; see also Chapter 13).

The postearthquake increase in suspended sediment load in rivers decreases rapidly within a decade or less after an earthquake, consistent with decreasing frequency of new landslides (Marc et al., 2015). Sediment concentrations associated with the 1999 Chi Chi earthquake returned to preearthquake levels within 7 years of the earthquake (Hovius et al., 2011). Similar time scales were observed in the Wenchuan earthquake, with the total mass of sediment transported in the suspended load being 19% (+52%, -9%) of the total generated by coseismic landsliding (Wang et al., 2015). Geomorphic analysis of the volumetric connectivity of Wenchuan coseismic landslides to the fluvial network, defined as the volume of landslide material that intersects with a channel network, showed a weak correlation with suspended sediment (Li et al., 2016; West et al., 2014). The lack of a spatial correlation was interpreted as being related to unquantified sources of hillslope sediment entering the channel network. Similar deficits in mass between coseismic landsliding and suspended load have been observed in the Chi Chi (Hovius et al., 2011) and Kanto (Koi et al., 2008) events. However, rapid erosion of landslide sediment associated with the 1970 Madang earthquake led to an estimated 50% of coseismic landslide debris being evacuated within 6 months of the earthquake (Pain & Bowler, 1973). This event differs from other examples in that the landslides associated with the earthquake were mostly shallow, described as 'a "skin"... removed from the entire slope' (Pain & Bowler, 1973), suggesting that a preponderance of finer-grained material was deposited in this event.

By contrast, the bedload transport of earthquake-derived sediment is less well understood than the transport of suspended-load and wash-load sediment and is limited to a few specific observations. After the 1999 Chi Chi earthquake, a detailed analysis of the Peikang River basin demonstrated that landslide-derived sediment caused up to 4 m of aggradation within the channel. The sediment pulse generated by the earthquake smothered exposed bedrock. Evacuation rate estimates of hundreds of years suggested that the sediment pulse may limit bedrock incision across most or all of the interseismic cycle of ~500 years (Yanites et al., 2010a). Where this stored sediment enters uplifted bedrock portions of these river networks, it can then provide tools for rapid bedrock incision and knickpoint retreat (Cook et al., 2013). This effect suggests a complex interaction between the channel network and bedload immediately after an earthquake. Detailed repeat LiDAR observations of the bedload pulse caused by a large coseismic landslide associated with the 2016 Kaikōura earthquake demonstrated the complexity of bedload sediment transport of coseismic debris (Tunnicliffe et al., 2024). This study noted pulsed phases of sediment aggradation and degradation in the channel related to the delivery of sediment by erosion of the landslide material. The result was a "wave" of sediment that translated and diffused downstream of the source landslides. In contrast to these studies, the 2016 Gorkha earthquake has generated no clear bedload pulse in the major river systems of the Melamchi-Indrawati and Bhote Koshi (Graf et al., 2024). Utilising analysis of floodplain width derived from repeated satellite imagery, there was no obvious pulse of sediment derived from coseismic landslide deposits; instead, there was a single debris flow of glacial debris (unrelated to the earthquake) that was responsible for rapid widening and 12 m of aggradation in the Melamchi catchment (Fig. 9.6). A similar lack of aggradation can be seen on the major river systems close to the epicentral areas of the 2008 Wenchuan and 2005 Kashmir earthquakes (Fig. 9.6). The Min Jiang, which drains the epicentral area of the Wenchuan earthquake, has experienced major inputs of sediment from postearthquake debris flows (Fan et al., 2018) and has evidence of fluvial aggradation in tributary catchments (Harvey et al., 2022), yet there has been no major evidence of accumulation in the channel network.

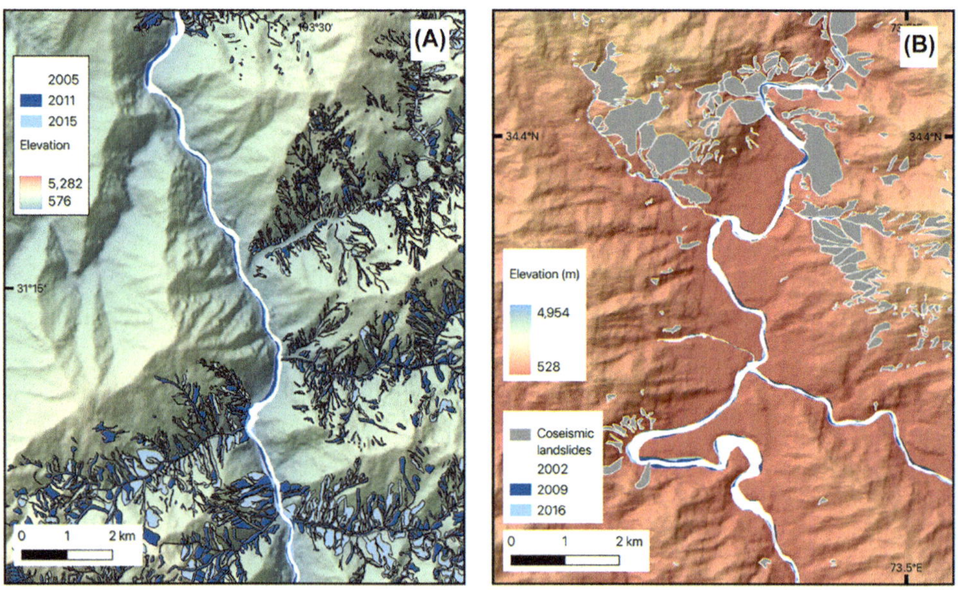

Figure 9.6 Bedload responses of major river systems to earthquake- derived sediment inputs. In each case, a polygon representing the width of the floodplain was calculated using repeat satellite imagery of the major river system before and after each earthquake. (A) Map of the Min Jiang close to the epicentral area of the 2008 Wenchuan earthquake in China. There is no consistent change in the width of the floodplain associated with increased sediment loading despite numerous large debris flow events. (B) Map of the Jhelum and Neelum Rivers close to the epicentre of the 2005 Kashmir earthquake in Pakistan. In none of these examples is a large pulse of floodplain sediment observed within the major river systems. Coseismic landslides were mapped by Basharat et al. (2021). Map created on an SRTM-derived basemap.

It appears that much of this sediment moves rapidly through the river network and is stored in downstream reservoirs (Li et al., 2025). This is likely due to the high stream power of the Min and the delivery of smaller grain sizes that are incorporated into the suspended load of the river. Similarly, the Neelum and Jhelum Rivers that drain the epicentral area of the Kashmir earthquake show neither aggradation nor evidence of a sediment pulse associated with large coseismic landslides directly entering the river (Fig. 9.6B). Similarly, tributary channels show little evidence of aggradation. The observations of bedload transport from these few recent earthquakes highlight the complexity of the relationship between sediment transport from hillslopes to channels and within channels after an earthquake. In these decadal-scale observations, significant aggradation occurs in smaller to medium-scale catchments that can lead to slow rates of evacuation over hundreds to thousands of years, while major rivers show little evidence of the impact of earthquake sediment (Fig. 9.6).

^{10}Be analysis of river sediments deposited by floods after an earthquake has provided further insight into the time scales for reworking of landslide debris (West et al., 2014; Wilkinson et al., 2022). Coseismic landsliding erodes bedrock and soil from slopes, depositing sediment (that is sourced from depth, having low cosmogenic concentration) into the fluvial system. Where preearthquake cosmogenic concentrations are known, the net reduction in concentration after the earthquake reflects the mass of low-concentration sediment sourced from bedrock and deep hillslope sediment fills and transported into the fluvial system. After the Wenchuan earthquake, ^{10}Be concentrations in river sediment decreased in proportion with the areal concentration of coseismic landslides, providing a potential mechanism for tracking the rate of sediment input from earthquakes into the fluvial system (West et al., 2014). Similar work after the Kaikōura earthquake did not record significant postearthquake changes in the ^{10}Be concentrations in river sand, which was attributed to a lack of connection between landslides and the fluvial system (Wilkinson et al., 2022). Overall, the analysis of the sediment load generated by earthquakes points to increased sediment fluxes driven by the spatial distribution, depth, and channel connectivity of the coseismic landslides and modified by the frequency of intense precipitation events.

The differing responses of the bedload and suspended load within postearthquake landscapes reflect the complex chain of geomorphic processes that control how and where sediment can enter the fluvial system (Yanites et al., 2025). Ultimately, the rate at which sediment is exported from orogens following an earthquake depends on the location of sediment deposition within the channel network and the efficacy of hillslope and fluvial processes acting on that sediment. Earthquake-triggered landslides are more commonly triggered close to ridges because of seismic amplification (Densmore & Hovius, 2000). As a result, landslide material can be deposited across the hillslopes and in the fluvial network where it can be remobilised by a combination of hillslope and fluvial processes (Francis et al., 2022).

The sediment budget method applied to recent earthquakes provides insight into how and where earthquake sediment moves through and is exported from catchments in the several decades after a large earthquake (e.g. Francis et al., 2022; Pearce & Watson, 1986). For this review, we have summarised and compiled simple short-term (<50 years) sediment budgets for six historical earthquakes (1929 Murchison earthquake, New Zealand; 1970 Madang earthquake, Papua New Guinea; 1999 Chi Chi earthquake, Taiwan; 2005 Kashmir Earthquake, Pakistan; 2008 Wenchuan earthquake, China; and 2016 Gorkha earthquake, Nepal) (Fig. 9.7). For each of these earthquakes, we compared the calculated mass of sediment generated by earthquake-triggered landslides and by earthquake-affected rainfall-triggered landslides (*sensu* Hovius et al., 2011; Jones et al., 2021) to different geomorphic and hydrologic estimates of sediment export from channels; the length of the record of sediment export sets the time scale for the budget. Where possible, we compiled other evidence for sediment fluxes between hillslopes and the channel network. Except for the 1970 Madang earthquake, the mass of sediment that was exported from the orogen was <10% of the mass generated by earthquake-triggered landslides. All these examples are in areas of high precipitation rates in either a tropical or monsoonal climate and are unlikely to be limited by a lack of water. As discussed above, the Madang earthquake appears to be dominated by shallow, fine-grained landslides that were readily mobilised by fluvial processes.

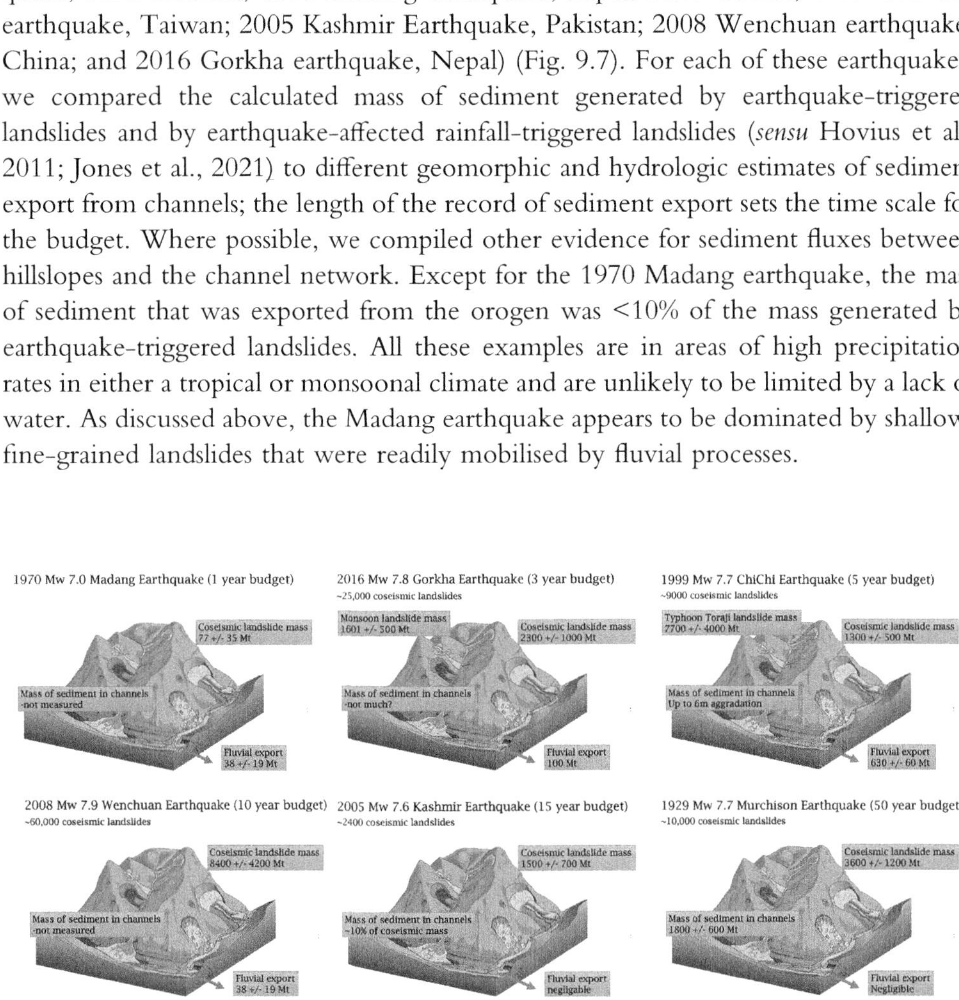

Figure 9.7 Representative sediment budget for major earthquakes compiled from literature sources. The main limitation on the development of sediment budgets for large earthquakes is the measurement of the total fluvial export, which often depends on the existence of gauging stations and is generally limited to the suspended load fraction. Here, the time scale of each budget represents the total sediment exported from the fluvial system as a function of time since the earthquake. The Kashmir and Murchison earthquakes have negligible export as there is a lack of gauged data, but it appears that sediment remains stored in landslides and channel deposits.

The most detailed sediment budget has been calculated for the 2008 Wenchuan earthquake, where high-resolution postearthquake imagery allowed for the observation of specific hillslope processes driving sediment transport from hillslope to channels and from low-order to higher-order channels (Francis et al., 2022). For this earthquake, sediment was almost entirely generated by earthquake shaking, and debris flows were the major mechanism by which sediment was transported to channels and through the upper parts of the fluvial network. The low sediment transport capacity in the lower-order streams limited the rate of fluvial export of sediment, particularly when compared with debris flows. Hence, despite being fluvially connected, the sediment flux from these lower-order streams was dictated by the frequency and magnitude of debris flows that traversed them (Francis et al., 2022). Francis et al.'s (2022) estimate of approximately 15% of landslide sediment export to the Min River is consistent with accumulation in the downstream Zipingpu reservoir (Li et al., 2025), suggesting that the additional bedload from coseismic landsliding is transported effectively through the major rivers of the Longmenshan. Large changes to the fluvial network were prevalent in the first 5 years after the Chi Chi earthquake, which was a relatively small coseismic landslide event (1300 ± 500 Mt; Marc et al., 2015), but was followed by a large monsoonal landsliding event that has been attributed to seismic preconditioning of the slopes (estimated at 5000–13,000 Mt by Yanites et al., 2010b). The combination of coseismic landsliding and landslides associated with Typhoon Toraji caused aggradation of up to 6 m in the channel network, with an estimated evacuation time of ~500 years, close to the return period of earthquakes with the same magnitude as Chi Chi (Yanites et al., 2011; Yanites et al., 2010a). The 5-year postearthquake record of suspended sediment from catchments draining the landslide-affected areas corresponded to a loss of 660 ± 63 Mt of sediment (Hovius et al., 2011). No corresponding estimate of bedload export was made, although Yanites et al., (2010a) estimated bedload as approximately 10% of the suspended load and within the uncertainty bounds of that estimate. The Wenchuan and Chi Chi earthquakes represent the largest coseismic and postseismic landslide events for which we have detailed postearthquake records. Each postearthquake landscape was subject to a different set of cascading processes, with Wenchuan dominated by debris flows, whereas Chi Chi postseismic landslides entered the channel network and were largely transported by fluvial processes. In both cases, there was a significant increase in the frequency and magnitude of geomorphic events, and sediment fluxes increased rapidly immediately after the earthquake before returning to preearthquake rates within a decade. Despite this, the total loss of sediment (both suspended load and bedload) from the orogen was less than 15% of the total landslide mass generated by the earthquake event within a decade of the earthquake. How, where, and at what rate the remaining sediment is released remains an open question.

The smaller coseismic landslide events associated with the Gorkha, Kashmir, and Murchison earthquakes have had the smallest impacts on the fluvial network. Suspended sediment yield after the Gorkha earthquake saw a small increase in 2016, but largely remained at preearthquake rates in 2015 and 2017 (Märki et al., 2021). A small debris flow response was recorded (Kincey et al., 2021), but this caused little change in the width of the floodplains, an indicator of the volume of sediment stored in large rivers (Graf et al., 2024). Similarly, despite an increase in turbidity in the Neelum River after the Kashmir earthquake (Das et al., 2007), little postearthquake activity has been recorded in coseismic landslides (Khattak et al., 2010) or within the channel network (Fig. 9.7). Finally, a sediment budget of the Murchison earthquake demonstrated that postearthquake modification of coseismic landslides was minimal and that much of the earthquake debris remained within the landscape on hillslopes and preserved in landslide dams (Pearce & Watson, 1986). All of these events have had small or negligible effects on the channel network, suggesting that the volume of an earthquake-triggered landslide event has a significant effect on the geomorphic response to that event (Wang et al., 2015). Smaller events still generate sediment, but in each of these examples, the sediment has remained either on hillslopes due to inefficient hillslope sediment transport processes or preserved within the upper parts of the channel network as landslide or landslide dam material. The result is a patchwork of thicker and thinner sediment along hillslopes where this landslide sediment is preserved. These thicker patches of sediment may be a form of sediment 'bomb' (Cook et al., 2021) and contribute to enhanced rates of large debris flows in these catchments when a sufficiently intense rain event occurs.

The grain size of the landslide material delivered to the channel network may also affect both the nature of the postearthquake geomorphic cascade and preservation of sediment within the channel network. The high rates of export from the Madang earthquake may be related to the shallow nature of landslides associated with that event that delivered relatively fine material to the channel network (Pain & Bowler, 1973). The grain size distribution of the landslide material within the channel network of each of these earthquakes affects the ability of the channel network to export that material. In the Wenchuan earthquake, debris flows were more efficient at eroding the range of grain sizes deposited in the channel network than fluvial processes (Harvey et al., 2024).

9.3 Long time scale fluvial response to cumulative events

The temporal and spatial evolution of active faulting over longer time scales ($>10^4$ years) exerts a fundamental control on topography, surface processes and sediment routing in continental settings (e.g. Allen, 2017; Cowie et al., 2006; Gawthorpe & Leeder, 2000; Pechlivanidou et al., 2018; Watkins et al., 2020; Whittaker & Walker, 2015). Relative uplift drives the development of topography and relief until an erosional steady state is achieved, which variously reflects the intensity of surface processes, rock mass strength, and the size, spacing, and slip rate of the faults (e.g. Cowie et al., 2006; Densmore et al., 2004, 2005, 2007; Roda-Boluda & Whittaker, 2017; Zondervan,

Stokes et al., 2020; Zondervan, Whittaker et al., 2020). In this section, we focus on river response to normal and thrust faulting over longer time scales than considered above, as these two fault types can lead to significant vertical motion of the Earth's surface (by >1 km) over million-year time scales (Allen & Allen, 2013).

It is crucial to stress that whenever we think of the longer-term erosional response of rivers to faulting, we are, by definition, time-integrating tectonic displacement and fluvial activity across multiple earthquake cycles and events (e.g. Cowie et al., 2006). Understanding these responses is therefore challenging because details of, for example, multiple river discharge or storm events, earthquake time series, sediment supply changes, or indeed detailed channel geometry adjustments are mostly unknowable over time scales that may exceed 10^6 years (Whittaker, 2012). This means that scaling up from short-term observations and approaches to longer time scales is not easy. So while fluvial landscapes can act as a tectonic tape recorder (e.g. Mudd et al., 2022; Pechlivanidou et al., 2019; Whipple, 2004; Whittaker & Boulton, 2012), decoding this record accurately requires that we can simplify long-term incision models while at the same time retaining sufficient fidelity to capture the key elements of the time-integrated river response to faulting.

Fortunately, much progress has been made in solving this problem in the last 25 years from both numerical modelling and observational perspectives (e.g. Cowie et al., 2006; He et al., 2021; Roda-Boluda et al., 2018; Tucker & Whipple, 2002; Whipple & Tucker, 2002; Whittaker, 2012). The response of rivers to active normal faulting was particularly the focus of early advances in terms of landscape evolution modelling over longer time periods, because such faults are relatively simple to constrain in terms of their along- and across-strike uplift profiles (e.g. Whittaker et al., 2007b). The stream power model has become the most widely used rule to parameterise river incision in these studies (Leonard et al., 2023; Whipple & Tucker, 1999): in its simplest form it expresses river incision rate E as a power-law function of an upstream drainage area A and local slope S:

$$E = KA^m S^n \tag{9.2}$$

where m and n are positive constant exponents, and K represents other factors including bedrock erodibility, discharge-drainage area scaling, and potentially sediment supply effects (Tucker & Whipple, 2002; Whipple & Tucker, 2002); these factors are discussed in further detail below. The fundamental appeal of the stream power 'family' of erosion rules for time-integrated landscape evolution studies mainly lies in the fact that the primary inputs, slope and channel area, are easy to obtain from DEMs or to model, without requiring detailed knowledge of incisional dynamics or climate in the past. If erosion rate balances a fault-driven uplift rate, U (i.e. at a topographic steady state), then the effects of tectonics are translated directly into parameter E. Consequently, Eq. 9.2, as well as numerous studies from field and modelling perspectives, show that channel steepness k_s, is expected to scale with fault-driven uplift rates, such as measures of footwall-component throw rate in a normal fault setting (e.g. Attal et al., 2008; Roda-Boluda & Whittaker, 2017; Roda-Boluda et al., 2018). That essentially means that when time-integrated fault slip rates are larger, channels crossing such faults are steeper at the same drainage area, all else such as rock type being equal (Fig. 9.8A,B).

A key parameter is n, the slope exponent in Eq. 9.2 (e.g. Royden & Taylor Perron, 2013). A classic 'unit stream power model' implies $n = 1$ and $m = 0.5$, allowing erosion rates to be proportional to channel gradient upstream of active faults. This simple model is physically tractable (for instance, we can relate measures such as energy expenditure on the bed, or basal shear stress, directly to channel gradient), while other authors have argued for different values of n based on global compilations which include areas of active faulting (e.g. Harel et al., 2016). As an extreme example for rivers crossing active normal faults in the Gulf of Corinth (Greece), Gallen and Fernández-Blanco (2021) in a DEM-based study suggested that a best-fit value of n may be up to 7 for four eastern catchments by inverting river long profiles. The authors recognised that this high n might result from their best-fit model being biased towards the downstream part of the profiles. In contrast, for the same region, Pechlivanidou et al. (2019) employed $n = 1$ in their *pyBadlands* model to successfully reproduce the sediment volume archived in the Corinth Basin when spatially variable erodibilities were used. These studies show that determining the parameters (or indeed the applicability) of the stream power erosion rule in areas of active faulting from modelling or DEM-based approaches alone is not easy, and may complicate the use of the channel steepness index in order to determine rock uplift rates. This is important because, when $n > 1$, the landscape system displays shock-wave behaviour (e.g. Roydon & Taylor Perron, 2013) and information about active faulting may be lost, as we discuss later.

While it is important to stress that channels crossing active faults, whether reverse or normal, may have reached topographic steady state, it is the transient response of rivers to a change in tectonic boundary conditions that has been the mainstay of research into long-term fluvial response to active faulting in the last 20 years (Hopkins & Dawers, 2015; Kent et al., 2021; Tucker & Whipple, 2002; Whipple & Tucker, 2002; Whittaker et al., 2007b; Whittaker, 2012; Zhou et al., 2024; Zondervan, Stokes et al., 2020; Zondervan, Whittaker et al., 2020). Both theory and numerical modelling have demonstrated that for channels incising bedrock and crossing and active normal faults, and whose erosional dynamics can be approximated by this stream power family of incision laws, knickpoints will migrate upstream in footwall catchments in response to the initiation of faulting or a change in relative base-level, such as a change in fault slip rate (Attal et al., 2008; Gasparini et al., 2006; He et al., 2021; Tucker & Whipple, 2002; Whipple & Tucker, 2002; Wobus et al., 2006) (Fig. 9.9A).

While many numerical modelling studies have looked at the broad generation of knickpoints, surprisingly few have directly and specifically addressed realistic

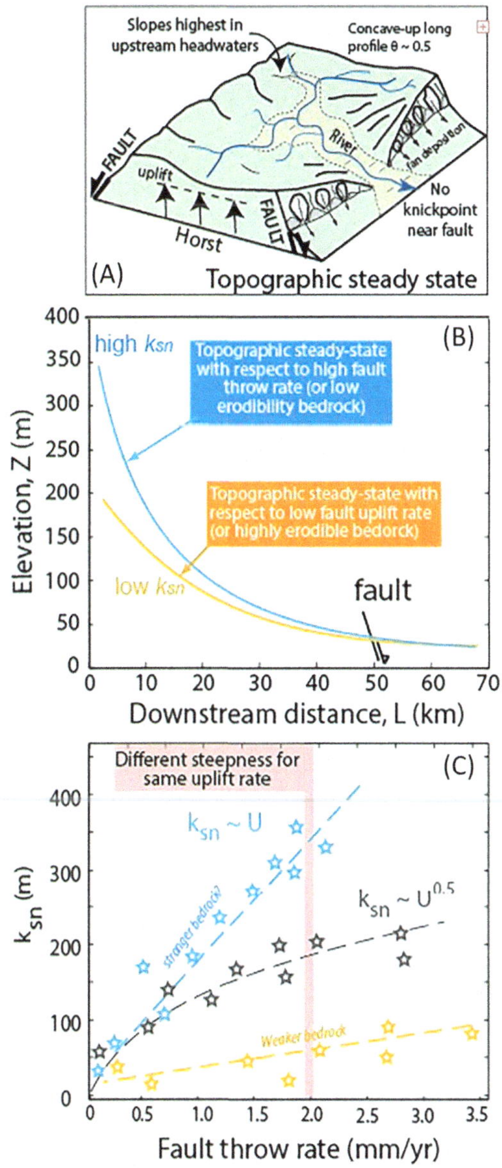

Figure 9.8 (A) Schematic diagram of a river catchment eroding a normal fault horst block at topographic steady state. (B) Example of a concave-up river long profile for high fault throw rates (blue) and low fault throw rates (orange), leading to high and low channel steepness indices, respectively. (C) Channel steepness index and how it may vary as a function of fault throw rate for a linear scaling between k_{sn} and uplift rate/erosion rate (blue line), consistent with $n = 1$ in a stream power rule and for nonlinear scaling (black line), which would be consistent with $n \neq 1$. Note that reducing bedrock erodibility also affects channel steepness (orange line).

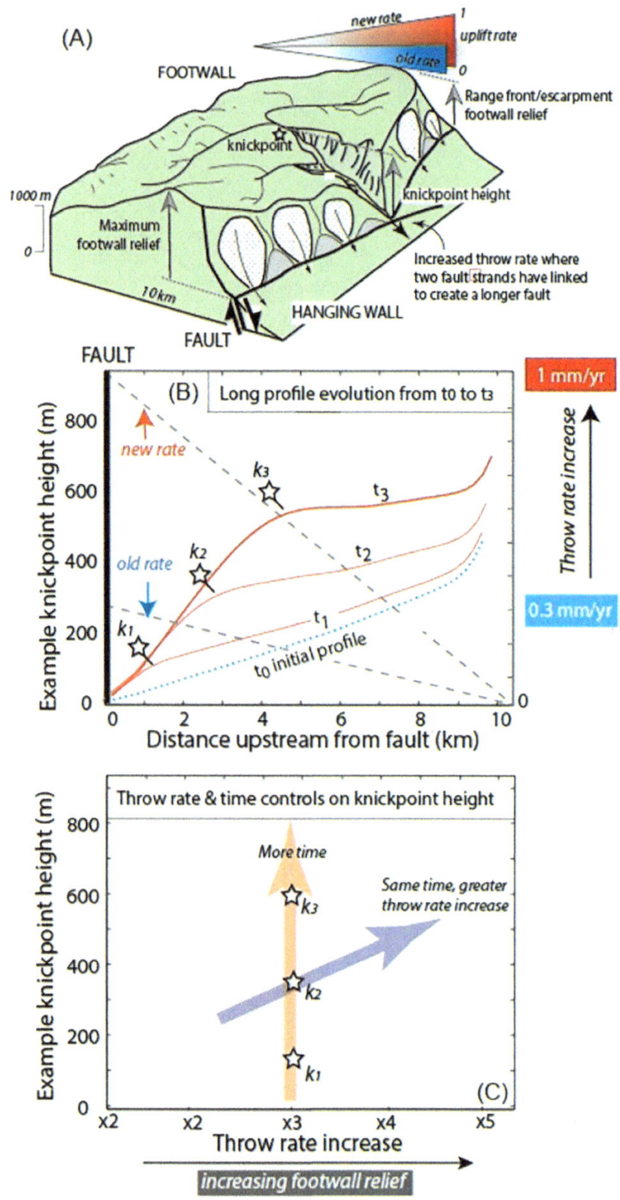

Figure 9.9 (A) Schematic diagram of a river catchment eroding a normal fault block and undergoing a transient response to a change in fault throw rate by a factor of three. The knickpoint illustrated (white star) represents the furthest the wave of incision has propagated into the footwall block. (B) Example of river long profile evolution over time for the scenario depicted in (A). At time t_0, the river has a concave-up long profile and was in a topographic steady state. Following the change in fault slip rate, a knickpoint (star) propagates upstream and grows in elevation (t_1-t_3). (C) Knickpoint height as a function of fault throw rate increase. The elevation of the knickpoint, measured from where the river crosses the fault, grows with time (orange arrow). For any single time since a fault throw rate increase, the elevation of the knickpoint would also depend on the degree of fault acceleration (purple arrow).

configurations of active faulting. For instance, Attal et al. (2008) modelled river response of an individual catchment to active normal faulting in Italy, incorporating fault slip rate and channel geometry. Their results showed that the elevation and height of knickpoints above the fault scaled directly with the magnitude of the fault slip rate increase and the time elapsed since the change of slip rate (Fig. 9.9B,C). Cowie et al. (2006), at the scale of an entire normal fault array, coupled a physically-driven fault growth and interaction model to a landscape evolution model to explore the coupling between tectonics and river behaviour at a regional scale. Their results showed how active faulting controlled both the location of the main drainage divide and the size of the footwall catchments that develop along an evolving basin-bounding normal fault. They also illustrated how fluvial incision varied markedly upstream of active structures and showed how modelled fault growth directly influenced the incision rate of rivers that drained into the neighbouring hanging-wall basin and thus their likelihood of being 'defeated' by active faulting. More recently, He et al. (2021) modelled landscape response to the linkage of two normal faults. The results demonstrated how the fluvial landscape responded in time and space to the changes in uplift patterns that accompanied fault linkage. For instance, both the steepest topography and the most elevated erosion rates migrated from the centres of the former fault segments to the linkage zone, while the shape of the drainage divide also evolved dynamically.

Ultimately, observational field studies in areas of active faulting have allowed us to test and validate these model predictions and have enabled us to constrain plausible time scales for river responses to active faulting. Working in the central Apennines of Italy, Whittaker et al. (2007a), Whittaker et al. (2007b), and Whittaker et al. (2008) studied active normal faults that had undergone a well-constrained slip rate acceleration at c.0.8 million years ago, and were among the first to show that rivers crossing these structures had knickpoints that could be directly related to this change in slip rate. In contrast, rivers crossing active faults that had maintained a constant slip rate for c.3 million years appeared to have concave-up, steady state river long profiles, suggesting that any knickpoints generated 3 million years ago had already propagated through the catchments. Consequently, these authors concluded that the landscape response time to active faulting in hard carbonate bedrock was long—between 1 and 3 million years for that part of Italy. This study also argued that a field-derived discharge-based unit stream power model ($n = 1$) broadly constrained the distribution of incision upstream of the fault as long as channel width was expressly considered (see also below). These results are important because they illustrate that the time-integrated effects of multiple earthquakes over hundreds of thousands of years could, to first order, resemble numerical model predictions of knickpoint migration upstream of active faults in response to constant uplift rates (e.g. Attal et al., 2008; Fig. 9.9). These authors also validated the prediction that knickpoint elevations, measured from above the point where the fault intersected the river, scaled both with fault throw rate and with the time elapsed since the tectonic perturbation. Similar observations have subsequently been made for rivers crossing active normal faults around the world, including fault arrays in Turkey (Boulton & Whittaker, 2009; Kent et al., 2016; Kent et al., 2017), Greece (Çal et al., 2024; Zondervan, Stokes et al., 2020; Zondervan, Whittaker et al., 2020; Zhou et al., 2024), China (e.g. He et al., 2021), the United States (Hopkins & Dawers, 2015), and the East African rift (Wedmore et al., 2022) amongst many others. Consequently, the spatial distribution, elevation, and characteristics of knickpoints are now recognised as critical markers of active faulting in modern landscapes (e.g. Gallen & Wegmann, 2017; Hopkins & Dawers, 2015; Roda-Boluda et al., 2019) as well as in preserved ancient settings, such as incised palaeo-landscapes imaged in seismic reflection surveys of fault blocks in the Norwegian North Sea (e.g. Wrona et al., 2023). Examples of river long profiles showing knickpoints upstream of well-documented active normal faults in Calabria, Italy, and the Sperchios graben of Greece are shown in Fig. 9.10A and B, respectively.

Knickpoints may record important information about the evolution of fault segments that is not available from time-averaged stratigraphic data sets, such as the temporal evolution of fault interaction. In particular, a number of field studies demonstrate how knickpoints upstream of faults serve as detailed markers of successive fault growth and linkage phases (Roda-Boluda & Whittaker, 2017; Whittaker & Walker, 2015; Whittaker et al., 2008; Zhou et al., 2024) particularly where multiple knickpoints can be seen in river long profiles along the strike of faults. For instance, knickpoints located on rivers draining across former fault segment boundaries can be used to constrain the magnitude and timing of slip rate enhancement during fault array interaction (Boulton & Whittaker, 2009; Kent et al., 2017; Roda-Boluda & Whittaker, 2017; Whittaker & Walker, 2015) (Figs. 9.10, 9.11).

As active faults link and lengthen, strain localisation can lead to the development of larger, higher slip-rate structures by segment linkage, while in contrast 'unsuccessful' fault segments located across strike may become inactive assuming regional strain rates do not change (Cowie & Roberts, 2001; Cowie et al., 1993). In fact, the throw rate enhancement associated with the linkage of fault segments can be estimated from a geometric calculation considering the prelinkage length of the segments and the distance to the nearest tip of the newly-linked structure (Cowie & Roberts, 2001; Whittaker & Walker, 2015) (Fig. 9.11). At the same time, fault throw rates postlinkage can be reconciled with the height and distribution of knickpoints along the strike in the fault footwall, particularly where multiple knickpoints might exist (e.g. Fig. 9.11B), enabling sophisticated insights into the history of fault growth from active faults (e.g. Roda-Boluda & Whittaker, 2017, 2018). Yet because faulting can lead to drainage reorganisation (Duffy et al., 2015; Geurts et al., 2018), great care must be taken to evaluate this effect when considering landscape responses at a fault array scale. While the above studies mostly focus on normal faults, similar observations can be made in areas of

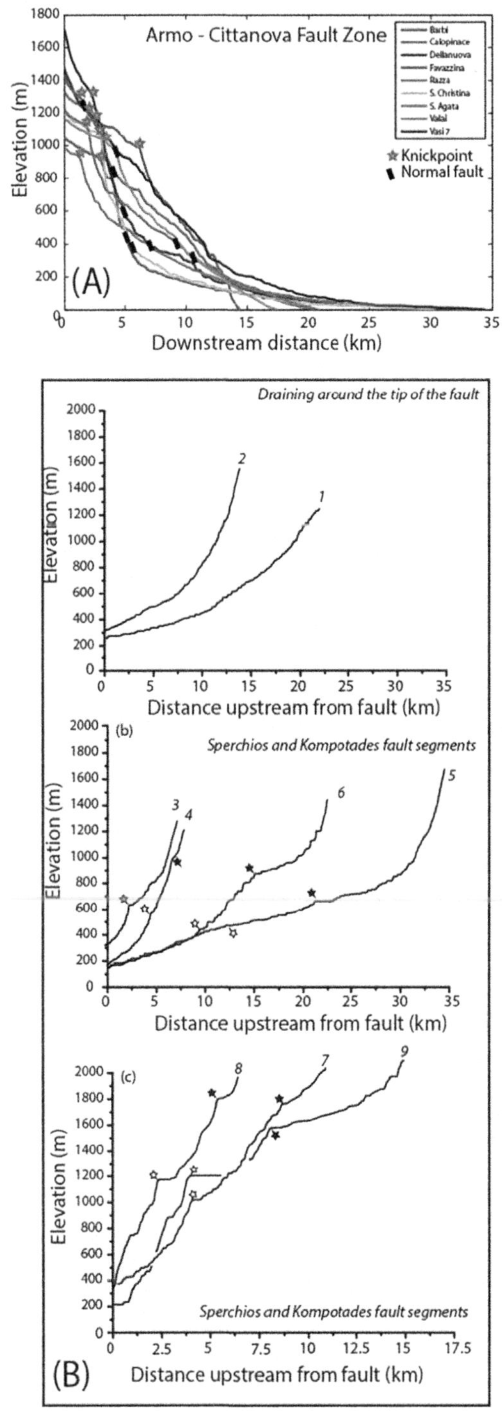

Figure 9.10 Examples of river long profiles crossing active normal faults in Italy and Greece. (A) Selected long profiles along the Cittanova -Armo faults in Calabria, Italy. Knickpoints are highlighted as dots. The rivers are selected from the localities shown in Fig. 9.11A. (B) River long profiles crossing the Sperchios and Kompotades fault segments in the North Gulf of Evia region, Greece. These profiles mostly show two sets of knickpoints (stars on rivers 3 -9, lower panels), while rivers that drain around the tip of the Sperchios fault segment (rivers 1 and 2, top panel) are smoothly concave-up. The location of these fault segments is shown in Fig. 9.11B.

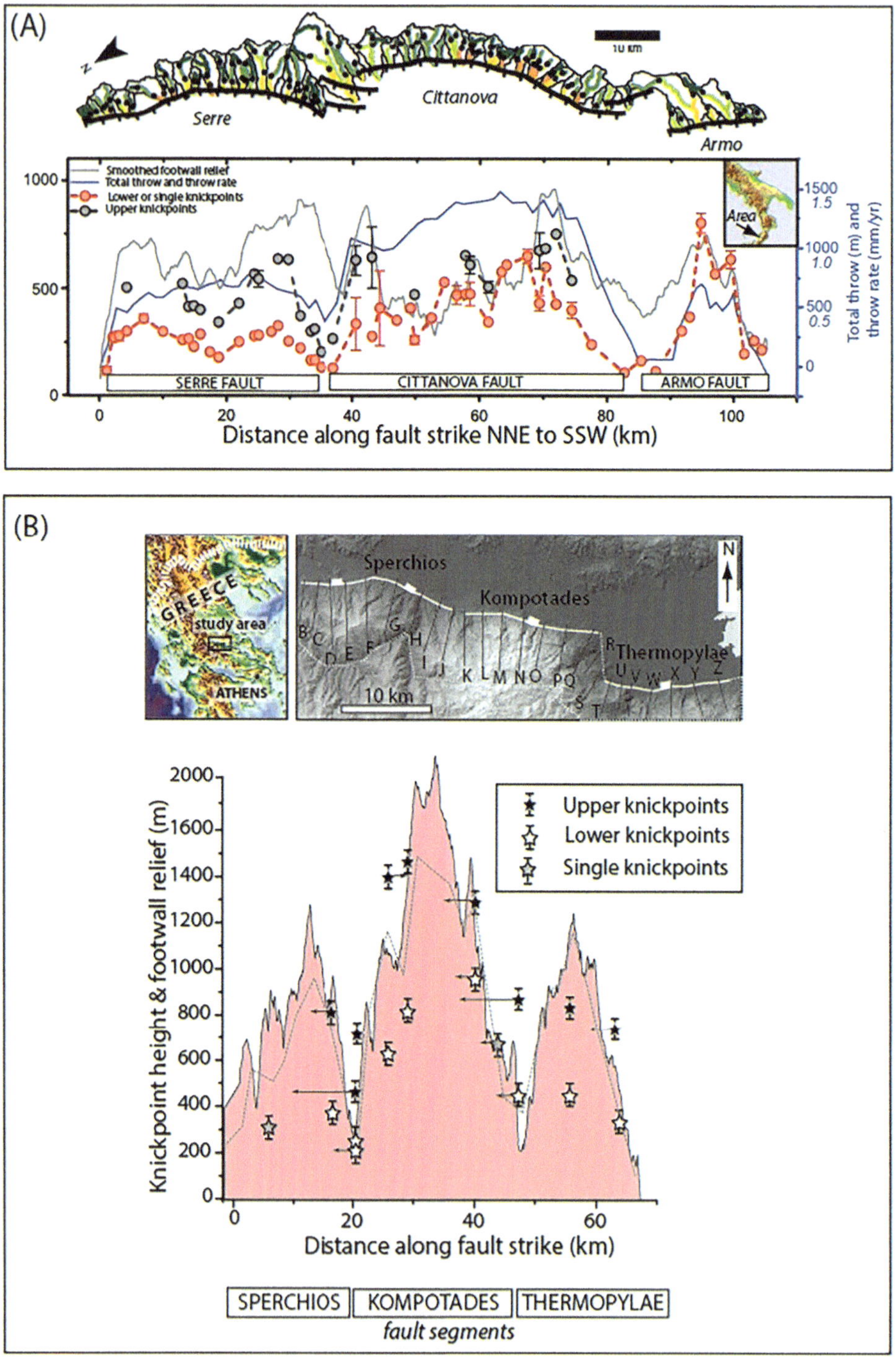

Figure 9.11 Examples of two studies that have used knickpoints to trace the growth and interaction of normal faults (Roda-Boluda & Whittaker, 2017; Whittaker & Walker, 2015). (A) Knickpoints documented

(continued)

active thrusting (e.g. Kirby & Whipple, 2012). For instance, the classic work of Lavé and Avouac (2001) in the Siwalik Hills of the frontal Himalaya clearly showed that rivers increased their gradient and adjusted their planform to increase bed shear stresses in response to fault bend folding, creating prominent knickpoints in the channels. Similar observations have subsequently been made for thrust faults in Taiwan (e.g. Yanites et al., 2010b), the Zagros Mountains of Iran (e.g. Oveisi et al., 2009; Ramsey et al., 2008), Morocco (Boulton et al., 2014), New Zealand (Formento-Trigilio et al., 2003), and the contractional domain of the Italian Apennines (e.g. Wegmann & Pazzaglia, 2009), amongst many others. In such studies, deformed fill or strath terraces have proven to be particularly useful markers for tracking river response to thrusting (Holbrook & Schumm, 1999; Zondervan et al., 2022), particularly given that fault bend or fault propagation folding associated with thrusting leads to more complex deformation and uplift patterns than are typically observed with normal faults (e.g. Miller et al., 2007).

It is implicit in the stream power erosion rule that the knickpoint migration celerity, V, can be expressed as:

$$V = \Psi A^m S^{n-1} \tag{9.3}$$

where Ψ represents other controls on knickpoint retreat rate, such as sediment flux. It is this velocity which ultimately controls the tempo of landscape evolution and the time scale over which fault-bounded catchments can record tectonics. In a theoretical sense, the position of these knickpoints upstream of the fault record the farthest a wave of fault-related erosion has propagated upstream (e.g. Kent et al., 2017; Whittaker & Boulton, 2012). For a unit stream power model ($m = 0.5$, $n = 1$), slope does not matter, and landscape response times fundamentally depend on drainage area. These predictions

upstream of the Serre, Cittanova and Armo faults in Calabria, Italy. In map view (upper panel), green colours document channel segments with a lower steepness index and yellow and red colours document channels with a higher steepness index, while black dots represent knickpoints. Adapted from Roda-Boluda and Whittaker, 2017. When these knickpoints are plotted along strike (lower panel), an upper set (grey) traces the footwall relief of the faults, implying these knickpoints relate to fault initiation at c.1 million years ago. A lower set (pink), which is also found at segment boundaries, is interpreted to represent an increase in fault slip rate due to later linkage of the fault segments. Only one set of knickpoints is seen on the Armo fault, suggesting that this structure has not changed its slip rate. (B) In the Sperchios Graben, central Greece (location shown in upper panel), an upper set of knickpoints on rivers crossing the active faults (black) broadly reflects the footwall relief. A lower set (white) is interpreted to represent the interaction of the three fault segments, which are clearly expressed in the footwall relief profile. The fault interaction event occurred by 0.9 million years ago at the latest. *Adapted from Whittaker, A.C., & Walker, A.S. (2015). Geomorphic constraints on fault throw rates and linkage times: Examples from the Northern Gulf of Evia, Greece. Journal of Geophysical Research: Earth Surface, 120, 137–158. https://doi.org/10.1002/2014jf003318.*

have been validated for moderate-sized rivers crossing active faults in Greece, Italy and Turkey (e.g. Whittaker & Boulton, 2012; Whittaker & Walker, 2015), with drainage areas of 10s to 100s of square kilometres, and where knickpoints have been shown to migrate at millimetres to centimetres a year. Bedrock lithology, in terms of erodibility, is encompassed in the parameter K (Eq. 9.2), and consequently, when bedrock erodibility is higher, knickpoint migration rates are faster (Kent et al., 2021; Whittaker & Boulton, 2012; Zhou et al., 2024; Zondervan, Whittaker et al., 2020). Values of K for these studies are typically of the order of 10^{-7} to 10^{-5}/year, implying landscape-scale response times to active normal faulting from 10^5 to 10^7 years. Significantly, for footwall catchments that obey the unit stream power rule and which also obey Hack scaling, response times to faulting are scale invariant: this is because faster knickpoint celerities envisaged by Eq. 9.3 are neatly balanced by longer trunk stream lengths (c.f. Whittaker, 2012). Nevertheless, there is evidence to suggest that $n \neq 1$ for rivers crossing faults. For instance, Whittaker and Boulton (2012) showed that apparent bedrock erodibility K appeared to scale with fault slip rate, meaning that rivers crossing active faults in Greece and Italy had proportionally faster landscape response times where rock uplift rates are higher, which the authors tentatively related to dynamic channel width effects (see below). Yet if $n > 1$, then 'consuming knickpoints' can be generated (Roydon & Taylor Perron, 2013). In this case, driven by faster migration velocities on steeper slopes, fault-driven knickpoints can 'catch up' with each other. Where this situation occurs, information about past active faulting will be lost from the fluvial archive.

Moreover, two factors important to constraining river responses to active faulting are absent from simple versions of a stream power erosion rule— namely, channel hydraulic geometry and long-term sediment supply. Field studies as well as numerical models show that both are important. Field data show that dynamic channel narrowing can be a key mechanism of knickpoint retreat and fluvial response to active faulting, particularly for bedrock rivers (Attal et al., 2008; Hopkins & Dawers, 2015; Kent et al., 2021; Turowski et al., 2009; Yanites, 2018;Zondervan, Stokes et al., 2020; Zondervan, Whittaker et al., 2020). Several studies have illustrated that rivers upstream of active faults do not display conventional hydraulic scaling (in other words, their width cannot be described as a simple power-law function of upstream drainage area; e.g. Duvall et al., 2004; Whittaker et al., 2007b; Yanites et al., 2010b). Instead, rivers appear to narrow either in conjunction with, or as a response to, steepening, which enables bedrock shear stresses and hence incisional capacity to be enhanced (e.g. Amos & Burbank, 2007). These responses have been linked to the achievement of a channel gradient threshold (e.g. Hopkins & Dawers, 2015) as well as sediment input (Turowksi et al., 2009; Yanites, 2018), and may lessen the sensitivity of channel slope to rock uplift rates (Whittaker et al., 2007a). Unsurprisingly, different formulations of river erosion rules make different predictions of width response to active faulting (e.g. Turowski et al., 2009; Yanites, 2018) and some models and observations indicate that reduced channel widths can be maintained even after the passage of a knickpoint (Kent et al., 2021), while in other cases, widening is the response (Baynes et al., 2022). While no simple erosion rules adequately capture these dynamic hydraulic geometry effects, it remains extremely challenging to incorporate them into fault evolution scenarios over million-year time scales, because we usually do not have constraints on channel geometry changes that have taken place in the past (Yanites, 2018; Zondervan, Whittaker et al., 2020). Such effects do, however, provide one means to explain nonlinear scaling between fault-driven uplift rates and channel steepness in areas of active faulting.

In a similar way, river response to fault growth can be modulated by sediment supply in fault-bounded catchments. Sediment moving through channels has long been hypothesised to modulate incision (e.g. Gasparini et al., 2006, 2007; Sklar & Dietrich, 2006; Whipple & Tucker, 2002). Where sediment supply is optimised to drive bedrock incision, bedload sediment in the channel can act as 'tools' to abrade and detach bedrock, leading to the enhancement of channel erosion rates (e.g. Sklar & Dietrich, 2004; Yanites, 2018); in contrast, too much sediment covers the bedrock and inhibits incision (e.g. Sklar & Dietrich, 2004). Numerical and analogue models clearly show that these effects can significantly modulate how active faulting is expressed in fluvial landscapes (e.g. Cowie et al., 2006; Gasparini et al., 2006), while it is patently clear from the modern studies described in this chapter that faulting and earthquakes fundamentally drive sediment supply from hillslopes to channels. Yet, field studies quantifying the relative effect of sediment supply on channel incision over time periods of 10^4 to 10^6 years have been sparse (c.f. Hobley et al., 2011). One notable exception is the work of Cowie et al. (2008), in which the authors compared four footwall rivers in Italy and Greece crossing active normal faults with similar slip rates to isolate the effect of long-term sediment supply. The study showed clearly how bed shear stresses near the active faults did not increase in proportion to fault throw rate, and thus the results provided evidence for both the 'tools' and 'cover' effects. Moreover, such work showed that channel steepening and knickpoint generation upstream of active faults is dampened in some circumstances where sediment supply plays an important role in governing long-term landscape evolution, and ignoring this effect could lead to an underestimation of fault activity. Similar outcomes were noted by Kent et al. (2021) for active faults in Turkey, although these authors noted that variations in bedrock erodibility could also have explained their results. Cowie et al. (2006) provided field confirmation of the hypothesis that the expression of active faulting in fluvial landscapes differs when sediment supply is a significant proportion of long-term transport capacity, but few studies have attempted to address field-based perspectives of long-term landscape evolution upstream of active faults, and this remains an under-researched area.

Formal quantitative methods to invert transient river long profiles for the faulting history at a fault array scale have now been successfully developed (e.g. Goren et al.,

2014; Quye-Sawyer et al., 2020), although their outputs are of course sensitive to the assumptions incorporated in the inversion scheme. Regardless of these limitations, this type of approach is especially powerful where the syn-tectonic stratigraphic archive of the faulting is not exposed (e.g. Roda-Boluda & Whittaker, 2017). It is also important to consider that the relatively local, river-related, fault-block-scale geomorphological patterns described above are superimposed on the much larger-scale fault network-controlled, 'basin-and-range-style' relief (e.g. Gawthorpe & Leeder, 2000) and consequently, at the scale of a fault array, river response to tectonics will be a combination of multiple faults with different evolutionary histories (Cowie et al., 2006).

9.4 Outstanding research challenges

The discussion in the sections above indicates that there are a number of outstanding research questions on the interactions between active faulting and fluvial erosion in time and space that need to be addressed. We explore four of these themes in greater detail in the next section, highlighting critical knowledge gaps and potential research directions.

9.4.1 What proportion of earthquake-generated sediment is preserved within orogens across earthquake cycles?

The long-term (>10^5 years) dynamic evolution of earthquake-derived sediment in landscapes remains an area of open research. Based on recent earthquake sediment budgets, many studies have inferred that earthquake sediment may be preserved and affect the evolution of orogens across thousands of years (Dai et al., 2021; Li et al., 2017). However, these inferences assume a simple linear or constant rate of postearthquake sediment export. The potential for sediment to be stored across earthquake cycles has been explored in only a small number of studies. Using a 0-D bucket-style model, Francis et al. (2020) demonstrated that the sediment generated by earthquake-triggered landslides is largely eroded from the orogen between large earthquakes; however, when large earthquakes are closely spaced in time, sediment can accumulate in the orogen. When considering channel evolution across multiple seismic cycles, the rate of sediment supply from hillslopes remains a key unknown in understanding the legacy of earthquake sediment in the fluvial system (Croissant et al., 2019). These initial studies have demonstrated that earthquakes can generate significant volumes of sediment, largely on hillslopes and away from the channel network. The total bedrock erosion by earthquake-triggered landslides across multiple earthquake cycles could be up to 50% of the total long-term denudation of orogens (Francis et al., 2020). The location of earthquake-triggered landslide sediment deposition is related to the position of the steepest topography (Jones et al., 2021) and earthquake parameters, particularly the ground acceleration caused by passing seismic waves (Gorum et al., 2011). Hence, sediment generated by earthquakes is not simply related to incision, nor is it consistent with landslide generation via coupled hillslope-channel processes (Egholm et al., 2013). The cascade of geomorphic processes that deliver that sediment from hillslopes to the channel further complicates the rate and location of earthquake-derived sediment flux into channels (Francis et al., 2022). The effects of the sediment generated by earthquakes on the longer-term evolution of channels and landscapes remain a significant open question. While not questioning current models of long-term orogenic evolution *per se*, it is likely that areas with more frequent earthquakes will have complicated catchment evolution histories.

Furthermore, the heavy-tailed nature of earthquake-triggered landslide volume distributions and their spatial properties mean that sediment generated by earthquakes can become isolated from efficient hillslope or fluvial transport. There are countless examples of large earthquake-triggered landslides preserved within orogens (Chen et al., 2022; Korup & Clague, 2009; Korup et al., 2007). These landslides are preserved and affect the local fluvial responses for 10^1–10^4 years (Korup, 2006; Ouimet et al., 2007). After noting that most large rivers that drain active orogens are not choked with sediment, Croissant et al. (2017) demonstrated that channel narrowing increases sediment transport capacity and enhances erosion of the landslide mass. When scaled to multiple landslides, the hydrodynamic feedbacks within the channel are complicated by the frequency of upstream landslide deposits, which affects the rate of removal of downstream landslides (Croissant et al., 2019). Large landslides that dam rivers reduce upstream bed and water surface slopes via sedimentation and create local knickpoints that steepen the channel until the landslide material is eroded. In the steep landscapes of High Mountain Asia, landslide dam deposits of greater than 400 m thickness have been recorded, with dozens of landslide dams adding a major transient effect to river profile evolution in these mountains (Ouimet et al., 2007). Pervasive landslide-dammed lake sediment, possibly of mid-Pleistocene age (Xigeda Formation; Yang et al., 2022), is preserved within steep, rapidly eroding sections of the Middle Yangtze system, suggesting that local disruptions to channel properties may persist over 10^4–10^6 years. Evacuation time scales of this coarse landslide sediment ultimately impinge on rivers' ability to erode bedrock and set the rate of base-level lowering in these orogens. These studies highlight the need for a more detailed and systematic understanding of the impact of landslide dams in the orogen-scale evolution of catchments.

9.4.2 Can knickpoints be used as palaeoseismic indicators?

Robust palaeoseismic catalogues are important for constraining earthquake return periods and the size of events prior to detailed historic accounts, or prior to the instrumental period, for seismic hazard assessments. The potential to link individual knickpoints to earthquakes or earthquake clusters is a tantalising idea and an outstanding challenge. Although the recent earthquakes in Taiwan and New Zealand described above demonstrate that knickpoints do form as a result of significant vertical coseismic offsets, these features are likely to be relatively short-lived, persisting for 10s of years (Cook et al., 2013). Reports of knickpoints correlated to earthquakes along small streams in China (Sun et al., 2016; Wei et al., 2015) are problematic owing to the very small catchment sizes (below the threshold drainage areas where the stream power model will be valid), raising questions as to what is actually being preserved and measured in these cases. Further research on river systems known to have been affected by earthquakes, long-term monitoring, and the identification of new locations where the formation of knickpoints can be directly related to fault offsets are required across different tectonic and climatic settings. Being able to confidently identify small-scale (<10 m) knickpoints within the stochastic variability of the bedrock channel in the form of riffles and steps will be necessary as DEM resolutions increase to avoid erroneous labelling of erosional features as knickpoints, while the use of complementary modelling (e.g. Steer et al., 2019) and correlations across multiple channels would increase the robustness of conclusions.

9.4.3 Independent dating of knickpoint retreat

The rate of knickpoint retreat is generally estimated based upon assumed or known timings of the forcing event (i.e. Berlin & Anderson, 2007; Bishop et al., 2005; Boulton & Whittaker, 2009; Crosby & Whipple, 2006; Kent et al., 2017; Loget & Van Den Driessche, 2009; Whittaker & Boulton, 2012) or by correlating it to deposits (e.g. river terraces) of known date (Finnegan, 2013; Liu et al., 2024). Yet, few studies have directly dated knickpoint retreat. For example, Mackey et al. (2014) used in situ ^{3}He cosmogenic exposure dating to date the rate of waterfall retreat on Kauai, Hawaii, through samples of the upper reaches of the channel bedrock combined with samples from strath terraces, cliff faces and boulders below the knickpoint. A similar approach using in situ ^{10}Be cosmogenic nuclides was taken by Jakica et al. (2011) to date the top of knickpoints along a slowly eroding escarpment in Western Australia. More studies that obtain direct measurements of knickpoint retreat are necessary to calibrate numerical models and test assumptions, as well as to develop robust empirical relationships that can be applied in the absence of datable material or causative events.

9.4.4 Bridging the scales

The final research frontier that we highlight around understanding the fluvial response to coseismic drivers is the ability to integrate across time scales from individual events to multiple earthquake cycles. This is a challenging task, as has been highlighted in all the sections above; while remaining unaccomplished, it is hindering the ability to use fluvial landforms as palaeoseismic indicators through the shredding of the geomorphic signal preserved in river long profiles and hillslopes in much the way that sediment transport shreds the climatic and tectonic signals preserved in the stratigraphic record (Jerolmack & Paola, 2010).

The nonlinearity of processes such as the periodicity and stochasticity of earthquake timing and recurrence and uncertainty in fundamental relationships, such as that between geomorphic metrics like catchment slope or k_{sn} and uplift rate, variously reported as being linear (Boulton et al., 2024; Roda-Boluda et al., 2019) and nonlinear (e.g. Ouimet et al., 2009), contribute to the complexity and inherent uncertainties. Issues around constraining key parameters such as K (Eq. 9.2), which not only combines multiple factors including precipitation/discharge and rock strength/erodibility, but also depends in its dimensions on the stream power model chosen and which can span many orders of magnitude (e.g. Kirby & Whipple, 2001; Wobus et al., 2006), introduce further problems. While increasingly sophisticated modelling approaches (e.g. Goren et al., 2014; Quye-Sawyer et al., 2020) can provide new insights, there are still relatively few comprehensive field investigations of the long-term landscape and fluvial response of rivers to active faults, especially in regions of complex bedrock or high sediment supply. Therefore developing new field 'laboratories' and coupling field observations with modelling should be key areas for future research.

9.5 Conclusions

In this chapter, we have synthesised existing knowledge and highlighted recent advances in the use of rivers to understand active tectonics and their potential as proxies for seismic records. Although the physical processes underpinning how rivers respond to long- and short-term tectonic drivers are now fairly well understood, supported by numerous empirical data on the relationships among various fluvial metrics and erosion and uplift, significant challenges remain in explicitly linking fluvial geomorphology to individual events or earthquake sequences, and even in confidently deriving uplift rates or histories from river morphology. Additionally, while recent large earthquakes have provided case studies on the volume, grain size and distribution of coseismic and postseismic sediment supply into rivers, a significant amount of variability has been observed, especially in the grain size and volume of transported material. Therefore while existing research supports general increases in sediment flux with earthquake magnitude, the time scale of this perturbation and the geomorphic impacts are still debated. Indeed, integrating across short to long time scales is a key issue in understanding the fluvial response to fault offset, uplift and sediment loading, and while we recognise that the fluvial system encodes information on seismic cycles and rock uplift, there is still much work to be done to allow us to fully decode this information.

Acknowledgements

We thank the editors for inviting us to participate in this volume.

References

Aktas, Y. D., So, E., & Johnson, C. (2023). The Turkiye earthquake sequence of February 2023: A longitudinal study report by EEFIT.

Allen, P. A. (2017). *Sediment routing systems: The fate of sediment from source to sink.* Cambridge University Press.

Allen, P. A., & Allen, J. R. (2013). *Basin analysis: Principles and application to petroleum play assessment.* John Wiley & Sons.

Amos, C. B., & Burbank, D. W. (2007). Channel width response to differential uplift. *Journal of Geophysical Research: Earth Surface, 112*, F02010. https://doi.org/10.1029/2006JF000672.

Andrés-Martínez, M., Pérez-Gussinyé, M., Armitage, J., & Morgan, J. P. (2019). Thermomechanical Implications of Sediment Transport for the Architecture and Evolution of Continental Rifts and Margins. *Tectonics, 38*(2), 641–665. https://doi.org/10.1029/2018tc005346.

Attal, M., Tucker, G. E., Whittaker, A. C., Cowie, P. A., & Roberts, G. P. (2008). Modeling fluvial incision and transient landscape evolution: Influence of dynamic channel adjustment. *Journal of Geophysical Research: Earth Surface, 113*. https://doi.org/10.1029/2007JF000893.

Basharat, M., Riaz, M. T., Jan, M. Q., Xu, C., & Riaz, S. (2021). A review of landslides related to the 2005 Kashmir Earthquake: implication and future challenges. *Natural Hazards, 108*(1), 1–30. https://doi.org/10.1007/s11069-021-04688-8.

Baynes, E. R. C., Lague, D., Steer, P., & Davy, P. (2022). Dynamic bedrock channel width during knickpoint retreat enhances undercutting of coupled hillslopes. *Earth Surface Processes and Landforms, 47*, 3629–3640. https://doi.org/10.1002/esp.5477.

Berlin, M. M., & Anderson, R. S. (2007). Modeling of knickpoint retreat on the Roan Plateau, western Colorado. *Journal of Geophysical Research, 112*. https://doi.org/10.1029/2006jf000553.

Bishop, P., Hoey, T. B., Jansen, J. D., & Lexartza Artza, I. (2005). Knickpoint recession rate and catchment area: The case of uplifted rivers in Eastern Scotland. *Earth Surface Processes and Landforms, 30*, 767–778. https://doi.org/10.1002/esp.1191.

Boulton, S. J., Rodés, Á., Fabel, D., Alçiçek, M. C., & Whittaker, A. C. (2024). Complex erosional response to uplift and rock strength contrasts in transient river systems crossing an active normal fault revealed by 10Be and 26Al cosmogenic nuclide analyses. *Earth Surface Processes and Landforms, 49*, 1428–1450. https://doi.org/10.1002/esp.5778.

Boulton, S. J., & Stewart, I. S. (2013). Holocene coastal notches in the Mediterranean region: Indicators of palaeoseismic clustering? *Geomorphology*.

Boulton, S. J., Stokes, M., & Mather, A. E. (2014). Transient fluvial incision as an indicator of active faulting and PlioQuaternary uplift of the Moroccan High Atlas. *Tectonophysics*.

Boulton, S. J., & Whittaker, A. C. (2009). Quantifying the slip rates, spatial distribution and evolution of active normal faults from geomorphic analysis: Field examples from an oblique-extensional graben, southern Turkey. *Geomorphology, 104*, 299–316.

Çal, Ç., Boulton, S. J., & Mildon, Z. (2024). Structural and geomorphological constraints on the activity of the Sparta Fault (Greece). *Journal of the Geological Society, 181*, jgs2024–jgs2066. https://doi.org/10.1144/jgs2024-066.

Chamberlain, E. L., Goodbred, S. L., Steckler, M. S., Wallinga, J., Reimann, T., Akhter, S. H., et al. (2024). Cascading hazards of a major Bengal basin earthquake and abrupt avulsion of the Ganges River. *Nature Communication, 15*, 4975. https://doi.org/10.1038/s41467-024-47786-4.

Chen, Y.-G., Chen, W.-S., Lee, J.-C., Lee, Y.-H., Lee, C.-T., Chang, H.-C., & Lo, C.-H. (2001). Surface Rupture of 1999 Chi-Chi Earthquake Yields Insights on Active Tectonics of Central Taiwan. *Bulletin of the Seismological Society of America, 91*(5), 977–985. https://doi.org/10.1785/0120000721.

Chen, C. M., Shyu, J. B. H., Tsui, H. K., & Hsieh, Y. C. (2022). Preservation and transportation of large landslide deposits under decadal and millennial timescales in the Taiwan orogenic belt. *Geomorphology, 415*. https://doi.org/10.1016/j.geomorph.2022.108402.

Chen, H.-E., Chiu, Y.-Y., Cheng, C.-Y., & Chen, S.-C. (2024). Knickpoints and fixed points: The evolution of fluvial morphology under the combined effect of fault uplift and dam obstruction on a soft bedrock river. *Earth Surface Dynamics, 12*, 1329–1346. https://doi.org/10.5194/esurf-12-13292024.

Chilton, K. D., & Spotila, J. A. (2022). Uncovering the controls on fluvial bedrock erodibility and knickpoint expression: A high-resolution comparison of bedrock properties between knickpoints and non-knickpoint reaches. *Journal of Geophysical Research: Earth Surface, 127*, e2021JF006511. https://doi.org/10.1029/2021JF006511.

Cook, K. L. (2017). An evaluation of the effectiveness of low-cost UAVs and structure from motion for geomorphic change detection. *Geomorphology, 278*, 195–208. https://doi.org/10.1016/j.geomorph.2016.11.009.

Cook, K. L., Rekapalli, R., Dietze, M., Pilz, M., Cesca, S., Rao, N. P., et al. (2021). Detection and potential early warning of catastrophic flow events with regional seismic networks. *Science, 374*, 87–92. https://doi.org/10.1126/science.abj1227.

Cook, K. L., Turowski, J. M., & Hovius, N. (2014). River gorge eradication by downstream sweep erosion. *Nature Geosciences, 7*, 682–686. https://doi.org/10.1038/ngeo2224.

Cook, K. L., Turowski, J. M., & Hovius, N. (2013). A demonstration of the importance of bedload transport for fluvial bedrock erosion and knickpoint propagation. *Earth Surface Processes and Landforms, 38*, 683–695. https://doi.org/10.1002/esp.3313.

Cowie, P. A., Whittaker, A. C., Attal, M., Roberts, G., Tucker, G. E., & Ganas, A. (2008). New constraints on sediment-fluxdependent river incision: Implications for extracting tectonic signals from river profiles. *Geology, 36*, 535–538. https://doi.org/10.1130/G24681A.1.

Cowie, P. A., Attal, M., Tucker, G. E., Whittaker, A. C., Naylor, M., Ganas, A., et al. (2006). Investigating the surface process response to fault interaction and linkage using a numerical modelling approach. *Basin Research, 18*, 231–266. https://doi.org/10.1111/j.1365-2117.2006.00298.x.

Cowie, P. A., & Roberts, G. P. (2001). Constraining slip rates and spacings for active normal faults. *Journal of Structural Geology, 23*, 1901–1915. https://doi.org/10.1016/S0191-8141(01)00036-0.

Cowie, P. A., Vanneste, C., & Sornette, D. (1993). Statistical physics model for the spatiotemporal evolution of faults. *Journal of Geophysical Research: Solid Earth, 98*, 21809–21821. https://doi.org/10.1029/93JB02223.

Croissant, T., Lague, D., Steer, P., & Davy, P. (2017). Rapid post-seismic landslide evacuation boosted by dynamic river width. *Nature Geosciences, 10*, 680–684. https://doi.org/10.1038/ngeo3005.

Croissant, T., Steer, P., Lague, D., Davy, P., Jeandet, L., & Hilton, R. G. (2019). Seismic cycles, earthquakes, landslides and sediment fluxes: Linking tectonics to surface processes using a reduced-complexity model. *Geomorphology, 339*, 87–103. https://doi.org/10.1016/j.geomorph.2019.04.017.

Crosby, B. T., & Whipple, K. X. (2006). Knickpoint initiation and distribution within fluvial networks: 236 waterfalls in the Waipaoa River, North Island, New Zealand. *Geomorphology, 82*, 16–38. https://doi.org/10.1016/j.geomorph.2005.08.023.

Dadson, S. J., Hovius, N., Chen, H., Dade, W. B., Lin, J. C., Hsu, M. L., et al. (2004). Earthquake-triggered increase in sediment delivery from an active mountain belt. *Geology, 32*, 733–736. https://doi.org/10.1130/G20639.1.

Dai, L., Scaringi, G., Fan, X., Yunus, A. P., Liu-Zeng, J., Xu, Q., et al. (2021). Coseismic debris remains in the orogen despite a decade of enhanced landsliding. *Geophysical Research Letters, 48*. https://doi.org/10.1029/2021GL095850.

D'Arcy, M., Roda-Boluda, D. C., & Whittaker, A. C. (2017). Glacial-interglacial climate changes recorded by debris flow fan deposits, Owens Valley, California. *Quaternary Science Reviews, 169*, 288–311. https://doi.org/10.1016/j.quascirev.2017.06.002.

Das, J. D., Saraf, A. K., & Panda, S. (2007). Satellite data in a rapid analysis of Kashmir earthquake (October 2005) triggered landslide pattern and river water turbidity in and around the epicentral region. *International Journal of Remote Sensing, 28*, 1835–1842. https://doi.org/10.1080/01431160600954720.

Densmore, A. L., Dawers, N. H., Gupta, S., & Guidon, R. (2005). What sets topographic relief in extensional footwalls? *Geology, 33*, 453–456. https://doi.org/10.1130/G21440.1.

Densmore, A. L., Dawers, N. H., Gupta, S., Guidon, R., & Goldin, T. (2004). Footwall topographic development during continental extension. *Journal of Geophysical Research: Earth Surface, 109*. https://doi.org/10.1029/2003JF000115.

Densmore, A. L., Gupta, S., Allen, P. A., & Dawers, N. H. (2007). Transient landscapes at fault tips. *Journal of Geophysical Research, 112*. https://doi.org/10.1029/2006jf000560.

Densmore, A. L., & Hovius, N. (2000). Topographic fingerprints of bedrock landslides. *Geology, 28*, 371–374. https://doi.org/10.1130/0091-7613(2000)28<371:TFOBL>2.0.CO;2.

DiBiase, R. A., Whipple, K. X., Lamb, M. P., & Heimsath, A. M. (2015). The role of waterfalls and knickzones in controlling the style and pace of landscape adjustment in the western San Gabriel Mountains, California. *GSA Bulletin, 127*, 539–559. https://doi.org/10.1130/B31113.1.

Duffy, B., Quigley, M., Barrell, D. J. A., Van Dissen, R., Stahl, T., Leprince, S., et al. (2013). Fault kinematics and surface deformation across a releasing bend during the 2010 MW 7.1 Darfield, New Zealand, earthquake revealed by differential LiDAR and cadastral surveying. *GSA Bulletin, 125*, 420–431. https://doi.org/10.1130/B30753.1.

Duffy, O. B., Brocklehurst, S. H., Gawthorpe, R. L., Leeder, M. R., & Finch, E. (2015). Controls on landscape and drainage evolution in regions of distributed normal faulting: Perachora Peninsula, Corinth Rift, Central Greece. *Basin Research, 27*, 473–494. https://doi.org/10.1111/bre.12084.

Duvall, Alison, Kirby, Eric, & Burbank, Douglas (2004). Tectonic and lithologic controls on bedrock channel profiles and processes in coastal California. *Journal of Geophysical Research: Earth Surface, 109*(F345), 27. https://doi.org/10.1029/2003JF000086.

Egholm, D. L., Knudsen, M. F., & Sandiford, M. (2013). Lifespan of mountain ranges scaled by feedbacks between landsliding and erosion by rivers. *Nature, 498*, 475–478. https://doi.org/10.1038/nature12218.

Ewiak, O., Victor, P., & Oncken, O. (2015). Investigating multiple fault rupture at the Salar del Carmen segment of the Atacama Fault System (northern Chile): Fault scarp morphology and knickpoint analysis. *Tectonics, 34*, 187–212. https://doi.org/10.1002/2014TC003599.

Fan, R. L., Zhang, L. M., Wang, H. J., & Fan, X. M. (2018). Evolution of debris flow activities in Gaojiagou Ravine during 2008–2016 after the Wenchuan earthquake. *Engineering Geology, 235*, 1–10. https://doi.org/10.1016/j.enggeo.2018.01.017.

Fernández-Blanco, D., de Gelder, G., Lacassin, R., & Armijo, R. (2020). Geometry of Flexural Uplift by Continental Rifting in Corinth, Greece. *Tectonics, 39*(1). https://doi.org/10.1029/2019tc005685.

Finnegan, N. J. (2013). Interpretation and downstream correlation of bedrock river terrace treads created from propagating knickpoints. *Journal of Geophysical Research: Earth Surface, 118*, 54–64. https://doi.org/10.1029/2012JF002534.

Finnegan, N. J., & Balco, G. (2013). Sediment supply, base level, braiding, and bedrock river terrace formation: Arroyo Seco, California, USA. *GSA Bulletin, 125*, 1114–1124. https://doi.org/10.1130/B30727.1.

Flint, J. J. (1974). Stream gradient as a function of order, magnitude, and discharge. *Water Resources Research, 10*, 969–973. https://doi.org/10.1029/WR010i005p00969.

Formento-Trigilio, M. L., Burbank, D. W., Nicol, A., Shulmeister, J., & Rieser, U. (2003). River response to an active fold-and-thrust belt in a convergent margin setting, North Island, New Zealand. *Geomorphology, 49*, 125–152. https://doi.org/10.1016/S0169-555X(02)00167-8.

Francis, O., Fan, X., Hales, T., Hobley, D., Xu, Q., & Huang, R. (2022). The fate of sediment after a large earthquake. *Journal of Geophysical Research: Earth Surface, 127*, e2021JF006352-e2021JF006352. https://doi.org/10.1029/2021JF006352.

Francis, O. R., Hales, T. C., Hobley, D. E. J., Fan, X., Horton, A. J., Scaringi, G., et al. (2020). The impact of earthquakes on orogen-scale exhumation. *Earth Surface Dynamics, 8*, 579–593. https://doi.org/10.5194/esurf8-579-2020.

Gallen, S. F., & Fernández-Blanco, D. (2021). A new data-driven bayesian inversion of fluvial topography clarifies the tectonic history of the corinth rift and reveals a channel steepness threshold. *Journal of Geophysical Research: Earth Surface, 126*, e2020JF005651. https://doi.org/10.1029/2020JF005651.

Gallen, S. F., & Wegmann, K. W. (2017). River profile response to normal fault growth and linkage: An example from the Hellenic forearc of south-central Crete, Greece. *Earth Surface Dynamics, 5*, 161–186. https://doi.org/10.5194/esurf-5-161-2017.

Gasparini, N. M., Bras, R. L., & Whipple, K. X. (2006). Numerical modeling of non–steady-state river profile evolution using a sediment-flux-dependent incision model. In S. D. Willett, N. Hovius, M. T. Brandon, & D. M. Fisher (Eds.). *Tectonics, climate, and landscape evolution* (pp. 0). Geological Society of America. https://doi.org/10.1130/2006.2398(08).

Gasparini, N. M., Whipple, K. X., & Bras, R. L. (2007). Predictions of steady state and transient landscape morphology using sediment-flux-dependent river incision models. *Journal of Geophysical Research: Earth Surface, 112*. https://doi.org/10.1029/2006JF000567.

Gawthorpe, R. L., & Leeder, M. R. (2000). Tectono-sedimentary evolution of active extensional basins. *Basin Research, 12*, 195–218. https://doi.org/10.1111/j.1365-2117.2000.00121.x.

Geurts, A. H., Cowie, P. A., Duclaux, G., Gawthorpe, R. L., Huismans, R. S., Pedersen, V. K., et al. (2018). Drainage integration and sediment dispersal in active continental rifts: A numerical modelling study of the central Italian Apennines. *Basin Research, 30*, 965–989. https://doi.org/10.1111/bre.12289.

Goren, L., Fox, M., & Willett, S. D. (2014). Tectonics from fluvial topography using formal linear inversion: Theory and applications to the Inyo Mountains, California. *Journal of Geophysical Research: Earth Surface, 119*, 1651–1681. https://doi.org/10.1002/2014JF003079.

Gorum, T., Fan, X., van Westen, C. J., Huang, R. Q., Xu, Q., Tang, C., et al. (2011). Distribution pattern of earthquakeinduced landslides triggered by the 12 May 2008 Wenchuan earthquake. *Geomorphology, 133*, 152–167. https://doi.org/10.1016/j.geomorph.2010.12.030.

Graf, E. L. S., Sinclair, H. D., Attal, M., Gailleton, B., Adhikari, B., & Baral, B. R. (2024). Geomorphological and hydrological controls on sediment export in earthquake-affected catchments in the Nepal Himalaya. *Earth Surface Dynamics, 12*, 135–161. https://doi.org/10.5194/esurf-12-135-2024.

Hack, J. T. (1957). *Studies of longitudinal stream profiles in Virginia and Maryland*. Geological Survey Professional Paper.

Harel, M. A., Mudd, S. M., & Attal, M. (2016). Global analysis of the stream power law parameters based on worldwide 10Be denudation rates. *Geomorphology*. https://doi.org/10.1016/j.geomorph.2016.05.035.

Harvey, E. L., Hales, T. C., Hobley, D. E. J., Liu, J., & Fan, X. (2022). Measuring the grain-size distributions of mass movement deposits. *Earth Surface Processes and Landforms, 47*, 1599–1614. https://doi.org/10.1002/esp.5337.

Harvey, E. L., Kincey, M. E., Rosser, N. J., Gadtaula, A., Collins, E., Densmore, A. L., et al. (2024). Review of landslide inventories for Nepal between 2010 and 2021 reveals data gaps in global landslide hotspot. *Nature Hazards*. https://doi.org/10.1007/s11069-024-07013-1.

Haviv, I., Enzel, Y., Whipple, K. X., Zilberman, E., Matmon, A., Stone, J., et al. (2010). Evolution of vertical knickpoints (waterfalls) with resistant caprock: Insights from numerical modeling. *Journal of Geophysical Research: Earth Surface, 115*. https://doi.org/10.1029/2008JF001187.

He, Z., & Ma, B. (2015). Holocene paleoearthquakes of the Daqingshan fault detected from knickpoint identification and alluvial soil profile. *Journal of Asian Earth Sciences, 98*, 261–271. https://doi.org/10.1016/j.jseaes.2014.11.025.

He, C., Yang, C.-J., Turowski, J. M., Rao, G., Roda-Boluda, D. C., & Yuan, X.-P. (2021). Constraining tectonic uplift and advection from the main drainage divide of a mountain belt. *Nature Communication, 12*, 544. https://doi.org/10.1038/s41467-020-20748-2.

Hobley, D. E. J., Sinclair, H. D., Mudd, S. M., & Cowie, P. A. (2011). Field calibration of sediment flux dependent river incision. *Journal of Geophysical Research: Earth Surface, 116*. https://doi.org/10.1029/2010JF001935.

Holbrook, J., & Schumm, S. A. (1999). Geomorphic and sedimentary response of rivers to tectonic deformation: A brief review and critique of a tool for recognizing subtle epeirogenic deformation in modern and ancient settings. *Tectonophysics, 305*, 287–306. https://doi.org/10.1016/S00401951(99)00011-6.

Hopkins, M. C., & Dawers, N. H. (2015). Changes in bedrock channel morphology driven by displacement rate increase during normal fault interaction and linkage. *Basin Research, 27*, 43–59. https://doi.org/10.1111/bre.12072.

Hovius, N., Meunier, P., Lin, C. W., Chen, H., Chen, Y. G., Dadson, S., et al. (2011). Prolonged seismically induced erosion and the mass balance of a large earthquake. *Earth and Planetary Science Letters, 304*, 347–355. https://doi.org/10.1016/j.epsl.2011.02.005.

Huang, M.-H., Fielding, E. J., Liang, C., Milillo, P., Bekaert, D., Dreger, D., & Salzer, J. (2017). Coseismic deformation and triggered landslides of the 2016 M_w 6.2 Amatrice earthquake in Italy. *Geophysical Research Letters, 44*(3), 1266–1274. https://doi.org/10.1002/2016gl071687.

Jakica, S., Quigley, M. C., Sandiford, M., Clark, D., Fifield, L. K., & Alimanovic, A. (2011). Geomorphic and cosmogenic nuclide constraints on escarpment evolution in an intraplate setting, Darling Escarpment, Western Australia. *Earth Surface Processes and Landforms, 36*, 449–459. https://doi.org/10.1002/esp.2058.

Jerolmack, D. J., & Paola, C. (2010). Shredding of environmental signals by sediment transport. *Geophysical Research Letters, 37*, 1–5. https://doi.org/10.1029/2010GL044638.

Jones, J. N., Boulton, S. J., Stokes, M., Bennett, G. L., & Whitworth, M. R. Z. (2021). 30-year record of Himalaya masswasting reveals landscape perturbations by extreme events. *Nature Communications, 12*, 1–15. https://doi.org/10.1038/s41467-021-26964-8.

Kent, E., Boulton, S. J., Stewart, I. S., Whittaker, A. C., & Alçiçek, M. C. (2016). Geomorphic and geological constraints on the active normal faulting of the Gediz (Alaşehir) Graben, Western Turkey. *Journal of the Geological Society, 173*, 666–678. https://doi.org/10.1144/jgs2015-121.

Kent, E., Boulton, S. J., Whittaker, A. C., Stewart, I. S., & Cihat Alçiçek, M. (2017). Normal fault growth and linkage in the Gediz (Alaşehir) Graben, Western Turkey, revealed by transient river long-profiles and slopebreak knickpoints. *Earth Surface Processes and Landforms, 42*, 836–852. https://doi.org/10.1002/esp.4049.

Kent, E., Whittaker, A. C., Boulton, S. J., & Alçiçek, M. C. (2021). Quantifying the competing influences of lithology and throw rate on bedrock river incision. *GSA Bulletin, 133*, 1649–1664. https://doi.org/10.1130/B35783.1.

Khattak, G. A., Owen, L. A., Kamp, U., & Harp, E. L. (2010). Evolution of earthquake-triggered landslides in the Kashmir Himalaya, northern Pakistan. *Geomorphology, 115*, 102–108. https://doi.org/10.1016/j.geomorph.2009.09.035.

Kincey, M. E., Rosser, N. J., Robinson, T. R., Densmore, A. L., Shrestha, R., Pujara, D. S., et al. (2021). Evolution of coseismic and post-seismic landsliding after the 2015 Mw 7.8 gorkha earthquake, Nepal. *Journal of Geophysical Research: Earth Surface, 126*, e2020JF005803. https://doi.org/10.1029/2020JF005803.

Kirby, E., & Whipple, K. (2001). Quantifying differential rock-uplift rates via stream profile analysis. *Geology, 29*, 415–418. https://doi.org/10.1130/0091-7613(2001)029<0415:qdrurv>2.0.co;2.

Kirby, E., & Whipple, K. X. (2012). Expression of active tectonics in erosional landscapes. *Journal of Structural Geology, 44*, 54–75. https://doi.org/10.1016/j.jsg.2012.07.009.

Koi, T., Hotta, N., Ishigaki, I., Matuzaki, N., Uchiyama, Y., & Suzuki, M. (2008). Prolonged impact of earthquake-induced landslides on sediment yield in a mountain watershed: The Tanzawa region, Japan. *Geomorphology, 101*, 692–702. https://doi.org/10.1016/j.geomorph.2008.03.007.

Korup, O. (2006). Effects of large deep-seated landslides on hillslope morphology, western Southern Alps, New Zealand. *Journal of Geophysical Research: Earth Surface, 111*. https://doi.org/10.1029/2004JF000242.

Korup, O., & Clague, J. J. (2009). Natural hazards, extreme events, and mountain topography. *Quaternary Science Reviews, 28*, 977–990. https://doi.org/10.1016/j.quascirev.2009.02.021.

Korup, O., Clague, J. J., Hermanns, R. L., Hewitt, K., Strom, A. L., & Weidinger, J. T. (2007). Giant landslides, topography, and erosion. *Earth and Planetary Science Letters, 261*, 578–589. https://doi.org/10.1016/j.epsl.2007.07.025.

Langridge, R., Ries, W., Litchfield, N., Villamor, P., Van Dissen, R., Barrell, D., et al. (2016). The New Zealand active faults database. *New Zealand Journal of Geology and Geophysics, 59*, 86–96. https://doi.org/10.1080/00288306.2015.1112818.

Langridge, R. M., Clark, K. J., Almond, P., Baize, S., Howell, A., Kearse, J., et al. (2023). Late Holocene earthquakes on the Papatea Fault and its role in past earthquake cycles, Marlborough, New Zealand. *New Zealand Journal of Geology and Geophysics, 66*, 317–341. https://doi.org/10.1080/00288306.2022.2117829.

Langridge, R. M., Rowland, J., Villamor, P., Mountjoy, J., Townsend, D. B., Nissen, E., et al. (2018). Coseismic rupture and preliminary slip estimates for the papatea fault and its role in the 2016 Mw 7.8 Kaikōura, New Zealand, earthquake. *Bulletin of the Seismological Society of America, 108*, 1596–1622. https://doi.org/10.1785/0120170336.

Lavé, J., & Avouac, J. P. (2001). Fluvial incision and tectonic uplift across the Himalayas of central Nepal. *Journal of Geophysical Research, 106*, 26561. https://doi.org/10.1029/2001JB000359.

Lee, Y.-H., Lu, S.-T., Shih, T.-S., Hsieh, M.-L., & Wu, W.-Y. (2005). Structures Associated with the Northern End of the 1999 Chi-Chi Earthquake Rupture, Central Taiwan: Implications for Seismic-Hazard Assessment. *Bulletin of the Seismological Society of America, 95*(2), 471–485. https://doi.org/10.1785/0120020170.

Leonard, J. S., Whipple, K. X., & Heimsath, A. M. (2023). Isolating climatic, tectonic, and lithologic controls on mountain landscape evolution. *Science Advances, 9*, eadd8915. https://doi.org/10.1126/sciadv.add8915.

Li, G., West, A. J., Densmore, A. L., Hammond, D. E., Jin, Z., Zhang, F., et al. (2016). Connectivity of earthquake-triggered landslides with the fluvial network: Implications for landslide sediment transport after the 2008 Wenchuan earthquake. *Journal of Geophysical Research: Earth Surface, 121*, 703–724. https://doi.org/10.1002/2015JF003718.

Li, G., West, A. J., Densmore, A. L., Jin, Z., Zhang, F., Wang, J., et al. (2017). Earthquakes drive focused denudation along a tectonically active mountain front. *Earth and Planetary Science Letters, 472*, 253–265. https://doi.org/10.1016/j.epsl.2017.04.040.

Li, G. K., West, A. J., Jin, Z., Qiu, H., Zhang, F., Wang, J., Hammond, D. E., Densmore, A. L., Hilton, R. G., Dong, S., Atwood, A., Fischer, W. W., & Lamb, M. P. (2025). Large riverbed sediment flux sustained for a decade after an earthquake. *Nature, 644*, 398–403. https://doi.org/10.1038/s41586-025-09354-8.

Lin, G. W. (2018). Variations of fluvial sediment transport after large earthquakes: Field study in Taiwan catchments. *Water (Switzerland), 10*. https://doi.org/10.3390/w10121836.

Liu, Y., Wang, J.-K., & Zhang, J.-F. (2024). Quaternary evolution of the Jinshan Gorge and Yellow River: Insight from numerical modelling and terrace correlation. *Earth Surface Processes and Landforms, 49*, 2567–2582. https://doi.org/10.1002/esp.5843.

Loget, N., & Van Den Driessche, J. (2009). Wave train model for knickpoint migration. *Geomorphology, 106*, 376–382. https://doi.org/10.1016/j.geomorph.2008.10.017.

Mackey, B. H., Scheingross, J. S., Lamb, M. P., & Farley, K. A. (2014). Knickpoint formation, rapid propagation, and landscape response following coastal cliff retreat at the last interglacial sea-level highstand: Kaua'i, Hawai'i. *GSA Bulletin, 126*, 925–942. https://doi.org/10.1130/B30930.1.

Malatesta, L. C., & Lamb, M. P. (2017). Formation of waterfalls by intermittent burial of active faults. *GSA Bulletin, 130*, 522–536. https://doi.org/10.1130/B31743.1.

Marc, O., Hovius, N., Meunier, P., Uchida, T., & Hayashi, S. (2015). Transient changes of landslide rates after earthquakes. *Geology, 43*, 883–886.

Märki, L., Lupker, M., France-Lanord, C., Lavé, J., Gallen, S., Gajurel, A. P., et al. (2021). An unshakable carbon budget for the Himalaya. *Nature Geosciences, 14*, 745–750. https://doi.org/10.1038/s41561-021-00815-z.

McEwan, E., Stahl, T., Howell, A., Langridge, R., & Wilson, M. (2023). Coseismic river avulsion on surface rupturing faults: Assessing earthquake-induced flood hazard. *Sciences Advances, 9*, eadd2932. https://doi.org/10.1126/sciadv.add2932.

Meghraoui, M., Jaegy, R., Lammali, K., & Albarède, F. (1988). Late Holocene earthquake sequences on the El Asnam (Algeria) thrust fault. *Earth and Planetary Science Letters, 90*, 187–203. https://doi.org/10.1016/0012-821X(88)90100-8.

Miller, S. R., Slingerland, R. L., & Kirby, E. (2007). Characteristics of steady state fluvial topography above fault-bend folds. *Journal of Geophysical Research: Earth Surface, 112*. https://doi.org/10.1029/2007JF000772.

Mudd, S. M., Roda-Boluda, D. C., Goren, L., & Clubb, F. J. (2022). Beyond the long profile. In J. Shroder, F. (Jack), & E. Wohl (Vol. Eds.), *Treatise on geomorphology: vol. 6.1*, (pp. 22–52). Elsevier. https://doi.org/10.1016/b978-0-12-818234-5.00026-2.

Olive, J.-A., Behn, M. D., & Malatesta, L. C. (2014). Modes of extensional faulting controlled by surface processes. *Geophysical Research Letters, 41*(19), 6725–6733. https://doi.org/10.1002/2014gl061507.

Ouimet, W. B., Whipple, K. X., & Granger, D. E. (2009). Beyond threshold hillslopes: Channel adjustment to base-level fall in tectonically active mountain ranges. *Geology, 37*, 579–582. https://doi.org/10.1130/G30013A.1.

Ott, R. F., Gallen, S. F., Caves Rugenstein, J. K., Ivy-Ochs, S., Helman, D., Fassoulas, C., Vockenhuber, C., Christl, M., & Willett, S. D. (2019). Chemical Versus Mechanical Denudation in Meta-Clastic and Carbonate Bedrock Catchments on Crete, Greece, and Mechanisms for Steep and High Carbonate Topography. *Journal of Geophysical Research: Earth Surface, 124*(12), 2943–2961. https://doi.org/10.1029/2019jf005142.

Ouimet, W. B., Whipple, K. X., Royden, L. H., Sun, Z., & Chen, Z. (2007). The influence of large landslides on river incision in a transient landscape: Eastern margin of the Tibetan Plateau (Sichuan, China). *GSA Bulletin, 119*, 1462–1476. https://doi.org/10.1130/B26136.1.

Oveisi, B., Lavé, J., Van Der Beek, P., Carcaillet, J., Benedetti, L., & Aubourg, C. (2009). Thick- and thin-skinned deformation rates in the central Zagros simple folded zone (Iran) indicated by displacement of geomorphic surfaces. *Geophysical Journal International, 176*, 627–654. https://doi.org/10.1111/j.1365-246X.2008.04002.x.

Pain, C. F., & Bowler, J. M. (1973). Denudation following the November 1970 earthquake at Madang, Papua New Guinea. *Zeitschrift Für Geomorphologie, 18*, 92–104.

Pearce, A. J., & Watson, A. J. (1986). Effects of earthquake-induced landslides on sediment budget and transport over a 50-yr period. *Geology, 14*, 1–4. https://doi.org/10.1130/00917613(1986)14<52:EOELOS>2.0.CO;2.

Pechlivanidou, S., Cowie, P. A., Duclaux, G., Nixon, C. W., Gawthorpe, R. L., & Salles, T. (2019). Tipping the balance: Shifts in sediment production in an active rift setting. *Geology, 47*, 259–262. https://doi.org/10.1130/G45589.1.

Pechlivanidou, S., Cowie, P. A., Hannisdal, B., Whittaker, A. C., Gawthorpe, R. L., Pennos, C., et al. (2018). Source-to-sink analysis in an active extensional setting: Holocene erosion and deposition in the Sperchios rift, central Greece. *Basin Research, 30*, 522–543. https://doi.org/10.1111/bre.12263.

Quigley, M., & Duffy, B. (2020). Effects of earthquakes on flood hazards: A case study from christchurch, New Zealand. *Geosciences, 10*, 114. https://doi.org/10.3390/geosciences10030114.

Quigley, M. C., Hughes, M. W., Bradley, B. A., van Ballegooy, S., Reid, C., Morgenroth, J., et al. (2016). The 2010–2011 Canterbury earthquake sequence: Environmental effects, seismic triggering thresholds and geologic legacy. *Tectonophysics, 672–673*, 228–274. https://doi.org/10.1016/j.tecto.2016.01.044.

Quye-Sawyer, J., Whittaker, A. C., & Roberts, G. G. (2020). Calibrating fluvial erosion laws and quantifying river response to faulting in Sardinia, Italy. *Geomorphology, 370*, 107388. https://doi.org/10.1016/j.geomorph.2020.107388.

Ramsey, L. A., Walker, R. T., & Jackson, J. (2008). Fold evolution and drainage development in the Zagros mountains of Fars province, SE Iran. *Basin Research, 20*, 23–48. https://doi.org/10.1111/j.13652117.2007.00342.x.

Roda-Boluda, D. C., D'Arcy, M., McDonald, J., & Whittaker, A. C. (2018). Lithological controls on hillslope sediment supply: Insights from landslide activity and grain size distributions. *Earth Surface Processes and Landforms, 43*, 956–977.

Roda-Boluda, D. C., D'Arcy, M., Whittaker, A. C., Gheorghiu, D. M., & Rodés, Á. (2019). 10Be erosion rates controlled by transient response to normal faulting through incision and landsliding. *Earth and Planetary Science Letters, 507*, 140–153. https://doi.org/10.1016/j.epsl.2018.11.032.

Roda-Boluda, D. C., & Whittaker, A. C. (2017). Structural and geomorphological constraints on active normal faulting and landscape evolution in Calabria, Italy. *Journal of the Geological Society, 174*, 701–720. https://doi.org/10.1144/jgs2016-097.

Rothman, S. D., Scheingross, J. S., McCoy, S. W., & Dow, H. W. (2023). Impacts of spontaneous waterfall development on bedrock river longitudinal profile morphology. *Journal of Geophysical Research: Earth Surface, 128*, e2022JF007057. https://doi.org/10.1029/2022JF007057.

Royden, L., & Taylor Perron, J. (2013). Solutions of the stream power equation and application to the evolution of river longitudinal profiles. *Journal of Geophysical Research: Earth Surface, 118*, 497–518. https://doi.org/10.1002/jgrf.20031.

Sklar, L. S., & Dietrich, W. E. (2004). A mechanistic model for river incision into bedrock by saltating bed load. *Water Resources Research, 40*(6). https://doi.org/10.1029/2003wr002496.

Sklar, L. S., & Dietrich, W. E. (2006). The role of sediment in controlling steady-state bedrock channel slope: Implications of the saltation-abrasion incision model. *Geomorphology, 82*, 58–83. https://doi.org/10.1016/j.geomorph.2005.08.019.

Snyder, N. P., Whipple, K. X., Tucker, G. E., & Merritts, D. J. (2000). Stream profiles in the Mendocino triple junction region, northern California. *GSA Bulletin, 112*, 1250–1263. https://doi.org/10.1130/00167606(2000)112<1250:lrttfd>2.3.co;2.

Stahle, D. W., VanArsdale, R. B., & Cleaveland, M. K. (1992). Tectonic signal in baldcypress trees at reelfoot lake, tennessee. *Seismological Research Letters, 63*, 439–447. https://doi.org/10.1785/gssrl.63.3.439.

Steer, P., Croissant, T., Baynes, E., & Lague, D. (2019). Statistical modelling of coseismic knickpoint formation and river response to fault slip. *Earth Surface Dynamics,* 7, 681–706. https://doi.org/10.5194/esurf-7-681-2019.

Stock, J., & Dietrich, W. E. (2003). Valley incision by debris flows: Evidence of a topographic signature. *Water Resources Research, 39*(4). https://doi.org/10.1029/2001wr001057.

Sun, C., Wan, T., Xie, X., Shen, X., & Liang, K. (2016). Knickpoint series of gullies along the Luoyunshan Piedmont and its relation with fault activity since late Pleistocene. *Geomorphology, 268*, 266–274. https://doi.org/10.1016/j.geomorph.2016.06.026.

Tolorza, V., Mohr, C. H., Carretier, S., Serey, A., Sepúlveda, S. A., Tapia, J., et al. (2019). Suspended sediments in chilean rivers reveal low postseismic erosion after the maule earthquake (Mw 8.8) during a severe drought. *Journal of Geophysical Research: Earth Surface, 124*, 1378–1397. https://doi.org/10.1029/2018JF004766.

Tucker, G. E., & Whipple, K. X. (2002). Topographic outcomes predicted by stream erosion models: Sensitivity analysis and intermodel comparison. *Journal of Geophysical Research: Solid Earth, 107*, ETG 1–16. https://doi.org/10.1029/2001JB000162.

Tunnicliffe, J., Howarth, J., & Massey, C. (2024). Controls on fluvial sediment evacuation following an earthquake-triggered landslide: Observations from LiDAR time series. 10, eadi5560.

Turowski, J. M., Lague, D., & Hovius, N. (2009). Response of bedrock channel width to tectonic forcing: Insights from a numerical model, theoretical considerations, and comparison with field data. *Journal of Geophysical Research: Earth Surface, 114*. https://doi.org/10.1029/2008JF001133.

Wang, N., Han, Z., Li, X., Chen, G., Wang, X., & Lu, H. (2015). Tectonic uplift of Mt. Lushan indicated by the steepness indices of the river longitudinal profiles. *Acta Gepgraphica Sinica, 70*, 1516–1525. https://doi.org/10.11821/dlxb201509013.

Watkins, S. E., Whittaker, A. C., Bell, R. E., Brooke, S. A. S., Ganti, V., Gawthorpe, R. L., et al. (2020). Straight from the source's mouth: Controls on field-constrained sediment export across the entire active Corinth Rift, central Greece. *Basin Research, 32*, 1600–1625. https://doi.org/10.1111/bre.12444.

Wedmore, L. N. J., Turner, T., Biggs, J., Williams, J. N., Sichingabula, H. M., Kabumbu, C., et al. (2022). The luangwa rift active fault database and fault reactivation along the southwestern branch of the East African Rift. *Solid Earth, 13*, 1731–1753. https://doi.org/10.5194/se-13-1731-2022.

Wegmann, K. W., & Pazzaglia, F. J. (2009). Late quaternary fluvial terraces of the romagna and marche apennines, Italy: Climatic, lithologic, and tectonic controls on terrace genesis in an active orogen. *Quaternary Science Reviews, 28*, 137–165. https://doi.org/10.1016/j.quascirev.2008.10.006.

Wei, Z., Bi, L., Xu, Y., & He, H. (2015). Evaluating knickpoint recession along an active fault for paleoseismological analysis: The Huoshan Piedmont, Eastern China. *Geomorphology, 235*, 63–76. https://doi.org/10.1016/j.geomorph.2015.01.013.

West, A. J., Hetzel, R., Li, G., Jin, Z., Zhang, F., Hilton, R. G., et al. (2014). Dilution of 10 Be in detrital quartz by earthquake-induced landslides : Implications for determining denudation rates and potential to provide insights into landslide sediment dynamics. *Earth and Planetary Science Letters, 396*, 143–153. https://doi.org/10.1016/j.epsl.2014.03.058.

Whipple, K. X. (2004). Bedrock rivers and the geomorpjology of active orogens. *Annual Review of Earth & Planetary Sciences, 32*, 151–185. https://doi.org/10.1146/annurev.earth.32.101802.120356.

Whipple, K. X., & Tucker, G. E. (2002). Implications of sediment-flux-dependent river incision models for landscape evolution. *Journal Geophys Research Solid Earth, 107*. https://doi.org/10.1029/2000JB000044.

Whipple, K. X., & Tucker, G. E. (1999). Dynamics of the stream-power river incision model: Implications for height limits of mountain ranges, landscape response timescales, and research needs. *Journal of Geophysical Research: Solid Earth, 104*, 17661–17674. https://doi.org/10.1029/1999JB900120.

Whittaker, A. C. (2012). How do landscapes record tectonics and climate? *Lithosphere, 4*. https://doi.org/10.1130/RF.L003.1.

Whittaker, A. C., Attal, M., Cowie, P.a, Tucker, G. E., & Roberts, G. (2008). Decoding temporal and spatial patterns of fault uplift using transient river long profiles. *Geomorphology, 100*, 506–526. https://doi.org/10.1016/j.geomorph.2008.01.018.

Whittaker, A. C., & Boulton, S. J. (2012). Tectonic and climatic controls on knickpoint retreat rates and landscape response times. *Journal of Geophysical Research: Earth Surface, 117*.

Whittaker, A. C., Cowie, P. A., Attal, M., Tucker, G. E., & Roberts, G. P. (2007a). Bedrock channel adjustment to tectonic forcing: Implications for predicting river incision rates. *Geology, 35*, 103–106. https://doi.org/10.1130/G23106A.1.

Whittaker, A. C., Cowie, P. A., Attal, M., Tucker, G. E., & Roberts, G. P. (2007b). Contrasting transient and steady-state rivers crossing active normal faults: New field observations from the central apennines, Italy. *Basin Research, 19*, 529–556. https://doi.org/10.1111/j.1365-2117.2007.00337.x.

Whittaker, A. C., & Walker, A. S. (2015). Geomorphic constraints on fault throw rates and linkage times: Examples from the Northern Gulf of Evia, Greece. *Journal of Geophysical Research: Earth Surface, 120*, 137–158. https://doi.org/10.1002/2014jf003318.

Wilkinson, C., Stahl, T., Jones, K., Fujioka, T., Fink, D., & Norton, K. P. (2022). Delayed 10Be dilution in detrital quartz following extensive coseismic landsliding: A 2016 Kaikōura earthquake case study. *Earth and Planetary Science Letters, 581*, 117392. https://doi.org/10.1016/j.epsl.2022.117392.

Wobus, C., Whipple, K. X., Kirby, E., Snyder, N., Johnson, J., Spyropolou, K., et al. (2006). Tectonics from topography: Procedures, promise, and pitfalls. In S. D. Willett, N. Hovius, M. T. Brandon, & D. M. Fisher (Vol. Eds.), *Tectonics, climate, and landscape evolution: vol. 398*Geological Society of America. https://doi.org/10.1130/2006.2398(04) p. 0–0.

Wrona, T., Whittaker, A. C., Bell, R. E., Gawthorpe, R. L., Fossen, H., Jackson, C. A.-L., et al. (2023). Rift kinematics preserved in deep-time erosional landscape below the northern North Sea. *Basin Research, 35*, 744–761. https://doi.org/10.1111/bre.12732.

Yang, Y., Xu, S., Binnie, S. A., Ritter, B., Cui, L.-F., Cao, Z.-P., et al. (2022). Late Cenozoic locally landslide-dammed lakes across the Middle Yangtze River. *Geomorphology, 413*, 108366. https://doi.org/10.1016/j.geomorph.2022.108366.

Yanites, B. J. (2018). The dynamics of channel slope, width, and sediment in actively eroding bedrock river systems. *Journal of Geophysical Research: Earth Surface, 123*, 1504–1527. https://doi.org/10.1029/2017JF004405.

Yanites, B. J., Clark, M. K., Roering, J. J., West, A. J., Zekkos, D., Baldwin, J. W., et al. (2025). Cascading land surface hazards as a nexus in the Earth system. *Science, 388*, eadp9559. https://doi.org/10.1126/science.adp9559.

Yanites, B. J., Tucker, G. E., Hsu, H. L., Chen, C. C., Chen, Y. G., & Mueller, K. J. (2011). The influence of sediment cover variability on long-term river incision rates: An example from the Peikang River, central Taiwan. *Journal of Geophysical Research: Earth Surface, 116*, 1–13. https://doi.org/10.1029/2010JF001933.

Yanites, B. J., Tucker, G. E., Mueller, K. J., & Chen, Y.-G. (2010a). How rivers react to large earthquakes: Evidence from central Taiwan. *Geology, 38*, 639–642. https://doi.org/10.1130/G30883.1.

Yanites, B. J., Tucker, G. E., Mueller, K. J., Chen, Y. G., Wilcox, T., Huang, S. Y., et al. (2010b). Incision and channel morphology across active structures along the Peikang River, central Taiwan: Implications for the importance of channel width. *Bulletin of the Geological Society of America, 122*, 1192–1208. https://doi.org/10.1130/B30035.1.

Zhou, Z., Whittaker, A. C., Bell, R. E., & Hampson, G. J. (2024). Unravelling tectonic and lithological effects on transient landscapes in the Gulf of Corinth, Greece. *Basin Research, 36*, e12901. https://doi.org/10.1111/bre.12901.

Zondervan, J. R., Stokes, M., Boulton, S. J., Telfer, M. W., & Mather, A. E. (2020). Rock strength and structural controls on fluvial erodibility: Implications for drainage divide mobility in a collisional mountain belt. *Earth and Planetary Science Letters, 538*, 116221. https://doi.org/10.1016/j.epsl.2020.116221.

Zondervan, J. R., Stokes, M., Telfer, M. W., Boulton, S. J., Mather, A. E., Buylaert, J.-P., et al. (2022). Constraining a model of punctuated river incision for Quaternary strath terrace formation. *Geomorphology, 414*, 108396. https://doi.org/10.1016/j.geomorph.2022.108396.

Zondervan, J. R., Whittaker, A. C., Bell, R. E., Watkins, S. E., Brooke, S. A. S., & Hann, M. G. (2020). New constraints on bedrock erodibility and landscape response times upstream of an active fault. *Geomorphology, 351*, 106937. https://doi.org/10.1016/J.GEOMORPH.2019.106937.

CHAPTER 10

Glacial processes and impacts

Maximillian Van Wyk de Vries[1,2]
[1]Department of Geography, University of Cambridge, Cambridge, United Kingdom,
[2]Department of Earth Sciences, University of Cambridge, Cambridge, United Kingdom

10.1 Introduction and context

Glaciers and earthquakes frequently coexist in tectonically active mountain belts, where crustal deformation drives seismicity and high elevations allow snow and ice to accumulate. Glaciers form wherever annual snow accumulation exceeds ablation, spanning environments from the equator (e.g. the Volcán Cayambe ice cap, Ecuador) to the poles (e.g. the Antarctic ice sheet), and have occupied much larger areas during past glacial periods. As a result, interactions between seismic processes and the cryosphere have occurred repeatedly through Earth's history and across a wide range of climatic and tectonic settings. Interactions between glaciers and earthquakes take two broad forms: seismic shaking can perturb glacier systems and trigger a range of associated hazards, while glacier mass changes can modify stresses acting on seismogenic faults (Fig. 10.1). This chapter examines these primary interactions, alongside a range of secondary processes that link earthquakes, cryospheric dynamics and landscape evolution. The discussion is constrained to processes occurring on Earth and excludes purely glaciological seismic events (icequakes).

Research has long recognised that large-magnitude earthquakes trigger a range of secondary hazards, including cryospheric hazards. Some of the earliest documented evidence of this is provided by detailed field surveys following the 1899 Yakutat Bay earthquakes (M 8–8.5), Alaska, which described widespread glacier fracturing, ice avalanching, enhanced crevassing and sudden changes in glacier dynamics attributed to intense seismic shaking (Tarr et al., 1912). Subsequent historical inventories confirmed that earthquakes can initiate snow avalanches across a broad range of magnitudes and distances, establishing that seismic shaking is an effective destabilising mechanism in snow-covered terrain (Podolskiy et al., 2010a). Large earthquakes can trigger ice- and rock-ice avalanches that can transform into highly mobile debris flows with catastrophic impacts, as exemplified by the 1970 M_w 7.9 Ancash earthquake in Peru (Ericksen et al., 1970). Later geomorphic and remote-sensing studies documented exceptionally mobile rock-ice avalanches and glacier collapses in steep mountain environments, including the 2002 Kolka–Karmadon event, where large volumes of rock and glacier ice were mobilised in a single failure sequence (Huggel et al., 2005; Kotlyakov et al., 2004),

Geomorphic Impacts of Earthquakes
DOI: https://doi.org/10.1016/B978-0-443-40375-0.00005-0

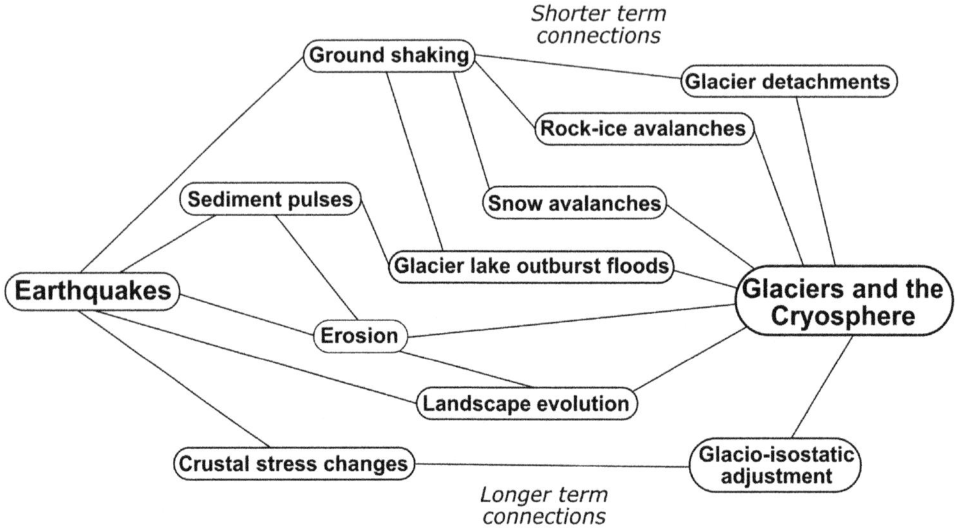

Figure 10.1 Conceptual systems diagram illustrating coupling between earthquakes and the cryosphere across time scales.

although a direct seismic link was rarely established. The 2015 M_w 7.8 Gorkha earthquake (Chapter 15) further reinforced this understanding, with extensive coseismic landslides, snow and rock-ice avalanches observed across the high Himalaya, highlighting the sensitivity of steep mountainous systems to seismic disturbance (Fujita et al., 2017; Kargel et al., 2016). Although early hazard frameworks often assumed a strong link between earthquakes and glacial lake outburst floods (GLOFs), recent systematic assessments indicate that directly earthquake-triggered GLOFs are comparatively rare, with earthquakes more commonly influencing ongoing cryosphere-driven instability and multi-hazard cascades (Chen et al., 2023; Wood et al., 2024).

Ongoing cryospheric hazard research continues to investigate which physical thresholds and preconditioning factors determine when and how earthquakes trigger failures. A unifying concept emerging from recent work is that many glaciers and periglacial systems remain close to mechanical or thermal failure thresholds, making them particularly sensitive to transient external forcing such as seismic shaking (e.g. Kääb et al., 2021; Podolskiy et al., 2010a). While the capacity for strong ground motion to initiate snow and ice avalanches is well established in historical inventories (Ericksen et al., 1970; Fujita et al., 2017; Podolskiy et al., 2010a), research continues into which intrinsic material properties and landscape geometries modulate this response (e.g. Evans, 2021), particularly as permafrost degradation under climate warming weakens periglacial slopes and lowers failure thresholds (Gruber & Haeberli, 2007; Haeberli et al., 2017). Glacier geometry and basal hydrology are recognised as critical controls on dynamic instability, yet there is limited consensus on how thickness, slope and subglacial water pressures interact with seismic wavefields to govern failure and cascade potential (Faillettaz et al., 2015; Gnyawali et al., 2020). Ongoing debate over the frequency of catastrophic moraine-dam failures in inventories (Allen et al., 2019) highlights the need to integrate long-term cryospheric preconditioning with short-term seismic forcing when assessing earthquake-related cryospheric hazard. Detailed observations of recent events, such as the (nonseismic) May 2025 Blatten glacier detachment in Switzerland (e.g. Büntgen et al., 2025; Islam et al., 2025) and the December 2025 M_w 7.0 Hubbard Glacier earthquake and landslides/avalanches in Alaska, USA (USGS, 2025), will further enrich our mechanical understanding of glacier and rock-ice collapses and stability thresholds. The advent of global, subweekly revisit time and medium (≤10 m) resolution satellite imagery, alongside extensive global seismic networks, substantially improves the probability of detecting future interactions and the completeness of resulting inventories.

For glacier to earthquake links, recognition that ice-sheet loss and other cryospheric mass changes can influence tectonic and seismic behaviour predates modern geodetics and geophysical data. As early as the late 19th century, geomorphological observations in northern Europe and North America documented anomalous fault scarps, displaced shorelines and differential uplift in formerly glaciated regions following deglaciation, suggesting a link between ice-sheet melting and crustal stress changes (Jäckli, 1965; Steffen et al., 2021). These observations laid the groundwork for later mechanistic interpretations of postglacial faulting. Over the past 50 years, a range of studies have formalised the link between glacio-isostatic adjustment (GIA) and intraplate stress perturbations (e.g. Peltier & Andrews, 1976; Wu & Hasegawa, 1996), leading to systematic criteria for identifying glacially induced faults in formerly glaciated landscapes (Lagerbäck, 1992) and linking spatial patterns of seismicity to postglacial stress evolution. Quantitative modelling developed from the 1970s to the 1990s demonstrated that the viscoelastic response of the lithosphere-mantle system to ice-load removal can, in some cases, generate stress changes of a similar magnitude to tectonic loading, with stresses migrating through time as rebound proceeds (Peltier & Andrews, 1976; Steffen et al., 2021; Wu & Hasegawa, 1996). More recent work has integrated LiDAR-derived fault inventories, refined palaeoseismic chronologies and demonstrated via stress modelling that glacial loading and unloading have exerted a persistent influence on fault slip histories and modern seismicity across a range of formerly glaciated regions (Steffen et al., 2021).

In recent publications, research on glacially triggered faulting and intraplate seismicity particularly focuses on the magnitude and persistence of cryosphere-driven stress perturbations relative to background tectonic and local stress sources, and the extent to which these impacts are observable for modern (anthropogenic) glacier mass loss. GIA modelling shows that ice unloading can generate stress changes of tens of MPa, sufficient to reactivate favourably oriented preexisting faults, but these stresses are transient, depth-dependent and sensitive to Earth rheology, lithospheric structure and ice-history assumptions, making robust prediction of fault reactivation in space and time a major challenge (Grollimund & Zoback, 2001; Peltier & Andrews, 1976; Steffen & Steffen, 2021; Steffen et al., 2021; Wu & Hasegawa, 1996). For context, a stress change of tens of MPa is comparable to, or larger than, typical earthquake stress drops (of order 1 to 10 MPa), highlighting that rebound-related stresses can be mechanically relevant where faults are favourably oriented. GIA provides clear examples of large-magnitude stress modulation, but it is also spatially restricted and transient. Attribution of present-day seismicity in formerly glaciated regions remains debated: while some studies invoke continuing rebound control, focal mechanisms, stress observations and updated models suggest that residual GIA stresses are often modest at the regional scale, with even distal plate-boundary forces and other loads commonly of similar or greater magnitude (e.g. Gudmundsson, 1999; Steffen et al., 2021). This uncertainty is compounded by evidence that faulting is not always restricted to a narrow deglacial window in time, with radiocarbon, geomorphic and LiDAR-based studies indicating episodic rupture extending into the mid to late Holocene and variable segment timing, complicating fault age attribution and the distinction between glacially modulated slip and other sources (Lagerbäck, 1992; Mikko et al., 2015; Olesen, 1988; Steffen et al., 2021). Fluids are recognised as playing a key role in modulating fault rupture timing, but hydromechanical processes remain poorly constrained; unloading-related pore-pressure changes may influence fault stability over broad areas, but they are rarely incorporated alongside evolving Coulomb stress in regional assessments (Steffen et al., 2021). Finally, debate continues over the spatial extent of seismicity, with candidate faults proposed near ice margins and outside former ice limits in Europe, evidence that the Laurentide ice sheet modified seismicity in the New Madrid seismic zone, USA (Grollimund & Zoback, 2001) and indirect evidence from Canadian lake-sediment records highlights the need for consistent inventories and direct constraints on fault-zone properties (Steffen et al., 2021). Recent work on the Antarctic ice sheet has identified possible interactions (Barbosa et al., 2025), but local observations remain scarce, and both future ice loss volumes and fault sensitivities are uncertain and require further research. Addressing these uncertainties across regions will require complementary constraints on fault architecture, stress and hydrogeology at depth, motivating long-term monitoring efforts (Steffen et al., 2021).

This chapter examines interactions between earthquakes and glaciers across all environments, but examples particularly focus on tectonically active, glaciated mountain regions, where seismic and cryospheric processes coevolve across time scales from seconds to millennia. To keep the scope constrained, purely glaciological seismic phenomena (e.g. icequakes) are excluded unless they demonstrably contribute to broader cryospheric or multihazard impacts. In addition, glacio-seismic and glacio-tectonic processes observed or inferred elsewhere in the solar system are beyond the scope of this chapter. Here, we explore three core topics: (1) how earthquakes influence glacier stability and trigger cryospheric hazards such as snow, ice and rock-ice avalanches and

GLOFs; (2) how glaciers and cryospheric mass changes modulate crustal stresses, fault slip and seismicity through loading, unloading and hydrological processes; and (3) how earthquake-glacier interactions propagate across time scales and how climate change modifies the boundary conditions under which they operate. These issues form the core of the following three sections. The chapter adopts an explicitly integrated perspective, as earthquake–cryosphere interactions cannot be understood through isolated process studies alone but require coupling seismic, glaciological, geomorphological and natural hazard studies. Building on historical and mechanistic context, it discusses cascading processes and sediment-tectonic coupling, cryospheric feedbacks on seismicity, and concludes with a synthesis of key uncertainties, hazard implications and priorities for future research.

10.2 Cryospheric hazards and multihazard cascades

This section focuses on seismic and cryospheric hazards not as an end in themselves, but as a lens to explore how and where earthquakes and glaciers interact. The most common interactions occur when near-threshold cryospheric systems are perturbed by seismic forcing, making the associated hazards a sensitive indicator of coupling between ground motion, ice, snow and sediment. Establishing this context is useful prior to examining how cryospheric mass changes feed back into crustal stress and seismicity later in Section 10.3. In conventional seismic risk frameworks, the consequences of earthquake-related ground shaking are primarily framed in terms of damage to engineered structures. The same ground motion, however, equally impacts natural systems. Shaking intensity, rupture geometry, distance and local site and topographic effects (e.g. topographically amplified shaking) generate large, rapidly applied inertial forces capable of destabilising rock and ice, with widespread coseismic landsliding a key consequence of earthquakes and with a density, volume and mobility scaling with peak ground acceleration and shaking duration (e.g. Jibson et al., 2006; Roback et al., 2018). In glaciated and seasonally snow-covered terrain, these forces act on snow, ice and rock-ice interfaces, enabling direct triggering of snow, ice and rock-ice avalanches (Podolskiy et al., 2010a, b) and contributing to sediment mobilisation during major earthquakes, primarily via landsliding (Kargel et al., 2016). To fully understand seismic hazard in glaciated regions, it is therefore necessary to understand glacier-earthquake interactions. In the following paragraphs, we will explore baseline cryospheric hazard processes and broader multihazard interactions, before considering detailed case studies of seismically modulated glacial hazards.

Many cryospheric systems exist close to mechanical and thermal stability thresholds, such that relatively small perturbations can produce abrupt and high-magnitude responses (e.g. Faillettaz et al., 2015; Kääb et al., 2021). These responses include nonlinear seasonal accelerations and surges that are present across a wide range of glaciers, but also complete failure and collapse in glacier detachments (e.g. Kääb et al., 2021). Glaciers, permafrost-affected rock slopes and ice- or moraine-dammed lakes are all capable of sudden failure when critical thresholds are exceeded, initiating potentially hazardous ice and rock-ice avalanches, debris flows and GLOFs (Clague, 2013). Glacier thinning and retreat progressively remove lateral and basal support from adjacent hillslopes, promoting debris-flow-driven rockfalls and large rock avalanches, while warming-induced permafrost degradation weakens fracture networks (Gruber & Haeberli, 2007; Haeberli et al., 2017). At the same time, glacier downwasting favours the formation and growth of ice- and moraine-dammed lakes, many of which are structurally weak and susceptible to GLOF formation, producing floods far larger and more damaging than meteorological events of comparable catchment size (Allen et al., 2019; Clague & Evans, 2000; Emmer & Cochachin, 2013). Hydrological processes can also influence these hazards by elevating pore pressures, accelerating glacier flow and governing the initiation and magnitude of subglacial floods (e.g. Faillettaz et al., 2015). These systems are also conditioned by longer-term climatic change, bringing many closer to failure thresholds and making them increasingly sensitive to seismic shaking.

The interaction between seismic and cryospheric processes cannot be viewed in isolation, but must be considered as a component of a broader multihazard and cascading hazard system. This multihazard perspective emphasises that earthquakes, landslides, floods and cryospheric hazards are linked through triggering, compounding and sequential interactions, forming dynamic hazard networks rather than simple binary interaction pairs (Gill & Malamud, 2014, 2016; Van Wyk de Vries, 2025). In glaciated regions, seismic shaking may concurrently initiate snow, ice, or rock-ice avalanches, destabilise moraine or ice dams, trigger landslides and redistribute large sediment volumes into valleys and lakes, also raising the likelihood of secondary flooding, landslides, debris flows, or outburst events. Conversely, climate-driven cryospheric change modifies boundary conditions for seismic impacts by altering slope stability, ice volume, lake inventories and sediment availability, both increasing the likelihood and complexity of cascading failures when earthquakes occur and modifying their dominant characteristics through time, complicating the understanding and management of the associated risk. These interactions are characteristic of many high-impact events (e.g. Sattar et al., 2025; Shugar et al., 2021), with hazard outcomes shaped by coupled cryospheric, tectonic, climatic and anthropogenic processes and requiring integrated management and risk mitigation approaches (Gill & Malamud, 2014, 2016; Van Wyk de Vries, 2025). This coupling of transient seismic forcing with evolving cryospheric state in a multihazard system underpins the case study events examined below.

Earthquake-induced destabilisation of moraine- or ice-dammed lakes has long been regarded as a plausible trigger of GLOFs, either through direct weakening of dams or indirectly via earthquake-triggered landslides and ice avalanches that generate displacement waves and dam overtopping (Clague & Evans, 2000; Lützow et al., 2023;

Veh and Wang, 2025). Particularly prior to the advent of satellites and widespread seismic networks, a seismic trigger was often hypothesised for GLOFs observed primarily through their distal impacts. However, empirical support for this mechanism is limited, with the 1970 M_w 7.9 Ancash earthquake in the Cordillera Blanca, Peru, being a rare example of such a case. This earthquake directly triggered at least six GLOFs from moraine-dammed lakes in adjacent valleys, primarily via coseismic mass movements entering lakes, initiating dam overtopping and causing rapid drainage (Wood et al., 2024). Beyond this exceptional case, regional analyses indicate that earthquake-related GLOFs are rare: of 59 earthquakes between 1900 and 2021 whose effects intersected glacial lakes in the Tropical Andes, only the aforementioned Ancash earthquake led to observed GLOFs (Wood et al., 2024). While historical earthquake and GLOF inventories are highly incomplete, in the modern era all medium to large earthquakes are automatically catalogued. While GLOFs and lake drainage events are not yet routinely catalogued on a global scale, the large majority are easily detectable from both optical and radar satellite imagery, and the paucity of modern 'earthquake-triggered GLOF' observations strongly suggests that, while plausible, they are a rare occurrence and not a key driver of GLOF risk in most instances (Lützow et al., 2023; Veh and Wang, 2025). Despite being a secondary or speculative GLOF trigger in most cases, earthquakes have other well-documented hazard interactions with the cryosphere.

Earthquake shaking can directly destabilise steep glacierised slopes, triggering the failure of rock-ice cliffs, hanging glaciers, or snow- and ice-mantled rock walls and producing mixed avalanches with high mobility and severe downstream impacts. This process was again observed in the 1970 M_w 7.9 Ancash earthquake in Peru's Cordillera Blanca, as well as the GLOFs discussed earlier, seismic shaking destabilised a perched rock and ice mass on Nevado Huascarán, generating a rock-ice avalanche that rapidly entrained additional snow, firn, ice and debris and transformed into a rapid and long-runout debris flow that destroyed much of the town of Yungay, causing several thousand fatalities (Evans et al., 2009; Wood et al., 2024). The presence of ice both within the flow and in the initial high-elevation runout zone reduced internal and basal friction, enabling exceptional runout and extensive valley-floor inundation (Fig. 10.2). A prior nonseismic partial glacier detachment had occurred in a nearby location 8 years prior, in 1962, and the potential for further collapses had been recognised (Carey, 2010), although not in concert with seismic shaking.

More recently, the 2015 M_w 7.8 Gorkha earthquake triggered a wide suite of rock, snow and ice avalanches across Nepal and Tibet, including the deadly Langtang rock-ice avalanche in central Nepal (Fig. 10.3; Fujita et al., 2017; Kargel et al., 2016; Lacroix, 2016). There, strong horizontal and vertical accelerations acted on a snow and ice-loaded rockwall near the seasonal snowline, where winter snow and low temperatures alongside the shaking promoted brittle failure and high initial mobility (Fujita et al., 2017; Kargel et al., 2016). The failure evolved as a multistage rock-ice avalanche,

Figure 10.2 The 1970 Huascarán earthquake-avalanche cascade, Cordillera Blanca, Peru. (A) Oblique view of Nevado Huascarán showing the source area of the earthquake-triggered ice-rock avalanche (Ericksen et al., 1970). (B) Map view of the avalanche impact and inundation area, including the burial of the town of Yungay (Servicio Aerofotogramétrico Nacional; Mergili et al., 2018). The image is approximately 5 km across, Rh = Ranrahirca. (C) Recent view of the Huascarán massif, highlighting the re-vegetated deposit in the foreground and landslide scar in the background (I.R. Viajero, May 2019).

transitioning into a high-velocity granular flow and powerful air blast that caused widespread loss of life and destroyed infrastructure (Gnyawali et al., 2020; Kargel et al., 2016). Similar, though generally smaller-scale, ice-rich rock avalanches have been documented following large earthquakes in Alaska (Jibson et al., 2006), while inventories confirm that seismic triggering of snow avalanches is widespread, underscoring the sensitivity of cryospheric mass movements to transient ground motion (Podolskiy et al., 2010a). This seismic triggering of mass flows is critical to consider for three reasons: (1) seismic ground shaking may trigger failure of areas that would otherwise have remained stable, (2) the wide spatial footprint of strong ground motion during large earthquakes may drive concurrent failure of rock, ice and snow leading to highly mobile mass-flow mixtures, reaching further than any individual flow might, and

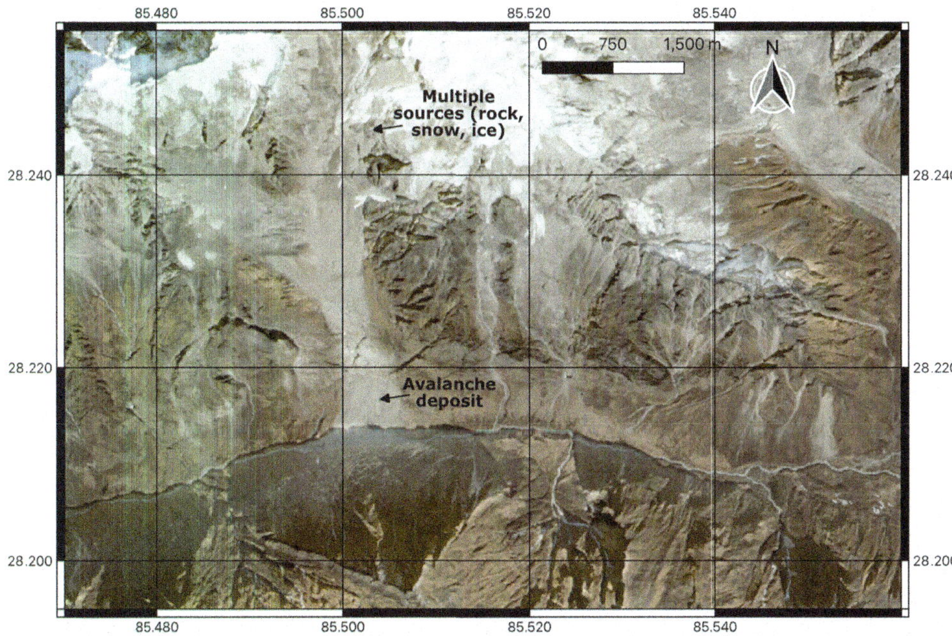

Figure 10.3 Post event Sentinel-2 (28 December 2015) satellite image of the Langtang Valley ice-rock avalanche triggered by the 25 April 2015 Mw 7.8 Gorkha earthquake, Nepal (e.g. Lacroix, 2016). The deposit extends from steep glacierised source areas into the valley floor, illustrating coupling between seismic ground motion, cryospheric slope instability and downstream sediment redistribution.

(3) seismically triggered cryospheric hazards are typically concurrent with other seismic hazards (e.g. landslides, building collapses), complicating postdisaster response.

The interaction of seismic shaking with glaciers often manifests most clearly through multihazard cascades, both instantaneous and over a period of time after the event. The 2002 Kolka–Karmadon catastrophe (Caucasus) is widely interpreted as a coupled rock-ice system failure in which a large rock/ice avalanche from the Dzhimarai-Khokh wall impacted and entrained much of Kolka Glacier, producing an exceptionally mobile ice-rock avalanche and downstream debris-flow sequence (Huggel et al., 2005; Kotlyakov et al., 2004). No earthquake, beyond the seismic signal of this glacier detachment itself, was recorded at the time of failure. However, seismic shaking in the weeks prior to failure was identified as a likely trigger for supraglacial landslides that loaded the glacier surface with rock debris (Kotlyakov et al., 2004) and potentially preconditioned the system for later failure. The exact role of seismic forcing is debated in this event, and other events show that glacier detachments can occur without prior earthquakes (Kääb et al., 2021), yet this highlights the possibility of strongly time-lagged multihazard interactions through preconditioning. A different example of long-tail hazard is Lake Sarez, Tajikistan, formed in 1911 when an earthquake triggered the Usoi landslide and dammed the Murghab River. Large natural dams can remain metastable for decades to centuries and then fail through earthquake shaking, overtopping by impulse waves from large slope failures or avalanches into the lake, GLOF floodwaves, internal erosion, or progressive deformation, with downstream consequences extending far beyond the initial trigger (Alford et al., 2000; Fan et al., 2020; Schuster and Alford, 2004). As such, Lake Sarez acts as a long-term and potentially catastrophic amplifier to future cryospheric hazards in its glacierized basin despite its seismic origin being more than a century ago. More broadly, both earthquakes and glacier change erode, mobilise and deposit large volumes of sediment across mountainous landscapes, leading to persistent hazard and interacting across a wide range of time scales, as also discussed in the final section.

Across these cases, earthquake-driven cryospheric hazard is best treated as a state-dependent problem: seismic shaking can, and in some cases does act as an immediate trigger, but the greatest impacts often arise where strong ground motion intersects preconditioned slope, glacier and lake systems, propagating through hazard cascades which may persist long after the mainshock. Indeed, due to the complexity of mountain systems, the indirect link between past seismic primers and eventual hazards may often be missed. For instance, earthquake-generated barrier lakes demonstrate how single coseismic landslides can restructure the hazard profile of an entire glacierised basin (Schuster and Alford, 2004), introducing a new and potentially catastrophic failure threshold. This coupling motivates both the need for seismic processes to be considered in cryospheric hazard assessments and vice versa, and for both to be embedded in fully integrated multihazard assessments. Having established how seismic shaking interacts with glacier systems, the next section examines the opposite process, how changes in ice volume can modulate seismicity and its consequences.

10.3 Cryospheric influence on seismicity

This section explores the mechanisms by which glacier and broader cryosphere mass variability can influence crustal stress, fault behaviour and seismicity across a range of spatial and temporal scales. First, the physical basis for this glacier-earthquake coupling is outlined, focusing on stress changes induced by surface load changes and the sensitivity of critically stressed faults to modest perturbations. The observational and modelling approaches that integrate seismic and glaciological data to quantify the extent and magnitude of cryosphere-driven stress transfer are then reviewed. Subsequent subsections discuss evidence for stress modulation at seasonal to interannual time scales, before examining postglacial rebound and paleoseismic records, and finally considering modern ice loss as a potential driver of seismicity. The section then considers case studies from Fennoscandia (Fig. 10.4) and from modern settings such as Greenland and Patagonia to illustrate both long-term and contemporary expressions of cryospheric control on seismicity. This is concluded with a discussion of the implications for seismic hazard in

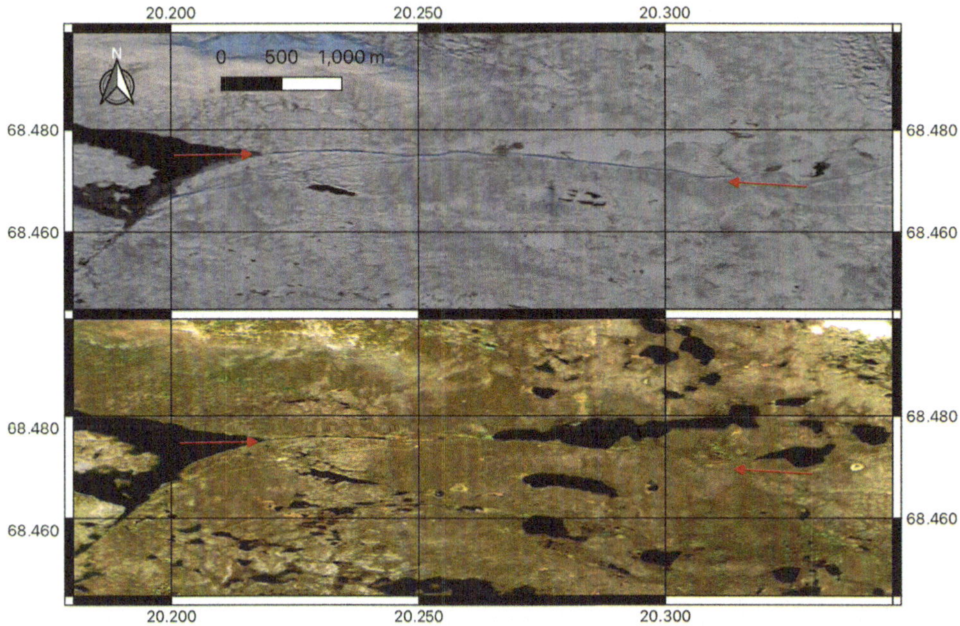

Figure 10.4 Optical Sentinel-2 imagery of the Pärvie postglacial fault, northern Sweden, shown under (upper) summer and (lower) winter conditions. Red arrows highlight the fault scarp.

rapidly deglaciating regions, alongside remaining uncertainties that limit attribution and forecasting of glacier-modulated seismic hazard.

Rapid changes in ice and snow loading drive lithospheric stress changes by modifying vertical normal stress and modifying horizontal deviatoric stresses, each of which may generate new faults or, more commonly, promote slip on favourably oriented preexisting structures (Peltier & Andrews, 1976; Steffen et al., 2021; Wu & Hasegawa, 1996). Ice-sheet unloading (and, with opposite sign, loading) produces spatially and temporally heterogeneous stress perturbations, with Coulomb stress changes that may be locally comparable in magnitude to long-term tectonic forcing (Hampel et al., 2007; Hetzel & Hampel, 2005). Because many crustal faults exist close to failure, even modest cryosphere-driven stress changes of order 0.1–1 MPa can advance or delay rupture timing by altering the stress distribution on critically stressed fault planes (Bettinelli et al., 2008; Hetzel & Hampel, 2005; Steffen et al., 2021). Numerical and analytical models further demonstrate that even nonoptimally oriented faults can be reactivated under glacial loading and unloading due to rotation of the principal stress axes and transient increases in horizontal stress during rebound, helping to explain the range of postglacial faulting geometries and timings observed in formerly glaciated continental shield regions (Steffen et al., 2021). Collectively, these mechanisms frame glacier crustal load changes as stress modulators that primarily influence where and when earthquakes occur by altering the timing of failure on existing fault systems. Importantly, this implies that any resulting seismic risk is primarily tied to alteration of rupture position and timing on potentially active faults that already exist.

Constraints from observations and numerical models of ice-sheet-earthquake coupling have advanced primarily to test whether glacier-driven stress changes are large enough to influence earthquake occurrence. In formerly glaciated regions, notably Fennoscandia and eastern North America, geomorphic and palaeoseismic evidence indicates a phase of elevated intraplate seismicity following deglaciation. Fault scarps are observed offsetting glacial landforms, glacio-fluvial deposits and late-Quaternary shorelines, demonstrating that significant brittle deformation postdated ice retreat rather than reflecting inherited structures; importantly, the seismic magnitude implied by these faults is larger than generally expected within the background tectonic regime (Adams, 1989; Lagerbäck, 1992; Olesen, 1988). Trenching and sedimentological studies show that many of these scarps record surface-rupturing earthquakes in late-glacial to early Holocene time, expressed through displaced bedrock, seismically disturbed tills, liquefaction features and widespread mass movements in unconsolidated sediments (Adams, 1989). High-resolution LiDAR mapping has since refined and expanded regional fault inventories, revealing the continuity, segmentation and spatial extent of postglacial fault scarps that crosscut glacial deposits and landforms across large areas of northern Sweden (Mikko et al., 2015). Together, these geomorphic, stratigraphic and remote-sensing observations provide a robust if spatially biased observational basis for postglacial intraplate faulting, motivating more detailed investigation of individual fault systems in well-preserved regions (Steffen et al., 2021). More recently, this geological field evidence is complemented by extensive seismic monitoring networks and satellite and airborne geodetic techniques.

Continuous GNSS observations directly capture uplift and horizontal deformation associated with ice unloading, while gravimetric data constrain the magnitude and spatial pattern of surface mass change, together providing improved boundary conditions for modelling stress evolution in the lithosphere (Steffen & Wu, 2011; Steffen et al., 2021; Wu & Hasegawa, 1996). These datasets are increasingly combined with seismic catalogues, numerical modelling and the paleoseismic constraints discussed above to explore the full range of glacier-earthquake interactions. Coupled numerical models further show that predicted stress magnitudes and geometries are sensitive to a broad suite of assumptions, including the load history and lithosphere–asthenosphere rheology, reinforcing the need for high-quality observational constraints to reduce nonuniqueness and, where relevant, improve attribution confidence (Steffen et al., 2021). Collectively, these approaches provide a broad evidence base for past and present interactions; for a fuller understanding, it is useful to complement this with data on shorter-term load changes.

A small but growing literature documents seismic impacts from surface load changes substantially smaller and shorter-lived than those expected from glacier mass variation, reinforcing the case that these cryospheric load changes provide an important local to regional control on seismicity (e.g. Hainzl et al., 2006). Seasonal to interannual variations in snow, ice and hydrological loading generate measurable elastic deformation and stress perturbations, providing potential low-magnitude, high-frequency cryosphere-earthquake coupling (Bettinelli et al., 2008; Hainzl et al., 2006). GPS observations in northeastern Japan demonstrate coherent annual vertical and horizontal deformation driven by winter snow loading, confirming that subannual cryospheric loads measurably modulate crustal stress (Heki, 2001). Outside of the cryosphere, seasonal monsoon-driven water storage provides an analogue mechanism and has been proposed as a modulator of earthquake timing in the Himalaya, with inferred stress changes sufficient to advance or delay rupture on critically stressed faults without altering the underlying tectonic regime (Bettinelli et al., 2008). Comparable behaviour was observed in California, with seasonal hydrological loading inducing kPa-scale Coulomb stress variations and corresponding changes in seismicity (Johnson et al., 2017). Rainfall-triggered earthquakes, attributed to poroelastic stress transfer and pore-pressure diffusion, provide further evidence that even small transient surface forcing can measurably influence seismicity (Hainzl et al., 2006). The following paragraphs first explore the best available field evidence for these interactions from Scandinavian fault systems, and then consider modern evidence from regions of rapid ice loss.

Fault systems surrounding the former Fennoscandian ice sheet provide the most detailed and best-documented evidence for postglacial intraplate faulting. Major structures such as the Pärvie (Fig. 10.4), Stuoragurra (Fig. 10.5), Burträsk and Lansjärv faults form laterally continuous scarps tens to more than 100 km long, offsetting a range of precisely dated glacial deposits, shorelines and late-Quaternary landforms and indicating large-magnitude ($M_w \geq 7$) surface-rupturing earthquakes shortly after the collapse of the Fennoscandian ice sheet (Lagerbäck, 1992; Olesen, 1988; Steffen et al., 2021). In recent years, this dataset has been complemented by extensive, high-resolution LiDAR surveys that capture fault scarp locations, geometries and offset magnitudes in unprecedented detail (Mikko et al., 2015). Improving on earlier aerial photographs and satellite-based mapping, these LiDAR data confirm observations that many scarps crosscut glacial sediments rather than reflecting inherited bedrock topography (Mikko et al., 2015). Trenching and stratigraphic investigations provide direct palaeoseismic evidence for late-glacial to early Holocene rupture, including faulted tills, displaced glacio-fluvial and lacustrine sediments, and associated seismically triggered mass movements (Olesen, 1988; Steffen et al., 2021). A combination of the spatial association of these fault systems with the former ice sheet, precise geochronological dating and robust relative age models, coupled with sedimentary and remote-sensing evidence, and a good understanding of the baseline tectonic regime, leads to very high confidence that ice-sheet-scale unloading caused fault reactivation and large-magnitude earthquakes (Steffen et al., 2021). Given the substantial ground shaking

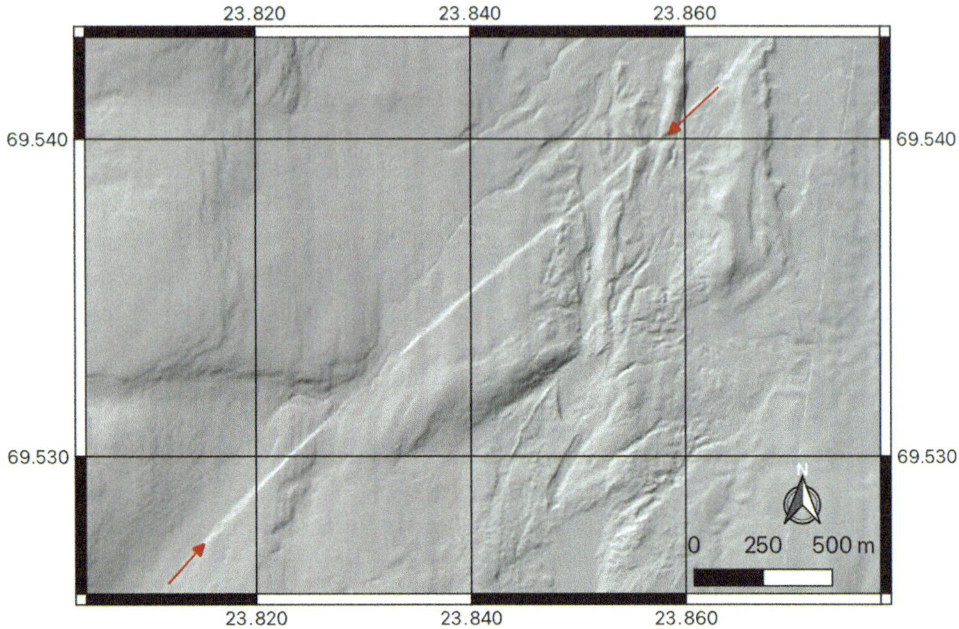

Figure 10.5 Lidar hillshade image (sun azimuth 315 degrees, altitude 40 degrees) of the Máze-Stuoragurra postglacial fault system, northern Norway. The fault scarp is expressed as a sharp, laterally continuous break in slope cutting across both glacial and postglacial landforms.

generated by such earthquakes and the presence of major population centres around rapidly shrinking glaciers, a key follow-on question is the extent to which such seismicity is observed today.

Greenland and Patagonia are home to some of the world's most rapid ice loss (Foresta et al., 2018; Khan et al., 2015; Van Wyk de Vries et al., 2023), and are projected to lose even greater ice volumes over the course of the 21st century (Bravo et al., 2021; Briner et al., 2020). They thus provide smaller-scale contemporary analogues to post-LGM Fennoscandia and possible glacial modulation of crustal deformation and seismicity (Ammirati et al., 2024; Olsen & Nettles, 2017). In Greenland, geodetic observations document rapid elastic and viscoelastic uplift near retreating ice margins, while seismic records show increased rates of localised microseismicity near major outlet glaciers (Olsen & Nettles, 2017; Steffen et al., 2021). A key challenge in Greenland is separating possible deglaciation-triggered crustal earthquakes from seismicity at the ice-bed interface driven by glacial basal sliding (e.g. Ekström et al., 2006), and indeed recent evidence suggests that englacial icequakes are a substantial driver of internal deformation of ice streams (Fichtner et al., 2025). In southern Patagonia, a dense seismic network deployed around the Southern Patagonian Icefield has revealed clusters of shallow earthquakes concentrated around regions of rapid ice thinning (Richter et al., 2016), with focal mechanisms indicating upper-crustal faulting distinct from deeper subduction-related seismicity (Ammirati et al., 2024). These observations coincide with some of the highest measured surface uplift rates globally, driven by rapid ice loss (Lange et al., 2014; Richter et al., 2016), and demonstrate how contemporary unloading can be a first-order driver of seismicity even in plate-boundary-adjacent settings. Together, the Greenland and Patagonia case studies provide preliminary evidence that current glacier mass loss is triggering earthquakes in both tectonically active and quiescent settings, yet much uncertainty remains on the spatial extent, temporal variability and maximum magnitude of these events.

On a global scale, stress changes and seismicity associated with glacier unloading are extremely limited compared to those driven by tectonic processes, but may be locally important for the spatial and temporal clustering of shallow seismicity (Steffen et al., 2021). Observations from Patagonia, Alaska and Iceland indicate that this modulation is spatially concentrated near zones of rapid ice loss and surface uplift, implying that glacier mass loss may modify seismic hazard, but there is little consensus as to the importance of this for overall risk management (Ammirati et al., 2024; Árnadóttir et al., 2009; Rollins et al., 2021). From a hazard perspective, this finding supports the value of time-dependent monitoring and scenario-based assessment near population centres and infrastructure in high-altitude and high-latitude environments, particularly proximal to areas of rapid glacial mass loss. Despite locally strong qualitative and case-based evidence, quantifying the wider contribution of cryospheric unloading to present-day seismic hazard remains challenging due to the short temporal overlap between high-resolution glaciological observations and instrumental seismic records, difficulties in separating cryosphere-driven modulation from tectonic, magmatic and hydrological forcing, and sensitivity of stress change simulations to poorly constrained model parameters such as subsurface rheology (Steffen & Wu, 2011; Steffen et al., 2021; Wu & Hasegawa, 1996). These uncertainties and data gaps currently limit the translation of future glacier volume change forecasts into changes in seismic magnitude and frequency, with changes in the probability of dangerous or damaging ground shaking for potentially exposed people and infrastructure even more uncertain. Despite this, current evidence of large-magnitude past events and mechanistic links is clear enough that postglacial seismicity should be captured in the risk assessments of many ice-adjacent regions, and ideally account for a maximum plausible earthquake magnitude in design criteria.

Collectively, the evidence reviewed here clearly indicates that glacier mass loss and associated surface load and crustal stress changes can modulate seismicity across a range of time scales. The clearest and largest-magnitude evidence of this is preserved in postglacial fault scarps (Steffen et al., 2021), yet evidence increasingly points towards a measurable seismic signal of present-day ice loss (Ammirati et al., 2024). The extent to which this meaningfully alters seismic risk for communities in high-altitude and high-latitude settings remains an open question, yet this provides a clear mechanistic bridge from anthropogenic climate change to ice loss and increased seismicity. Together with evidence for modern glacial modulation of magma generation rates (e.g. Auriac et al., 2013), this highlights that even these geohazards, typically considered independent of human activity, cannot be fully decoupled from it. For the broader ice-earthquake discussion, the key implication is that cryospheric load changes are most likely to modify where, when and how large earthquakes occur on existing fault systems, rather than whether areas are seismically active at all. The next section explores the longer timescale interactions between glaciers and seismicity through the lens of sediment connectivity and landscape evolution.

10.4 Sediment connectivity and climate-tectonic coupling

The movement and storage of sediment connect the short-timescale earthquake-cryosphere cascades discussed earlier with the longer-term coevolution of erosion, relief and deformation. Therefore earthquakes and glaciation interact not only through short-term failures but also by reorganising sediment connectivity and controlling whether material is evacuated or stored (Fan et al., 2019; Harries et al., 2021). In glaciated mountain belts, this connectivity is conditioned by hillslope-glacier coupling and the capacity of glaciers and proglacial rivers to act as debris conveyors (Scherler et al., 2011). Large earthquakes can also generate sediment pulses, particularly via landsliding, that temporarily exceed background fluvial and glacial fluxes, altering downstream transport conditions (Dadson et al., 2004; Fan et al., 2019). The geomorphic impact is commonly prolonged because coseismic and paraglacial sources are repeatedly reworked, sustaining elevated sediment yields for years to decades (e.g. Morin et al., 2018). Cryospheric change can further amplify this by modulating slope stability, shaking intensity and sediment availability, adding further sediment pulses to the system and making the case for earthquakes, glaciation and postevent sediment reworking to be treated as a coupled sediment network problem rather than isolated drivers (e.g. Densmore et al., 1998; McColl & Davies, 2013; McColl et al., 2012). With earthquakes playing both a constructive and destructive role in mountain building (e.g. Hovius et al., 2011), and glaciers providing an additional control on landscape erosion, it follows that long-term feedback loops may also exist between the two. The following paragraphs explore how sediment routing and landscape evolution can drive coupled changes in seismicity and glaciation.

Mass removal through erosion can contribute to isostatic and flexural responses that modify rock uplift patterns and, in some settings, the distribution and rate of fault slip, linking sediment export to deformation and seismicity. Earthquakes are central to this feedback because they both build relief and generate large, spatially focused erosion (primarily via landsliding), raising sediment export for years to decades as coseismic material is progressively connected to sediment transport networks (Dadson et al., 2004; Fan et al., 2019; Hovius et al., 2011; Malamud et al., 2004; Marc et al., 2019). In glaciated mountain belts, glacial downwasting and meltwater routing can therefore establish the baseline unloading rate, with seismic sediment pulses intermittently accelerating denudation. Together, the balance between mass removal and rock uplift determines how strongly overall deformation and orogenesis can be modulated by erosional unloading, in turn influencing seismicity (Brozović et al., 1997; Fan et al., 2019; Herman et al., 2021; Li et al., 2017). These same processes alter glaciation by determining the elevation of mountain peaks, with higher (and therefore colder) mountains generally supporting larger glaciers. Building on this, the following paragraph explores the glacial buzzsaw hypothesis, which considers possible feedback between overall topographic elevation and glaciation.

The glacial buzzsaw hypothesis proposes that, by its efficiency, glacial erosion imposes an effective upper limit on mountain elevation by preferentially lowering terrain above the equilibrium line altitude (ELA), such that overall elevation and peak heights become glacially and climatically constrained even where rock uplift continues (Brozovíc et al., 1997; Egholm et al., 2009; Herman et al., 2021; Mitchell & Montgomery, 2006). This links to the sediment connectivity ideas above as it emphasises that long-term landscape evolution is not just a function of climate and rock uplift but also of how sediment is eroded and removed from the system, creating space for episodic drivers such as earthquakes to influence long-term outcomes without dominating them. The 'height limit' thus emerges from sediment production and evacuation; through erosion, glaciers generate large sediment fluxes that must be routed through proglacial fluvial pathways, and the effectiveness and intermittency of that routing determine whether erosion can keep pace with uplift (Egholm et al., 2009; Herman et al., 2021). The extent of this erosive 'buzzsaw' and whether it can be reliably distinguished from tectonic relief controls are subject to active debate: some studies support an ELA-linked limit in specific settings, whereas others emphasise uplift-rate dependence, spatially variable glacial efficiency and the influence of nonglacial hillslope processes, such as topographically amplified shaking, that can complement or overwhelm glacial downwasting (Egholm et al., 2009; Herman et al., 2021; Mitchell & Montgomery, 2006; Prasicek et al., 2020). Other studies have, conversely, noted that glaciation can reduce long-term denudation rates (Thomson et al., 2010). Earthquakes enter this broader system in two ways: as an episodic but powerful source of sediment, and as a driver of orogenesis at convergent margins (Fan et al., 2019; Hovius et al., 2011). Long-term fault slip rates and seismicity may therefore be modified by erosional patterns, while episodic earthquake-landslide erosion pulses introduce a nonelevation-dependent sediment source that generally acts to reduce relief. In both cases, sediment routing and evacuation rate act as important long-term controls on landscape change (Fan et al., 2019; Herman et al., 2021; Hovius et al., 2011).

Climate sets the overall background conditions in which earthquake- and cryosphere-driven sediment pulses are generated, mobilised, evacuated and/or stored; it therefore determines the extent to which short-lived sediment cascades, as discussed in the hazard section above, accumulate into longer-term topographic change. Through its control on glacier extent (through ELA altitude in particular) and thermal regime, and by altering meltwater, climate changes the rate and spatial focus of erosion and sediment production (Cook et al., 2020; Koppes & Montgomery, 2009; Koppes et al., 2015; Norris et al., 2025; Patton et al., 2022). At the catchment scale, climate-driven variability in runoff and extremes determines whether this material is exported, accumulates, or remobilises additional sediment, or is buffered in transient storage (Chakraborty, 2023; Harries et al., 2021; Morin et al., 2018).

Together, these lines of evidence highlight several key points around long-term glacier-earthquake coupling. First, both glaciers and earthquakes are important sediment producers in mountainous areas, with both contributing to the same highly coupled sediment network and determining the overall erosion rate. Sediment removal, remobilisation and storage from each influence the other, and both exert an important influence on overall landscape evolution (Dadson et al., 2004; Fan et al., 2019; Hovius et al., 2011). Second, overall erosion and denudation patterns, with glaciation and seismicity both being first-order controls, feed back in turn into overall orogenesis and deformation patterns, setting the overall distribution of faults and their slip rates (Brozovíc et al., 1997; Fan et al., 2019; Herman et al., 2021; Li et al., 2017). Fault distribution and slip rate determine where earthquakes occur and how large they are. Third, overall sediment mobility, evacuation rates and orogenesis determine mountain belt hypsometry. Due to the elevation-dependence of temperature, higher mountains can host larger glaciers (all else equal), feeding back into glaciation extent (Brozovíc et al., 1997; Egholm et al., 2009; Herman et al., 2021; Mitchell & Montgomery, 2006; Prasicek et al., 2020). Fourth, climate acts as an important control on this overall coupled system, in particular by modulating glaciation and precipitation-driven sediment mobility (Chakraborty, 2023; Harries et al., 2021; Morin et al., 2018; Norris et al., 2025; Patton et al., 2022). Climate modulates not only glacier extent and discharge rates but also the rate and intermittency of sediment transfer, and therefore influences whether earthquake-driven sediment pulses amplify or dampen longer-term erosional efficacy (Chakraborty, 2023; Cook et al., 2020; Harries et al., 2021; Koppes & Montgomery, 2009; Koppes et al., 2015; Norris et al., 2025; Patton et al., 2022). Where export is efficient, sustained denudation can feed back on rock uplift and fault behaviour, linking sediment routing to seismicity (Densmore et al., 1998; Hetzel & Hampel, 2005; Li et al., 2017; Marc et al., 2019). Overall, the climate-glacier-earthquake sediment connectivity system and feedbacks therein set the baseline for the shorter-term process interactions. Seismic triggering of cryospheric multihazard chains and glacial fault modulation, discussed in

Sections 10.2 and 10.3, emphasise the coupling between these processes across a wide range of time scales.

10.5 Conclusions

Earthquake-cryosphere coupling operates bidirectionally and across time scales. Short-lived interactions can arise when seismic shaking perturbs near-threshold glacial and periglacial systems, producing mass-movement cascades. Over medium deglacial timescales, glacier mass loss and associated crustal unloading can exert a first-order control on fault activation patterns and slip. On longer geological time scales, the same coupling influences landscape evolution because earthquakes and glaciers jointly control where sediment is produced, how long it is stored and how rapidly it is exported, which in turn condition orogenesis and fault slip patterns. While operating across time scales, these processes are closely coupled, and this chapter therefore explores hazards, stress modulation and sediment transfer as linked expressions of the same system.

1. How do earthquakes influence glacier stability and cryospheric hazards? Many glacial and periglacial systems naturally exist close to internal stability thresholds, and seismic shaking can drive them to failure. This is most commonly observed through earthquake-triggered mass movements, which include landslides but also snow avalanches, rock-ice avalanches and glacier detachments. Indirect and lagged multi-hazard interactions, for instance, through glacial and seismic sediment pulses, provide another major glacial-seismic connection. In the first section, a range of relevant case studies is discussed, showcasing these direct mechanistic links, indirect multihazard connections, and broader implications for risk management in glaciated areas.
2. How do glaciers and cryospheric mass changes modulate stresses, fault slip and seismicity? Glacier mass change modifies the loading of the crust, driving changes in both horizontal and vertical stress fields. Altogether, this enables deglaciation to drive faulting in the upper crust, although this primarily occurs through reactivation of existing, favourably oriented fault networks. Integrated sedimentary and remote-sensing datasets from postglacial landscapes show evidence for M_W 7+ earthquakes and widespread induced mass movements, large enough to pose a substantial seismic hazard. Evidence of seismic modulation from much smaller and shorter-lived surface and hydrological load changes provides additional mechanistic support for glacial modulation of fault systems. Modern evidence from rapidly deglaciating settings shows evidence for induced seismicity, although the extent to which this alters overall seismic risk remains an open question.
3. How do earthquake-glacier interactions propagate across time scales, and how does climate change modify the boundary conditions under which they operate? Both earthquakes and glaciers are major sediment producers, with the overall sediment system exhibiting a range of interactions between the two. Glaciers provide a more

consistent, elevation-dependent erosion-transfer-storage profile, which tends to increase relief, while earthquakes drive more episodic erosion through induced mass movements. The overall long-term erosion and denudation profile balances orogenic mountain building, feeding back into the distribution and slip rate on active fault networks (determining overall seismicity) and mountain hypsometry (determining the extent of glaciation). Climate acts as an additional component of this system, modulating overall glacier extent and erosional efficiency alongside tectonic processes.

Taken together, the chapter shows that earthquakes and glaciers interact not only through short-timescale cascades but also by jointly regulating sediment production, storage and export, thereby shaping relief, erosion patterns and deformation over longer timescales. Earthquakes matter to long-term topography not because they are frequent, but because they intermittently reorganise sediment connectivity and inject large sediment pulses into glacio-fluvial routing systems. Glaciers matter not only as erosional agents but as dynamic conveyors and buffers that determine whether sediment is evacuated or stored, influencing the efficiency of erosional unloading and, in turn, are shaped by elevation feedbacks.

Four priorities emerge across this chapter, namely that there is a need for: (1) improved event-chain and process inventories that capture delayed and compound consequences (not just triggers), paired with consistent observation of postevent trajectories; (2) closer quantitative coupling between cryospheric mass change, stress modelling and fault properties, including better constraints on rheology and ice history; (3) direct constraints on fault-zone structure, fluids and in-situ stress in key postglacial fault systems, for example, motivated by long-term monitoring initiatives that aim to improve attribution of cryosphere-modulated seismicity; and (4) stronger integration of multihazard frameworks that explicitly link seismic, cryospheric and sediment-driven processes to risk reduction for high-mountain communities and infrastructure projects, enabling coupled assessment of cascading impacts, long-tail hazards and evolving risk under climate change.

References

Adams, J. (1989). Postglacial faulting in eastern Canada: Nature, origin and seismic hazard implications. *Tectonophysics, 163*, 323–331. https://doi.org/10.1016/0040-1951(89)90267-9.

Alford, D., Cunha, S. F., & Ives, J. D. (2000). Lake Sarez, Pamir Mountains, Tajikistan: Mountain hazards and development assistance. *Mountain Research and Development, 20*, 20–23. https://doi.org/10.1659/0276-4741(2000)020[0020:LSPMTM]2.0.CO;2 publisher: International Mountain Society.

Allen, S. K., Zhang, G., Wang, W., Yao, T., & Bolch, T. (2019). Potentially dangerous glacial lakes across the Tibetan Plateau revealed using a large-scale automated assessment approach. *Science Bulletin, 64*, 435–445. https://doi.org/10.1016/j.scib.2019.03.011.

Ammirati, J.-B., Azúa, K., Pastén-Araya, F., Richter, A., Wiens, D. A., Flores, M. C., Ruiz, S., Guzmán-Marín, P., Lanza, F., & Sielfeld, G. (2024). Fault reactivation linked to rapid ice-mass removal from the Southern Patagonian Icefield (48–52°S). *Tectonophysics, 880*, 230–320. https://doi.org/10.1016/j.tecto.2024.230320.

Árnadóttir, T., Lund, B., Jiang, W., Geirsson, H., Björnsson, H., Einarsson, P., & Sigurdsson, T. (2009). Glacial rebound and plate spreading: Results from the first countrywide GPS observations in Iceland. *Geophysical Journal International, 177*, 691–716. https://doi.org/10.1111/j.1365-246X.2008.04059.x.

Auriac, A., Spaans, K. H., Sigmundsson, F., Hooper, A., Schmidt, P., & Lund, B. (2013). Iceland rising: Solid earth response to ice retreat inferred from satellite radar interferometry and visocelastic modeling. *Journal of Geophysical Research: Solid Earth, 118*, 1331–1344. https://doi.org/10.1002/jgrb.50082.

Barbosa, I., Steffen, R., Steffen, H., Seidel, E., Gohl, K., & Hübscher, C.: (2025). Circum-antarctic glacially induced fault reactivation since the last glacial maximum. *Tectonics, 44*. e2024TC008 464, https://doi.org/10.1029/2024TC008464, _eprint https://agupubs.onlinelibrary.wiley.com/doi/pdf/10.1029/2024TC008464.

Bettinelli, P., Avouac, J.-P., Flouzat, M., Bollinger, L., Ramillien, G., Rajaure, S., & Sapkota, S. (2008). Seasonal variations of seismicity and geodetic strain in the Himalaya induced by surface hydrology. *Earth and Planetary Science Letters, 266*, 332–344. https://doi.org/10.1016/j.epsl.2007.11.021.

Bravo, C., Bozkurt, D., Ross, A. N., & Quincey, D. J. (2021). Projected increases in surface melt and ice loss for the Northern and Southern Patagonian Icefields. *Scientific Reports, 11*. https://doi.org/10.1038/s41598-021-95725-w publisher: Nature Publishing Group.

Briner, J. P., Cuzzone, J. K., Badgeley, J. A., Young, N. E., Steig, E. J., Morlighem, M., Schlegel, N.-J., Hakim, G. J., Schaefer, J. M., Johnson, J. V., Lesnek, A. J., Thomas, E. K., Allan, E., Bennike, O., Cluett, A. A., Csatho, B., de Vernal, A., Downs, J., Larour, E., & Nowicki, S. (2020). Rate of mass loss from the Greenland Ice Sheet will exceed Holocene values this century. *Nature, 586*, 70–74. https://doi.org/10.1038/s41586-020-2742-6 publisher: Nature Publishing Group.

Brozovíc, N., Burbank, D. W., & Meigs, A. J. (1997). Climatic limits on landscape development in the northwestern Himalaya. *Science (New York, N.Y.), 276*, 571–574. https://doi.org/10.1126/science.276.5312.571 publisher: American Association for the Advancement of Science.

Büntgen, U., Oppenheimer, C., Farinotti, D., Nahtz, T., & Esper, J. (2025). The 2025 Blatten disaster in the Swiss Alps followed exceptional warming and highlights the vulnerability of people and heritage in glaciated landscapes. *Communications Earth & Environment, 6*, 994. https://doi.org/10.1038/s43247-025-02994-8 publisher: Nature Publishing Group.

Carey, M. (2010). *In the shadow of melting glaciers: Climate change and andean society*. Oxford University Press google-Books-ID: JiH_YkVGqNkC.

Chakraborty, A. (2023). How erosion shapes dynamic Quaternary mountain environments: A review. *Quaternary Science Advances, 9*. https://doi.org/10.1016/j.qsa.2023.100071 100 071.

Chen, N., Liu, M., Allen, S., Deng, M., Khanal, N. R., Peng, T., Tian, S., Huggel, C., Wu, K., Rahman, M., & Somos-Valenzuela, M. (2023). Small outbursts into big disasters: Earthquakes exacerbate climate-driven cascade processes of the glacial lakes failure in the Himalayas. *Geomorphology, 422*, 108–539. https://doi.org/10.1016/j.geomorph.2022.108539.

Clague, J. J. (2013). Cryospheric hazards. *Geology Today, 29*, 73–79. https://doi.org/10.1111/gto.12005, _eprint https://onlinelibrary.wiley.com/doi/pdf/10.1111/gto.12005.

Clague, J. J., & Evans, S. G. (2000). A review of catastrophic drainage of moraine-dammed lakes in British Columbia. *Quaternary Science Reviews, 19*, 1763–1783. https://doi.org/10.1016/S0277-3791(00)00090-1.

Cook, S. J., Swift, D. A., Kirkbride, M. P., Knight, P. G., & Waller, R. I. (2020). The empirical basis for modelling glacial erosion rates. *Nature Communications, 11*, 759. https://doi.org/10.1038/s41467-020-14583-8 publisher: Nature Publishing Group.

Dadson, S. J., Hovius, N., Chen, H., Dade, W. B., Lin, J.-C., Hsu, M.-L., Lin, C.-W., Horng, M.-J., Chen, T.-C., Milliman, J., & Stark, C. P. (2004). Earthquake-triggered increase in sediment delivery from an active mountain belt. *Geology, 32*, 733–736. https://doi.org/10.1130/G20639.1.

Densmore, A. L., Ellis, M. A., & Anderson, R. S. (1998). Landsliding and the evolution of normal-fault-bounded mountains. *Journal of Geophysical Research: Solid Earth, 103*, 203–215. 219, https://doi.org/10.1029/98JB00510, _eprint https://agupubs.onlinelibrary.wiley.com/doi/pdf/10.1029/98JB00510.

Egholm, D. L., Nielsen, S. B., Pedersen, V. K., & Lesemann, J.-E. (2009). Glacial effects limiting mountain height. *Nature, 460*, 884–887. https://doi.org/10.1038/nature08263 publisher: Nature Publishing Group.

Ekström, G., Nettles, M., & Tsai, V. C. (2006). Seasonality and increasing frequency of greenland glacial earthquakes. *Science (New York, N.Y.), 311*, 1756–1758. https://doi.org/10.1126/science.1122112 publisher: American Association for the Advancement of Science.

Emmer, A., & Cochachin, A. (2013). The causes and mechanisms of moraine-dammed lake failures in the Cordillera Blanca, North American Cordillera, and Himalayas. *AUC GEOGRAPHICA, 48*, 5–15. https://karolinum.cz/casopis/auc-geographica/rocnik-48/cislo-2/clanek-2301.

Ericksen, G. E., Plafker, G., & Concha, J. F. (1970). *Preliminary report on the geologic events associated with the May 31, 1970, Peru earthquake*. U.S. Geological Survey, google-Books-ID: wfJeAAAAMAAJ.

Evans, I. S. (2021). Glaciers, rock avalanches and the 'buzzsaw' in cirque development: Why mountain cirques are of mainly glacial origin. *Earth Surface Processes and Landforms, 46*, 24–46. https://doi.org/10.1002/esp.4810, _eprint https://onlinelibrary.wiley.com/doi/pdf/10.1002/esp.4810.

Evans, S. G., Bishop, N. F., Fidel Smoll, L., Valderrama Murillo, P., Delaney, K. B., & Oliver-Smith, A. (2009). A re-examination of the mechanism and human impact of catastrophic mass flows originating on Nevado Huascarán, Cordillera Blanca, Peru in 1962 and 1970. *Engineering Geology, 108*, 96–118. https://doi.org/10.1016/j.enggeo.2009.06.020.

Faillettaz, J., Funk, M., & Vincent, C. (2015). Avalanching glacier instabilities: Review on processes and early warning perspectives. *Reviews of Geophysics, 53*, 203–224. https://doi.org/10.1002/2014RG000466, _eprint https://agupubs.onlinelibrary.wiley.com/doi/pdf/10.1002/2014RG000466.

Fan, X., Scaringi, G., Korup, O., West, A. J., Van Westen, C. J., Tanyas, H., Hovius, N., Hales, T. C., Jibson, R. W., Allstadt, K. E., Zhang, L., Evans, S. G., Xu, C., Li, G., Pei, X., Xu, Q., & Huang, R. (2019). Earthquake-induced chains of geologic hazards: patterns, mechanisms, and impacts. *Reviews of Geophysics, 57*, 421–503. https://doi.org/10.1029/2018RG000626, _eprint https://agupubs.onlinelibrary.wiley.com/doi/pdf/10.1029/2018RG000626.

Fan, X., Dufresne, A., Siva Subramanian, S., Strom, A., Hermanns, R., Tacconi Stefanelli, C., Hewitt, K., Yunus, A. P., Dunning, S., Capra, L., Geertsema, M., Miller, B., Casagli, N., Jansen, J. D., & Xu, Q. (2020). The formation and impact of landslide dams – State of the art. *Earth-Science Reviews, 203*, 103–116. https://doi.org/10.1016/j.earscirev.2020.103116.

Fichtner, A., Hofstede, C., Kennett, B. L. N., Svensson, A., Westhoff, J., Walter, F., Ampuero, J.-P., Cook, E., Zigone, D., Jansen, D., & Eisen, O. (2025). Hidden cascades of seismic ice stream deformation. *Science (New York, N.Y.), 387*, 858–864. https://doi.org/10.1126/science.adp8094 publisher: American Association for the Advancement of Science.

Foresta, L., Gourmelen, N., Weissgerber, F., Nienow, P., Williams, J. J., Shepherd, A., Drinkwater, M. R., & Plummer, S. (2018). Heterogeneous and rapid ice loss over the Patagonian Ice Fields revealed by CryoSat-2 swath radar altimetry. *Remote Sensing of Environment, 211*, 441–455. https://doi.org/10.1016/j.rse.2018.03.041.

Fujita, K., Inoue, H., Izumi, T., Yamaguchi, S., Sadakane, A., Sunako, S., Nishimura, K., Immerzeel, W. W., Shea, J. M., Kayastha, R. B., Sawagaki, T., Breashears, D. F., Yagi, H., & Sakai, A. (2017). Anomalous winter-snow-amplified earthquake-induced disaster of the 2015 Langtang avalanche in Nepal. *Natural Hazards and Earth System Sciences, 17*, 749–764. https://doi.org/10.5194/nhess-17-749-2017 publisher: Copernicus GmbH.

Gill, J. C., & Malamud, B. D. (2014). Reviewing and visualizing the interactions of natural hazards. *Reviews of Geophysics, 52*, 680–722. https://doi.org/10.1002/2013RG000445, _eprint https://agupubs.onlinelibrary.wiley.com/doi/pdf/10.1002/2013RG000445.

Gill, J. C., & Malamud, B. D. (2016). Hazard interactions and interaction networks (cascades) within multi-hazard methodologies. *Earth System Dynamics, 7*, 659–679. https://doi.org/10.5194/esd-7-659-2016 publisher: Copernicus GmbH.

Gnyawali, K. R., Xing, A., & Zhuang, Y. (2020). Dynamic analysis of the multi-staged ice–rock debris avalanche in the Langtang valley triggered by the 2015 Gorkha earthquake, Nepal. *Engineering Geology, 265*. https://doi.org/10.1016/j.enggeo.2019.105440 105 440.

Grollimund, B., & Zoback, M. D. (2001). Did deglaciation trigger intraplate seismicity in the New Madrid seismic zone? *Geology, 29*, 175–178. https://doi.org/10.1130/0091-7613(2001)029<0175:DDTISI>2.0.CO;2.

Gruber, S., & Haeberli, W. (2007). Permafrost in steep bedrock slopes and its temperature-related destabilization following climate change. *Journal of Geophysical Research: Earth Surface, 112*.

https://doi.org/10.1029/2006JF000547, _eprint https://agupubs.onlinelibrary.wiley.com/doi/pdf/10.1029/2006JF000547.

Gudmundsson, A. (1999). Postglacial crustal doming, stresses and fracture formation with application to Norway. *Tectonophysics, 307*, 407–419. https://doi.org/10.1016/S0040-1951(99)00107-9.

Haeberli, W., Schaub, Y., & Huggel, C. (2017). Increasing risks related to landslides from degrading permafrost into new lakes in de-glaciating mountain ranges. *Geomorphology, 293*, 405–417. https://doi.org/10.1016/j.geomorph.2016.02.009.

Hainzl, S., Kraft, T., Wassermann, J., Igel, H., & Schmedes, E. (2006). Evidence for rainfall-triggered earthquake activity. *Geophysical Research Letters, 33*. https://doi.org/10.1029/2006GL027642, _eprint https://agupubs.onlinelibrary.wiley.com/doi/pdf/10.1029/2006GL027642.

Hampel, A., Hetzel, R., & Densmore, A. L. (2007). Postglacial slip-rate increase on the Teton normal fault, northern Basin and Range Province, caused by melting of the Yellowstone ice cap and deglaciation of the Teton Range? *Geology, 35*, 1107–1110. https://doi.org/10.1130/G24093A.1.

Harries, R. M., Gailleton, B., Kirstein, L. A., Attal, M., Whittaker, A. C., & Mudd, S. M. (2021). Impact of climate on landscape form, sediment transfer and the sedimentary record. *Earth Surface Processes and Landforms, 46*, 990–1006. https://doi.org/10.1002/esp.5075, _eprint https://onlinelibrary.wiley.com/doi/pdf/10.1002/esp.5075.

Heki, K. (2001). Seasonal modulation of interseismic strain buildup in northeastern Japan driven by snow loads. *Science (New York, N.Y.), 293*, 89–92. https://doi.org/10.1126/science.1061056 publisher: American Association for the Advancement of Science.

Herman, F., De Doncker, F., Delaney, I., Prasicek, G., & Koppes, M. (2021). The impact of glaciers on mountain erosion. *Nature Reviews Earth & Environment, 2*, 422–435. https://doi.org/10.1038/s43017-021-00165-9 publisher: Nature Publishing Group.

Hetzel, R., & Hampel, A. (2005). Slip rate variations on normal faults during glacial–interglacial changes in surface loads. *Nature, 435*, 81–84. https://doi.org/10.1038/nature03562 publisher: Nature Publishing Group.

Hovius, N., Meunier, P., Lin, C.-W., Chen, H., Chen, Y.-G., Dadson, S., Horng, M.-J., & Lines, M. (2011). Prolonged seismically induced erosion and the mass balance of a large earthquake. *Earth and Planetary Science Letters, 304*, 347–355. https://doi.org/10.1016/j.epsl.2011.02.005.

Huggel, C., Zgraggen-Oswald, S., Haeberli, W., Kääb, A., Polkvoj, A., Galushkin, I., & Evans, S. G. (2005). The 2002 rock/ice avalanche at Kolka/Karmadon, Russian Caucasus: Assessment of extraordinary avalanche formation and mobility, and application of QuickBird satellite imagery. *Natural Hazards and Earth System Sciences, 5*, 173–187. https://doi.org/10.5194/nhess-5-173-2005 publisher: Copernicus GmbH.

Islam, N., Carrivick, J. L., Coulthard, T., Westoby, M., Dunning, S., & Gindraux, S. (2025). A growing threat of multi-hazard cascades highlighted by the Birch Glacier collapse and Blatten landslide in the Swiss Alps. *Geology Today, 41*, 200–205. https://doi.org/10.1111/gto.12526, _eprint https://onlinelibrary.wiley.com/doi/pdf/10.1111/gto.12526.

Jibson, R. W., Harp, E. L., Schulz, W., & Keefer, D. K. (2006). Large rock avalanches triggered by the M 7.9 Denali Fault, Alaska, earthquake of November 2002. *Engineering Geology, 83*, 144–160. https://doi.org/10.1016/j.enggeo.2005.06.029.

Johnson, C. W., Fu, Y., & Bürgmann, R. (2017). Seasonal water storage, stress modulation, and California seismicity. *Science (New York, N.Y.), 356*, 1161–1164. https://doi.org/10.1126/science.aak9547 publisher: American Association for the Advancement of Science.

Jäckli, H. C. A. (1965). Pleistocene glaciation of the swiss alps and signs of postglacial differential uplift. In J. R. H. E. Wright, & D. G. Frey (Vol. Eds.), *International studies on the quaternary: Papers prepared on the occasion of the VII Congress of the International Association for Quaternary Research Boulder: 84*, (pp. 1965). Colorado: Geological Society of America. https://doi.org/10.1130/SPE84p153 p. 0.

Kargel, J. S., Leonard, G. J., Shugar, D. H., Haritashya, U. K., Bevington, A., Fielding, E. J., Fujita, K., Geertsema, M., Miles, E. S., Steiner, J., Anderson, E., Bajracharya, S., Bawden, G. W., Breashears, D. F., Byers, A., Collins, B., Dhital, M. R., Donnellan, A., Evans, T. L., ... Young, N. (2016). Geomorphic and geologic controls of geohazards induced by Nepal's 2015 Gorkha earthquake. *Science (New York, N.Y.), 351*, aac8353. https://doi.org/10.1126/science.aac8353 publisher: American Association for the Advancement of Science.

Khan, S. A., Aschwanden, A., Bjørk, A. A., Wahr, J., Kjeldsen, K. K., & Kjær, K. H. (2015). Greenland ice sheet mass balance: A review. *Reports on Progress in Physics, 78.* https://doi.org/10.1088/0034-4885/78/4/046801 publisher: IOP Publishing.

Koppes, M., Hallet, B., Rignot, E., Mouginot, J., Wellner, J. S., & Boldt, K. (2015). Observed latitudinal variations in erosion as a function of glacier dynamics. *Nature, 526,* 100–103. https://doi.org/10.1038/nature15385 publisher: Nature Publishing Group.

Koppes, M. N., & Montgomery, D. R. (2009). The relative efficacy of fluvial and glacial erosion over modern to orogenic timescales. *Nature Geoscience, 2,* 644–647. https://doi.org/10.1038/ngeo616 publisher: Nature Publishing Group.

Kotlyakov, V. M., Rototaeva, O. V., & Nosenko, G. A. (2004). The September 2002 Kolka Glacier Catastrophe in North Ossetia, Russian federation: Evidence and analysis. *Mountain Research and Development, 24,* 78–83. https://doi.org/10.1659/0276-4741(2004)024[0078:TSKGCI]2.0.CO;2 publisher: International Mountain Society.

Kääb, A., Jacquemart, M., Gilbert, A., Leinss, S., Girod, L., Huggel, C., Falaschi, D., Ugalde, F., Petrakov, D., Chernomorets, S., Dokukin, M., Paul, F., Gascoin, S., Berthier, E., & Kargel, J. S. (2021). Sudden large-volume detachments of low-angle mountain glaciers – more frequent than thought? *The Cryosphere, 15,* 1751–1785. https://doi.org/10.5194/tc-15-1751-2021 publisher: Copernicus GmbH.

Lacroix, P. (2016). Landslides triggered by the Gorkha earthquake in the Langtang valley, volumes and initiation processes. *Earth, Planets and Space, 68*(46). https://doi.org/10.1186/s40623-016-0423-3.

Lagerbäck, R. (1992). Dating of late quaternary faulting in northern Sweden. *Journal of the Geological Society, 149,* 285–291. https://doi.org/10.1144/gsjgs.149.2.0285 publisher: The Geological Society of London.

Lange, H., Casassa, G., Ivins, E. R., Schröder, L., Fritsche, M., Richter, A., Groh, A., & Dietrich, R. (2014). Observed crustal uplift near the Southern Patagonian Icefield constrains improved viscoelastic Earth models. *Geophysical Research Letters, 41,* 805–812. https://doi.org/10.1002/2013GL058419.

Li, G., West, A. J., Densmore, A. L., Jin, Z., Zhang, F., Wang, J., Clark, M., & Hilton, R. G. (2017). Earthquakes drive focused denudation along a tectonically active mountain front. *Earth and Planetary Science Letters, 472,* 253–265. https://doi.org/10.1016/j.epsl.2017.04.040.

Lützow, N., Veh, G., & Korup, O. (2023). A global database of historic glacier lake outburst floods. *Earth System Science Data, 15,* 2983–3000. https://doi.org/10.5194/essd-15-2983-2023 publisher: Copernicus GmbH.

Malamud, B. D., Turcotte, D. L., Guzzetti, F., & Reichenbach, P. (2004). Landslides, earthquakes, and erosion. *Earth and Planetary Science Letters, 229,* 45–59. https://doi.org/10.1016/j.epsl.2004.10.018.

Marc, O., Behling, R., Andermann, C., Turowski, J. M., Illien, L., Roessner, S., & Hovius, N. (2019). Long-term erosion of the Nepal Himalayas by bedrock landsliding: The role of monsoons, earthquakes and giant landslides. *Earth Surface Dynamics,* 7, 107–128. https://doi.org/10.5194/esurf-7-107-2019 publisher: Copernicus GmbH.

McColl, S. T., & Davies, T. R. H. (2013). Large ice-contact slope movements: Glacial buttressing, deformation and erosion. *Earth Surface Processes and Landforms, 38,* 1102–1115. https://doi.org/10.1002/esp.3346, _eprint https://onlinelibrary.wiley.com/doi/pdf/10.1002/esp.3346.

McColl, S. T., Davies, T. R. H., & McSaveney, M. J. (2012). The effect of glaciation on the intensity of seismic ground motion. *Earth Surface Processes and Landforms, 37,* 1290–1301. https://doi.org/10.1002/esp.3251, _eprint https://onlinelibrary.wiley.com/doi/pdf/10.1002/esp.3251.

Mergili, M., Frank, B., Fischer, J.-T., Huggel, C., & Pudasaini, S. P. (2018). Computational experiments on the 1962 and 1970 landslide events at Huascarán (Peru) with r.avaflow: Lessons learned for predictive mass flow simulations. *Geomorphology, 322,* 15–28. https://doi.org/10.1016/j.geomorph.2018.08.032.

Mikko, H., Smith, C. A., Lund, B., Ask, M. V., & Munier, R. (2015). LiDAR-derived inventory of post-glacial fault scarps in Sweden. *GFF, 137,* 334–338. https://doi.org/10.1080/11035897.2015.1036360 publisher: Taylor & Francis _eprint: https://doi.org/10.1080/11035897.2015.1036360.

Mitchell, S. G., & Montgomery, D. R. (2006). Influence of a glacial buzzsaw on the height and morphology of the Cascade Range in central Washington State, USA. *Quaternary Research, 65,* 96–107. https://doi.org/10.1016/j.yqres.2005.08.018.

Morin, G. P., Lavé, J., France-Lanord, C., Rigaudier, T., Gajurel, A. P., & Sinha, R.: (2018). Annual sediment transport dynamics in the narayani basin, central Nepal: assessing the impacts of erosion processes in the annual sediment budget. *Journal of Geophysical Research: Earth Surface, 123*, 2341–2376. https://doi.org/10.1029/2017JF004460, _eprint https://agupubs.onlinelibrary.wiley.com/doi/pdf/10.1029/2017JF004460.

Norris, S. L., Gosse, J. C., Millan, R., Mouginot, J., Rabatel, A., Morlighem, M., Bolton, M. S. M., & Alley, R. B. (2025). Drivers of global glacial erosion rates. *Nature Geoscience, 18*, 732–739. https://doi.org/10.1038/s41561-025-01747-8 publisher: Nature Publishing Group.

Olesen, O. (1988).The Stuoragurra Fault, evidence of neotectonics in the Precambrian of Finnmark, northern Norway, https://nynjg.geologi.no/publications/the-stuoragurra-fault-evidence-of-neotectonics-in-the-precambrian-of-finnmark-northern-norway/,

Olsen, K. G., & Nettles, M. (2017). Patterns in glacial-earthquake activity around Greenland, 2011–13. *Journal of Glaciology, 63*, 1077–1089. https://doi.org/10.1017/jog.2017.78.

Patton, H., Hubbard, A., Heyman, J., Alexandropoulou, N., Lasabuda, A. P. E., Stroeven, A. P., Hall, A. M., Winsborrow, M., Sugden, D. E., Kleman, J., & Andreassen, K. (2022). The extreme yet transient nature of glacial erosion. *Nature Communications, 13*, 7377. https://doi.org/10.1038/s41467-022-35072-0 publisher: Nature Publishing Group.

Peltier, W. R., & Andrews, J. T. (1976). Glacial-Isostatic adjustment—I. The forward problem. *Geophysical Journal International, 46*, 605–646. https://doi.org/10.1111/j.1365-246X.1976.tb01251.x.

Podolskiy, E., Nishimura, K., Abe, O., & Chernous, P. (2010a). Compiling an inventory of earthquake-induced snow avalanches, 1899-2010. *Letter [Avalanche Subcommittee of Japanese Society of Snow and Ice], 45*, 8–9.

Podolskiy, E. A., Nishimura, K., Abe, O., & Chernous, P. A. (2010b). Earthquake-induced snow avalanches: I. Historical case studies. *Journal of Glaciology, 56*, 431–446. https://doi.org/10.3189/002214310792447815.

Prasicek, G., Hergarten, S., Deal, E., Herman, F., & Robl, J. (2020). A glacial buzzsaw effect generated by efficient erosion of temperate glaciers in a steady state model. *Earth and Planetary Science Letters, 543*, 116–350. https://doi.org/10.1016/j.epsl.2020.116350.

Richter, A., Ivins, E., Lange, H., Mendoza, L., Schröder, L., Hormaechea, J. L., Casassa, G., Marderwald, E., Fritsche, M., Perdomo, R., Horwath, M., & Dietrich, R. (2016). Crustal deformation across the Southern Patagonian Icefield observed by GNSS. *Earth and Planetary Science Letters, 452*, 206–215. https://doi.org/10.1016/j.epsl.2016.07.042.

Roback, K., Clark, M. K., West, A. J., Zekkos, D., Li, G., Gallen, S. F., Chamlagain, D., & Godt, J. W. (2018). The size, distribution, and mobility of landslides caused by the 2015 M_w 7.8 Gorkha earthquake, Nepal. *Geomorphology, 301*, 121–138. https://doi.org/10.1016/j.geomorph.2017.01.030.

Rollins, C., Freymueller, J. T., & Sauber, J. M. (2021). Stress promotion of the M_w~7.8 fairweather fault earthquake and others in southeast Alaska by glacial isostatic adjustment and inter-earthquake stress transfer. *Journal of Geophysical Research: Solid Earth, 126*. e2020JB020 411, https://doi.org/10.1029/2020JB020411, _eprint https://agupubs.onlinelibrary.wiley.com/doi/pdf/10.1029/2020JB020411.

Sattar, A., Cook, K. L., Rai, S. K., Berthier, E., Allen, S., Rinzin, S., de Vries, M. V. W., Haeberli, W., Kushwaha, P., Shugar, D. H., Emmer, A., Haritashya, U. K., Frey, H., Rao, P., Gurudin, K. S. K., Rai, P., Rajak, R., Hossain, F., Huggel, C., ... Bhat, S. Y. (2025). The Sikkim flood of October 2023: Drivers, causes, and impacts of a multi-hazard cascade. *Science (New York, N.Y.), 387*, eads2659. https://doi.org/10.1126/science.ads2659 publisher: American Association for the Advancement of Science.

Scherler, D., Bookhagen, B., & Strecker, M. R. (2011). Hillslope-glacier coupling: The interplay of topography and glacial dynamics in High Asia. *Journal of Geophysical Research: Earth Surface, 116*. https://doi.org/10.1029/2010JF001751, _eprint https://agupubs.onlinelibrary.wiley.com/doi/pdf/10.1029/2010JF001751.

Schuster, R., & Alford, D. (2004). Usoi landslide dam and lake Sarez, Pamir mountains, Tajikistan. *Environmental & Engineering Geoscience, 10*, 151–168. https://doi.org/10.2113/10.2.151.

Shugar, D. H., Jacquemart, M., Shean, D., Bhushan, S., Upadhyay, K., Sattar, A., Schwanghart, W., McBride, S., de Vries, M. V. W., Mergili, M., Emmer, A., Deschamps-Berger, C., McDonnell, M.,

Bhambri, R., Allen, S., Berthier, E., Carrivick, J. L., Clague, J. J., Dokukin, M., ... Westoby, M. J. (2021). A massive rock and ice avalanche caused the 2021 disaster at Chamoli, Indian Himalaya. *Science (New York, N.Y.), 373*, 300–306. https://doi.org/10.1126/science.abh4455 publisher: American Association for the Advancement of Science.

Steffen, H., & Wu, P. (2011). Glacial isostatic adjustment in Fennoscandia—A review of data and modeling. *Journal of Geodynamics, 52*, 169–204. https://doi.org/10.1016/j.jog.2011.03.002.

Steffen, H., Olesen, O., & Sutinen, R. (Eds.). (2021). *Glacially-triggered faulting*Cambridge: Cambridge University Press. https://doi.org/10.1017/9781108779906.

Steffen, R., & Steffen, H. (2021). Reactivation of non-optimally orientated faults due to glacially induced stresses. *Tectonics, 40*. e2021TC006 853, https://doi.org/10.1029/2021TC006853, _eprint https://agupubs.onlinelibrary.wiley.com/doi/pdf/10.1029/2021TC006853.

Tarr, R. S., Martin, L., & Gilbert, G. K. (1912). The earthquakes at Yakutat Bay, Alaska, in September, 1899, Tech. Rep. 69, U.S. Geological Survey, https://doi.org/10.3133/pp69, iSSN: 2330-7102 Publication Title: Professional Paper.

Thomson, S. N., Brandon, M. T., Tomkin, J. H., Reiners, P. W., Vásquez, C., & Wilson, N. J. (2010). Glaciation as a destructive and constructive control on mountain building. *Nature, 467*, 313–317. https://doi.org/10.1038/nature09365 publisher: Nature Publishing Group.

USGS: 2025 M7.0 Hubbard glacier earthquake-triggered landslides and snow avalanches | U.S. Geological Survey. 2025. https://www.usgs.gov/programs/landslide-hazards/science/2025-m70-hubbard-glacier-earthquake-triggered-landslides-and.

Van Wyk de Vries, M. (2025). All hazards are multi-hazards, few of them are natural. *Npj Natural Hazards, 2*, 18. https://doi.org/10.1038/s44304-025-00071-w publisher: Nature Publishing Group.

Van Wyk de Vries, M., Romero, M., Penprase, S. B., Ng, G.-H. C., & Wickert, A. D. (2023). Increasing rate of 21st century volume loss of the Patagonian Icefields measured from proglacial river discharge. *Journal of Glaciology, 69*, 1187–1202. https://doi.org/10.1017/jog.2023.9.

Veh, G., Wang, B. G., Zirzow, A., Schmidt, C., Lützow, N., Steppat, F., Zhang, G., Vogel, K., Geertsema, M., Clague, J. J., & Korup, O. (2025). maller glacier lake outburst floods despite worldwide growth in lake area. *Nature Water, 3*, 271–283. https://doi.org/10.1038/s44221-025-00388-w publisher: Nature Publishing Group.

Wood, J. L., Harrison, S., Wilson, R., Emmer, A., Kargel, J. S., Cook, S. J., Glasser, N. F., Reynolds, J. M., Shugar, D. H., & Yarleque, C. (2024). Shaking up assumptions: Earthquakes have rarely triggered Andean glacier lake outburst floods. *Geophysical Research Letters, 51*. e2023GL105 578, https://doi.org/10.1029/2023GL105578, _eprint https://agupubs.onlinelibrary.wiley.com/doi/pdf/10.1029/2023GL105578.

Wu, P., & Hasegawa, H. S. (1996). Induced stresses and fault potential in eastern Canada due to a realistic load: A preliminary analysis. *Geophysical Journal International, 127*, 215–229. https://doi.org/10.1111/j.1365-246X.1996.tb01546.x.

CHAPTER 11

Postseismic sediment and geochemical fluxes

Jin Wang[1] **and Gen K. Li**[2]
[1]State Key Laboratory of Loess Science, Institute of Earth Environment, Chinese Academy of Sciences, Xi'an, P.R. China,
[2]Department of Earth Science, University of California, Santa Barbara, CA, United States

11.1 Introduction

Earthquakes represent a powerful geological force that shapes the Earth's surface, with impacts extending far beyond the fault rupture and continuing well after the cessation of shaking. During and shortly after earthquakes, ground shaking and seismic wave propagation can trigger widespread landslides (Keefer, 1984; Malamud et al., 2004), dislocating soil and rock from hillslopes and facilitating riverine sediment export (Wang et al., 2015). Earthquake-induced landslides and subsequent sediment transport exert profound effects on Earth's surface processes, significantly altering sediment (Hovius et al., 2011; Howarth et al., 2012) and geochemical fluxes, including organic carbon erosion, degradation and burial (Howarth et al., 2025; Wang et al., 2016). This chapter reviews research on the impact of earthquakes on sediment dynamics, the carbon cycle, gas emissions, and metal release. Understanding postseismic sediment and geochemical fluxes – their magnitudes, patterns and compositions – is crucial for assessing the role of earthquakes as a devastating hazard as well as a primary driver of landscape evolution and geochemical cycles.

11.2 Earthquake impacts on sediment transport

While the duration of ground shaking during a large earthquake is typically brief (seconds to minutes), the seismic energy released is enormous. This shaking, coupled to seismic wave propagation, often induces widespread hillslope failures, or landsliding, whose numbers and magnitudes generally increase with earthquake magnitude (Keefer, 1984; Malamud et al., 2004; see Chapter 7). Landslides supply large quantities of detrital material to river systems, facilitating sediment transport in earthquake-impacted landscapes. This seismically induced sediment cascade – a chain of sediment fluxes through interconnected geomorphic domains (e.g. hillslopes and rivers) and processes (e.g. transport and deposition) – often propagates across time scales from coseismic to interseismic periods. During and immediately after hillslope failure, some landslides, especially those connected to river channels, rapidly deliver sediment into river networks and boost riverine sediment fluxes (Li et al., 2016; Wang et al., 2015).

Geomorphic Impacts of Earthquakes
DOI: https://doi.org/10.1016/B978-0-443-40375-0.00016-5

Over subsequent months to years, this landsliding may cause significant landform change within the river network, for example, by altering channel morphology (Tunnicliffe et al., 2024), aggrading channel beds to enhance overbank flood hazards (Korup et al., 2004) or migrating drainage divides (Dahlquist et al., 2018). Over longer time scales (10s–1000s of years), earthquake-induced landslides and associated geomorphic processes can be a major player in mountain belt evolution (Densmore & Hovius, 2000; Wang et al., 2020), governing orogenic erosion rates (Li et al., 2017) and sediment fluxes to lowlands and coastal environments (Howarth et al., 2012; Howarth et al., 2025). Previous studies have relied on hydrological gauging and surveys downstream of the epicentral area (Hovius et al., 2011; Li et al., 2025; Wang et al., 2015) to quantify postseismic sediment fluxes. Recent advances in remote sensing (Francis et al., 2022), high-resolution topographic data (Tunnicliffe et al., 2024), geochemical tracing techniques (Wang et al., 2020; West et al., 2014) and numerical modelling (Croissant et al., 2017) have improved researchers' capacity to quantify the sediment cascade and geomorphic consequences following large earthquakes, providing new insights into the sedimentary responses to seismic events.

11.2.1 Volumes of landslides generated by earthquakes

The impact of earthquakes on sediment transport is primarily shaped by coseismic landslides and postseismic debris flows (Wang et al., 2015). Coseismic landslides are among the most conspicuous and hazardous geomorphic consequences of large earthquakes. Strong ground shaking reduces slope stability by increasing pore-water pressures, fracturing bedrock and dislodging weathered material. In steep terrain, even moderate earthquakes can trigger extensive slope failures, while large-magnitude events often produce tens of thousands of landslides across wide regions (Keefer, 1984; Marc et al., 2016). The total volume of coseismic landslides provides a primary estimate of their contribution to sediment routing systems. Both landslide area and volume scale positively, to first order, with earthquake magnitude (Malamud et al., 2004), with landslide density (in terms of landslide number or landslide-affected area per unit landscape area) increasing closer to fault ruptures (Bloom et al., 2021; Parker et al., 2011). Regions experiencing higher-magnitude earthquakes and located closer to the rupture zone tend to produce greater landslide volumes, leading to larger increases in sediment flux. Over the period of 1975–2025, 403 onshore earthquakes with magnitudes greater than M_w 7.0 have been recorded globally (Fig. 11.1). Applying the empirical relationship between earthquake magnitude and total coseismic landslide volume reported by Malamud et al. (2004), we estimate that these events have produced approximately 170 km^3 of landslide material – an amount equivalent to nearly half of the global fluvial suspended sediment load exported during this timespan (Milliman & Farnsworth, 2011). We note that this represents a simplified exercise as the appropriate

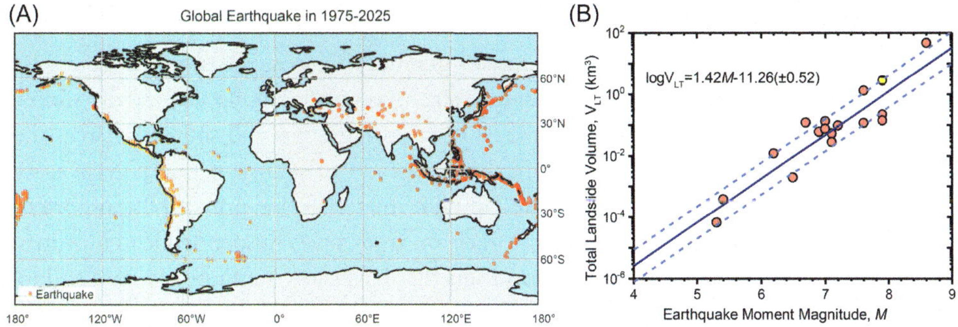

Figure 11.1 Global earthquakes greater than M_w 7 and the relationship between earthquake magnitude and landslide volume over the period of 1975–2025. (A) Earthquakes (red dots). (B) Estimated total coseismic landslide volume as a function of earthquake moment magnitude. Orange circles represent landslide volumes associated with 16 events reported in Malamud et al. (2004). The solid line is the least-squares best fit to the data, with dashed lines indicating standard deviations from the best fit. The yellow circle highlights the landslide volume for the 2008 Wenchuan earthquake (Li et al., 2014). Map lines delineate study areas and do not necessarily depict accepted national boundaries. *(A, B) Sourced from the USGS (https://earthquake.usgs.gov/earthquakes/search/). Modified from Malamud, B. D., Turcotte, D. L., Guzzetti, F., & Reichenbach, P. (2004). Landslides, earthquakes, and erosion.* Earth and Planetary Science Letters, 229***,*** *45–59.*

scaling relationship between magnitude and total volume may vary across different regions and environmental conditions (Li et al., 2017; Marc et al., 2016). Nonetheless, this first-order estimate emphasises the importance of earthquake-induced landslides in sediment transport at a global scale.

At local to regional scales, coseismic landslide volumes are affected by geomorphic, geological and climatic factors (Graf et al., 2024; Rault et al., 2019). For instance, the 2008 M_w 7.9 Wenchuan earthquake in China generated more than 60,000 landslides (Li et al., 2014; 2016), exceeding those triggered by the 2015 M_w 7.8 Gorkha earthquake in Nepal (Roback et al., 2018; Valagussa et al., 2021). This discrepancy likely reflects the extensive preseismic landsliding that consumes unstable slopes in the monsoon-dominated Himalayas of central Nepal (Petley et al., 2007). Landslide size is not only governed by ground motion but also by local lithology, topography and tectonic stress fields (Li & Moon, 2021; Valagussa et al., 2019). Moreover, seismic shaking can cause rock damage and regolith loosening, and as a result, landslides often continue occurring in the years following the mainshock, prolonging the seismic impact on sediment flux (see Chapters 15–16; Fan et al., 2018; Kincey et al., 2021; Marc et al., 2015). Often boosted by landslide input, postseismic sediment fluxes are modulated by hydrological and climatic variabilities (Chapter 9), which could dampen or amplify the effect of earthquake-induced landslides (Howarth et al., 2025; Wang et al., 2015; Zhang et al., 2019). Continuous monitoring over longer time scales (>10s of years) is thus crucial for comprehending postseismic sediment dynamics.

11.2.2 Impacts of earthquakes on suspended sediment flux

One of the most immediate and widespread geomorphic consequences of large earthquakes is their impact on sediment transport. Strong ground shaking can rapidly mobilise vast volumes of granular materials composed of rock fragments and soil and deliver them into the sediment routing system, setting off a cascade of sediment transport processes. Once they enter the river system, those landslide materials are primarily transported as suspended load and bedload with distinct effects on landscapes. Suspended load is primarily composed of fine particles such as clays and silts that have slow settling velocities and long advection lengths. Thus suspended load can travel long distances (>10s–100s of km) downstream of the epicentral region and settle in lowland rivers and deltas, regulating magnitudes and patterns of sedimentation in those deposition zones (Dethier et al., 2022). In contrast, bedload consists of coarse grains that span a wide range of sizes from sands to cobbles and boulders, with significant influences on the morphodynamics of landslide-fuelled channels (Croissant et al., 2017), which in turn modulate mountain relief (Egholm et al., 2013). The understanding of these sediment loads and associated processes has significantly evolved over the last several decades. In the 1970s–80s, when continuous monitoring of water and sediment discharges was limited, studies mainly focused on landform changes as a consequence of sediment transport, revealing that landslide material could be stored in fluvial systems for long periods of time and could cause changes in channel morphology. For example, following large earthquakes in both Papua New Guinea and New Zealand, channels were buried by up to 10 m of debris, much of which remained stored for more than 50 years following the mainshock (Pearce & Watson, 1986). Studies have also documented downstream transfer of landslide sediment (Pearce & Watson, 1983, 1986), providing field evidence that earthquake-induced landslides govern sediment movements in postseismic periods.

In the 1990s–2000s, improvements in hydrological monitoring enabled more quantitative assessments. A landmark case was the 1999 M_w 7.7 Chi-Chi earthquake in Taiwan (Chapter 13), where, due to the dense network of hydrological monitoring stations, researchers were able to quantify postseismic suspended sediment fluxes at high resolution for the first time. However, direct comparisons of pre- and postseismic sediment concentrations are often confounded by rainfall and discharge variability. To address this, Dadson et al. (2004) and Hovius et al. (2011) applied the widely observed power-law relationship between water discharge and suspended sediment concentration, using the coefficient κ to represent unit or water discharge-normalised suspended sediment concentration (Fig. 11.2a). They found that, in rivers draining the epicentral region, the value of κ increased by a factor of four during the first 4 years after the earthquake. Extending their observations to 2007, Hovius et al. (2011) showed that landslide activities and suspended sediment transport rates were more than five times

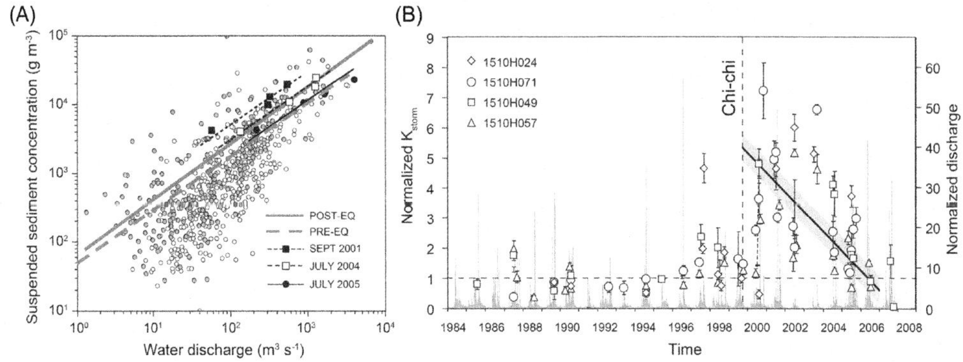

Figure 11.2 Hydrometric data for the Choshui River, Taiwan, before and after the 1999 Chi-Chi earthquake. (A) Water discharge and suspended sediment concentration were measured at the mainstream gauging station from 1987 to 2006. Pre- and postearthquake measurements are represented by open and filled grey circles, respectively. The lines depict the best-fit power laws to the preearthquake data (1987–99, grey dashed line), the floods of 2001, 2004 and 2005 (black lines) and the postearthquake measurements (grey solid line). (B) Time evolution of unit or water discharge-normalised suspended sediment concentration κ at four different gauging stations during major storm events. The black line shows linear least-squares best fit to the postearthquake unit suspended sediment concentrations, showing the recovery of earthquake impacts in 6 ± 0.8 year (1σ shown in grey). *Both panels adapted from Hovius, N., Meunier, P., Lin, C.-W., Chen, H., Chen, Y.-G., Dadson, S., Horng, M.-J., & Lines, M. (2011). Prolonged seismically induced erosion and the mass balance of a large earthquake.* Earth and Planetary Science Letters, 304**,** *347–355.*

greater than preseismic levels, with elevated fluxes persisting for approximately 6 years (Fig. 11.2b). Earthquake-induced landslides contributed nearly half of the suspended sediment exported during this period.

Some other major earthquakes have exhibited similar patterns. A few hours after the 2008 Wenchuan earthquake, suspended sediment concentrations in affected rivers increased more than 50-fold (Wang et al., 2015). Annual suspended sediment fluxes surged by a factor of 3–7 over the next 4 years, with the magnitude of increase correlating positively with upstream landslide density (Li et al., 2017; 2025; Wang et al., 2015). However, not all large earthquakes exhibit significant suspended sediment responses. For example, following the 2010 M_w 8.8 Maule earthquake in Chile, no significant increase in suspended sediment was observed in rivers draining landslide-affected areas, likely due to low rainfall in the postseismic period (Tolorza et al., 2019). After the 2015 M_w 7.8 Gorkha earthquake, suspended sediment yields did not show immediate changes that could be related to earthquakes, but peaked a year later, potentially reflecting delayed responses that are controlled by monsoonal rainfall (Märki et al., 2021). These observations showcase the diverse responses of suspended sediment flux to earthquakes, highlighting the importance of climate in regulating postseismic suspended sediment transport.

11.2.3 Postearthquake bedload sediment flux

In addition to supplying fine material to the river network, earthquake-triggered landslides can also deliver coarse debris (sand-sized and larger) that is primarily transported as bedload. That coarse landslide material can infill channels, causing upstream aggradation and downstream incision (Fan et al., 2016; Yanites et al., 2010). However, quantifying bedload responses has been challenging because direct measurement of bedload transport is inherently difficult (Croissant et al., 2017; Tunnicliffe et al., 2024).

To date, few studies have directly measured pre- and postseismic changes in bedload flux. As a result, researchers have often relied on indirect methods to constrain the bedload component. Digital elevation models (DEMs), satellite imagery and airborne LiDAR have allowed repeated assessments of hillslope and channel morphology changes, which provide useful information for constraining bedload transport. DEM- and imagery-based surveys of landslide inventories help to quantify landslide material movements, but carry major uncertainties from the need to apply empirical scaling relationships between landslide area and volume (Francis et al., 2022). Based on repeated image-based surveys of channels, Francis et al. (2022) estimated that in the decade following the 2008 Wenchuan earthquake, the induced seismic landslides delivered 77.9 ± 30.0 Mt of sediment (representing ~12% of the total landslide sediment mass in the catchment) to the Min Jiang River, with the remaining ~88% stored on hillslopes, suggesting centennial- to millennial-scale influences on sediment routing systems.

LiDAR offers a higher level of precision: comparing sequential scans allows quantifying erosion, deposition and bedload transport directly. For example, studies in New Zealand have demonstrated that postseismic bedload transport is highly contingent on boundary conditions such as channel geometry and bed composition (Tunnicliffe et al., 2024). Importantly, channel narrowing can temporarily enhance transport capacity (Croissant et al., 2017), whereas boulder armouring increases bed roughness and slows vertical incision (Tunnicliffe et al., 2024). These feedback mechanisms demonstrate the importance of water-sediment-channel interplays in predicting bedload sediment fluxes and transport pathways.

Additional insights can be gained from depositional sinks such as man-made reservoirs (Li et al., 2025; Zhang et al., 2019) and marginal seas (Mountjoy et al., 2018). For example, following the 2008 Wenchuan earthquake, riverbed sediment fluxes increased by more than 20 times, composing ~65% of the total postseismic sediment load. The river sediment budget indicates that in the decade following the 2008 Wenchuan earthquake, about 9% of the total landslide-derived material was evacuated (Fig. 11.3; Li et al., 2025), which aligns with DEM-based estimates (Francis et al., 2022). Importantly, fluvial evacuation rates showed little decline during this period, indicating the long-lasting, decadal-scale influence of landslide debris on bedload sediment fluxes (Li et al., 2025).

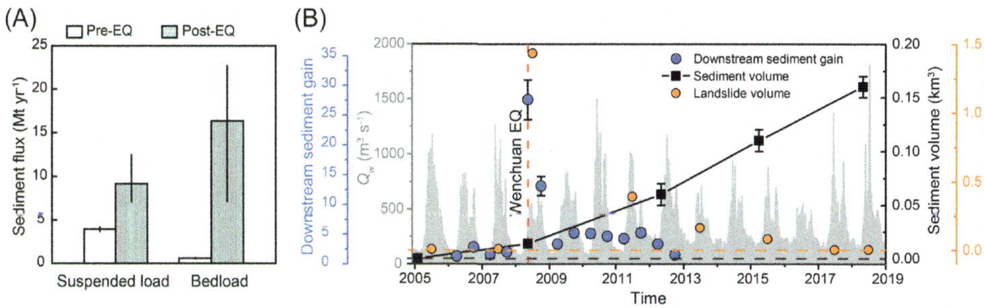

Figure 11.3 Fluvial sediment flux and landslide volume before and after the 2008 Wenchuan earthquake. (A) Comparison of total suspended load and bedload fluxes before (white bars) and after (grey bars) the earthquake. Error bars represent the 16th to 84th percentiles of the Monte Carlo results. (B) Sediment volumes deposited in the downstream reservoir before and after the earthquake (black squares; Li et al., 2025), along with landslide volume data (orange circles) from Fan et al. (2019) and the ratios of downstream to upstream suspended sediment flux (blue circles) for the Fu Jiang affected by the earthquake-triggered landslides (Wang et al., 2015). Background grey bars show the monthly averaged water discharge of the Min Jiang upstream. *Adapted from Li, G. K., West, A. J., Jin, Z., Qiu, H., Zhang, F., Wang, J., Hammond, D. E., Densmore, A. L., Hilton, R. G., Dong, S., Atwood, A., Fischer, W. W., & Lamb, M. P. (2025). Large riverbed sediment flux sustained for a decade after an earthquake.* Nature, 644*, 398–403.*

Geochemical tracers provide complementary evidence of postseismic bedload transport. Cosmogenic ^{10}Be concentrations in quartz grains, which record cumulative near-surface exposure times, typically decline following large earthquakes as freshly exposed, ^{10}Be-depleted landslide material enters rivers. Postseismic ^{10}Be signatures have indicated <2 to >10x increases in sediment flux within the channel network after the 2008 Wenchuan earthquake (Wang et al., 2017; Wang et al., 2025; West et al., 2014). In some basins, ^{10}Be concentrations had nearly returned to preseismic values within a decade as vegetation recovered and sediment supply diminished (Wang et al., 2017), although these trends may diverge from reservoir sediment records (Li et al., 2025). The case of the Wenchuan earthquake highlights that the transport of landslide material after a seismic event may last over centennial time scales. Modelling studies from other places likewise yield divergent estimates of bedload transport rates and associated recovery time (Croissant et al., 2017; Tunnicliffe et al., 2024; Yanites et al., 2010), reflecting uncertainties in measuring riverine sediment transport capacity and its temporal evolution.

Overall, seismic impacts on bedload transport are modulated by a range of factors, including landslide volumes, landslide-channel connectivity, climate and water discharge. We speculate that, immediately after large earthquakes and landslide input, sediment yields from many river systems are likely to be limited by transport capacity ('transport-limited'). Intense rainfall could enhance the transport capacity through increasing water discharge and facilitating other sediment transport processes, such as debris flows, shifting the system to a 'supply-limited' regime where bedload transport rates depend on available landslide debris in river channels.

11.2.4 Time scales of earthquake-landslide influence on sediment flux

The storage of earthquake-derived landslide material within catchments has profound implications not only for sediment transport but also for cascading hazards, such as elevated flood risks due to channel aggradation. To better understand how earthquakes induce cascading hazards through landsliding, it is critical to determine the time scale over which rivers will evacuate landslide debris. This time scale, to first order, sets the duration over which earthquake-triggered landslides affect sediment flux. Comparing this time scale to the seismic recurrence interval also determines whether earthquake-derived landslide material can be completely exported before the next earthquake, which is key to understanding mountain belt evolution in seismically active regions (Li et al., 2019).

Given that large earthquakes tend to recur on time scales of centuries to millennia in many tectonically active regions, while sediment monitoring typically spans only decades, accurately constraining the landslide debris evacuation time scale is inherently challenging. A mechanistic understanding of the geomorphic processes involved in landslide sediment transport is thus needed to project long-term patterns from short-term observations. Past studies have measured the spatiotemporal pattern of postseismic sediment fluxes, which provides some baseline information on the pace at which rivers evacuate landslide debris. Rivers transport fine-grained materials from landslides as suspended load, and therefore the retention time of fine-grained materials can be inferred from changes in suspended sediment signals. Wang et al. (2015) compared the fine-grained fraction of landslide deposits and measured suspended sediment fluxes, and estimated that fine-grained materials could influence sediment flux for years to centuries. However, long-term evacuation rates still remain poorly constrained, with large uncertainties on the removal of bedload materials that are transported more slowly. Yanites et al. (2010) and Li et al. (2025) suggested that decades to centuries are required to evacuate coarse landslide materials based on field measurements. Modelling that accounts for channel morphodynamical changes (Croissant et al., 2017) has estimated that half the coarse fraction could be removed within 5–25 years, but consequent bed armouring may increase the time required (Tunnicliffe et al., 2024).

Stratigraphic records offer valuable insight into sedimentary responses over multiple earthquake cycles, although they may be located some way down the fluvial system from the main areas of coseismic and postseismic landsliding. For instance, based on sediment cores drilled in lakes adjacent to the Southern Alps, New Zealand, Howarth et al. (2012) documented sedimentation rates three times higher than background levels following past $M_w > 7.6$ earthquakes, with impacts lasting approximately 50 years.

Similarly, check-dam records in Japan indicate that sediment from earthquakes 80 years prior still influences present-day erosion (Koi et al., 2008). However, several challenges exist in interpreting seismic signals in sedimentary records: distinguishing seismic from other event signals is difficult, because both postseismic sediment pulses and storm events can generate similar types of deposit, such as turbidites (Howarth et al., 2025); sedimentary records can be dominated by nonseismic and nonclimatic signals, for example those documenting tectonic stress buildup (McSaveney et al., 2015); the sedimentary signals tend to be modified during propagation (Jerolmack & Paola, 2010; Romans et al., 2016); and the postseismic evolution of sediment flux does not necessarily follow a simple monotonic trajectory (Fig. 11.4; Tunnicliffe et al., 2024), but often exhibits a pulsed pattern modulated by aftershocks and episodes of intense rainfall (Francis et al., 2022), making it difficult to directly identify postseismic signals from sedimentary records and to define the precise mode and time scale of postseismic recovery.

Thus current estimates of the duration of the seismic influence on sediment flux vary widely, ranging from years to centuries. The estimates depend on methods and a range of factors, including landslide volumes, connectivity of landslide debris to the fluvial network, climate and geomorphic setting. A critical unresolved question is whether earthquake-induced landslide material can be fully evacuated within a single recurrence

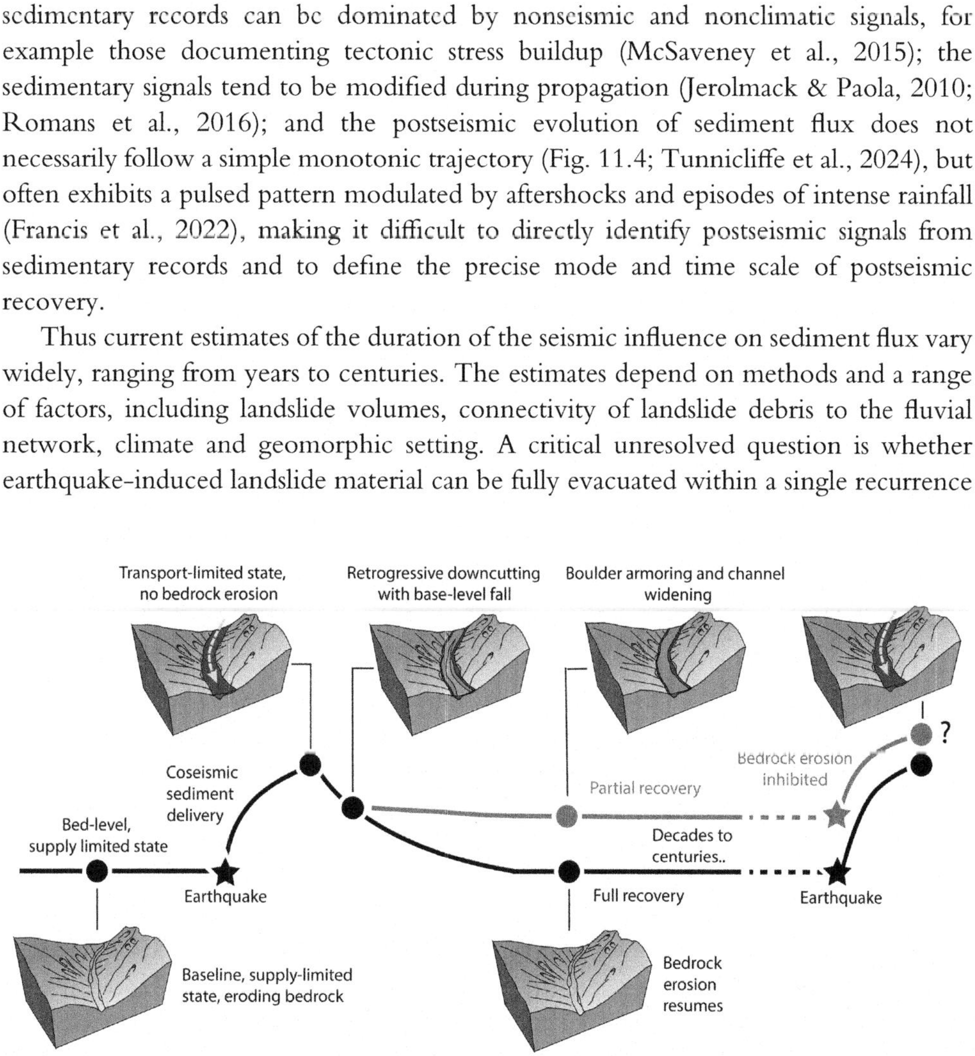

Figure 11.4 Schematic evolution of sediment transport rate (lines) and river channel morphology (block diagrams) following a large earthquake, relative to the occurrence of future seismic events. The precise temporal evolution of postearthquake sediment flux, the duration of seismic impacts and the recovery process are yet to be constrained. *Adapted from Tunnicliffe, J., Howarth, J., & Massey, C. (2024). Controls on fluvial sediment evacuation following an earthquake-triggered landslide: Observations from LiDAR time series.* Science Advances, 10, *eadi5560.*

interval. Answering this question is important for assessing the role of earthquakes in topographic development, as it determines the extent to which earthquake-induced landslide erosion counteracts coseismic rock uplift. Some studies tended to support the view that full evacuation is achievable: landslide-derived erosion rates appeared broadly consistent with regional erosion rates over different time scales (Li et al., 2017) and geomorphic surveys often reveal a lack of extensive relict deposits from previous large seismic events (Parker et al., 2011). Other studies suggest longer evacuation periods: vegetation recovery and stabilisation of landslide deposits can markedly reduce removal rates in the years following an earthquake (Francis et al., 2022; Lin et al., 2006), leading to extended sequestration of landslide material on hillslopes. Tunnicliffe et al. (2024) showed that the interval of earthquake-related sediment release does not align with the seismic recurrence interval, suggesting that subsequent earthquakes may remobilise legacy material deposited during prior seismic events. Future studies and continued monitoring are needed to improve and calibrate current models and to better constrain the evacuation time of earthquake-induced landslides.

11.2.5 Approaches to studying postseismic sediment flux

Despite the challenges posed by the low frequency of large earthquakes on human time scales, several methods have been developed over the past decades to study the impacts of large earthquakes and postseismic sediment fluxes, each with advantages and limitations.

(1) Geomorphic mapping and direct monitoring: Satellite imagery enables geomorphic mapping to construct single and multitemporal landslide inventories and to measure channel sedimentation caused by bedload transport (Francis et al., 2022; Graf et al., 2024). Landslide inventories provide essential data on landslide distribution, runout and deposition (Fan et al., 2019, Larsen & Montgomery, 2012), though they offer limited capacity to quantify fluxes. Advances in these techniques may also help infer suspended sediment dynamics, though high uncertainties remain. Airborne LiDAR offers high-precision topographic change detection, enabling direct quantification of erosion and deposition (Tunnicliffe et al., 2024). These efforts can be combined with direct field monitoring (e.g. sediment thickness measurements at a point) to estimate volumes over time and thus temporal volumetric changes in different domains of the system.

(2) River discharge monitoring: Hydrological networks that measure water discharge and sediment loads before and after earthquakes allow quantitative assessments of short-term impacts (Dadson et al., 2004; Tolorza et al., 2019; Wang et al., 2015), though distinguishing seismic from climatic influences remains challenging. Additionally, these measurements are overwhelmingly focused on suspended sediment load and so are necessarily incomplete.

(3) Sedimentary records: Reservoirs, lakes and marine basins act as repositories for earthquake-liberated landslide debris. Bathymetric surveys, coring, and dating provide valuable estimates of sediment budgets and deposition rates across events (Li et al., 2025).

(4) Geochemical tracers: Cosmogenic nuclides (e.g. in-situ ^{10}Be, isotopes of U, Sr; Li et al., 2024; West et al., 2014) and radiocarbon dating (Howarth et al., 2012; Slosson et al., 2022) are useful for measuring erosion rates and quantifying shifts in sediment sourcing following earthquakes.

(5) Numerical modelling: Process-based models simulate sediment delivery, transport and storage, integrating landslide supply and hydrological forcing (Croissant et al., 2017; Yanites et al., 2010). However, uncertainties remain due to the complexity of sediment transfer pathways that limit our understanding of modern processes governing postseismic sediment transport. It remains very challenging to apply these in a predictive sense, for example, to anticipate the cascading impacts that we would expect from a large earthquake.

11.3 Impacts of earthquakes on carbon cycles

Earthquake-triggered landslides have profound influences on not only sediment transport but also geochemical cycles. In this section, we focus on the carbon flux as a case study to discuss the seismic impacts on geochemical fluxes more generally. We choose carbon for two reasons: most existing studies have focused on carbon, and the carbon cycle represents a central process in global biogeochemical cycles with far-reaching influences on Earth's climate and the cycling of other key elements (such as nitrogen). That being said, the most immediate and conspicuous effect of large earthquakes arises from landslide-induced damage to vegetation and soils. Landslides result in the destruction of terrestrial vegetation and the mobilisation of soil organic carbon, disrupting ecosystems and regional carbon cycling. Moreover, the destruction of vegetation by landslides diminishes its ability to sequester atmospheric carbon, whereas the decomposition of landslide-eroded biomass releases CO_2 to the atmosphere via heterotrophic processes (Chambers et al., 2007; Kang et al., 2023). Studies show that disturbed soils, particularly those in deeper horizons that are now exposed to oxygen through landslides, could decompose and emit CO_2 at fast rates (Shi et al., 2025). Landslides, especially those involving bedrock, can also mobilise rock-derived organic carbon whose oxidation provides a long-term CO_2 source (Zondervan et al., 2023). The ultimate fate of landslide-mobilised organic carbon residing in soil, vegetation and rock – whether buried or weathered – and the balance between these terms set the net effect of earthquake-induced landslides on the carbon cycle (Croissant et al., 2021, Wang et al., 2016).

11.3.1 Earthquake effects on biospheric organic carbon erosion and burial

The amount of organic carbon mobilised by earthquakes can be substantial and in some cases, comparable to biogeochemical fluxes transported at much larger spatial scales than those of earthquake-affected regions. Garwood et al. (1979) investigated two shallow earthquakes in Panama in 1976 (M_L 6.7 and 7.0) and found that landslides denuded around 54 km^2 (12%) of forest cover across a 450 km^2 region. When these figures were extrapolated to a larger scale, it was suggested that such seismic events could lead to the destruction of over 10% of forest cover, demonstrating the potential for earthquakes to significantly alter regional biospheric carbon storage. More recent studies have focused on the 2008 Wenchuan earthquake. Liu et al. (2022) reported that approximately 90% of soil and vegetation organic carbon in the epicentral area was destroyed during the earthquake. Chen et al. (2009) estimated that Wenchuan-induced landslides mobilised approximately 13.6 Tg (1 Tg = 10^{12} g) of biospheric organic carbon, based on landslide mapping and spatial datasets of soil and forest organic carbon density. Their estimated amount is comparable to the annual particulate organic carbon load of the Amazon River (Richey et al., 1990), the largest river on Earth by water discharge volume. Other studies have provided varying estimates of the organic carbon mobilised during the Wenchuan earthquake. Zeng et al. (2016), using remote sensing and field surveys, estimated that forest biomass loss in the epicentral region was around 10.9 Tg C. In contrast, Ren et al. (2009), using a scalable geo-fluid model, estimated a significantly higher value of about 235 Tg C of vegetation destruction. These discrepancies primarily arise from their adopted methods and data sources and further studies are needed to reconcile the differences.

Other studies have estimated biomass loss due to storm-triggered landslides, which could provide a useful comparison against that due to earthquake-induced landslides. For example, Hurricane Katrina in 2005 resulted in the destruction of approximately 105 Tg of biomass carbon, which is equivalent to 50%–140% of the annual carbon sink of US forests (Chambers et al., 2007). In regions where both large earthquakes and high-magnitude storms occur (e.g. New Zealand), one would expect significant and efficient mobilisation of organic carbon from soils and vegetation (Howarth et al., 2025). Research efforts have also been aimed at improving methods. Advances in LiDAR technology have allowed for more precise estimates of landslide volumes and biomass losses. When combined with carbon inventory data collected at high spatial resolutions, these technologies can lead to more accurate and detailed assessments of seismic impacts on biospheric carbon fluxes (Hao et al., 2025).

Earthquake-induced vegetation and soil loss influence regional carbon budgets in two primary ways: (1) reduced net primary productivity (NPP) leads to a decrease in carbon uptake from the atmosphere (Yang et al., 2023) and (2) the accelerated decomposition of dead biomass and disturbed soils results in the release of CO_2 (Kang et al., 2023; Liu et al., 2022). While direct monitoring of postseismic CO_2 emissions is still limited, modelling studies suggest that microbial decomposition of landslide-mobilised organic carbon represents a major pathway for carbon release, competing with fluvial export of organic carbon (Croissant et al., 2021; Wang et al., 2016). Additionally, earthquake surface rupture can expose deep soils to atmospheric oxygen, significantly accelerating the decomposition of organic carbon and the subsequent release of CO_2. For example, even earthquakes that induce limited landslides on the Tibetan Plateau could generate hundreds of surface rupture segments, with lengths ranging from metres to kilometres, resulting in a significant cumulative CO_2 release flux (Shi et al., 2025).

Recent work has focused on processes that draw down CO_2 via (1) vegetation regrowth and soil re-establishment, which sequester atmospheric CO_2 (Croissant et al., 2021, Restrepo et al., 2009); and (2) particulate organic carbon burial in depositional environments, where mobilised organic matter may be preserved over geologic time scales (Frith et al., 2018; Howarth et al., 2025; Wang et al., 2016). Vegetation recovery can occur relatively rapidly. After the 1999 Chi-Chi earthquake, vegetation regrowth in the landslide-affected area reached about 59% within 2 years and nearly 90% within 6 years (Lin et al., 2005; Lin et al., 2006). Thirteen years after the 2008 Wenchuan earthquake, MODIS satellite data indicated that nearly half of the landslide-affected areas had recovered to preearthquake NPP levels (Yang et al., 2023). While vegetation tends to recover quickly, the soil regeneration rate remains much less well studied and is likely slow: the biospheric particulate organic carbon in alpine catchments can be millennia old (Wang et al., 2019), implying that, in theory, soil carbon reservoirs may require thousands of years to replenish, but this recovery may remain incomplete if interrupted by the recurrence of major earthquakes. Despite these advances, further research is needed to better depict the recovery trajectories of soil carbon pools and associated carbon fluxes following seismic events.

The influence of earthquake-induced landslides also extends to areas downstream of the epicentral region. Landslide-mobilised organic carbon could be exported from mountain catchments and buried downstream, potentially forming a long-term carbon sink (Hilton & West, 2020; Howarth et al., 2025). This biospheric carbon, derived from soils and vegetation, represents atmospheric carbon that was recently fixed by the biosphere. Its erosion, fluvial transport and burial provide an efficient pathway for carbon sequestration over geological time scales (Hilton & West, 2020). A thorough understanding of the amount and fate of this organic carbon is crucial for evaluating the role of seismic events in the global carbon cycle and their broader implications for climate change. In general, rapid and enhanced burial of biospheric organic carbon is expected in postseismic periods, mostly due to either enhanced sediment fluxes or increased sediment delivery to depositional (e.g., offshore) settings and increased frequency of turbidity currents (Blair & Aller, 2012; Thunell et al., 1999). Turbidity currents can increase the burial efficiency of organic carbon in anoxic or low-oxygen environments (Blair & Aller, 2012; Talling et al., 2024). Yet case studies offer varying

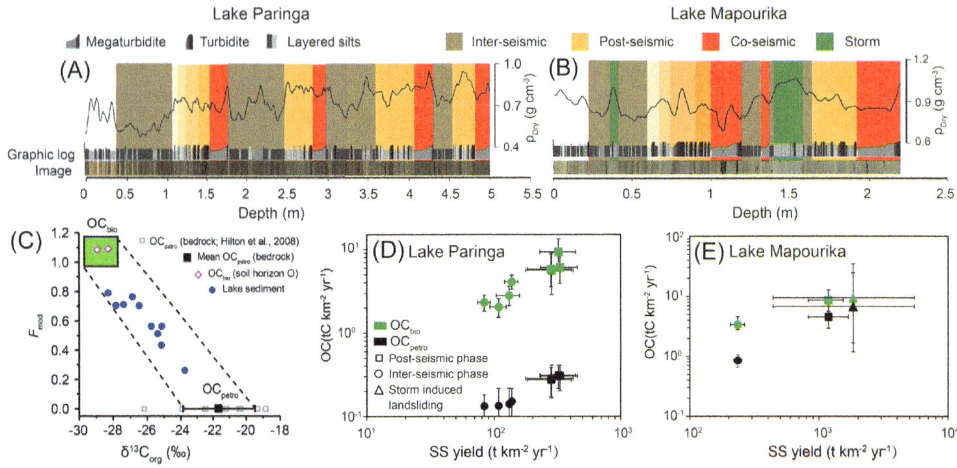

Figure 11.5 Quantitative analysis of pre- and postearthquake organic carbon sources and fluxes using lake sediment cores. (A) and (B) Core images, graphic logs and dry density measurements for Lakes Paringa and Mapourika, located along the Alpine Fault, South Island, New Zealand. Colours show inferred facies: interseismic (brown), coseismic (red), postseismic (orange), or storm-derived (green). (C) Mixing model of bedrock (OC_{petro}) and biospheric (OC_{bio}) organic carbon end members, based on the measured organic carbon isotopes (radiocarbon activity reported as fraction modern, F_{mod} and stable carbon isotopic ratio reported as $\delta^{13}C$), used to separate OC_{petro} and OC_{bio} in sediments from Lakes Paringa and Mapourika. (D) and (E) Relationship between suspended sediment, OC_{bio}, and OC_{petro} yields during postseismic, interseismic, and storm-driven landsliding phases in Lakes Paringa and Mapourika. Compared to interseismic phases, sedimentation rates and organic carbon burial efficiency increased during postseismic phases and storm-driven landslide events. *Modified after Howarth, J. D., Fitzsimons, S. J., Moody, A., Wang, J., Garnett, M. H., Croissant, T., Densmore, A. L., Howell, A., & Hilton, R. G. (2025). Long term carbon export from mountain forests driven by hydroclimate and extreme event driven landsliding.* Communications Earth & Environment, *6, 432.*

perspectives. After the 2008 Wenchuan earthquake, Wang et al. (2016) found that the particulate organic carbon supply from rivers draining areas with high landslide density increased severalfold. In contrast, following the 2015 Gorkha earthquake, despite landslide volumes approximately four times the interseismic sediment yield, no significant increases in riverine fluxes of sediment and particulate organic carbon were observed (Märki et al., 2021), and some rivers even exhibited reduced fluxes in 2015. Sedimentary archives provide additional insights: analyses of lake sediments from New Zealand's South Island have indicated that four major earthquakes along the Alpine Fault over the past millennium contributed approximately 40% of organic carbon export during that time period (Frith et al., 2018), with wetter climatic conditions doubling organic carbon burial rates (Fig. 11.5; Howarth et al., 2025). Similarly, Bao et al. (2018) observed episodic deposition of preaged organic carbon in deep-sea trenches linked to earthquake events.

In summary, the fate of landslide-mobilised biospheric organic carbon plays an important role in determining the effect of earthquakes on atmospheric CO_2. While short-term processes following earthquakes tend to favour CO_2 emissions, long-term processes such as burial and vegetation regrowth could cause CO_2 drawdown to sustain long-term carbon sequestration (Croissant et al., 2021). Besides modulating the biospheric organic carbon cycle, earthquakes also influence carbon fluxes through enhanced rock weathering, as discussed below.

11.3.2 Earthquake effects on rock weathering

Rock weathering influences atmospheric CO_2 concentrations over different time scales (Hilton & West, 2020). The main processes involved are:

(1) Silicate weathering: Silicate minerals react with water and CO_2 to form carbonic acid (HCO_3^-), which subsequently leads to the formation of calcium carbonate ($CaCO_3$), producing a CO_2 sink over 10^6-year time scales:

$$CaSiO_3 + 2CO_2 + 3H_2O \rightarrow Ca^{2+} + H_4SiO_4 + 2HCO_3^- \quad (11.1)$$

$$Ca^{2+} + 2HCO_3^- \rightarrow CaCO_3 + CO_2 + H_2O \quad (11.2)$$

(2) Carbonate weathering involving carbonic acid: Calcium carbonate ($CaCO_3$) reacts with water and CO_2 to form bicarbonate. After entering the marine system, HCO_3^- undergoes reverse reactions, which have no significant effect on atmospheric CO_2 beyond the carbonate compensation time, that is, $>10^4$ years:

$$CaCO_3 + CO_2 + H_2O \rightarrow Ca^{2+} + 2HCO_3^- \quad (11.3)$$

(3) Coupled pyrite-carbonate weathering: Pyrite undergoes oxidation in the presence of oxygen and water, producing sulphuric acid, which dissolves carbonate rocks and releases CO_2 into the atmosphere and this CO_2 source operates on time scales shorter than pyrite formation in marine environments (10^6–10^7years; Torres et al., 2014):

$$4FeS_2 + 15O_2 + 14H_2O \rightarrow 4Fe(OH)_3 + 8H_2SO_4 \quad (11.4)$$

$$CaCO_3 + H_2SO_4 \rightarrow CO_2 + H_2O + Ca^{2+} + SO_4^{2-} \quad (11.5)$$

(4) Oxidative weathering of rock-derived, petrogenic organic carbon as a CO_2 source: When sedimentary rocks are exhumed, organic carbon undergoes oxidation, facilitated by microbial activity, resulting in a CO_2 source over 10^6-year time scales:

$$CH_2O + O_2 \xrightarrow{\text{enzymes}} CO_2 + H_2O \qquad (11.6)$$

Tectonic rock uplift is widely regarded as a major driver of silicate weathering by exposing fresh bedrock surfaces, which accelerates CO_2 consumption (Hilton & West, 2020). This process is one of the primary mechanisms regulating atmospheric CO_2 concentrations and climate over geological time scales, forming the basis of the prevailing 'uplift-weathering' hypothesis that links mountain-building events to global cooling and glacial cycles (Raymo & Ruddiman, 1992). As frequent and sudden tectonic disturbances, earthquakes offer a unique opportunity to investigate how tectonic activity impacts weathering and climate.

A few studies have made direct comparisons of chemical weathering rates before and after seismic events. Notably, Jin et al. (2016) conducted such an assessment in the Min Jiang catchment in the Longmen Shan mountains of China, in the area that was affected by the 2008 Wenchuan earthquake. Preearthquake studies of chemical weathering in this region had been conducted, allowing pre- versus postseismic comparisons. Their analyses revealed that total solute fluxes increased significantly after the earthquake, with a fourfold increase in Na/Ca ratios and largely elevated $^{87}Sr/^{86}Sr$. Similarly, the silicate-derived alkalinity and associated CO_2 consumption rate rose by a factor of approximately 4 (Jin et al., 2016). In the same study area and using multiple isotope tracers (U, Sr, Be), Li et al. (2024) demonstrated that postseismic chemical weathering was concentrated within landslide deposits, rather than the conventional weathering zone in soils, likely due to the exposure of abundant fresh, reactive mineral surfaces created by mass wasting. Emberson et al. (2015, 2018) also reported that landslides accelerate chemical weathering processes. However, following the 2015 Gorkha earthquake, no significant increase in silicate weathering fluxes was detected, likely due to already high background weathering rates under monsoonal conditions and limited changes in postseismic physical erosion rates (Märki et al., 2021).

As introduced earlier, rock weathering encompasses carbon source and sink processes across different time scales. Carbonate weathering by carbonic acid consumes atmospheric CO_2 on $<10^4$-year time scales (the carbonate compensation time scale in the oceans) but has a neutral effect on CO_2 over 10^6-year time scales, because the dissolved carbonate species ultimately precipitate in the oceans with CO_2 released back into the atmosphere. Pyrite oxidation produces sulphuric acid, which reacts with carbonates and releases CO_2 into the atmosphere and this pyrite-carbonate-coupled CO_2 source operates on time scales shorter than the pyrite formation time scale in the oceans (10^6–10^7 years) (Torres et al., 2014). Organic carbon in sedimentary rocks, or petrogenic organic carbon, represents a large, ancient carbon pool, whose oxidative weathering constitutes a major long-term CO_2 source (Zondervan et al., 2023). At the global scale, these weathering processes are strongly influenced by physical erosion rates (Hilton & West, 2020). Because the reaction kinetics of carbonate, pyrite and petrogenic organic carbon weathering are commonly considered to be faster than those of silicate minerals (Hilton & West, 2020), we expect earthquake-triggered landslides – by exposing fresh rock and enhancing erosion – to significantly increase these fluxes. This inference is supported by several case studies utilising field data, geochemical analyses and numerical modelling. Emberson et al. (2018) combined seepage chemistry, catchment discharge data and modelling to show that bedrock landsliding in New Zealand and Taiwan can accelerate rock weathering, with all weathering reactions contributing to a net carbon source. Bufe et al. (2024) analysed catchments in Taiwan, New Zealand and Sichuan, finding that tectonically active regions function as carbon sinks under moderate erosion rates, but shift to net carbon sources when erosion rates are extremely high and landslides are frequent.

Beyond the above understanding, the role of earthquake-triggered landslides in the long-term carbon cycle remains to be better quantified, especially in settings with varying climatic, geologic and tectonic conditions. We recommend future research on systematic assessments of seismic impacts on rock weathering fluxes across these various environmental contexts. Resolving the effect of seismic landslides on the long-term carbon cycle could also contribute to the fundamental question of whether tectonically active regions act as net carbon sinks or sources, and under what conditions.

11.3.3 Earthquake effects on greenhouse gas emissions

Fault activity and rupture during earthquakes can also lead to the release of subsurface CO_2 into the atmosphere, particularly in carbonate-rich regions where crustal reservoirs store significant amounts of CO_2 released through the metamorphism of subducted carbonates (Chiodini et al., 2020). Elevated CO_2 emissions have been documented after the 2008 Wenchuan and 2015 Gorkha earthquakes (Girault et al., 2018; Zheng et al., 2013), as well as smaller events, such as an M_L 3.5 earthquake in the Czech Republic in 2014 (Fischer et al., 2017). Chiodini et al. (2020) observed that CO_2 emissions correlate with the progression of seismic sequences over decadal time scales. Over million-year time scales, plate collisions drive magmatic and metamorphic degassing, producing cumulative CO_2 outgassing which may have contributed significantly to long-term climate regulation (Guo et al., 2021). Additionally, carbon monoxide (CO) anomalies have been detected before, during and after large earthquakes (Cui et al., 2024).

Methane (CH_4) is another critical greenhouse gas that could be released during earthquakes, particularly in regions where methane hydrates or natural gas reservoirs exist. Atmospheric CH_4 anomalies have been observed following the 2008 Wenchuan and 2013 Lushan earthquakes in China (Cui et al., 2024). Offshore earthquakes can also destabilise gas hydrate deposits. For instance, the 2010 M_w 8.8 Maule earthquake fractured hydrate-bearing sediments, creating migration pathways for free gas to escape into the water column (Geersen et al., 2016). Fischer et al. (2013) demonstrated that the 1945 M_w 8.1 Makran earthquake offshore of Pakistan triggered a methane release equivalent to 3.26×10^8 mol over 60 years. However, other events have shown negligible impacts; for instance, the 2021 M_s 6.0 Luxian earthquake in China did not result in detectable leakage in the adjacent shale gas fields (Liu et al., 2024).

While earthquake-induced greenhouse gas emissions are generally at local scales and short-lived, repeated large earthquakes in tectonically active regions may contribute significantly to long-term fluxes (Girault et al., 2018; Tamburello et al., 2018). This is especially relevant in areas with substantial gas reservoirs and active fault systems, where earthquakes may disturb those reservoirs to trigger considerable emissions with potential regional climate implications. We suggest that future research should prioritise (1) monitoring and quantifying gas flux changes under various seismic settings for a mechanistic, quantitative understanding of how earthquakes affect greenhouse gas emissions and (2) assessing the long-term environmental impacts of earthquake-triggered greenhouse gas releases.

11.4 Earthquake effects on other geochemical fluxes

In addition to their impact on the carbon cycle, earthquakes influence the release of various gases and elements that are crucial to the climate system and ecological environments. During seismic events, rock fracturing, friction and water-rock interactions often result in the release of hydrogen gas. Changes in pore pressure triggered by earthquakes can accelerate hydrogen accumulation and outgassing, particularly in seismically active regions where emissions often exhibit fluctuating pulses (Ito et al., 1999). Earthquakes can also enhance the degassing of radon, a radioactive gas, by altering fault and near-fault permeability (Ghosh et al., 2009; Mogro-Campero et al., 1980; Purtscheller et al., 1995). Radon has also been considered a potential indicator for earthquake forecasting (Hauksson, 1981). Although elevated radon concentrations do not directly predict seismic activity, their notable postseismic fluctuations provide valuable observational signals that may assist in understanding seismic processes. Degassing of other gases at active fault zones has been documented as well, including helium, alkanes, alkyl aromatic hydrocarbons, as well as highly volatile metals such as mercury, arsenic and antimony (Ito et al., 1999). The concentration anomalies of these substances are often associated with variations in regional stress fields and aftershock activity, as shown by Toutain and Baubron (1999).

In volcanically active regions, earthquakes can also influence volcanic degassing. Seismic waves have been shown to trigger volcanic eruptions, even at distances exceeding 1000 km from the epicentre (Bayramov et al., 2025). Earthquakes may also accelerate the release of volcanic gases such as SO_2, CO_2 and He (Hsu et al., 2010). Postseismic changes in the connectivity between magma chambers and fault systems can also enhance deep gas release, potentially affecting atmospheric composition, regional climate and local environments. Some other studies have reported decreases in SO_2 concentrations following earthquakes, likely due to the dominance of degassing processes in more viscous magmas or reduced permeability after seismic shaking (Avouris et al., 2017).

The influence of earthquakes on metal and nutrient fluxes deserves particular attention as well. Those fluxes liberated by earthquakes may transiently enhance primary productivity, with subsequent effects on biogeochemical cycles. Specifically, earthquake-triggered landslides can facilitate rock weathering, not only affecting carbon fluxes but also leading to the release of essential nutrients such as nitrogen and phosphorus (Buendía et al., 2010). These elements, in turn, can stimulate aquatic ecosystem productivity, increase organic carbon burial and contribute to atmospheric CO_2 drawdown (Sigman & Haug, 2003). Groundwater chemistry also exhibits changes before and after seismic events, modulating the magnitude and composition of riverine solute fluxes in earthquake-impacted regions (Rosen et al., 2018; Skelton et al., 2014). In marine sediments and estuarine zones, earthquake-induced sediment resuspension and fluid mixing can release trace metals (e.g. Fe, Mn) and nutrients (e.g. P, Si, N), significantly modulating biological productivity (Hung & Ho, 2014; Xue et al., 2025). Nonetheless, seismic impacts on metal and nutrient fluxes remain much less studied than those on carbon fluxes, pointing to important avenues for future research.

11.5 Conclusions

Over the past several decades, a growing body of research has provided compelling evidence of the significant impacts that earthquakes have on sediment and geochemical fluxes. Notable examples include the 1999 Chi-Chi earthquake, the 2008 Wenchuan earthquake, the 2015 Gorkha earthquake and various historical seismic events in New Zealand. Several important patterns have emerged from these case studies. Notably, large earthquakes often induce widespread landslides that destabilise rock and soil structures. Some of the material generated by these landslides is directly transported into river channels, immediately affecting sediment transport (Mountjoy et al., 2018; Wang et al., 2015), while other deposits remain on hillslopes, requiring postseismic processes such as debris flows to mobilise them (Dahlquist & West, 2019; Francis et al., 2022; Hovius et al., 2011; Kincey et al., 2023). Earthquakes lower the thresholds for debris flow initiation, leading to an increase in postseismic debris flow activities (Huang & Fan, 2013). Postseismic sediment transport is influenced both by coseismic landslides and environmental conditions, including hydrology, climate and topography (Tunnicliffe et al., 2024; Wang et al., 2015; Zhang et al., 2019). Fine-grained materials may remain stored in epicentral regions for years to centuries before being exported to downstream areas (Wang et al., 2015), while coarser materials, transported as bedload, evolve in more complex ways and may persist for centennial to millennial time scales (Croissant et al., 2017; Li et al., 2025; Tunnicliffe et al., 2024). More field observations are needed to better assess the evacuation time scale of landslide-derived sediments (in comparison to the seismic recurrence interval) and their long-term impacts on landscape systems.

Earthquake-induced landslides can also disrupt the terrestrial biosphere and associated carbon storage, reducing the short-term carbon sequestration capacity of ecosystems while potentially enhancing the degradation of biospheric organic matter (Chen et al., 2009). Consequently, earthquakes may act as short-term carbon sources. However, these landslides simultaneously increase the erosion and mobilisation of biospheric organic carbon, facilitating the transport and burial of particulate organic carbon in rivers (Howarth et al., 2025; Wang et al., 2016). Coupled with rapid postseismic vegetation recovery, these processes can ultimately allow earthquakes to induce and sustain long-term carbon sinks (Croissant et al., 2021).

Besides influencing the biosphere, earthquake-induced landslides can enhance rock weathering fluxes (Jin et al., 2016; Li et al., 2024) by exposing fresh mineral surfaces, with potential influences on the climate system over geological time scales. Case studies suggest that earthquake-induced landslides likely accelerate the weathering of carbonates, silicates, pyrite, and rock-derived organic carbon (Hilton & West, 2020) and the net carbon balance associated with earthquake-driven rock weathering may depend on the sensitivity of these processes to erosion rates (Bufe et al., 2024).

Despite these insights, several important questions remain to be better understood, pointing to future research avenues:

- What sets the magnitude of impacts on sediment and geochemical fluxes from seismic events? Some studies have found limited effects of earthquakes on sediment and geochemical fluxes, such as the 2015 M_w 7.8 Gorkha earthquake in Nepal (Märki et al., 2021) and the 2010 M_w 8.8 Maule earthquake in Chile (Tolorza et al., 2019). We speculate that the effects of earthquakes in those regions may be dampened by high rates of interseismic processes or weak climatic forcing, but understanding these trade-offs requires future research.
- How do sediment and geochemical fluxes evolve after large earthquakes in space and time? We suggest continuous monitoring and systematic collection of sediment and river water samples to assess the spatiotemporal dynamics of those postseismic fluxes. Sedimentary archives offer valuable records as well to reconstruct those variations in the past.
- What is the role of climate in modulating postseismic sediment and geochemical fluxes? Continued long-term monitoring, coupled with interpretation of sedimentary records, in seismically active regions under diverse climatic contexts, would yield valuable insights.

- What methods can be adopted or developed to better understand the seismic impact on sediment and geochemical fluxes? Given the interdisciplinary nature of this task, we recommend a combination of methods, including field studies, geochemical analyses, sediment budgeting, remote sensing techniques (e.g. LiDAR) and process-based models for a comprehensive assessment of how earthquakes impact sediment transport and geochemical cycles. Understanding postseismic sediment and geochemical fluxes would provide valuable insights into landscape evolution, geochemical cycling and hazard mitigation in seismically active regions.

Acknowledgements

J.W. acknowledges funding from the National Natural Science Foundation of China (42322305, 42221003). G.K.L. thanks the support from the Hellman Family Foundation.

References

Avouris, D. M., Carn, S. A., & Waite, G. P. (2017). Triggering of volcanic degassing by large earthquakes. *Geology, 45*, 715–718.

Bao, R., Strasser, M., Mcnichol, A. P., Haghipour, N., Mcintyre, C., Wefer, G., & Eglinton, T. I. (2018). Tectonically-triggered sediment and carbon export to the Hadal zone. *Nature Communications, 9*, 121.

Bayramov, Z., Viltres, R., Doubre, C., Maggi, A., Jolivet, R., & Rivera, L. (2025). Very-long-range dynamic triggering of mud volcano unrest and silent magnitude-6 fault slip. *Science (New York, N.Y.), 389*, 996–1002.

Blair, N. E., & Aller, R. C. (2012). The fate of terrestrial organic carbon in the marine environment. *Annual Review of Marine Science, 4*, 401–428.

Bloom, C. K., Howell, A., Stahl, T., Massey, C., & Singeisen, C. (2021). The influence of off-fault deformation zones on the near-fault distribution of coseismic landslides. *Geology, 50*, 272–277.

Buendía, C., Kleidon, A., & Porporato, A. (2010). The role of tectonic uplift, climate, and vegetation in the long-term terrestrial phosphorous cycle. *Biogeosciences, 7*, 2025–2038.

Bufe, A., Rugenstein, J. K. C., & Hovius, N. (2024). CO_2 drawdown from weathering is maximized at moderate erosion rates. *Science (New York, N.Y.), 383*, 1075–1080.

Chambers, J. Q., Fisher, J. I., Zeng, H., Chapman, E. L., Baker, D. B., & Hurtt, G. C. (2007). Hurricane Katrina's carbon footprint on U.S. Gulf Coast forests. *Science (New York, N.Y.), 318*, 1107.

Chen, H., Wu, N., Yuan, X., Gao, Y., & Zhu, D. (2009). Aftermath of the Wenchuan earthquake. *Frontiers in Ecology and the Environment, 7*, 72.

Chiodini, G., Cardellini, C., Di Luccio, F., Selva, J., Frondini, F., Caliro, S., Rosiello, A., Beddini, G., & Ventura, G. (2020). Correlation between tectonic CO_2 earth degassing and seismicity is revealed by a 10-year record in the Apennines, Italy. *Science Advances, 6*, eabc2938.

Croissant, T., Hilton, R. G., Li, G. K., Howarth, J., Wang, J., Harvey, E. L., Steer, P., & Densmore, A. L. (2021). Pulsed carbon export from mountains by earthquake-triggered landslides explored in a reduced-complexity model. *Earth Surface Dynamics, 9*, 823–844.

Croissant, T., Lague, D., Steer, P., & Davy, P. (2017). Rapid post-seismic landslide evacuation boosted by dynamic river width. *Nature Geoscience, 10*, 680–684.

Cui, Y., Huang, J., Zeng, Z., & Zou, Z. (2024). CO emissions associated with three major earthquakes occurring in diverse tectonic environments. *Remote Sensing, 16*, 480.

Dadson, S. J., Hovius, N., Chen, H., Dade, W. B., Lin, J. C., Hsu, M. L., Lin, C. W., Horng, M. J., Chen, T. C., & Milliman, J. (2004). Earthquake-triggered increase in sediment delivery from an active mountain belt. *Geology, 32*, 733–736.

Dahlquist, M. P., & West, A. J. (2019). Initiation and runout of post-seismic debris flows: Insights from the 2015 Gorkha earthquake. *Geophysical Research Letters, 46*, 9658–9668.

Dahlquist, M. P., West, A. J., & Li, G. (2018). Landslide-driven drainage divide migration. *Geology, 46*, 403–406.

Densmore, A. L., & Hovius, N. (2000). Topographic fingerprints of bedrock landslides. *Geology, 28*, 371–374.

Dethier, E. N., Renshaw, C. E., & Magilligan, F. J. (2022). Rapid changes to global river suspended sediment flux by humans. *Science (New York, N.Y.), 376*, 1447–1452.

Egholm, D. L., Knudsen, M. F., & Sandiford, M. (2013). Lifespan of mountain ranges scaled by feedbacks between landsliding and erosion by rivers. *Nature, 498*, 475–478.

Emberson, R., Galy, A., & Hovius, N. (2018). Weathering of reactive mineral phases in landslides acts as a source of carbon dioxide in mountain belts. *Journal of Geophysical Research: Earth Surface, 123*, 2695–2713.

Emberson, R., Hovius, N., Galy, A., & Marc, O. (2015). Chemical weathering in active mountain belts controlled by stochastic bedrock landsliding. *Nature Geoscience, 9*, 42.

Fan, N., Nie, R., Wang, Q., & Liu, X. (2016). Dramatic undercutting of piedmont rivers after the 2008 Wenchuan Ms 8.0 Earthquake. *Scientific Reports, 6*, 37108.

Fan, X., Domènech, G., Scaringi, G., Huang, R., Xu, Q., Hales, T. C., Dai, L., Yang, Q., & Francis, O. (2018). Spatio-temporal evolution of mass wasting after the 2008 M_w 7.9 Wenchuan earthquake revealed by a detailed multi-temporal inventory. *Landslides, 15*, 2325–2341.

Fan, X., Scaringi, G., Domènech, G., Yang, F., Guo, X., Dai, L., He, C., Xu, Q., & Huang, R. (2019). Two multi-temporal datasets that track the enhanced landsliding after the 2008 Wenchuan earthquake. *Earth System Science Data, 11*, 35–55.

Fischer, D., Mogollón, J. M., Strasser, M., Pape, T., Bohrmann, G., Fekete, N., Spiess, V., & Kasten, S. (2013). Subduction zone earthquake as potential trigger of submarine hydrocarbon seepage. *Nature Geoscience, 6*, 647–651.

Fischer, T., Matyska, C., & Heinicke, J. (2017). Earthquake-enhanced permeability–evidence from carbon dioxide release following the M_L 3.5 earthquake in West Bohemia. *Earth and Planetary Science Letters, 460*, 60–67.

Francis, O., Fan, X., Hales, T., Hobley, D., Xu, Q., & Huang, R. (2022). The fate of sediment after a large earthquake. *Journal of Geophysical Research: Earth Surface, 127* e2021JF006352.

Frith, N. V., Hilton, R. G., Howarth, J. D., Gröcke, D. R., Fitzsimons, S. J., Croissant, T., Wang, J., Mcclymont, E. L., Dahl, J., & Densmore, A. L. (2018). Carbon export from mountain forests enhanced by earthquake-triggered landslides over millennia. *Nature Geoscience, 11*, 772–776.

Garwood, N. C., Janos, D. P., & Brokaw, N. (1979). Earthquake-caused landslides: A major disturbance to tropical forests. *Science (New York, N.Y.), 205*, 997–999.

Geersen, J., Scholz, F., Linke, P., Schmidt, M., Lange, D., Behrmann, J. H., Völker, D., & Hensen, C. (2016). Fault zone controlled seafloor methane seepage in the rupture area of the 2010 Maule earthquake. *Central Chile. Geochemistry, Geophysics, Geosystems, 17*, 4802–4813.

Ghosh, D., Deb, A., & Sengupta, R. (2009). Anomalous radon emission as precursor of earthquake. *Journal of Applied Geophysics, 69*, 67–81.

Girault, F., Adhikari, L. B., France-Lanord, C., Agrinier, P., Koirala, B. P., Bhattarai, M., Mahat, S. S., Groppo, C., Rolfo, F., & Bollinger, L. (2018). Persistent CO_2 emissions and hydrothermal unrest following the 2015 earthquake in Nepal. *Nature Communications, 9*, 2956.

Graf, E. L. S., Sinclair, H. D., Attal, M., Gailleton, B., Adhikari, B. R., & Baral, B. R. (2024). Geomorphological and hydrological controls on sediment export in earthquake-affected catchments in the Nepal Himalaya. *Earth Surface Dynamics, 12*, 135–161.

Guo, Z., Wilson, M., Dingwell, D. B., & Liu, J. (2021). India-Asia collision as a driver of atmospheric CO_2 in the Cenozoic. *Nature Communications, 12*, 3891.

Hao, M., Jin, Z., Li, P., Song, Y., & Cao, G. (2025). Detection of tree destruction induced by heavy rainfall in the afforested loess catchment of China. *CATENA, 248*, 108573.

Hauksson, E. (1981). Radon content of groundwater as an earthquake precursor: Evaluation of worldwide data and physical basis. *Journal of Geophysical Research: Solid Earth, 86*, 9397–9410.

Hilton, R. G., & West, A. J. (2020). Mountains, erosion and the carbon cycle. *Nature Reviews Earth & Environment, 1*, 284–299.

Hovius, N., Meunier, P., Lin, C.-W., Chen, H., Chen, Y.-G., Dadson, S., Horng, M.-J., & Lines, M. (2011). Prolonged seismically induced erosion and the mass balance of a large earthquake. *Earth and Planetary Science Letters, 304*, 347–355.

Howarth, J. D., Fitzsimons, S. J., Moody, A., Wang, J., Garnett, M. H., Croissant, T., Densmore, A. L., Howell, A., & Hilton, R. G. (2025). Long term carbon export from mountain forests driven by hydroclimate and extreme event driven landsliding. *Communications Earth & Environment, 6*, 432.

Howarth, J. D., Fitzsimons, S. J., Norris, R. J., & Jacobsen, G. E. (2012). Lake sediments record cycles of sediment flux driven by large earthquakes on the Alpine fault. *New Zealand. Geology, 40*, 1091–1094.

Hsu, S.-C., Huang, Y.-T., Huang, J.-C., Tu, J.-Y., Engling, G., Lin, C.-Y., Lin, F.-J., & Huang, C.-H. (2010). Evaluating real-time air-quality data as earthquake indicator. *Science of The Total Environment, 408*, 2299–2304.

Huang, R., & Fan, X. (2013). The landslide story. *Nature Geoscience, 6*, 325–326.

Hung, J. J., & Ho, C. Y. (2014). Typhoon- and earthquake-enhanced concentration and inventory of dissolved and particulate trace metals along two submarine canyons off southwestern Taiwan. *Estuarine, Coastal and Shelf Science, 136*, 179–190.

Ito, T., Nagamine, K., Yamamoto, K., Adachi, M., & Kawabe, I. (1999). Preseismic hydrogen gas anomalies caused by stress-corrosion process preceding earthquakes. *Geophysical Research Letters, 26*, 2009–2012.

Jerolmack, D. J., & Paola, C. (2010). Shredding of environmental signals by sediment transport. *Geophysical Research Letters, 37*. https://doi.org/10.1029/2010GL044638 L19401.

Jin, Z., West, A. J., Zhang, F., An, Z., Hilton, R. G., Yu, J., Wang, J., Li, G., Deng, L., & Wang, X. (2016). Seismically enhanced solute fluxes in the Yangtze River headwaters following the A.D. 2008 Wenchuan earthquake. *Geology, 44*, 47–50.

Kang, D., Li, Y., Ma, L., & Zou, S. (2023). Landslide scales affect soil organic carbon accumulation by influencing microbial decomposition of plant-derived carbon after earthquakes. *Ecological Indicators, 155*, 110949.

Keefer, D. K. (1984). Landslides caused by earthquakes. *Geological Society of America Bulletin, 95*, 406–421.

Kincey, M. E., Rosser, N. J., Densmore, A. L., Robinson, T. R., Shrestha, R., Singh Pujara, D., Horton, P., Swirad, Z. M., Oven, K. J., & Arrell, K. (2023). Modelling post-earthquake cascading hazards: Changing patterns of landslide runout following the 2015 Gorkha earthquake, Nepal. *Earth Surface Processes and Landforms, 48*, 537–554.

Kincey, M. E., Rosser, N. J., Robinson, T. R., Densmore, A. L., Shrestha, R., Pujara, D. S., Oven, K. J., Williams, J. G., & Swirad, Z. M. (2021). Evolution of coseismic and post-seismic landsliding after the 2015 M_w 7.8 Gorkha earthquake, Nepal. *Journal of Geophysical Research: Earth Surface, 126* e2020JF005803.

Koi, T., Hotta, N., Ishigaki, I., Matuzaki, N., Uchiyama, Y., & Suzuki, M. (2008). Prolonged impact of earthquake-induced landslides on sediment yield in a mountain watershed: The Tanzawa region, Japan. *Geomorphology, 101*, 692–702.

Korup, O., Mcsaveney, M. J., & Davies, T. R. H. (2004). Sediment generation and delivery from large historic landslides in the Southern Alps, New Zealand. *Geomorphology, 61*, 189–207.

Larsen, I. J., & Montgomery, D. R. (2012). Landslide erosion coupled to tectonics and river incision. *Nature Geoscience, 5*, 468–473.

Li, G., West, A. J., Densmore, A. L., Hammond, D. E., Jin, Z., Zhang, F., Wang, J., & Hilton, R. G. (2016). Connectivity of earthquake-triggered landslides with the fluvial network: Implications for landslide sediment transport after the 2008 Wenchuan earthquake. *Journal of Geophysical Research: Earth Surface, 121*, 703–724.

Li, G., West, A. J., Densmore, A. L., Jin, Z., Parker, R. N., & Hilton, R. G. (2014). Seismic mountain building: Landslides associated with the 2008 Wenchuan earthquake in the context of a generalized model for earthquake volume balance. *Geochemistry, Geophysics, Geosystems, 15*, 833–844.

Li, G., West, A. J., Densmore, A. L., Jin, Z., Zhang, F., Wang, J., Clark, M., & Hilton, R. G. (2017). Earthquakes drive focused denudation along a tectonically active mountain front. *Earth and Planetary Science Letters, 472*, 253–265.

Li, G., West, A. J., & Qiu, H. (2019). Competing effects of mountain uplift and landslide erosion over earthquake cycles. *Journal of Geophysical Research: Solid Earth, 124*, 5101–5133.

Li, G. K., & Moon, S. (2021). Topographic stress control on bedrock landslide size. *Nature Geoscience, 14*, 307–313.

Li, G. K., West, A. J., Jin, Z., Qiu, H., Zhang, F., Wang, J., Hammond, D. E., Densmore, A. L., Hilton, R. G., Dong, S., Atwood, A., Fischer, W. W., & Lamb, M. P. (2025). Large riverbed sediment flux sustained for a decade after an earthquake. *Nature, 644*, 398–403.

Li, L., Robinson, L. F., Li, G. K., Hedding, D. W., Xian, F., Xu, Z., Li, L., Ouyang, S., Li, T., & Li, G. (2024). Multi-isotopic constraints on the impacts of landslide on weathering and erosion in an active mountain range. *Earth and Planetary Science Letters, 626*, 118527.

Lin, W.-T., Chou, W.-C., Lin, C.-Y., Huang, P.-H., & Tsai, J.-S. (2005). Vegetation recovery monitoring and assessment at landslides caused by earthquake in Central Taiwan. *Forest Ecology and Management, 210*, 55–66.

Lin, W.-T., Lin, C.-Y., & Chou, W.-C. (2006). Assessment of vegetation recovery and soil erosion at landslides caused by a catastrophic earthquake: A case study in Central Taiwan. *Ecological Engineering, 28*, 79–89.

Liu, J., Fan, X., Tang, X., Xu, Q., Harvey, E. L., Hales, T. C., & Jin, Z. (2022). Ecosystem carbon stock loss after a mega earthquake. *CATENA, 216*, 106393.

Liu, Z., Chen, Z., Li, Y., Zhao, Z., Hong, S., Hu, L., Ma, L., Lu, C., Zhao, Y., He, H., Su, S., Zhao, Y., Shao, W., Cao, Z., & Wang, H. (2024). Shale gas leakage and fault activation: Insight from the 2021 Luxian M_S 6.0 earthquake, China. *Tectonophysics, 891*, 230530.

Malamud, B. D., Turcotte, D. L., Guzzetti, F., & Reichenbach, P. (2004). Landslides, earthquakes, and erosion. *Earth and Planetary Science Letters, 229*, 45–59.

Marc, O., Hovius, N., Meunier, P., Gorum, T., & Uchida, T. (2016). A seismologically consistent expression for the total area and volume of earthquake-triggered landsliding. *Journal of Geophysical Research: Earth Surface, 121*, 640–663.

Marc, O., Hovius, N., Meunier, P., Uchida, T., & Hayashi, S. (2015). Transient changes of landslide rates after earthquakes. *Geology, 43*, 883–886.

McSaveney, M., Cox, S., & Hancox, G. (2015). Increasing rock-avalanche frequency correlates with increasing seismic moment release in New Zealand's Southern Alps. *EGU General Assembly Conference Abstracts*, 8074.

Milliman, J. D., & Farnsworth, K. L. (2011). *River discharge to the coastal ocean: A global synthesis*. Cambridge University Press.

Mogro-Campero, A., Fleischer, R. L., & Likes, R. S. (1980). Changes in subsurface radon concentration associated with earthquakes. *Journal of Geophysical Research: Solid Earth, 85*, 3053–3057.

Mountjoy, J. J., Howarth, J. D., Orpin, A. R., Barnes, P. M., Bowden, D. A., Rowden, A. A., Schimel, A. C. G., Holden, C., Horgan, H. J., Nodder, S. D., Patton, J. R., Lamarche, G., Gerstenberger, M., Micallef, A., Pallentin, A., & Kane, T. (2018). Earthquakes drive large-scale submarine canyon development and sediment supply to deep-ocean basins. *Science Advances, 4*, eaar3748.

Märki, L., Lupker, M., France-Lanord, C., Lavé, J., Gallen, S., Gajurel, A. P., Haghipour, N., Leuenberger-West, F., & Eglinton, T. (2021). An unshakable carbon budget for the Himalaya. *Nature Geoscience, 14*, 745–750.

Parker, R. N., Densmore, A. L., Rosser, N. J., De Michele, M., Li, Y., Huang, R., Whadcoat, S., & Petley, D. N. (2011). Mass wasting triggered by the 2008 Wenchuan earthquake is greater than orogenic growth. *Nature Geoscience, 4*, 449–452.

Pearce, A. J., & Watson, A. (1983). Medium-term effects of two landsliding episodes on channel storage of sediment. *Earth Surface Processes and Landforms, 8*, 29–39.

Pearce, A. J., & Watson, A. J. (1986). Effects of earthquake-induced landslides on sediment budget and transport over a 50-yr period. *Geology, 14*, 52–55.

Petley, D. N., Hearn, G. J., Hart, A., Rosser, N. J., Dunning, S. A., Oven, K., & Mitchell, W. A. (2007). Trends in landslide occurrence in Nepal. *Natural Hazards, 43*, 23–44.

Purtscheller, F., Pirchl, T., Sieder, G., Stingl, V., Tessadri, T., Brunner, P., Ennemoser, O., & Schneider, P. (1995). Radon emanation from giant landslides of Kofels (Tyrol, Austria) and Langtang Himal (Nepal). *Environmental Geology, 26*(1), 32–38.

Rault, C., Robert, A., Marc, O., Hovius, N., & Meunier, P. (2019). Seismic and geologic controls on spatial clustering of landslides in three large earthquakes. *Earth Surface Dynamics,* 7, 829–839.

Raymo, M., & Ruddiman, W. F. (1992). Tectonic forcing of late Cenozoic climate. *Nature, 359*, 117–122.

Ren, D., Wang, J., Fu, R., Karoly, D. J., Hong, Y., Leslie, L. M., Fu, C., & Huang, G. (2009). Mudslide-caused ecosystem degradation following Wenchuan earthquake 2008. *Geophysical Research Letters, 36* L05401.

Restrepo, C., Walker, L. R., Shiels, A. B., Bussmann, R., Claessens, L., Fisch, S., Lozano, P., Negi, G., Paolini, L., Poveda, G., Ramos-Scharrón, C., Richter, M., & Velázquez, E. (2009). Landsliding and its multiscale influence on mountainscapes. *Bioscience, 59*, 685–698.

Richey, J. E., Hedges, J. I., Devol, A. H., Quay, P. D., Victoria, R., Martinelli, L., & Forsberg, B. R. (1990). Biogeochemistry of carbon in the Amazon River. *Limnology and Oceanography, 35*, 352–371.

Roback, K., Clark, M. K., West, A. J., Zekkos, D., Li, G., Gallen, S. F., Chamlagain, D., & Godt, J. W. (2018). The size, distribution, and mobility of landslides caused by the 2015 M_w 7.8 Gorkha earthquake, Nepal. *Geomorphology, 301*, 121–138.

Romans, B. W., Castelltort, S., Covault, J. A., Fildani, A., & Walsh, J. P. (2016). Environmental signal propagation in sedimentary systems across timescales. *Earth-Science Reviews, 153*, 7–29.

Rosen, M. R., Binda, G., Archer, C., Pozzi, A., Michetti, A. M., & Noble, P. J. (2018). Mechanisms of earthquake-induced chemical and fluid transport to carbonate groundwater springs after earthquakes. *Water Resources Research, 54*, 5225–5244.

Shi, P., Hu, X., Yang, H., Jiang, L., Ma, Y., Tang, H., Zhou, Q., Liu, F., & Liu, L. (2025). Earthquakes have accelerated the carbon dioxide emission rate of soils on the Qinghai-Tibet Plateau. *Global Change Biology, 31*, e70024.

Sigman, D. M., & Haug, G. H. (2003). The biological pump in the past. In H. Elderfield (Ed.). *Treatise on geochemistry*. Pergamon Press.

Skelton, A., Andrén, M., Kristmannsdóttir, H., Stockmann, G., Mörth, C.-M., Sveinbjörnsdóttir, Á., Jónsson, S., Sturkell, E., Guðrúnardóttir, H. R., Hjartarson, H., Siegmund, H., & Kockum, I. (2014). Changes in groundwater chemistry before two consecutive earthquakes in Iceland. *Nature Geoscience,* 7, 752–756.

Slosson, J. R., Hoke, G. D., & Lifton, N. (2022). Non-steady-state ^{14}C-^{10}Be and transient hillslope dynamics in steep high mountain catchments. *Geophysical Research Letters, 49* e2022GL100365.

Talling, P. J., Hage, S., Baker, M. L., Bianchi, T. S., Hilton, R. G., & Maier, K. L. (2024). The global turbidity current pump and its implications for organic carbon cycling. *Annual Review of Marine Science, 16*, 105–133.

Tamburello, G., Pondrelli, S., Chiodini, G., & Rouwet, D. (2018). Global-scale control of extensional tectonics on CO_2 earth degassing. *Nature Communications, 9*, 4608.

Thunell, R., Tappa, E., Varela, R., Llano, M., Astor, Y., Muller-Karger, F., & Bohrer, R. (1999). Increased marine sediment suspension and fluxes following an earthquake. *Nature, 398*, 233–236.

Tolorza, V., Mohr, C. H., Carretier, S., Serey, A., Sepúlveda, S. A., Tapia, J., & Pinto, L. (2019). Suspended sediments in Chilean rivers reveal low postseismic erosion after the Maule earthquake (M_w 8.8) during a severe drought. *Journal of Geophysical Research: Earth Surface, 124*, 1378–1397.

Torres, M. A., West, A. J., & Li, G. (2014). Sulphide oxidation and carbonate dissolution as a source of CO_2 over geological timescales. *Nature, 507*, 346–349.

Toutain, J.-P., & Baubron, J.-C. (1999). Gas geochemistry and seismotectonics: A review. *Tectonophysics, 304*, 1–27.

Tunnicliffe, J., Howarth, J., & Massey, C. (2024). Controls on fluvial sediment evacuation following an earthquake-triggered landslide: Observations from LiDAR time series. *Science Advances, 10*, eadi5560.

Valagussa, A., Frattini, P., Valbuzzi, E., & Crosta, G. B. (2021). Role of landslides on the volume balance of the Nepal 2015 earthquake sequence. *Scientific Reports, 11*, 3434.

Valagussa, A., Marc, O., Frattini, P., & Crosta, G. B. (2019). Seismic and geological controls on earthquake-induced landslide size. *Earth and Planetary Science Letters, 506*, 268–281.

Wang, J., Hilton, R. G., Jin, Z., Zhang, F., Densmore, A. L., Gröcke, D. R., Xu, X., Li, G., & West, A. J. (2019). The isotopic composition and fluxes of particulate organic carbon exported from the eastern margin of the Tibetan Plateau. *Geochimica et Cosmochimica Acta, 252*, 1–15.

Wang, J., Howarth, J. D., Mcclymont, E. L., Densmore, A. L., Fitzsimons, S. J., Croissant, T., Gröcke, D. R., West, M. D., Harvey, E. L., Frith, N. V., Garnett, M. H., & Hilton, R. G. (2020). Long-term patterns of hillslope erosion by earthquake-induced landslides shape mountain landscapes. *Science Advances, 6*, eaaz6446.

Wang, J., Jin, Z., Hilton, R. G., Zhang, F., Densmore, A. L., Li, G., & West, A. J. (2015). Controls on fluvial evacuation of sediment from earthquake-triggered landslides. *Geology, 43*, 115–118.

Wang, J., Jin, Z., Hilton, R. G., Zhang, F., Li, G., Densmore, A. L., Gröcke, D. R., Xu, X., & West, A. J. (2016). Earthquake-triggered increase in biospheric carbon export from a mountain belt. *Geology, 44*, 471–474.

Wang, W., Godard, V., Liu-Zeng, J., Scherler, D., Xu, C., Zhang, J., Xie, K., Bellier, O., Ansberque, C., De Sigoyer, J., & Team, A. (2017). Perturbation of fluvial sediment fluxes following the 2008 Wenchuan earthquake. *Earth Surface Processes and Landforms, 42*, 2611–2622.

Wang, W., Wang, W., Liu-Zeng, J., Godard, V., Zhang, J., Yang, J., Zheng, X., Wang, Z., Xie, K., & Li, M. (2025). Fifteen years observations of detrital ^{10}Be after the Wenchuan earthquake: Insights into landslide sediment transport following a large-magnitude seismic event. *Geomorphology, 482*, 109771.

West, A. J., Hetzel, R., Li, G., Jin, Z., Zhang, F., Hilton, R. G., & Densmore, A. L. (2014). Dilution of ^{10}Be in detrital quartz by earthquake-induced landslides: Implications for determining denudation rates and potential to provide insights into landslide sediment dynamics. *Earth and Planetary Science Letters, 396*, 143–153.

Xue, J., Li, B., Mao, Z., Ye, Z., Huang, T., Deng, Y., Hu, Z., Han, C., Wang, R., Wu, Q., Ji, M., Chen, R., Yang, H., & Huang, C. (2025). Lake state shifts after earthquakes in tectonically active regions. *Journal of Geophysical Research: Biogeosciences, 130* e2024JG008723.

Yang, Y., Fan, X., Wang, X., Lv, L., Zou, C., & Feng, Z. (2023). Net primary productivity changes associated with landslides induced by the 2008 Wenchuan Earthquake. *Land Degradation & Development, 34*, 1035–1050.

Yanites, B. J., Tucker, G. E., Mueller, K. J., & Chen, Y. G. (2010). How rivers react to large earthquakes: Evidence from central Taiwan. *Geology, 38*, 639–642.

Zeng, H., Lu, T., Jenkins, H., Negrón-Juárez, R. I., & Xu, J. (2016). Assessing earthquake-induced tree mortality in temperate forest ecosystems: A case study from Wenchuan, China. *Remote Sensing, 8*, 252.

Zhang, F., Jin, Z., West, A. J., An, Z., Hilton, R. G., Wang, J., Li, G., Densmore, A. L., Yu, J., Qiang, X., Sun, Y., Li, L., Gou, L., Xu, Y., Xu, X., Liu, X., Pan, Y., & You, C.-F. (2019). Monsoonal control on a delayed response of sedimentation to the 2008 Wenchuan earthquake. *Science Advances, 5*, eaav7110.

Zheng, G., Xu, S., Liang, S., Shi, P., & Zhao, J. (2013). Gas emission from the Qingzhu River after the 2008 Wenchuan earthquake, Southwest China. *Chemical Geology, 339*, 187–193.

Zondervan, J. R., Hilton, R. G., Dellinger, M., Clubb, F. J., Roylands, T., & Ogrič, M. (2023). Rock organic carbon oxidation CO_2 release offsets silicate weathering sink. *Nature, 623*, 329–333.

CHAPTER 12

On-fault geomorphic indicators of past earthquakes

Sally F. McGill
Department of Geological Sciences, California State University, San Bernardino, CA, United States

12.1 Introduction

Geomorphology has played a central role in the study of prehistoric earthquakes on active faults, including both strike-slip and dip-slip faults. Fault displacements during earthquakes typically produce fault scarps and related geomorphic features such as grabens, side-hill benches, triangular facets, wineglass canyons, shutter ridges, mole-tracks and laterally deflected, offset and beheaded channels. These geomorphic features can record the amount of strike-slip and dip-slip offset in past earthquakes, whether recent, historic, or prehistoric. The record of slip in prehistoric earthquakes that can be preserved within geomorphic features has provided crucial input to seismic hazard models and to our understanding of the behaviour of complex fault systems. For these reasons, many studies have investigated methods for interpreting the amount of slip from offset geomorphic features and how to accurately and precisely measure that slip and its uncertainties, as well as methods for inferring the number of earthquakes that contributed to the offset of a geomorphic feature.

Because many active faults have recurrence intervals that are longer than the period of historical record in their region, geomorphic records of past earthquakes provide important data to constrain seismic hazard models. Careful interpretation and reconstruction of the tectonic offset of dated landforms that have experienced multiple earthquakes can be used to calculate the average slip rate of the fault (e.g. Sieh & Jahns, 1984). Such geologic slip rates are important elements of seismic hazard models (e.g. Dawson et al., 2013; Field et al., 2014). Careful interpretation and compilation of landforms that have been offset in the past one to several earthquakes can also be used to infer the amount of slip per event in those earthquakes (e.g. Sieh, 1978). These results can then be used to infer the sizes of prehistoric earthquakes on a fault, as a guide for anticipating the range of plausible sizes of future earthquakes on the fault. Given dates for past events (typically from stratigraphic paleoseismic studies), probabilities of future earthquakes can also be inferred. This chapter will review and summarise some fault-related landforms and the methods and challenges of using them to infer the amount of slip in past earthquakes for the purposes of seismic hazard analysis and improved

Geomorphic Impacts of Earthquakes
DOI: https://doi.org/10.1016/B978-0-443-40375-0.00004-9

understanding of fault systems and the physics of earthquakes. Note that only on-fault indicators are considered; off-fault indicators such as coseismic landslides (Chapter 7) and liquefaction (Chapter 8) are dealt with elsewhere in this volume.

Previous chapters have described the processes by which landforms are created and modified as a result of fault rupture (Chapter 6) and earthquake shaking (Chapters 7, 8, and 9). Section 12.2 of this chapter will dive deeper into the processes of formation of some of the geomorphic features that form as a result of fault rupture, which were introduced in Chapter 6. Section 12.3 will summarise some considerations related to the accurate measurement of the amount of strike-slip and dip-slip offset of geomorphic features. Section 12.4 will explore how offset geomorphic markers have been used to estimate the amount of slip in prehistoric earthquakes. Section 12.5 will briefly summarise how studies of offset geomorphic markers can be used to infer the magnitude and timing of prehistoric earthquakes, and as an input to assessing seismic hazards, and will also show how studies of offset geomorphic features provide insight into the behaviour of complex fault systems and the physics of earthquakes.

12.2 Geomorphic features produced by past earthquakes

12.2.1 Map-view patterns for strike-slip, normal and thrust faults

Strike-slip faults (with horizontal slip, parallel to the strike of the fault) and dip-slip faults (with slip parallel to the dip direction of the fault plane) were introduced in Chapter 6. Here, we make some additional observations about their appearance as they cross-cut the land surface. Strike-slip fault planes are commonly (but not always) oriented vertically or nearly vertically (Anderson, 1905). Thus surficial fault rupture on strike-slip faults often cuts a planform straight line across any preexisting topography in the landscape (Fig. 12.1A). This contrasts with normal faults and thrust faults, which are typically nonvertical, resulting in surficial fault ruptures with map-view traces that are influenced by the local topography. For example, normal faults typically dip in the direction away from a mountain front at an angle of up to about 60 degrees (Anderson, 1905). Thus the trace of a normal fault along a range front will project to the ground surface farther into the mountains (farther upslope) in areas of high topography (e.g. along ridge crests that are perpendicular to the range being uplifted by the fault) and will project to the ground surface at a more valley-ward location (farther downslope) in areas of lower topography (e.g. within drainages that cross the normal fault). Thus normal faults typically have a fault trace that is concave toward the mountains where the fault crosses tributary valleys (Fig. 12.1B). In contrast, thrust faults dip in the direction beneath the mountains that they are uplifting (typically at an angle of about 30 degrees (Anderson, 1905)), so their traces typically intersect the ground surface farther valley-ward where they cross high topography (e.g. ridges) and are convex toward the mountains where they cross the lower topography within tributary valleys.

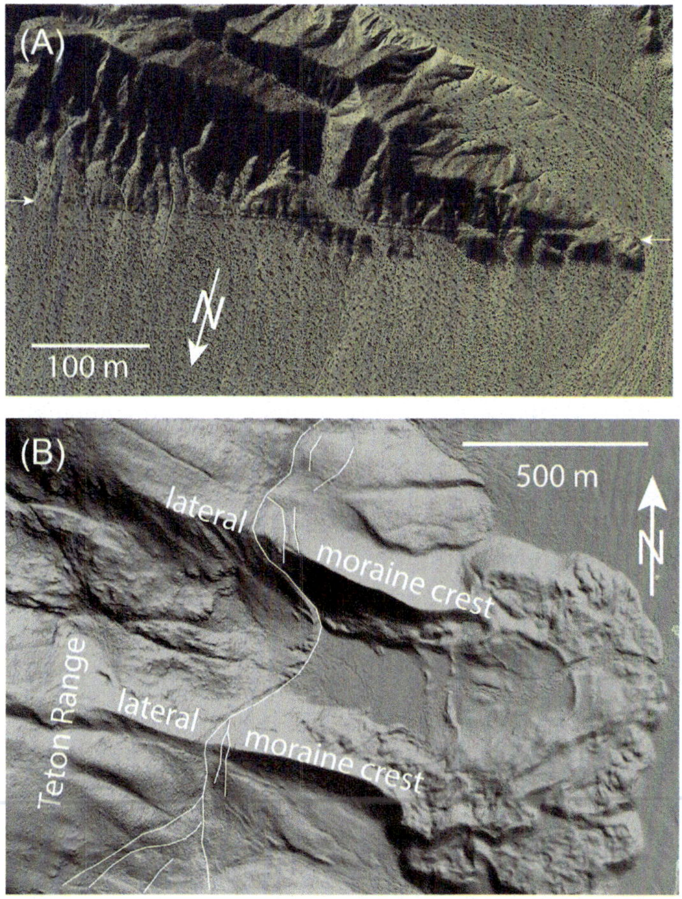

Figure 12.1 (A) Aerial photograph of the central Garlock fault, California, USA. Near-horizontal white arrows mark the fault trace. The vertical dip of the fault is apparent from the fault's linear trace as it traverses changing topography. Centre of image is approximately at N 35.5177°, W 117.3975°. Photo credit: U.S. Geological Survey (roll 1, frame 128, flight arranged by Malcolm Clark). (B) Hillshade representation of Lidar imagery along the Teton Fault, Wyoming, USA. White, curved lines mark the trace of this normal fault where it crosses a valley flanked by glacial moraines on the north and south. The eastward dip of the fault is apparent as the trace of the fault is concave toward the Teton Mountains (west of the fault), within the (low elevation) glaciated valley and is convex toward the Teton Mountains at the (higher elevation) moraine crests. Centre of image is approximately at N 43.7789°, W 110.7529°. Lidar survey conducted 9 July–4 August 2008 as part of the EarthScope Intermountain Seismic Belt LiDAR Project. Distributed by OpenTopography. https://doi.org/10.5069/G9VD6WCS. Accessed 10 November 2014. This material is based on services provided to the Plate Boundary Observatory by NCALM (https://ncalm.cive.uh.edu). PBO is operated by UNAVCO for EarthScope (http://www.earthscope.org) and supported by the National Science Foundation (No. EAR-0350028 and EAR-0732947).

12.2.2 Slip terminology for strike-slip and dip-slip faults

Most commonly, faults slip as a result of shear failure, in which case the slip vector lies within the plane of the fault and may include a strike-slip component (horizontal slip parallel to the strike of the fault plane), and a dip-slip component (slip that occurs parallel to the dip, or gradient of the fault plane) (Fig. 12.2). The dip-slip vector may be further resolved into its vertical component (throw) and its horizontal component (heave).

12.2.3 Fault scarps

Fault scarps are one of the most obvious types of geomorphic features produced by earthquakes. Fault scarps are abrupt, vertical, or steeply sloping, linear or curvilinear changes in elevation of the ground surface, which can form along any type of fault (strike-slip, normal, reverse, or oblique). The most classic examples and studies of fault scarps are along normal faults (e.g. Gilbert, 1890; Wallace, 1978). However, many of the terms, concepts and procedures for studying normal fault scarps can be applied to scarps that form along reverse and strike-slip faults as well.

12.2.3.1 Fault scarps along strike-slip faults

Strike-slip faults primarily have lateral slip, but there often may be a small component of vertical slip, which produces an obvious fault scarp along the surface rupture (e.g. Fig. 12.1A). A dip-slip component of motion may be substantial where there are bends or step-overs in strike-slip faults, but may also be present along long, straight sections of

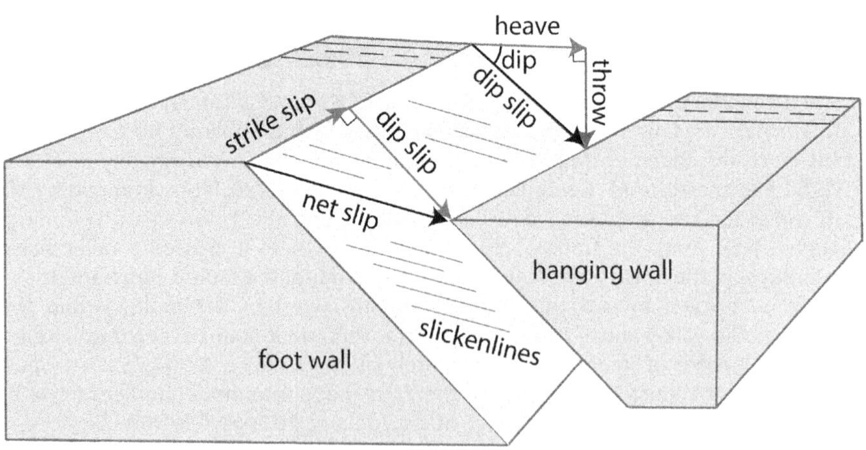

Figure 12.2 Block diagram illustrating the strike-slip and dip-slip components of the net slip vector, as well as the heave and throw components of the dip-slip vector. Slickenlines are striations on the fault plane, parallel to the net slip vector. For faults with nonvertical dips, the footwall is the block that lies beneath the fault plane, and the hanging wall is the block that overlies the fault plane. An offset road is illustrated on the ground surface.

a strike-slip fault. In the latter case, the Holocene scarps often face downhill, consistent with the longer-term sense of dip slip that has produced repeated uplift on one side of the fault (e.g. Fig. 12.1A). However, uphill-facing scarps can also form along strike-slip fault ruptures. In some cases, this is the result of lateral offset of a sloping ground surface, such as frequently occurs when alluvial fans are laterally offset (Fig. 12.3). In other cases, uphill-facing scarps are continuous across multiple drainages, suggesting that the component of dip-slip in the most recent earthquake(s) has been opposite to the deformation that created the broader scale topography (Fig. 12.4A,B). Where an uphill-facing scarp traverses a slope dominated by colluvial processes, colluvium eroded upslope from the fault will be deposited against the uphill-facing scarp (Fig. 12.4B),

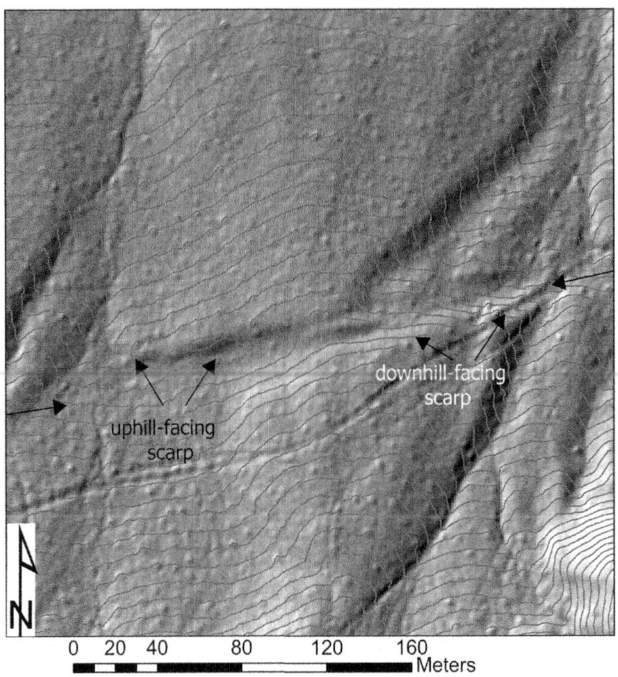

Figure 12.3 Lidar image of uphill-facing and downhill-facing scarps produced by left-lateral offset of an alluvial fan along the Garlock fault, illustrating how lateral offset of surfaces that have a component of slope that is parallel to the fault can produce fault scarps. Black arrows at the right and left edges of the image mark the location of the fault zone. Contour interval is 1 m. Ground surface generally slopes down to the north, but west edge of fan has a component of slope to the west and east edge has a component of slope to east. Centre of image is approximately at N 35.4821°, W 117.5466°. Lidar DEM from EarthScope Southern and Eastern California LiDAR Project. Distributed by OpenTopography. https://doi.org/10.5069/G9G44N6Q. Accessed 12 September 2008. *This material is based on services provided to the Plate Boundary Observatory by NCALM (*https://ncalm.cive.uh.edu/*). PBO is operated by UNAVCO for EarthScope (*http://www.earthscope.org*) and supported by the National Science Foundation (No. EAR-0350028 and EAR-0732947).*

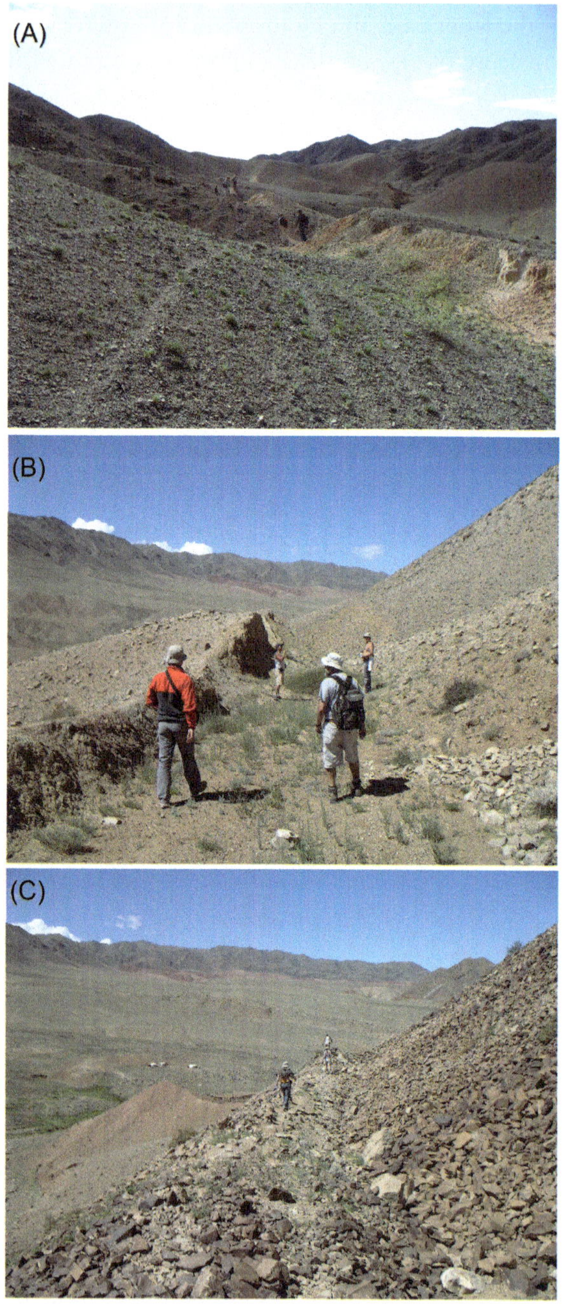

Figure 12.4 (A) and (B), Uphill-facing scarps formed in the 1957 Gobi-Altai earthquake along the Bogd fault, Mongolia. Note fine-grained sediment ponded behind the scarp in (B). (C), Side-hill bench along the 1957 Gobi-Altai earthquake rupture along the Bogd fault. *Photo credits: A, Sally McGill, 2005; B and C, Sally McGill, 2007.*

eventually filling in the trough and resulting in a side-hill bench (Fig. 12.4C), which is a common feature seen along prehistoric earthquake ruptures along strike-slip faults.

12.2.3.2 Fault scarps and grabens along normal faults

Although some portions of normal fault ruptures may have a single, simple fault scarp (e.g. Fig. 12.5A), the fault rupture is often more complex (Fig. 12.5B). Grove Karl Gilbert (1890) studied normal fault scarps along the western front of the Wasatch Range in Utah and observed that the array of fault scarps had complex and variable patterns, sometimes comprising a zone up to a few hundred metres wide. Within this zone were fault scarps facing both away from the mountain range and toward it, but always the scarp closest to the mountain front faced away from the mountains and typically had the largest amount of vertical slip (throw). The scarp with the largest throw is typically referred to as the primary scarp, and other scarps facing in the same direction are referred to as synthetic scarps, whereas those facing in the opposite direction are called antithetic scarps (Caskey, 1995). In addition, Gilbert (1890) found that the net throw on all of the faults within the zone was such that the footwall was uplifted relative to the valley.

Gilbert (1890) found that the scarp closest to the mountain front was usually 10–100 m valleyward of the contact between the bedrock of the mountain and alluvium of the valley and that its dip was usually 85–90 degrees, in contrast to the inferred dip of ~60 degrees within the bedrock below the alluvium that would be expected for normal faults (Anderson, 1905). Thus a prism of alluvium typically remains attached to and uplifted with the footwall of the fault, as shown by area <a b e> (the area enclosed

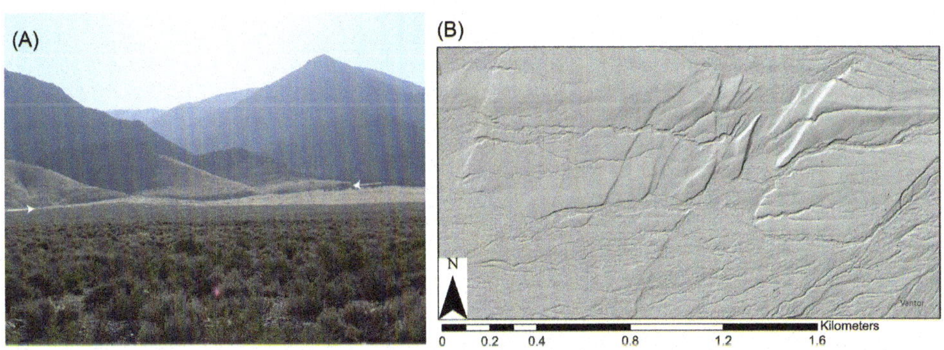

Figure 12.5 (A) Normal fault scarp along the base of the East Schell Range, Nevada. Photo credit: Rich Koehler. (B) Normal fault scarp array along the Panamint Valley fault west of Happy Canyon. Fault scarps trend roughly vertically in the image; flow on the alluvial fans is from right to left. Lidar hillshade from EarthScope Southern and Eastern California LiDAR Project. Distributed by OpenTopography. https://doi.org/10.5069/G9G44N6Q. Accessed 4 February 2026. *This material is based on services provided to the Plate Boundary Observatory by NCALM (https://ncalm.cive.uh.edu/). PBO is operated by UNAVCO for EarthScope (http://www.earthscope.org) and supported by the National Science Foundation (No. EAR-0350028 and EAR-0732947).*

between points a, b and e) in Fig. 12.6A, and Gilbert (1890) inferred that the steeply dipping scarp observed within alluvium at the surface must be listric (curved) (segment <b e> in Fig. 12.6A) and must root into the ~60 degrees dipping portion of the fault (segment <e y>) at depth. Assuming that the slip vector on the fault is parallel to the ~ 60 degrees dip of the fault at depth, motion of the fault will open a fissure along the more steeply dipping portion of the fault within the surficial alluvium (area <b e d> in Fig. 12.6A). Gilbert (1890) inferred that this fissure is quickly filled by the settling of one or both walls, via one of three different mechanisms. Fig. 12.6B shows closure of the fissure via gentle folding (warping) of the alluvial sediments in the prism that was originally located at <g d e> into the prism <g f e>. This mechanism results in a scarp, <b f>, that exaggerates the true throw, <b d>, on the fault. Fig. 12.6C shows closure of the fissure by settling of the hanging-wall side of the fissure along a fault, <h e>, that is antithetic to the primary fault. This forms a graben, <h i k b>, between the primary fault and the antithetic fault, the floor of which, <i k>, may slope in the direction opposite to the original slope of the alluvial surface, <h d>. Subtracting the height of the antithetic scarp, <h i>, from the height of the primary scarp, <b k>, will overestimate the true throw, <b d>, on the fault, unless the floor of the graben, <i k>, maintains the same slope as the original alluvial surface, <h d>, which is often not the case. Fig. 12.6D shows closure of the fissure by settling within the footwall side of the fissure, along a synthetic fault, <l e>, within the prism of alluvium, <a b e>, that is delineated in Fig. 12.6A. This results in two parallel scarps, <m d> and <l n>, both facing the valley. Summing the heights of these two scarps will overestimate the throw of the fault, <b d>, unless the original slope of the alluvial surface, <b l>, was

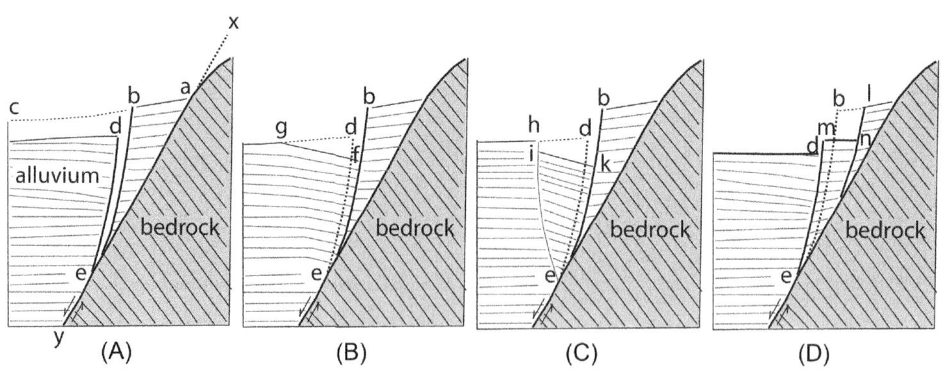

Figure 12.6 Cross-section sketches showing typical configurations of normal fault scarps in Utah, western USA. (A), initial configuration, immediately after earthquake. (B–D), three different mechanisms by which the initial fissure is commonly filled by collapse of alluvial deposits into the fissure. In each panel, dashed lines mark the former location of alluvial deposits and fissure walls. See text for discussion. *Redrafted from Gilbert, G. K. (1890). Lake Bonneville, U.S. Geological Survey Monograph, 1. US Government Printing Office, 438 pp., Figure 48).*

horizontal, and the surface, <m n>, of the block that was downdropped on the synthetic fault <l e> remains horizontal. Noting that fault scarps may represent repeated movements in a sequence of earthquakes over time, with different earthquakes sometimes rupturing different paths in the near-surface alluvium, Gilbert (1890) reasoned that the complex patterns of fault scarps that he observed could be formed by various combinations of these three deformation mechanisms.

Subsequent studies have provided support for Gilbert's (1890) interpretation that normal fault scarps that are near vertical at the surface often sole into bedrock faults that are dipping ~60 degrees or less. In laboratory sandbox models where normal faults are overlain by cohesionless sediment, the locus of slip within the sediment follows a near-vertical path to the surface, regardless of the dip of the fault within the underlying bedrock, and theoretical models show that this is a consequence of the dilational properties of the sediments as they deform (Cole & Lade, 1984).

Broad scarp arrays that are tens to hundreds of metres wide (e.g. Fig. 12.5B) may be underlain by a low-angle normal fault, and it is possible to infer the dip of the underlying low-angle fault by summing the throw and heave on each of the scarps (Axen et al., 1999). Because normal fault scarp arrays often include a graben, the resulting heave: throw ratio can be quite high, as the heave, h_i, across each of the fault scarps contributes to the total heave, H, whereas the throw, t_i, on the antithetic faults is subtracted from the throw on the synthetic faults to obtain the net throw, T. The dip of the underlying low-angle fault is calculated using Eq. 12.1 (Axen et al., 1999).

$$\theta = \tan^{-1}(\Sigma t_i / \Sigma h_i) \qquad (12.1)$$

12.2.3.3 Long-term growth of normal fault scarps with repeated earthquakes

12.2.3.3.1 Faceted spurs

Repeated earthquakes on a normal fault over 10^4–10^5 years can produce faceted spurs (Wallace, 1978) where a ridge spur that extends perpendicular to the crest of the range and to the range-front fault, intersects the fault (Figs. 12.7 and 12.8). The dip of the spurs is often less steep than the dip of the normal fault, indicating that faceted spurs are not simply exhumed fault planes. Tucker et al. (2011); Tucker et al. (2020) have shown that planar facets form when both the fault slip rate and the erosion rate are spatially and temporally constant. If the footwall contains a flat surface that is being uplifted, then the spurs between the fault-perpendicular drainages may be flat-topped and will produce a trapezoidal shape where they intersect the fault plane or the eroded fault plane (Wallace, 1978). However, with repeated rock uplift, the drainages will incise more deeply, and mass wasting along the walls of the drainages will cause the drainage walls to encroach more and more on the flat, upland surface, eventually reducing the spurs to sharp-crested ridges that form a triangular facet (Wallace, 1978) where they intersect the eroded fault plane (Figs. 12.7 and 12.8).

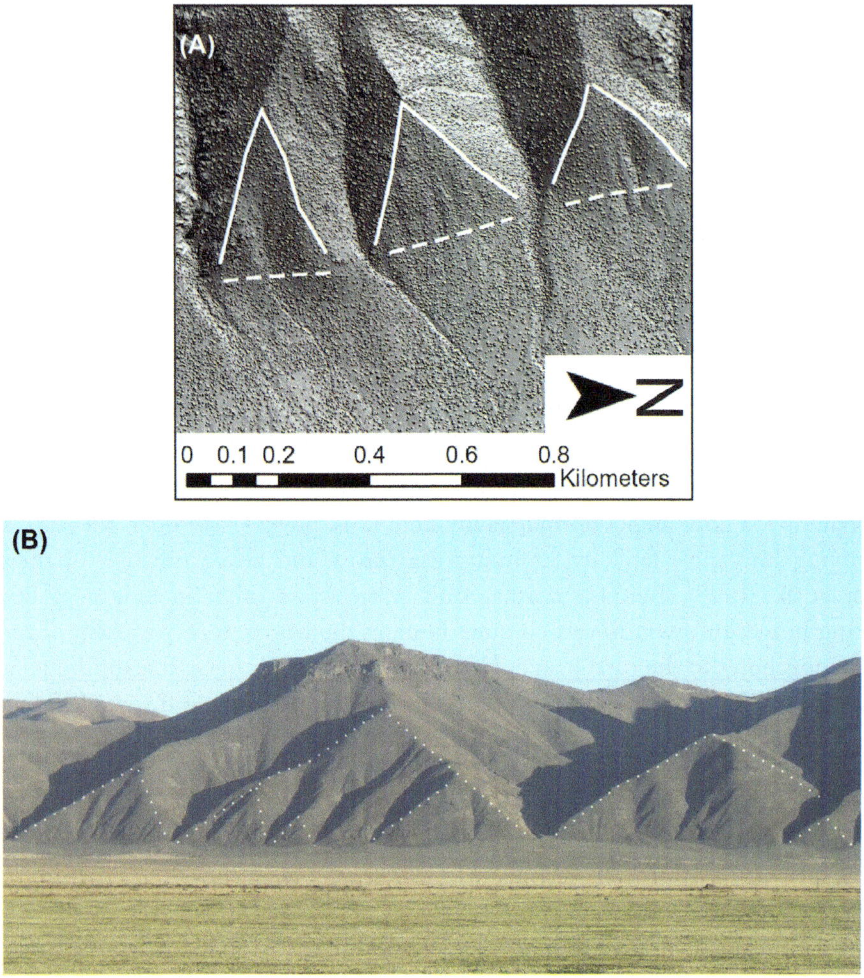

Figure 12.7 (A) Triangular facets (outlined with solid white lines) along the Toiyabe Range fault zone (marked with white dashed line), Nevada, USA. Centre of image approximately at N38.8352°, W117.2175°. Lidar hillshade image was created in OpenTopography.org from the USGS 3-D Elevation Program dataset NV Southern 1 D23. (B) Triangular facets (marked with dotted white lines) along the West Cortez Range, Nevada. *Photo by Rich Koehler.*

A number of landscape evolution modelling studies have investigated the factors that influence the size and shape of faceted spurs, including fault slip rate, fault dip, rates of rock weathering, downslope colluvial transport, bedrock landsliding, and fluvial erosion and transport. Densmore et al. (1998) highlighted the role of bedrock landsliding. Petit et al. (2009) found that for a given precipitation rate, a spectrum of mountain-front morphologies emerges depending on the relative rates of channel incision and hillslope diffusion. When hillslope diffusion is high, and channel bedrock is highly resistant to

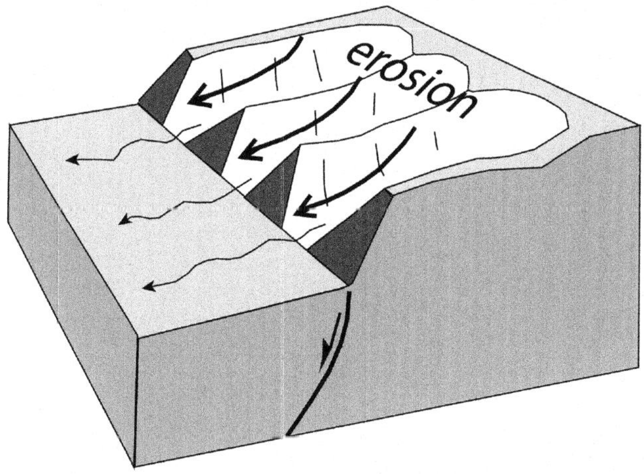

Figure 12.8 Illustration of the formation of triangular facets along a normal fault (Giletycz, 2015). *Redrafted from Giletycz, S. J., (2015). Landscape dynamics of the Hengchun Peninsula, Southern Taiwan, Ph.D. thesis, National Central University (Taiwan), 184 p.*

erosion, the scarp is smooth and undissected, whereas in the opposite conditions, many deep channels incise the mountain front. As conditions change from the former to the latter extreme, the formation of shallow incised channels, rounded facets and triangular facets is progressively favoured. Petit et al. (2009) also proposed a linear relationship between facet height and vertical throw rate. Hergarten (2024) found that the average size of facets is controlled by the ratio of hillslope erodibility and fluvial erodibility. Tucker et al. (2020) found, among other conclusions, that a break in slope at the transition from bedrock spur to colluvial wedge only develops where the rate of weathering is low relative to the rate of hillslope transport.

12.2.3.3.2 Wine-glass canyons

If a normal fault experiences a period of increased slip rate, erosional forces may not be able to compete with the rapid uplift. This can cause two adjacent triangular facets to begin to merge near their bases, as their bases are uplifted faster than the drainage between them can incise to maintain its previous gradient. This results in the formation of wineglass canyons, where the drainage between two spurs has been reduced to a very narrow, very steep slot canyon (analogous to the stem of a wine glass), compared to a much broader basin with a lower gradient farther upstream (representing the bowl of the wine glass) (Fig. 12.9). This broader, bowl-shaped upstream drainage area is inferred to have formed prior to the period of rapid fault slip during which the stem-shaped drainage cross-section formed. The term wineglass valley was coined by von Engeln (1942, p. 377) and was discussed by Wallace (1978), in the context of normal fault geomorphology, as a plan-view feature in which a basin is constricted at its mouth and

Figure 12.9 Wineglass canyon on the west side of the Tobin Range, Nevada, California (USA) is outlined by white dotted lines. White arrows mark the scarp from the 1915 Pleasant Valley earthquake rupture. *Photo by Rich Koehler.*

broad in its headwaters. More recent papers have used the term wineglass canyon and have focused on the front-view shape and slot-canyon aspects of wineglass canyons as indicators of renewed uplift or an increase in uplift rate along active normal faults (e.g. Fenton et al., 2003; Palumbo et al., 2004; Tortorici et al., 1995).

12.2.3.4 Fault scarps along reverse faults

Fault scarps produced by a reverse sense of slip have some unique characteristics because the hanging wall is faulted over the footwall, resulting, at least momentarily, in an overhanging fault scarp (Fig. 12.10). If the faulted material is cohesive enough, the overhanging scarp may remain preserved and visible for some time after the earthquake. However, in less cohesive material, the overhanging scarp may collapse during the earthquake itself (Fig. 12.11). Other complex fault scarp forms that have been observed in reverse faulting earthquakes are shown in Fig. 12.12 (Yu et al., 2010). Additional examples can be found for the 1968 M_w 6.5 Meckering and 1970 M_w 5.0 Calingiri earthquakes in Australia (Gordon & Lewis, 1980) and for the 1988 M_w 6.8 Spitak, Armenia earthquake (Philip et al., 1992).

12.2.3.5 Degradation of fault scarps over time

A fault scarp originally forms from fault slip during an earthquake. Along strike-slip and normal faults, the scarp may be vertical or steeply dipping immediately after the earthquake, with the surface of the scarp being the exhumed fault plane itself

Figure 12.10 (A) Overhanging right-lateral-reverse oblique-slip scarp along a restraining bend in the primary rupture of the 1992 Landers earthquake, southern California (USA). Photo credit: Sally McGill, 1992. (B) Overhanging scarp on a secondary thrust fault that slipped during the 1992 Landers earthquake. Field notebook and 15-cm ruler for scale. *Photo credit: Sally McGill, 1992.*

(Figs. 12.13A and 12.14A). However, the upper part of the exhumed fault plane spalls to produce a free-face at the top of the scarp and a debris slope that buries the base of the exhumed fault plane (Figs. 12.13B-12.15; Wallace (1977)). The debris slope rests at the angle of repose, typically 34 to 37 degrees (Wallace, 1977). Rainfall and local slope wash on the scarp remobilise and transport small particles from the debris slope to form a wash slope at the toe of the debris slope (Fig. 12.15). The dip of the wash slope ranges from the slope of the underlying surface up to 10 to 15 degrees (Wallace, 1977). In some locations, however, gullying may carry material farther away from the scarp instead of depositing it on the wash slope (Wallace, 1977). Collectively, the debris slope and the wash slope are referred to as a colluvial wedge. Over time, continued spallation further erodes the free face and enlarges the debris slope until it eventually buries the entire free face (Fig. 12.14C,D). As the free face retreats due to continued spallation, the sharp change in slope at the crest of the scarp, also known as the crestal break (Wallace, 1977), broadens over time, losing its sharpness (Fig. 12.14D,E). Over even longer time periods, the debris slope becomes reworked into a wash slope as surface runoff redistributes material from the upper part of the debris slope downslope onto the wash slope, leaving the entire scarp sloping at a shallower angle (Fig. 12.14E).

Figure 12.11 Hanging-wall collapse scarp along the Toromhon thrust, which ruptured in the 1957 Bogd earthquake, Mongolia. *Photo credit: Sally McGill, 2007.*

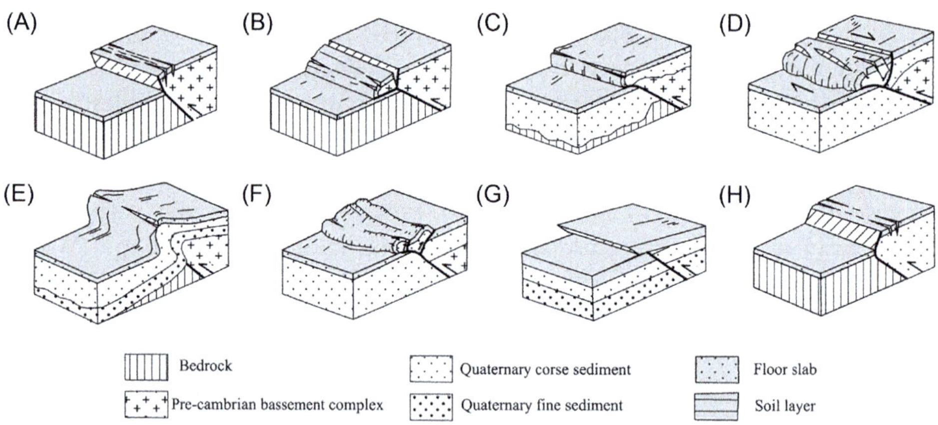

Figure 12.12 Examples of various morphologies of reverse fault scarps along the surface rupture of the 2008 Wenchuan earthquake (Yu et al., 2010). *From Yu, G., Xu, X., Klinger, Y., Diao, G., Chen, G., Feng, X., Li, C., Zhu, A., Yuan, R., Guo, T., & Sun, X. (2010). Fault-scarp features and cascading-rupture model for the M w 7.9 Wenchuan earthquake, eastern Tibetan plateau, China. Bulletin of the seismological Society of America, 100(5B), 2590–2614.*

Figure 12.13 Photos of fault scarp at the location with maximum vertical offset along the 1992 Landers earthquake rupture, California, were taken (A) four months after the earthquake and (B) about 10.75 years after the earthquake. In (B), note the development of a colluvial wedge, burying the base of the scarp, along with preservation of the upper portion of the free face. See Arrowsmith and Rhodes (1994) for a detailed analysis of degradation of fault scarps along this earthquake rupture in the first 11 months after the earthquake. *Photos by Mary Fagerson (A) and Anna Foutz (B).*

The rate of degradation of the scarp is highly dependent on the degree of induration of the material (e.g. alluvial sand and gravel versus indurated rock) as well as on climate, disturbance (such as by grazing or land use), and any subsequent fault rupture, which would superimpose a fresh scarp in the middle of an older degraded scarp. Wallace (1977) studied fault scarps in the semiarid western United States. Based on extrapolation from 23- to 62-year-old fault scarps from historic earthquakes, he inferred that the free face in that setting likely disappears after 1000–2000 years. Later investigators have suggested that, in a semiarid climate, scarp slopes in alluvium are reduced to the angle of repose within <1000 years (Hanks et al., 1984; Nash, 1986). Rounding of the crest of the scarp begins after the free face disappears and becomes significant by about 10,000 years, as indicated by the 1- to 5-m-wide rounded crestal zone in wave-cut cliffs from the most recent (~12 ka) high stand of Lake Lahontan in the western USA (Wallace, 1977). On scarps that are much older than 12 ka (potentially up to ~100,000 years old), he found maximum slopes of 8 to 9 degrees, much less than the angle of repose, suggesting that the debris slope had largely been reworked into a wash slope as surface runoff redistributed material downslope, leaving the entire scarp sloping at a shallower angle (Wallace, 1977).

Understanding the evolution over time of a fault scarp formed by a single earthquake enables one to begin to interpret scarps that have formed as a result of multiple earthquakes (Fig. 12.16). Fault scarps that represent the slip in more than one earthquake may have a rounded crest above the free face of the most recent earthquake

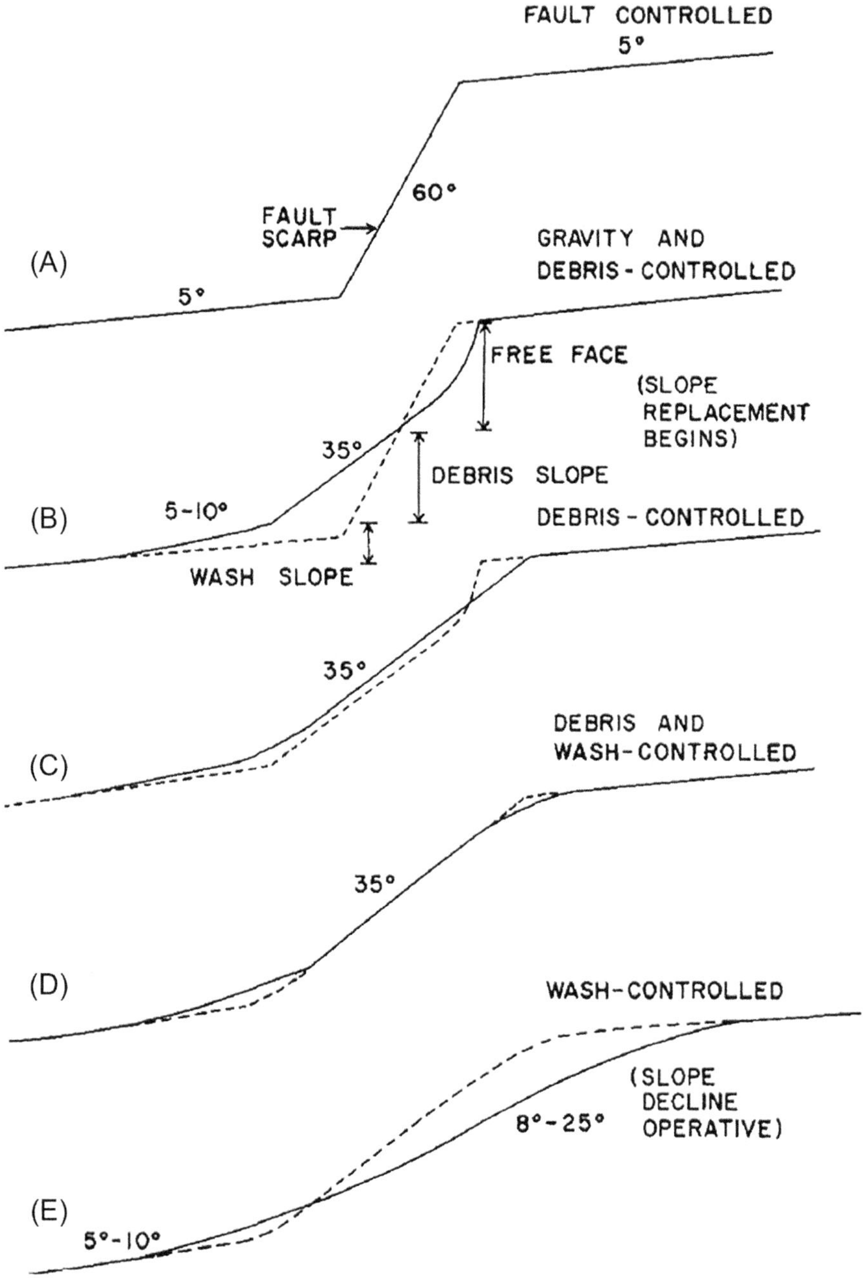

Figure 12.14 Evolution of a fault scarp over time. (A) Exhumed fault plane immediately after a normal faulting earthquake. (B–E) Shape of the fault scarp at progressively later times. In each panel, dashed line shows scarp configuration at the previous time step, and solid line shows the scarp profile at the current time step. See text for further explanation. *From Wallace, R.E. (1977). Profiles and ages of young fault scarps, north-central Nevada. Geological Society of America bulletin, 88(9), 1267–1281. https://doi.org/10.1130/0016-7606(1977)88<1267:PAAOYF>2.0.CO. Used with permission.*

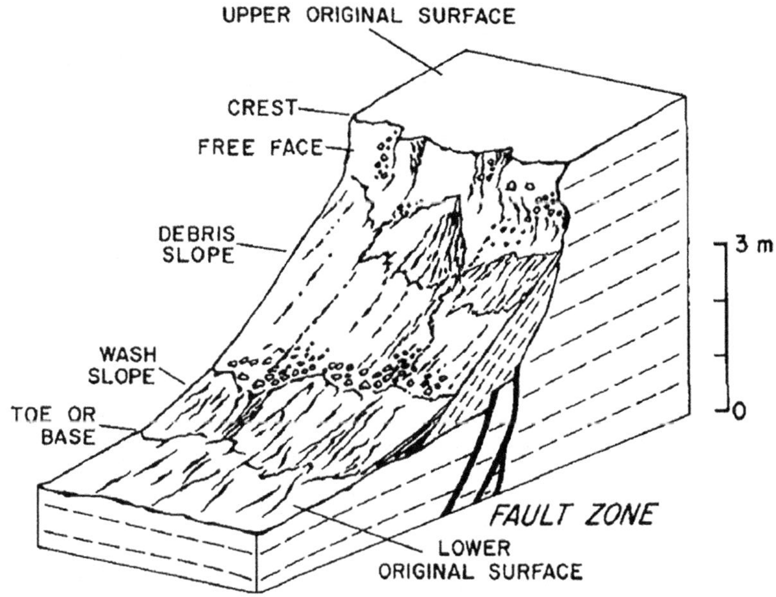

Figure 12.15 Early phase of fault scarp degradation, illustrating features of fault scarps discussed in text. *From Wallace, R.E. (1977). Profiles and ages of young fault scarps, north-central Nevada. Geological Society of America bulletin, 88(9), 1267–1281. https://doi.org/10.1130/0016-7606(1977)88<1267:PAAOYF>2.0.CO, GSAB. Used with permission.*

(Fig. 12.16C). In contrast, for a scarp that formed in a single earthquake, if the free face is preserved, it will extend all the way to the sharp crest of the scarp (Fig. 12.15). However, distinguishing a two-earthquake fault scarp from a single-earthquake fault scarp becomes difficult or impossible if the free face of the most recent earthquake is no longer preserved.

The degradation of fault scarps over time has been modelled with a diffusion equation so that the age of a single-earthquake scarp (or the rate of dip-slip on a fault with a multiearthquake scarp) can be estimated from a topographic profile across the scarp, along with knowledge of the diffusivity, κ (Avouac, 1993; Hanks et al., 1984; Hanks, 2000). Hanks (2000) suggests a diffusivity κ of ~ $0.01\ m^2/yr$ ($10\ m^2/ka$) for much of the Basin and Range province in the semiarid western U.S. Fig. 12.17 shows the modelled profile for a 5-m-high fault scarp after various periods of time, showing the gradual change in shape of the scarp, including decreasing scarp slope and decreasing sharpness of the scarp crest. A diffusion equation has been applied to modelling the ages of thrust fault scarps (Carretier et al., 2002; McPhillips & Scharer, 2023), and also to normal fault scarps (e.g. Hanks et al., 1984) and strike-slip fault scarps (e.g. Hilley et al., 2010), as well as to shoreline cliffs (e.g. Hanks et al., 1984) and to fluvial and alluvial terrace risers (e.g. Nash & Beaujon, 2006). Although thrust fault scarps start with a

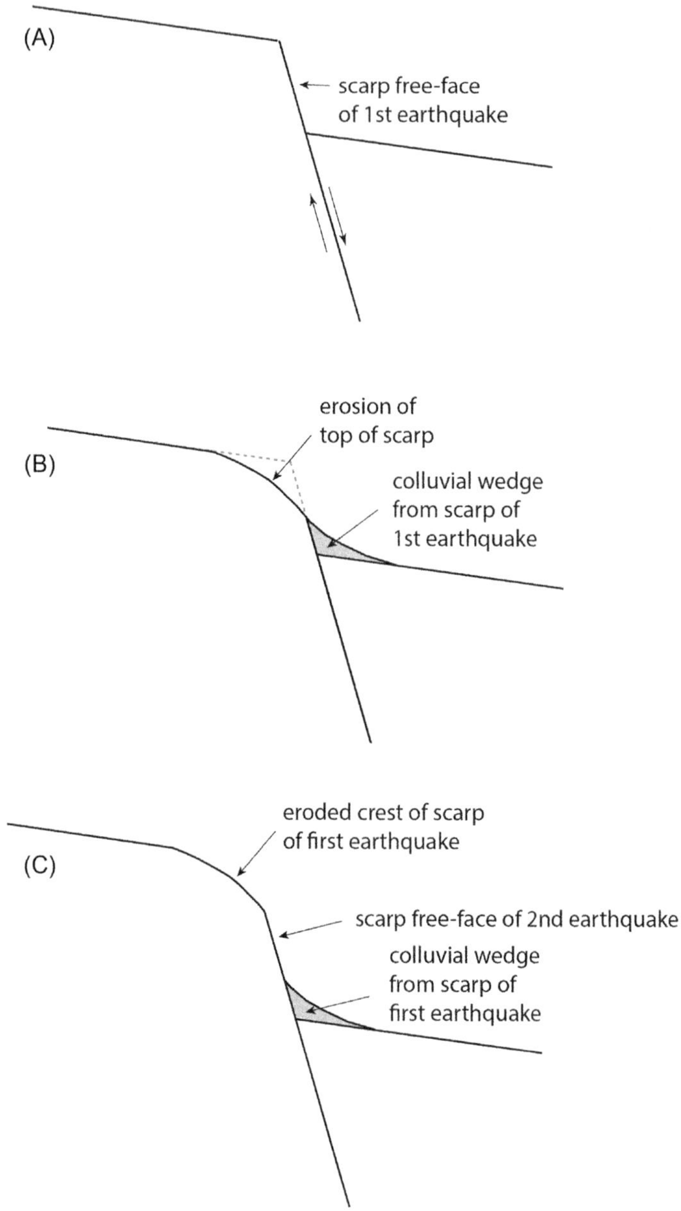

Figure 12.16 Cross-sectional illustrations of the evolution of a normal fault scarp formed by two earthquakes. (A) The first earthquake produces a scarp free-face. (B) Over decades or centuries, the scarp degrades via erosion of material from the top of the scarp and deposition of material in a colluvial wedge at the base of the scarp. (C) When the next earthquake offsets the ground surface, a new free-face forms, separating the rounded crest of the scarp from the first earthquake (above the new free-face) from the colluvial wedge at the base of the scarp from the first earthquake. With the continued passage of time, the free face of the more recent earthquake will degrade, creating a second colluvial wedge on top of the first.

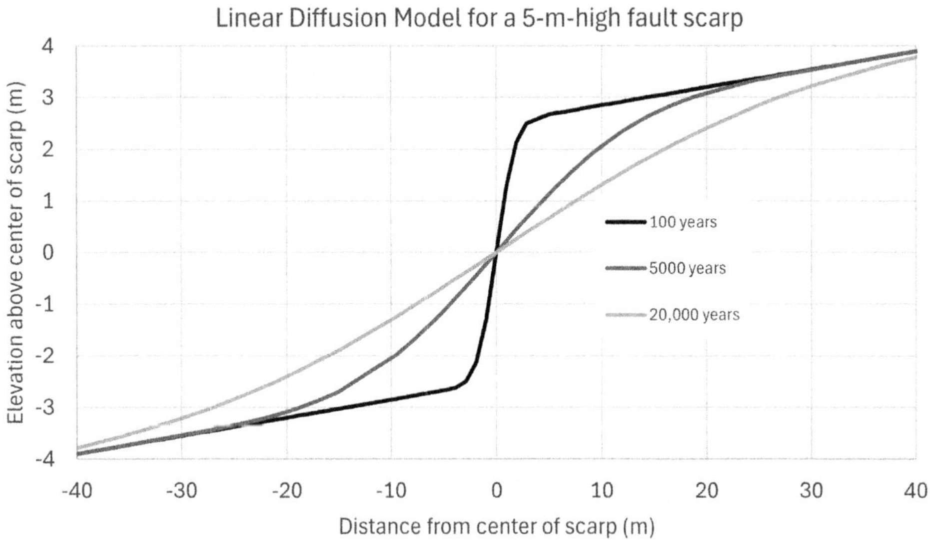

Figure 12.17 Scarp profile elevations (5x vertical exaggeration) versus distance for a 5-m high fault scarp offsetting a 2 degrees-sloping surface, modelled with a linear diffusion equation (Hanks et al., 1984), assuming a diffusivity of 0.01 m^2/yr.

shallower slope angle than high-angle normal fault scarps (~60 degrees) and strike-slip faults (90 degrees), diffusion modelling of thrust fault scarps is considered valid because all scarps in unconsolidated sediments, regardless of their initial slope, degrade to the angle of repose (~30–35 degrees) within a relatively short period of time, as mentioned above (Hanks et al., 1984; Nash, 1986). However, inaccurate ages will be obtained from diffusion modelling if the scarp being dated was formed by two separate fault strands whose degraded profiles have merged (Fig. 12.18; Nash & Beaujon, 2006).

12.2.4 Geomorphic features created by strike-slip faulting

12.2.4.1 En echelon *fractures and mole tracks*

Strike-slip fault ruptures with small lateral displacements are often recognised by an *en echelon* pattern of fractures (Fig. 12.19). This pattern has also been observed in laboratory models of faulting in which a layer of clay (up to a few cm thick) overlies two rigid boards that are moved in a strike-slip motion beneath the clay (Cloos, 1928; Riedel, 1929; Tchalenko, 1970), and these *en echelon* fractures are commonly referred to as Riedel shears. In earthquakes with small amounts of strike-slip, *en echelon* Riedel shears may be the only visible surficial expression of the underlying fault rupture (Fig. 12.19). For faults with right-lateral slip, the Riedel shears are oriented in a left-stepping pattern (Fig. 12.19). For left-lateral faults, the Riedel shears form a right-stepping pattern. Thus even in the absence of offset markers that directly indicate the sense of slip on an

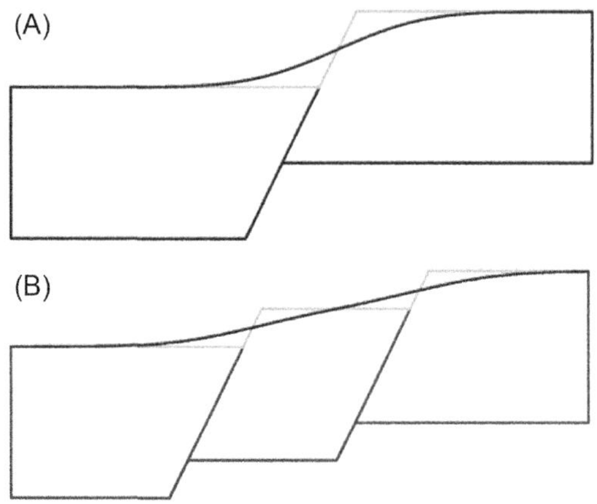

Figure 12.18 (A) Diffusion model for a fault scarp on a single fault strand. (B) Diffusion model for a compound fault scarp produced by two separate fault strands, which can no longer be distinguished at the surface. After the same time period, diffusion modelling would predict a shallower slope for the compound scarp, resulting in an erroneously old age, if it was modeled as a single scarp. *From Nash, D.B., & Beaujon, J.S. (2006). Modeling degradation of terrace scarps in Grand Teton National Park, USA. Geomorphology, 75(3-4), 400–407.*

Figure 12.19 Left-stepping *en echelon* fractures along the 2004 Parkfield earthquake rupture on the right-lateral San Andreas fault, central California (USA). *Photo credit: W. Bryant.*

earthquake surface rupture, these relationships can be used to infer left-lateral or right-lateral slip on the underlying fault.

If the lateral slip on the Riedel shears is sufficient (e.g. more than a few cm), uplift of sediment along thrust or reverse faults will occur at the steps between *en echelon* strands, forming a 'pop-up' structure (Fig. 12.20A). With larger amounts of lateral slip, these pop-ups combine to produce a surface rupture pattern described as a **mole track** (Fig. 12.20B,C). Open fissures can also form along strike-slip earthquake ruptures (Fig. 12.21). These usually fill with aeolian or colluvial sediments, probably within decades after the earthquake, but may perhaps remain recognisable as subtle linear depressions even after they are mostly filled.

12.2.4.2 Fault step-overs and associated geomorphic features

In addition to the small-scale fault step-overs that can occur along earthquake rupture traces (Fig. 12.22), when a larger-scale step exists between fault strands at depth, larger-scale accommodation features may develop at the surface. If the fault step is dilational (Fig. 12.23B,C), such that material is moving away from the step along both fault strands (that is, a right step in a right-lateral fault or a left step in a left-lateral fault), extension will occur within the region of fault overlap, resulting in a pull-apart basin (Fig. 12.24).

If the fault step is compressional (Fig. 12.23A,D), such that material is moving along both fault strands toward the region of overlap (a left step in a right-lateral fault or a right step in left-lateral fault), shortening will occur, producing uplifted topography in the area of overlap between the two fault strands. The uplifted topography is often referred to as a push-up, a pop-up, or a pressure ridge (if elongated). Pressure ridges can also form as a way of accommodating the transpression that results from along-strike changes in the dip of a strike-slip fault at depth (Hilley & Arrowsmith, 2008) or anywhere that the strike of the fault is not parallel to the slip vector (e.g. Yang et al., 2022).

When a step between two strands of a fault is fully connected at the surface by a primary through-going fault, the feature is usually referred to as either a releasing bend (if the bend accommodates a dilational step in the fault) or as a restraining bend (if it accommodates a compressional step in the fault). Mann (2007) provided a global catalogue and systematic categorisation of various types of bends and steps in strike-slip fault systems. Wesnousky (2006) analysed the frequency with which historical earthquake ruptures have terminated at, as opposed to propagating through, releasing and restraining steps of various sizes in strike-slip fault systems, while Zhu et al. (2024) conducted analogue and numerical modelling of the deformation at fault bends and steps.

12.2.4.3 Shutter ridges

Uphill-facing scarps that block or deflect one or more drainages are called shutter ridges. Although shutter ridges can form as a result of a component of dip-slip (see discussion of

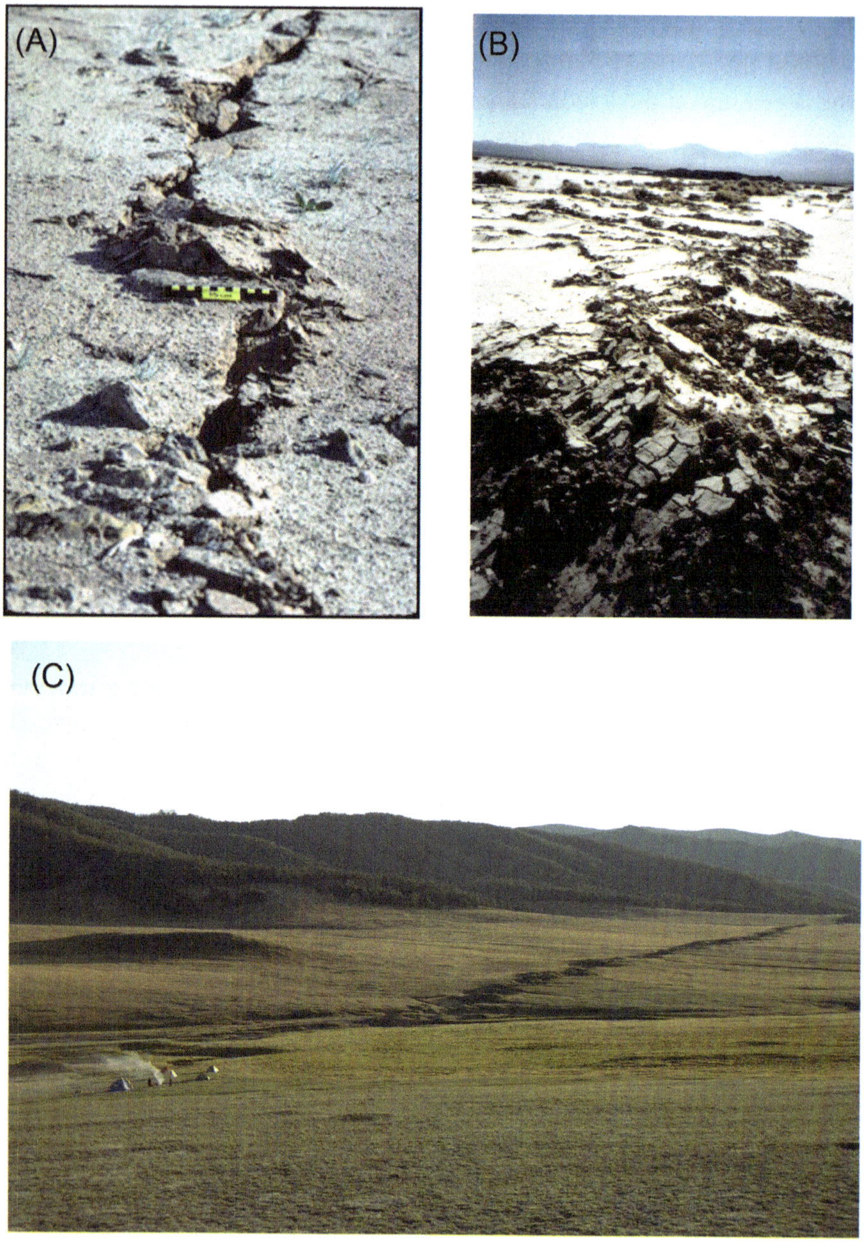

Figure 12.20 (A) Compressional 'pop-up' structure within a left-step between two right-lateral fault strands along the 1987 Superstition Hill earthquake rupture. (B) Mole-track-style surface deformation along the 1999, right-lateral, Hector Mine earthquake rupture. (C) Mole-track-style surface deformation produced by the 1905 earthquake rupture on the left-lateral Bulnay fault, still readily apparent a century later. *(A) Photo credit: Art Sylvester. (B) Photo credit: Sally McGill, 1999. (C) Photo credit: Jean-Francois Ritz.*

Figure 12.21 Open fissure that formed in the 1999 Hector Mine earthquake, which primarily had right-lateral slip. *Photo credit: Sally McGill (1999).*

uphill-facing scarps in Section 12.2.3.1), they also frequently form as a result of lateral slip, where a ridge, alluvial fan, colluvial wedge or other type of high ground is laterally faulted so as to obstruct an active drainage (Fig. 12.3). Fig. 12.25 shows two examples of shutter ridges that formed in a single earthquake, when fault-perpendicular ridges were left-laterally offset along the Bogd fault during the 1957 M_w 8.1 Gobi-Altay earthquake in Mongolia. Fig. 12.26 shows several examples of shutter ridges along the left-lateral Garlock fault, California, USA.

12.2.4.4 Laterally deflected, offset, and beheaded channels

Laterally offset geomorphic features are the most useful features for characterising prehistoric ruptures on strike-slip faults. These may include laterally offset or beheaded channels, and laterally offset landforms such as ridges, debris flows, terrace risers, or alluvial fans.

Figure 12.22 Small-scale pull-apart basin on a minor fault that formed in the 1992 Landers earthquake, southern California (USA). The opening of a small basin between right-stepping strands of this minor fault indicates that about 10 cm of right-lateral slip has occurred. *Photo by Sally McGill.*

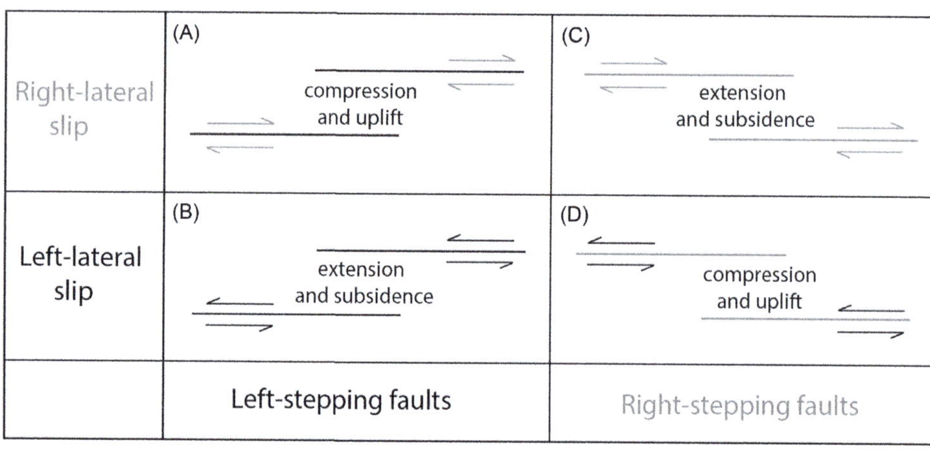

Figure 12.23 Diagrams illustrating the situations in which compression and uplift versus extension and subsidence occur at step-overs in strike-slip faults. See text for further explanation.

Wallace et al. (1968) used the term 'deflected stream' as a descriptive term to refer to a stream that is laterally bent or misaligned across a fault, where the upstream and downstream channel segments are still hydrologically connected to each other. In Wallace's (1968, p. 11–12) usage, the term 'deflected' does not indicate whether the channel's misalignment across the fault was accomplished by fault offset, by nontectonic processes (such as diversion of flow around an uphill-facing scarp, stream capture, or headward erosion of a channel along an easily erodible fault zone), or by a combination of tectonic and nontectonic processes. He referred to channels whose deflections are

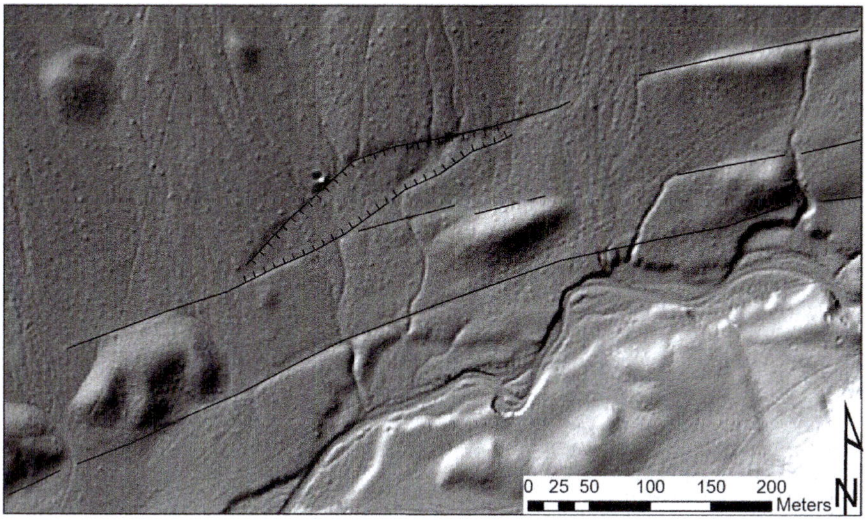

Figure 12.24 Pull-apart basin (outlined by hachures on the downdropped sides) that formed by a left-stepping transfer of left-lateral slip between strands of the Garlock fault. Centre of image is approximately at N35.4643°, W 117.6148°. Lidar DEM from EarthScope Southern and Eastern California LiDAR Project. Distributed by OpenTopography. https://doi.org/10.5069/G9G44N6Q. Accessed 12 September 2008. *This material is based on services provided to the Plate Boundary Observatory by NCALM (*https://ncalm.cive.uh.edu/*). PBO is operated by UNAVCO for EarthScope (*http://www.earthscope.org*) and supported by the National Science Foundation (No. EAR-0350028 and EAR-0732947).*

Figure 12.25 (A, B) Small shutter ridges formed by left-lateral offset of fault perpendicular ridges during the 1957 Gobi-Altai earthquake on the Bogd fault. *Photo credits: Sally McGill, 2005(a) and 2007(b).*

likely to be solely due to tectonic offset as 'true offsets' and those whose deflections may be partly or completely due to diversion of the channel around a preexisting topographic barrier as 'apparent offsets'. Sieh (1978), on the other hand, most often

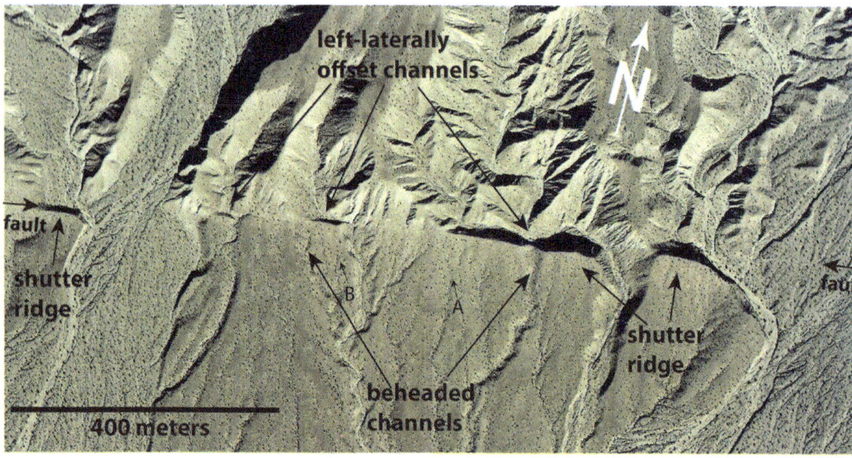

Figure 12.26 Aerial photograph of the central Garlock fault in Pilot Knob Valley, California, with various fault-related geomorphic features labelled. Stream flow direction is toward the bottom of the image. Centre of photo is approximately at N35.5606°, W 117.1990°. *Photo credit: U.S. Geological Survey, from 1:5000-scale, low sun-angle aerial photographs of the Garlock fault, flown January 17, 1976, under the coordination of Malcolm Clark.*

used the terms 'deflected', 'deflection' and 'deflections' where he interpreted the lateral separation of a channel along the fault to be at least partly due to diversion of flow around a preexisting topographic barrier.

In this chapter the term 'offset' channel is reserved for channels that Wallace et al. (1968) would call 'true offsets' – that is, for those deflected channels that can reasonably be inferred to have initially incised straight across the fault, so that the entire fault-parallel separation between the channel on the upstream versus downstream side of the fault can be attributed to lateral tectonic offset. When the terms 'deflection' or 'deflected' are used in this chapter, they are used as Wallace et al. (1968) used them – merely as describing the geometry of a channel that has a segment flowing along the fault trace, without interpreting whether that channel segment formed as a result of tectonic offset on the fault or as diversion around a topographic barrier along the fault (or by other nontectonic process). However, because of the ambiguity of the term 'deflected channel' in the literature, this chapter prefers to use the terms 'separated' and 'separation' when there is a need to describe the geometry of channels that are misaligned across the fault without yet committing to an interpretation of the cause of that misalignment. For channels in which it is clear that part or all of the deflection is due to diversion of the channel flow around a preexisting topographic barrier at the fault, this chapter uses the word 'diverted' to express this interpreted cause of the deflection.

Three examples of gullies or channels for which separation across the fault can clearly be entirely attributed to tectonic offset are shown in Fig. 12.27. In Fig. 12.27A, the gully remains clearly identifiable all the way to the fault, and, because the hydrologic

Figure 12.27 Examples of left-laterally separated drainages for which the entire separation can clearly be attributed to tectonic offset, thus warranting the label 'offset' drainage. (A) Small gully offset along the Bogd fault in the 1957 Gobi-Altai earthquake, Mongolia. Small backpack for scale where upstream gully segment intersects fault. Photo credit: Sally McGill, 2005. (B) Ephemeral channel that has likely been offset in at least one prehistoric earthquake in addition to being offset in the 1957 earthquake on the Bogd fault. Photo credit: Sally McGill, 2005. (C) Aerial photograph of a channel (outlined in black) offset about 15 m left-laterally in multiple prehistoric earthquakes along the central Garlock fault (marked with white arrows) in Pilot Knob Valley. Top of photo is to the NNW, and flow direction is toward the bottom of the image. Note that the width of the channel segment along the fault is narrower than the two segments north and south of the fault. See text for discussion. *Photo credit: U.S. Geological Survey, from 1:5000-scale, low sun-angle aerial photographs of the Garlock fault, flown January 17, 1976, under the coordination of Malcolm Clark.*

connection between the gully segments upstream and downstream from the fault has not been re-established after the earthquake, the intersection of the downstream section of the gully with the fault has not been significantly modified since the earthquake. In Fig. 12.27B,C, the hydrologic connection across the fault has not been disrupted, but the fault-parallel channel segment that flows along the fault is narrower, confirming that

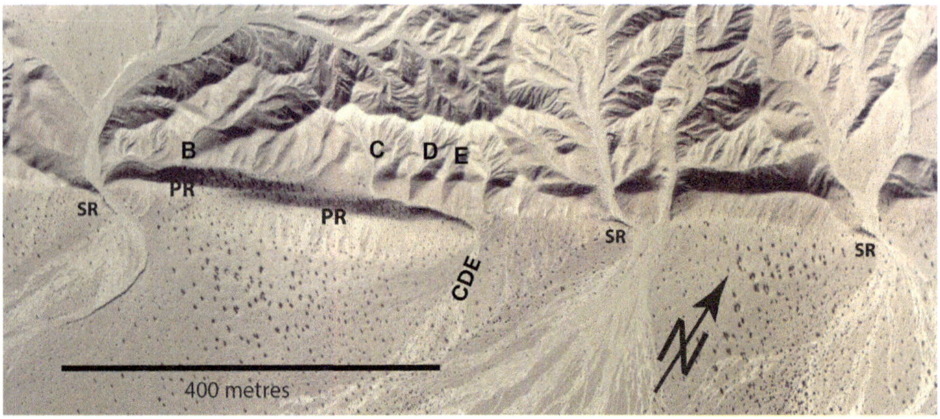

Figure 12.28 Aerial photograph of the left-lateral Garlock fault in Pilot Knob Valley, California, showing multiple shutter ridges. Flow direction of the major drainages is toward the bottom of the image. A fault-parallel pressure ridge (labeled 'PR') in the left half of the image blocks several drainages and diverts drainage B in a right-lateral direction, while diverting drainages C, D and E in a left-lateral direction. See text for further discussion of these drainages. The aforementioned pressure ridge itself appears to have been left-laterally offset 80–100 m. Several smaller shutter ridges, ranging in length from 10–25 m, are labelled 'SR'. These have formed as fault scarp colluvium that was shed from north to south across the fault has been left-laterally offset eastward, into the mouths of the adjacent drainages. *Photo credit: U.S. Geological Survey, from 1:5000-scale, low sun-angle aerial photographs of the Garlock fault, flown January 17, 1976, under the coordination of Malcolm Clark.*

it is younger than the channel segments up- and downstream from the fault and likely did not exist until after the channel was tectonically offset by its first experience of an earthquake.

Channels diverted around an obstacle along a strike-slip fault may be diverted either in the same or the opposite direction as the offset that would be produced by slip on a fault. As an example, in Fig. 12.28, channel B has been right-laterally deflected (no cause interpreted yet) around a pressure ridge, at the location of a (less active or inactive) strand of the left-lateral Garlock fault. However, this certainly does not indicate that right-lateral tectonic motion has occurred on this strand of the fault, because the channel geometry could have formed by channel flow being diverted around the shutter ridge. As another example, the upstream channel segments C, D and E in Fig. 12.28 are left-laterally separated from their combined downstream segment (CDE) across the active strand of the left-lateral Garlock fault. However, we cannot necessarily infer that the entire ~110 m of left-lateral separation of upstream channel C from downstream channel CDE, nor the ~70 m separation of channel D, nor the ~36 m separation of channel E, are the result of left-lateral slip along the Garlock fault. Instead, part or all of each separation may simply be the result of the pressure ridge existing prior to the formation of these channels and diverting their flow around the ridge, in a manner that happens to be a left-lateral deflection for channels C, D and E.

Wallace et al. (1968) used the term 'beheaded channel' to refer to a channel segment downstream from the fault that is no longer connected to its upstream source. Two beheaded channels are labelled in Fig. 12.26. This term is not applied to channels that could have originated solely on the downstream side of the fault and have experienced headward erosion that extends their channels upstream toward the fault (such as gullies A and B in Fig. 12.26). Rather, it is applied only to channels that are wide enough and deep enough at the fault that one may infer that the channel must have originated on the upstream side of the fault. For some beheaded channels, the source channel on the upstream side of the fault can be identified but is no longer hydrologically connected to the beheaded channel (e.g. Fig. 12.27A). For other beheaded channels, the source channel may not be readily identifiable, or multiple possible source channels may exist.

12.3 Measuring offsets of geomorphic features along active faults

Measurements of offsets of geomorphic features across active faults are necessary for estimating the amount of slip in recent earthquakes (e.g. DuRoss et al., 2020; Fletcher et al., 2014), in historical earthquakes (e.g. Beanland & Clark, 1994; Klinger et al., 2011; Sieh, 1978; Zielke, Arrowsmith, Grant Ludwig, et al., 2012), or in prehistoric earthquakes along an active fault (e.g. McGill & Sieh, 1991; Salisbury et al., 2012). Offset measurements of geomorphic markers that have experienced more than one earthquake can be used (in conjunction with information on the age of the marker) to measure the average slip rate of the fault (e.g. Sieh & Jahns, 1984). In certain cases, it may be possible to determine the number of earthquakes associated with an offset feature, making it possible to also estimate the average amount of slip in past surface-rupturing earthquakes at that location on the fault (e.g. Klinger et al., 2011; McGill & Sieh, 1991; Sieh, 1978). Fault slip rates and estimates of the amount of slip per earthquake are important inputs to seismic hazard models, and detailed studies of these parameters and their variations over time and space can provide insights into how complex fault systems behave, and also into the physics of earthquakes, as will be briefly explored in Section 12.5.

12.3.1 Measuring offsets on dip-slip faults

For a fault with pure dip-slip (whether the slip is normal or reverse), the total dip-slip is the vector sum of the vertical uplift (throw) and the horizontal extension or shortening (heave) (Fig. 12.2). In practice, the throw and the heave are usually not easy to measure directly, unless the fault scarp is very fresh and is within indurated material. Rather, what is most easily measurable in the field or on a topographic profile is the vertical separation of the ground surface displaced across the fault. The vertical separation is the vertical distance between projections of the preearthquake ground surface in the hanging wall and footwall of a fault (Caskey, 1995). Except in the case of a horizontal surface or a vertical fault, this vertical separation is different from the fault throw (Fig. 12.29). For faults that dip in the same direction as the surface slope (e.g. primary and synthetic scarps along a normal fault, see Section 12.2.3.2), vertical separation

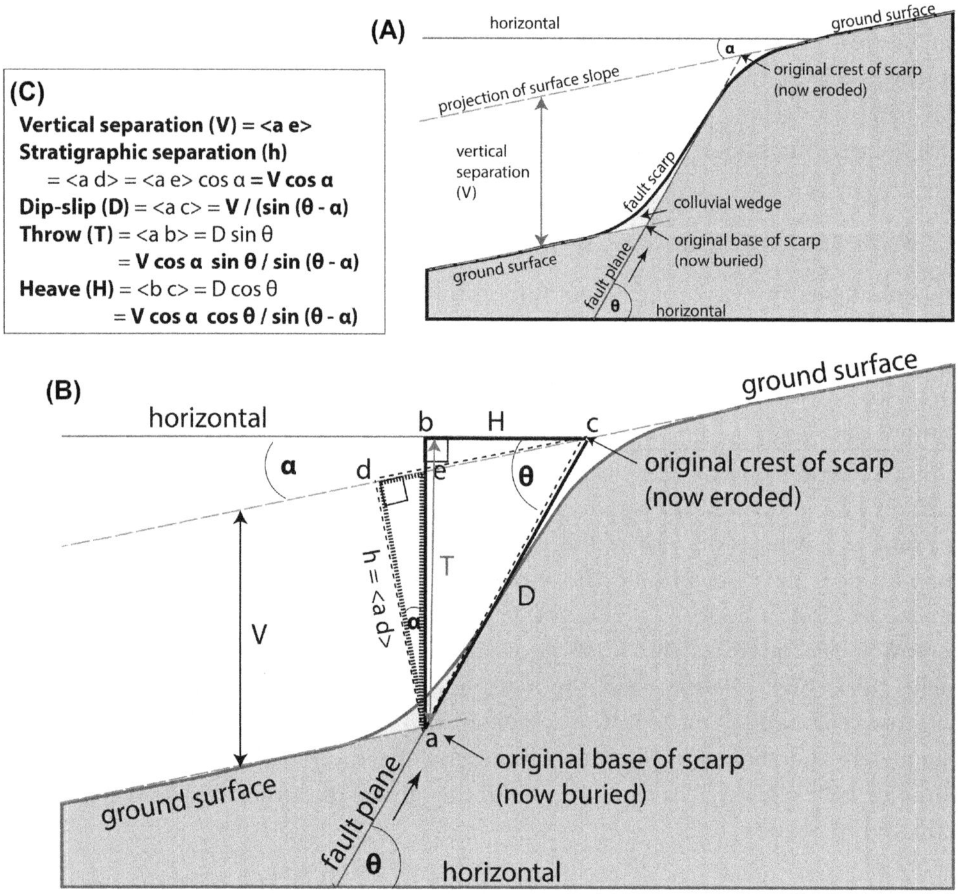

Figure 12.29 (A) Measurements of vertical separation, V, of an offset geomorphic surface across a normal fault, using the projection of the surface slope, *α*. (B) Definition of the three right triangles that are used in part C. These include <a b c>: solid black line connecting points a, b and c; <a d c>: thin, dashed black line; and <a d e>: thick, hachured, black line. (C) Trigonometric relationships (Caskey, 1995) that can be used to convert measured vertical separation (V = <a e> = the length between point a and point e), surface slope, α, and measured or assumed fault dip, *θ*, to the amount of dip-slip (D = <a c>), throw (T = <a b>), and heave (H = <b c>), on the fault. These equations assume that the surface slope away from the fault scarp is the same on each side of the fault. *Modified from Caskey, S.J. (1995). Geometric relations of dip slip to a faulted ground surface: New nomograms for estimating components of fault displacement. Journal of Structural Geology, 17(8), 1197–1202.*

underestimates the throw. For faults that dip in the opposite direction as the surface slope (e.g. thrust faults or antithetic normal faults), the vertical separation overestimates the throw (Caskey, 1995; Fig. 12.30). However, in combination with the original slope, α, of the offset surface, the vertical separation can be converted to the amount of dip-slip, throw and heave, if the fault dip, θ, is known, inferred or assumed (Caskey, 1995).

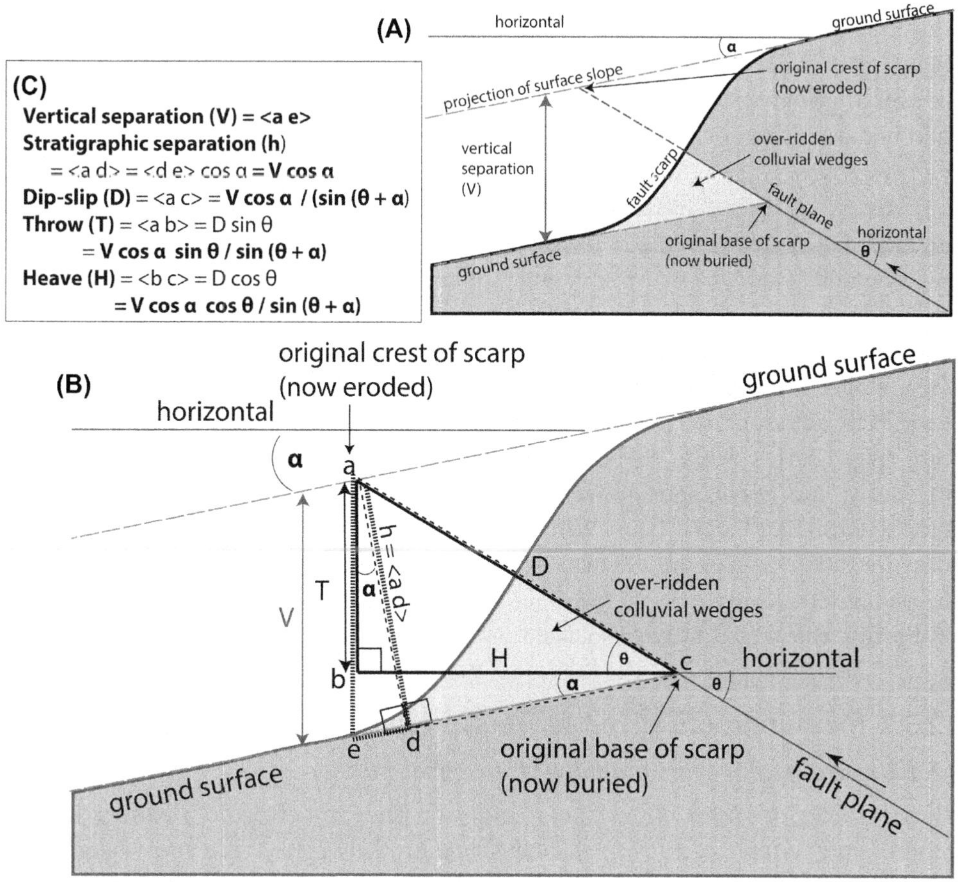

Figure 12.30 (A) Measurements of vertical separation, V, of an offset geomorphic surface across a thrust fault, using projection of the surface slope, α. (B) Definition of the three right triangles that are used in part C. These include <a b c>: solid black line, connecting points a, b and c; <a d c>: thin, dashed black line; and <a d e>: thick, hachured, black line). (C) Trigonometric relationships (Caskey, 1995) that can be used to convert measured vertical separation (V = <a e>), surface slope, α, and measured or assumed fault dip, θ, to the amount of dip-slip (D = <a c>), throw (T = <a b>), and heave (H = <b c>), on the fault. These equations assume that the surface slope away from the fault scarp is the same on each side of the fault. *Equations from Caskey, S.J. (1995). Geometric relations of dip slip to a faulted ground surface: New nomograms for estimating components of fault displacement. Journal of Structural Geology, 17(8), 1197–1202.*

In measuring the vertical separation of an offset surface from a profile across the fault, it is important to define the slope of the surface uphill from the fault, at a location that is uphill from the eroded crest of the scarp and from any tectonic warping near the fault (but no farther uphill than needed to find a representative planar surface) and to project this surface to a location above or beyond the toe of the scarp. Similarly, the slip of the surface downhill from scarp must be defined at a location downhill from the toe of the colluvial wedge and from any tectonic warping (but no farther downhill than needed) and then projected beneath the colluvial wedge toward the buried fault plane (Fig. 12.31). The vertical separation can then be measured between these two planar surfaces. If planar surfaces cannot be defined on each side of the fault, this method is not valid because there are no constraints on how the surfaces should be projected through the fault zone.

If the two projected surfaces have the same slope, then the vertical separation will be the same no matter where it is measured. However, because the surface slopes uphill and downhill from the scarp are often different, it is best to measure the vertical separation between the two projected planes as close as possible to their intersection with the fault plane, which is most likely located approximately at the middle of the scarp. Upper and lower bounds on the vertical separation can be obtained by measuring the vertical separation between the two projected planes at both the crest and toe of the scarp (Fig. 12.31; Caskey, 1995; McPhillips & Scharer, 2018). The calculations of dip-slip, throw and heave from the vertical slip also depend on the slope, α, of the offset surface (Figs. 12.29 and 12.30). When the surface slopes uphill and downhill from the scarp are different, separate calculations must be used based on separate values of α for the two different surface slopes to obtain maximum and minimum limits on the amount of dip-slip, throw and heave associated with the scarp as explained by Caskey (1995).

12.3.2 Measuring offsets on strike-slip faults

12.3.2.1 Evaluating the suitability of an offset marker

Measurements of lateral offsets along strike-slip faults involve the identification of a geomorphic feature, such as a channel bank or thalweg, terrace riser, debris flow, landslide deposit, or glacial moraine crest, that has been offset across the fault. For an offset marker to be useful, one must be able to confidently infer the preearthquake morphology of the marker (Zielke et al., 2015). This is necessary to ensure that the reported measurement accurately reflects the tectonic offset in the most recent earthquake (or the cumulative slip in a number of previous earthquakes) and is not biased by (1) diversion of a channel around an uphill-facing scarp, (2) postearthquake erosion of the marker, (3) stream piracy or (4) tectonic warping. The biases related to diversion of channels around uphill-facing scarps have been discussed in Section 12.2.4.4.

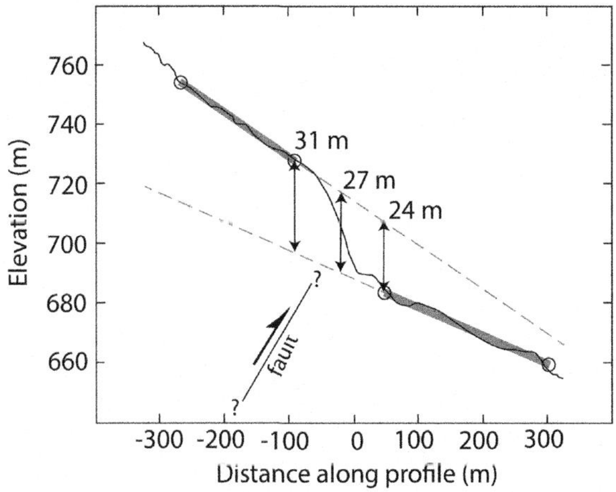

Figure 12.31 Example of a vertical separation measurement in a case where the slope of the alluvial surface upstream and downstream from the fault scarp is not the same (McPhillips and Scharer, 2018). Black line shows elevation data from a profile across a fault scarp constructed from a digital elevation model. Thick, solid grey lines bounded by open circles show the regions where a linear regression of the elevation data on the hanging wall and footwall defines the slopes of the fan surface that are projected across the fault zone (thin dashed grey line). Precise location of the fault within the width of the scarp is unknown, but is most likely near the centre. The preferred estimate of the vertical separation is measured at the mid-point between the hanging wall and footwall regression domains. The 2σ uncertainty was defined as the average difference between the vertical separation measurements at the top and base of the scarp. *Redrafted from McPhillips, D., & Scharer, K.M. (2018). Quantifying uncertainty in cumulative surface slip along the Cucamonga Fault, a crustal thrust fault in southern California. Journal of Geophysical Research: Solid Earth, 123(10), 9063–9083.*

12.3.2.1.1 Postearthquake erosion

Postearthquake erosion is particularly common for shutter ridges, where the offset material is faulted into the path of an active channel and is subject to erosion in each postearthquake precipitation event that produces flow in the channel (Figs. 12.26 and 12.28). Thus the length of a shutter ridge is likely to provide only a minimum estimate of its tectonic offset. Although the bank on the opposite side of the channel is protected from erosion on the downstream side of the fault, if the channel has widened by lateral erosion of that bank on the upstream side of the fault, this could also lead to an underestimate of the tectonic offset.

12.3.2.1.2 Stream piracy and channel aliasing

Offset channels are particularly prone to stream piracy. In cases where multiple closely-spaced drainages are incised into a surface that is old enough to have experienced many earthquakes, it may be difficult or impossible to determine which channel segments on opposite sides of the fault were originally connected across the fault, and there may have been multiple generations of stream capture, with the result

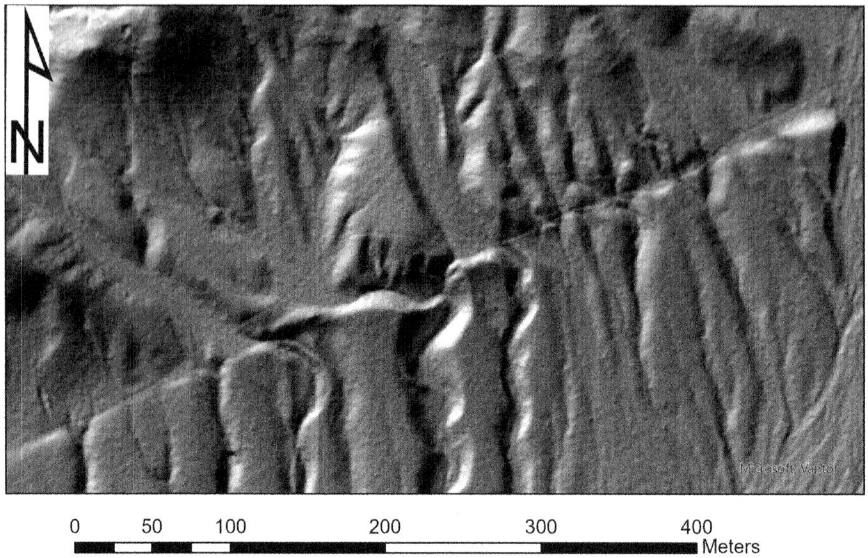

Figure 12.32 Closely spaced offset drainages along the central Garlock fault in the El Paso Mountains, California (USA), showing the difficulty in interpreting unique matches of upstream and downstream segments. Flow direction is toward the bottom of the image. Centre of image is approximately at N35.4394°, W117.6950. Lidar DEM from EarthScope Southern and Eastern California LiDAR Project. Distributed by OpenTopography. https://doi.org/10.5069/G9G44N6Q. Accessed 12 September 2008. *This material is based on services provided to the Plate Boundary Observatory by NCALM (*https://ncalm.cive.uh.edu/*). PBO is operated by UNAVCO for EarthScope (*http://www.earthscope.org*) and supported by the National Science Foundation (No. EAR-0350028 and EAR-0732947).*

that some of the downstream segments may have been connected to multiple different upstream sources at different times making it difficult or impossible to determine the tectonic offset of any of the channels with confidence (Fig. 12.32). This issue is called 'channel aliasing' (Reitman et al., 2023).

12.3.2.1.3 Tectonic warping near the fault

When tectonic warping is erroneously interpreted as a preearthquake bend in a channel, this leads to an underestimate of the tectonic offset of the marker. Coseismic lateral warping within a few metres or tens of metres of the fault zone has been documented by postearthquake surveys of human-made features that were known to be linear prior to the earthquake (e.g. Antoine et al., 2024; Dolan & Haravitch, 2014; Gold et al., 2015; Gold et al., 2021; Milliner et al., 2015; Rockwell et al., 2002; Treiman et al., 2002; Vallage et al., 2015; Zinke et al., 2014). Studies of prehistoric earthquake ruptures use offset natural features, such as ephemeral channels, in which case it is not possible to assume that the geomorphic feature was perfectly straight prior to the earthquake, making it difficult to distinguish whether latearal warping has occurred or not. Lateral

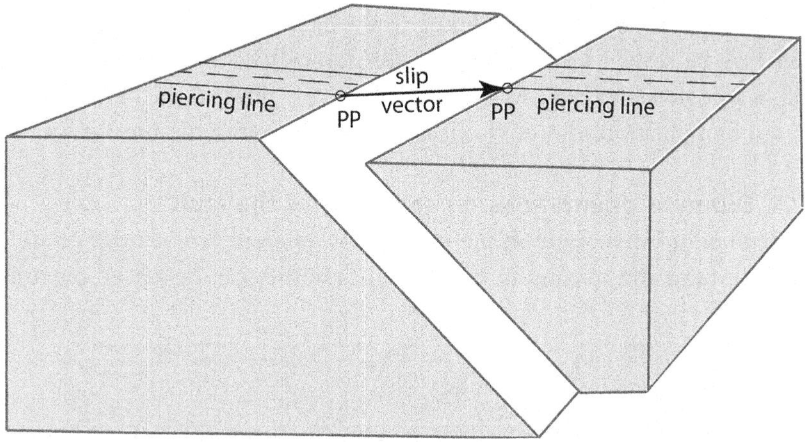

Figure 12.33 In this example, one edge of a road is selected as a piercing line, which intersects the fault at a piercing point (PP) on each side of the fault. Prior to slip on the fault, the two PPs were colocated. After slip on the fault, the slip vector extends from one PP to the other, within the plane of the fault.

warping and other off-fault deformations are more typical of immature faults with complex fault traces, whereas slip along mature faults is generally concentrated on the principal fault trace (Dolan & Haravitch, 2014; Ellis & Dunlap, 1988; Manighetti et al., 2021; McPhillips & Scharer, 2023).

12.3.2.2 Piercing points and piercing lines

In some cases, especially when measurements are made shortly after an earthquake along a fault zone comprising a single strand within relatively indurated material, a linear geomorphic feature that has been offset by the fault can be followed all the way to the fault and the precise location at which the feature intersects the fault plane can be identified on both sides of the fault. These linear features are called piercing lines, and their points of intersection with the fault plane are called piercing points (PPs) (Fig. 12.33).

12.3.2.2.1 Overprinting

When measuring offsets along a prehistoric fault rupture, however, the offset geomorphic markers often become more obscure as they approach the fault. They may be thoroughly broken up within a broad fault zone, tilted by secondary faults, buried by fault-scarp colluvium, and/or modified by postearthquake erosion (Lienkaemper, 2001; Ouchi, 2005; Sieh, 1978; Wallace et al., 1968). Zielke, Arrowsmith, Grant Ludwig, et al. (2012); Zielke et al. (2015) referred to such postearthquake modification of the offset geomorphic features within and near the fault zone as 'overprinting'. Where overprinting has occurred, the portion of the geomorphic feature whose preearthquake morphology remains clear (the piercing line) must be followed as close as possible toward the fault and then projected through the zone of overprinting to the fault plane (Fig. 12.34). Often the fault zone itself may be several metres wide but for purposes of the offset measurement, it is simplified as a single fault plane. As long as the piercing lines on the two sides of the fault have the same orientation, this does not introduce error into the measurement.

12.3.2.2.2 Different orientations on each side of the fault

When a channel or other geomorphic marker has different orientations on each side of the fault, minimum and maximum bounds on the offset can be placed by constructing

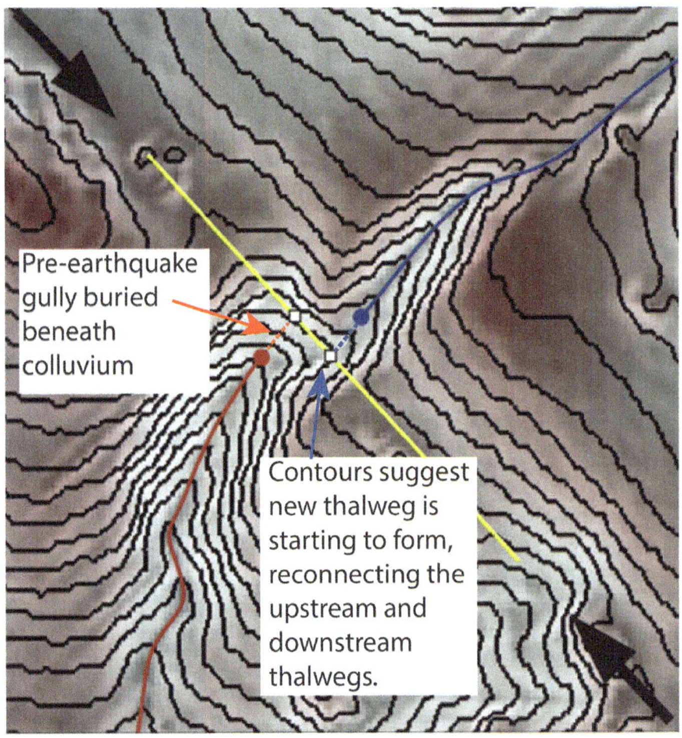

Figure 12.34 Blue and red curved lines mark a channel thalweg, as defined by the contour lines, upstream and downstream from the fault (respectively). The blue and red dots mark the points where the channel enters the zone of overprinting near the fault. The linear portion of the channel that is closest to the fault is projected (as a dashed line) to a simplified trace of the fault (yellow line) to obtain the piercing points (white squares). Notice that the thalweg that was once present between the red dot and the fault is no longer evident in today's contour lines, most likely because it has been buried by colluvium from the fault scarp. In addition, the shape of the contour lines immediately southwest of the fault suggests that postearthquake flow has begun to carve an incipient connection between the up- and downstream channel segments, further obscuring the channels' preearthquake morphology. *From Zielke, O., Klinger, Y., & Arrowsmith, J.R. (2015). Fault slip and earthquake recurrence along strike-slip faults—Contributions of high-resolution geomorphic data. Tectonophysics, 638, 43–62; their Fig. 12.8C).*

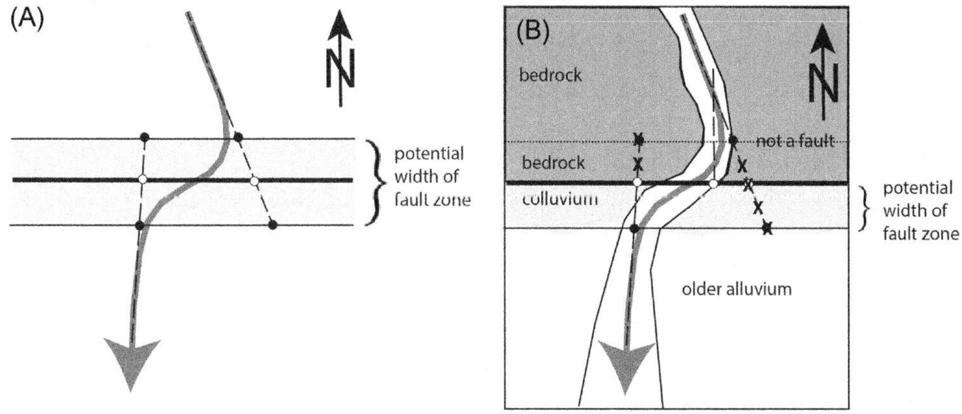

Figure 12.35 (A) Projection of a south-flowing stream channel through the zone of overprinting within the potential width of the fault zone. Thick grey line shows current channel thalweg. Thin, dashed black lines project the thalweg through the fault zone. White-filled circles are piercing points (PPs) where the projected piercing lines intersect the most likely position of the fault strand with the predominant offset, which are typically used to measure the preferred estimate of the channel offset. Black-filled circles are PPs where the projected piercing lines intersect the north and south boundaries of the fault zone, which would be used to measure minimum and maximum estimates of the offset, respectively. (B) If geologic field mapping reveals that bedrock (or alluvium that is older than the channel whose offset is being measured) is exposed in any area through which the channel had been projected based on remote observations, that projection is not valid (dashed lines marked with 'X's), and must be replaced with a valid projection that does not pass through material that is older than the channel that is offset.

two fault planes that bound the width of the fault zone and then projecting the channel on each side of the fault through the fault zone to the plane that defines the far side of the fault zone (Fig. 12.35A). The range of offsets defined by these two measurements will include the possibility that the change in orientation occurred anywhere within the fault zone.

When a channel bends as it approaches the fault, decisions must be made about which section of the channel should be used to define the orientation of the piercing line along which the channel will be projected to the fault through the zone of overprinting. It is often the case that the actual PP has been eroded or buried beneath fault-scarp colluvium, and the piercing lines need to be projected to the fault. However, piercing lines can only be validly projected through materials that are younger than the channel that has been offset. In some cases, field mapping may be required to determine the nature and relative age of the materials through which a piercing line is projected, to evaluate the validity of the projection (Fig. 12.35B). In addition to geologic mapping, excavations designed to follow a geomorphic marker to its actual intersection with the fault zone can also help to determine the preearthquake morphology of a geomorphic marker, where it has been buried by colluvium (e.g. Liu et al., 2004; Liu-Zeng et al., 2006; McGill & Sieh, 1993).

In Fig. 12.35B, fault-parallel excavation(s) across the valid projection line south of the fault could reveal whether a portion of the offset channel does or does not extend along that line to the fault, beneath the fault-scarp colluvium at the surface.

12.3.2.2.3 Measuring oblique slip

As noted above, there may also be a component of dip-slip along faults that are primarily strike-slip. For some offsets within indurated materials, the magnitude and orientation of the slip vector within the plane of the fault can be directly measured on the exhumed surface of the fault plane (Fig. 12.36). More commonly, however, especially if significant time has elapsed since the offset occurred, the exhumed fault plane may no longer be preserved, and it is easiest to measure the strike-slip and dip-slip components of the slip vector separately. In such cases, it is important to restore the lateral slip before using the methods described in Section 12.3.1 to measure the dip slip. If this is not done, the dip slip may be over- or underestimated (Fig. 12.37). Measurement of dip-slip after restoration of lateral offset may be accomplished by constructing a topographic profile along the piercing line (and omitting the dog-leg along the fault, where the piercing line is offset), or by subtracting the elevation of the PP on the downthrown side of the fault from the elevation of the PP on the upthrown side of the fault. Either method will provide the vertical separation of the piercing line

Figure 12.36 Measurement of the magnitude of the slip vector in the surface rupture of the 2010 El Mayor-Cucapah earthquake, Mexico. The slip vector (marked by a yellow tape measure) connects the two piercing points (white circles), where the piercing lines, defined by the intersection of the channel wall with the channel floor (marked with black lines), intersect the fault plane. *Photo credit: John Fletcher, 2010.*

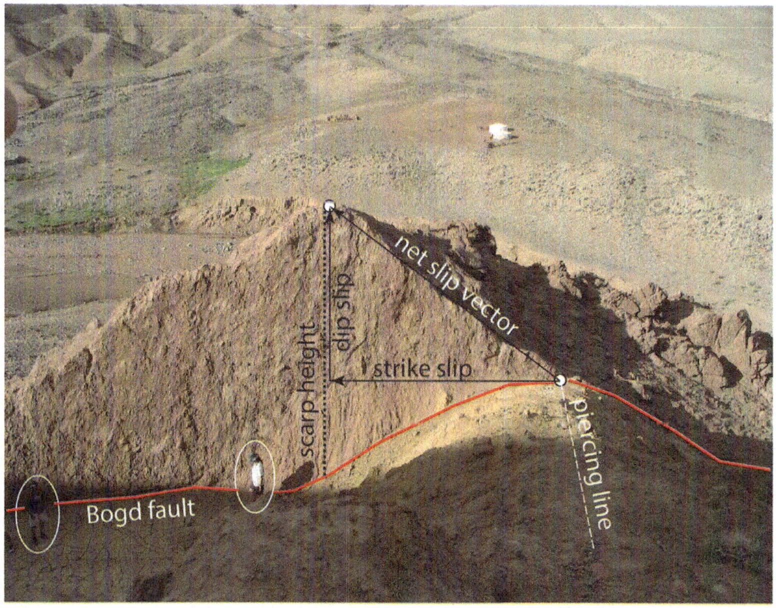

Figure 12.37 Left-lateral offset of a ridge during the 1957 Gobi-Altai earthquake, Mongolia. White ovals encircle persons for scale. Red line marks the trace of the Bogd fault. The piercing points (PPs) (white-filled circles with black outlines) are defined by the intersection of the ridge crest (the piercing line) on each side of the fault with the fault plane. The net slip vector connects the two PPs within the fault plane and can be decomposed into the strike-slip and dip-slip components of slip. A measurement of the scarp height beneath the PP on the left would overestimate the vertical slip in the 1957 earthquake. A measurement of the scarp height at the PP on the right would underestimate the vertical slip. Accurate measurement of the vertical slip can only be made after first restoring the lateral slip. *Photo credit: Sally McGill, 2005.*

across the fault, which can then be trigonometrically converted to dip-slip using the dip of the fault (see Section 12.3.1).

12.3.2.3 Technologies for measuring and documenting strike-slip offsets

12.3.2.3.1 Documenting strike-slip offsets

The methodologies for measuring and documenting lateral offsets have advanced from the use of tape measures in the field (McGill & Sieh, 1991; Sieh, 1978) to include the use of total stations (e.g. McGill & Sieh, 1991) and differential GPS (e.g. Rizza et al., 2011) to construct high-resolution topographic maps, visual analysis and optical image correlation of high-resolution satellite imagery (e.g. Gold et al., 2015; Klinger et al., 2005; Zinke et al., 2014), and digital elevation models (DEMs) constructed from optical satellite imagery (e.g. Antoine & Liu, 2025; Rockwell & Klinger, 2013), high-resolution aerial photography from balloons (Salisbury et al., 2018), unmanned aerial vehicles (e.g. Buckley et al., 2025; Burns, 2023), and from airborne, drone-borne, tripod-based, backpack-mounted and hand-held Lidar devices (e.g. Burns, 2023; Zielke, Arrowsmith, Grant Ludwig, et al., 2012). The advent of airborne Lidar, in particular, has revolutionised the ease with which large numbers of tectonically offset geomorphic features can be identified and measured (e.g. Zielke, Arrowsmith, Grant Ludwig, et al., 2012).

Zielke et al. (2015) provided an excellent summary of sources of topographic data and recommendations for use of various products derived from DEMs. Hillshade images and various other visualisations work well for identification of offset geomorphic markers, but contour lines derived from the DEM should be added when making offset measurements, to aid in accurate tracing of the marker shape and its projection onto the fault trace (Zielke et al., 2015). Remote measurements using DEMs and derivative products (hillshade images, slope maps, aspect maps and topographic contour lines) have the advantage of being able to view all parts of an offset feature at the same scale at the same time (Zielke et al., 2015). However, field observations and optical imagery provide valuable means to check interpretations made based on hillshade models (Zielke et al., 2015), particularly to evaluate the validity of projected piercing lines (Fig. 12.35).

12.3.2.3.2 Measuring: projection and back-slipping

Digital measurements can be made by using a digital measurement tool in a Geographic Information System or other software (Fig. 12.34), or by digitally slicing images along the fault and determining the amount of slip that must be restored along the fault to yield the most convincing preearthquake geometry for the offset geomorphic marker (e.g. Fig. 12.38). For measuring the offset of an alluvial fan from its source drainage, several authors have fit circles to the topographic contours on the fan to define the fan axis (Burns, 2023; Gold et al., 2015).

12.3.2.3.3 Semiautomated and automated measurements

Zielke and Arrowsmith (2012) developed a Matlab® graphical user interface, called LiDaCaoz (Lateral Displacement Calculator), for robust digital measurement of offset geomorphic markers based on cross-correlation of fault-parallel elevation profiles on both sides of a strike-slip fault. The profiles are projected onto the fault (which is assumed to have a vertical dip) using the user-defined orientation of the marker on each side of the fault. A test in which seven individuals independently used LiDiCaoz to measure the same 11 offset channels along the San Andreas fault showed that offset measurements by different users converged for simple, well-defined offset channels that are sharply offset across a narrow fault zone. However, results were more scattered for offset channels with more complex geometries.

Numerical landscape evolution simulations along a theoretical strike-slip fault (Reitman et al., 2019) have shown that within a set of >3000 automatically measured offsets from >135 simulations, individual measurements of the slip varied by ~30% and

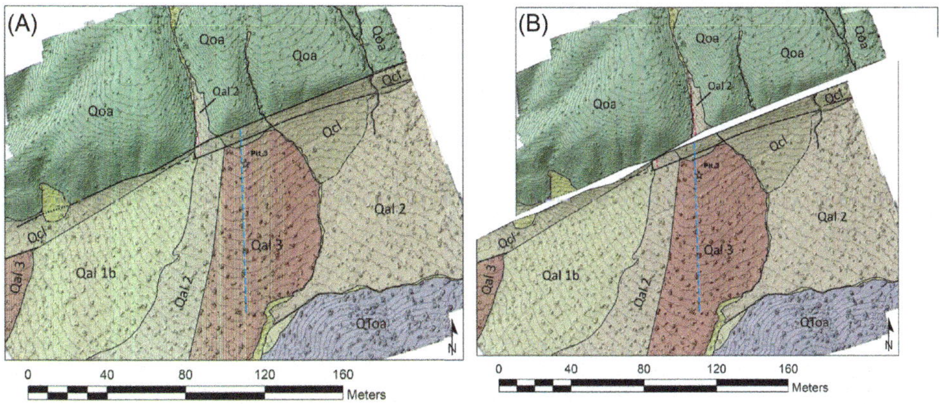

Figure 12.38 (A) Geologic map of a small alluvial fan (Qal3) offset from its source channel north of the Garlock fault. Blue dashed line marks the axis of the fan, as defined by the 0.5-m topographic contour lines. Pink dashed line marks the centre of the source channel, as measured between contour lines of equal height on each side of the channel. Samples for luminescence dating were collected from Pit 3. Underlying hillshade model and 0.5 m topographic contours were created in ArcGIS from a digital elevation model created using Agisoft Metashape software and drone-based photos. (B) Same geologic map with 23 m of left-lateral slip restored. *From Burns, J.E. (2023). Late holocene slip history of the central garlock fault, Mojave, Desert, California. San Bernardino: California State University, 137. M.S. thesis.*

the average of multiple offset measurements systematically underestimated modelled slip except when the fault zone is less than ~5 m wide, total slip is less than channel spacing (to avoid channel aliasing, see Section 12.3.2.1), and offsets are measured shortly after an earthquake.

12.4 Estimating the amount of slip in prehistoric earthquakes from offset geomorphic features

It is common practice after recent earthquakes (whether strike-slip or dip-slip) for investigators to measure the offsets along the rupture at as many different locations as possible and to plot these as a function of distance along the strike of the fault. For example, Reitman et al. (2022) compiled 39 published slip distributions for recent and historical strike-slip earthquake ruptures. Similar measurements have been made along faults for which the most recent rupture was prehistoric or was early enough in the historic period that the offsets were not measured in detail immediately after the earthquake, in an effort to characterise the magnitude of prehistoric earthquakes. Sieh (1978) pioneered this approach for strike-slip faults, building on earlier work by Wallace et al. (1968) and Clark (1970). Many subsequent investigations have continued to develop this approach (e.g. Klinger et al., 2000; Lin et al., 2020; Lindvall et al., 1989; McGill & Sieh, 1991; Reitman et al., 2024; Zielke et al., 2015; Zielke, Arrowsmith,

Grant Ludwig, et al., 2010, 2012). Although this section focuses on the application of this approach to strike-slip faults, many of the methods have been applied to dip-slip faults as well (e.g. Beanland & Clark, 1994; Salyards, 1985).

In the case of prehistoric ruptures (and early historic ruptures that are studied decades to a century or more after the earthquake), two additional challenges are faced in constructing a slip distribution along-strike: (1) the offset geomorphic features are more degraded than those observed shortly after an earthquake rupture, and are likely to have been modified by postearthquake erosion and deposition, and (2) it can be challenging to determine which offset geomorphic features have experienced only the most recent surface-rupturing earthquake and which ones have experienced the cumulative effects of two or more prehistoric earthquakes. The issue of estimating prehistoric fault slip from offset geomorphic markers that have been obscured by erosion or deposition has been addressed in Section 12.3. Here, we examine various methods of distinguishing the number of earthquakes that have contributed to the offset of a geomorphic marker.

12.4.1 Distribution of slip along strike

Typically, the distinction between geomorphic markers that have been offset in the most recent event only and those offset in multiple events is made by measuring all geomorphic markers with offsets that could reasonably represent the amount of slip in the past one to several earthquakes and plotting those offsets as a function of distance along the strike of the fault (Fig. 12.39A, e.g. Lindvall et al., 1989; McGill & Sieh, 1991; Rockwell et al., 1986; Sieh, 1978). The plot is then examined to see if the offset measurements cluster around discrete values. For example, if the amount of surficial offset was uniform along strike (or at least along kilometres-long stretches of the fault),

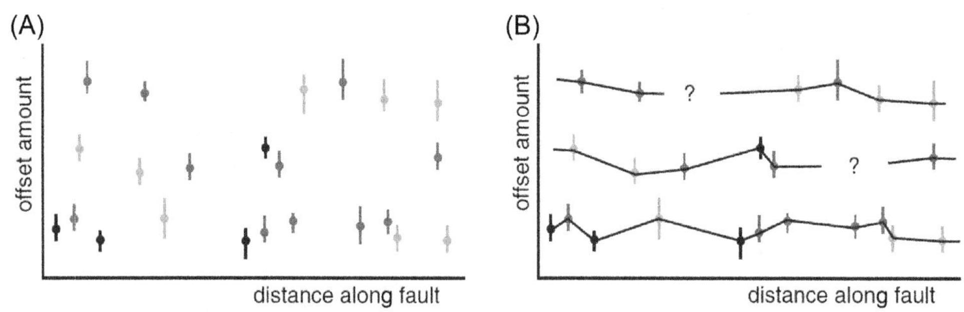

Figure 12.39 (A) Hypothetical plot of measurements of offset geomorphic markers as a function of distance along the fault. Lighter shaded symbols have lower quality ratings than darker symbols. (B) Same as A, but with interpretive lines connecting markers that are assumed to be offset in the most recent earthquake (lowest line), or in the past two or three earthquakes (middle and upper lines, respectively. *From Zielke, O., Klinger, Y., & Arrowsmith, J.R. (2015). Fault slip and earthquake recurrence along strike-slip faults—Contributions of high-resolution geomorphic data. Tectonophysics, 638, 43–62.*

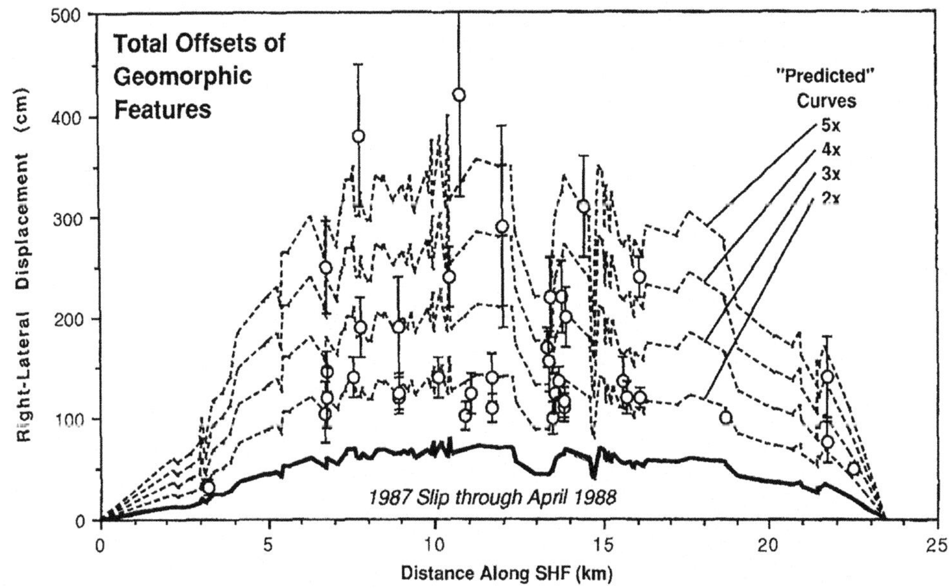

Figure 12.40 Small offsets of geomorphic features along the Supersition Hills fault show that repeated small offsets characterise ruptures on this fault. For simplicity, the slip distribution of the 1987 earthquake appears without the hundreds of data points from which it was constructed. Multiples of the 1987 slip distribution provide a reference for interpreting the number of earthquakes that may have contributed to the larger (and presumably older) offsets. *From Lindvall, S.C., Rockwell, T.K., & Hudnut, K.W. (1989). Evidence for prehistoric earthquakes on the Superstition Hills fault from offset geomorphic features. Bulletin of the Seismological Society of America, 79(2), 342–361.*

then the smallest offsets at any location along the fault likely represent the amount of slip in the most recent earthquake (see lowest connected line in Fig. 12.39B), and larger offsets would represent the cumulative slip in two or more earthquakes. If the penultimate event had a comparable amount of slip, then one might expect to find a set of geomorphic markers offset by approximately twice the amount of the smallest measured offsets (see middle connected line in Fig. 12.39B). Fig. 12.40 shows an example in which offset geomorphic features along the Superstition Hills fault, California (USA), were found to be offset in multiples of the amount of offset that occurred in the 1987 M_w 6.5 earthquake on that fault (Lindvall et al., 1989).

12.4.2 Cumulative offset probability density plots

To aid in distinguishing the number of discrete prehistoric slip events (earthquakes) represented by offset geomorphic markers along a fault, it is now common practice to construct a probabilistic histogram of the measurements by converting each measured offset and its uncertainties into a Gaussian probability density function and then

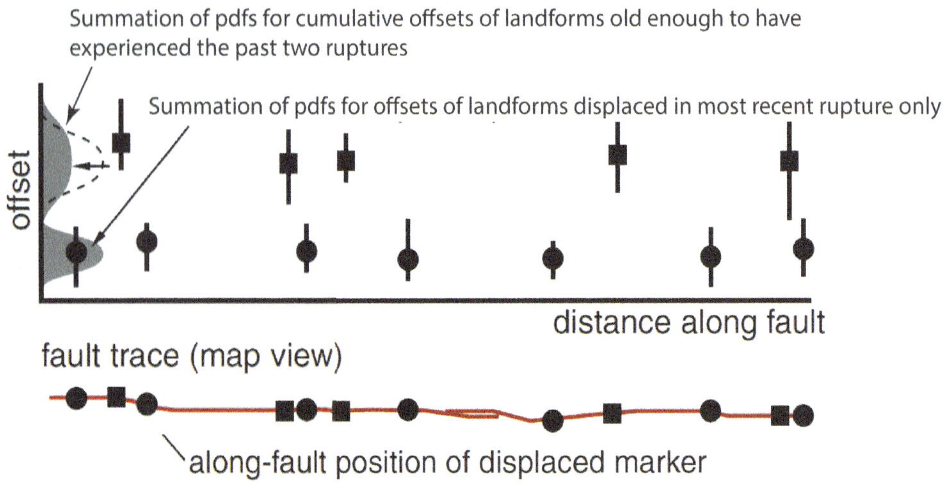

Figure 12.41 Hypothetical distribution along strike of the offsets of geomorphic markers, with Cumulative Offset Probability Density plotted on the vertical axis at left, showing that the offsets cluster around two discrete values. *From Zielke, O., Klinger, Y., & Arrowsmith, J.R. (2015). Fault slip and earthquake recurrence along strike-slip faults—Contributions of high-resolution geomorphic data. Tectonophysics, 638, 43–62.*

summing the probability density functions for all of the measured offsets along a given stretch of the fault (Fig. 12.41). McGill and Sieh (1991) first developed this approach when studying offset geomorphic features along the Garlock fault. Zielke et al. (2010) named the plot of stacked probability density functions a cumulative offset probability density (COPD) plot, and Zielke et al. (2015) presented a comprehensive summary of the method. The method has now been used in more than 20 studies (see Table 1 in Reitman et al., 2024). Two examples are shown in Fig. 12.42.

12.4.2.1 Limitations of one-dimensional cumulative offset probability density plots due to variable slip along strike

Of course, the COPD method assumes that the lateral offset in an individual earthquake is roughly constant along the length of the fault section, so that the probability density functions for individual offset measurements can be summed together regardless of their location along the fault section. However, McGill and Rubin (1999) measured densely spaced offsets along a 5-km length of the 1992, M_w 7.3, Landers, California earthquake, and found that the one-dimensional COPD plot had two peaks (Fig. 12.43), even though all of the offset markers were offset in the 1992 earthquake alone (McGill & Rubin, 1999). They found that the two peaks in the probability density plot were related to kilometre-scale lengths of fault that had different amounts of lateral offset in the 1992 earthquake.

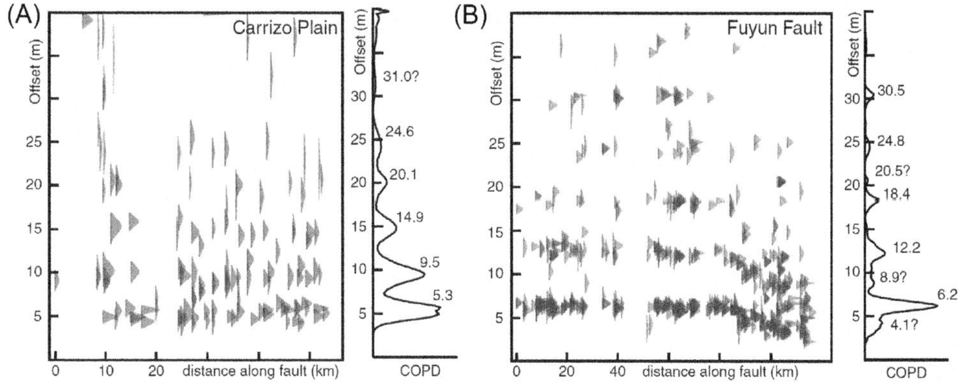

Figure 12.42 Offset measurements of geomorphic markers along the Carrizo Plain section of the San Andreas fault, USA (A), and the Fuyun fault China, (B), showing individual probability density functions for each measured offset, as wall as a one-dimensional Cumulative Offset Probability Density (COPD) plot for each fault. *Figure is from Zielke et al. (2015). Data are from Zielke et al. (2010), and Klinger et al. (2011).*

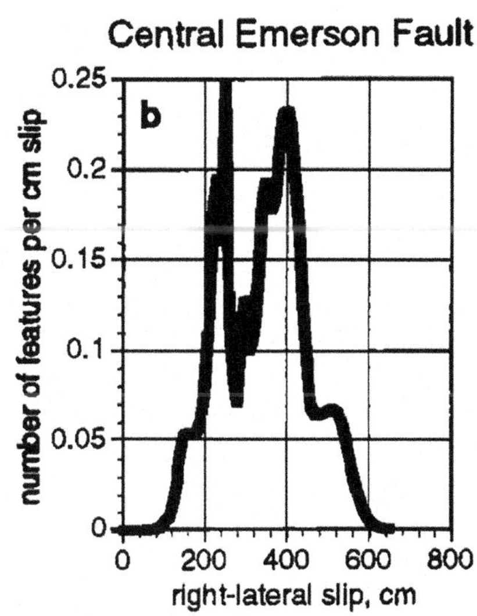

Figure 12.43 Cumulative offset probability density (COPD) function for a 5-km-long stretch of the 1992 Landers earthquake rupture, California (USA), showing that a bimodal COPD plot can be generated by a single earthquake. *From McGill, S.F., & Rubin, C.M. (1999). Surficial slip distribution on the central Emerson fault during the June 28, 1992, Landers earthquake, California. Journal of Geophysical Research: Solid Earth, 104(B3), 4811–4833.*

Zielke et al. (2015) showed that many historic earthquakes have had slip amounts that vary along strike at the scale of a few to tens of kilometres. These include the 1992 M_w 7.3 Landers (McGill & Rubin, 1999; Zachariasen & Sieh, 1995), the 1999 M_w 7.6

Izmit (Rockwell et al., 2002), the 2002 M_w 7.9 Denali (Haeussler et al., 2004), and the 2010 M_w 7.1 Darfield (Elliott et al., 2012; Quigley et al., 2012) earthquakes. This variability in slip can add ambiguity to the process of interpreting which offsets are small enough to be considered caused by the most recent earthquake alone versus which may represent the cumulative slip in two or more earthquakes. Reitman et al. (2024) found that approximately 20% of the 39 strike-slip earthquake ruptures that they compiled had a bimodal 1D COPD plot. Thus when interpreting 1D COPDs to distinguish multiple prehistoric earthquake ruptures, it is important to keep in mind that a single earthquake rupture can produce two or more peaks (Fig. 12.43).

To address this issue, more recent studies have used two-dimensional COPD plots (Fig. 12.44), in which the COPD is calculated separately along short sections of the fault, in order to allow for along-strike variability in the amount of slip in each paleo-earthquake (Bi et al., 2022; Chen et al., 2018; Haddon et al., 2016; Reitman et al., 2024; Ren et al., 2016; Zielke, Arrowsmith, Grant Ludwig, et al., 2012). A 2D COPD allows investigators to make choices about how to connect the various clusters of measurement values along strike for the most recent prehistoric earthquake and for any prior events.

12.4.2.2 Limitations of cumulative offset probability density plots due to the frequency of formation of new geomorphic features across the fault

The approach of using along-strike slip distributions and 1D or 2D COPD plots to estimate the amount of lateral slip in multiple prehistoric earthquakes along a strike-slip fault assumes that new geomorphic markers (such as channels or terrace risers) form more frequently than the occurrence of surface-rupturing earthquakes along the fault. Obviously, if no new geomorphic features form across the fault between two consecutive earthquakes, then only the slip in those two earthquakes combined (rather than individually) will be preserved in the geomorphic record. Thus some peaks in a COPD plot could represent two (or more) paleo-earthquakes.

Zielke et al. (2010) and Zielke et al. (2015) suggested that this may be the case along the San Andreas fault in the Carrizo Plain of central California. Analysis of lateral offsets of geomorphic features suggest that slip on this section of fault typically occurs in increments of 5.1 m. Given the 34-mm/yr slip rate of this portion of the fault, this would suggest an average recurrence interval of 140 ± 46 yr for surface-rupturing earthquakes (Zielke et al., 2010). However, investigations using fault-perpendicular excavations to document the timing of prehistoric earthquakes indicated an average recurrence interval of 88 ± 41 yr for such earthquakes (Akçiz et al., 2010). Zielke et al. (2015) suggested that the ~5 m increments of offset that they measured using geomorphic markers may each be dominated by slip from a single, large earthquake that ruptured the entire length and seismogenic depth of the Carrizo plain fault section

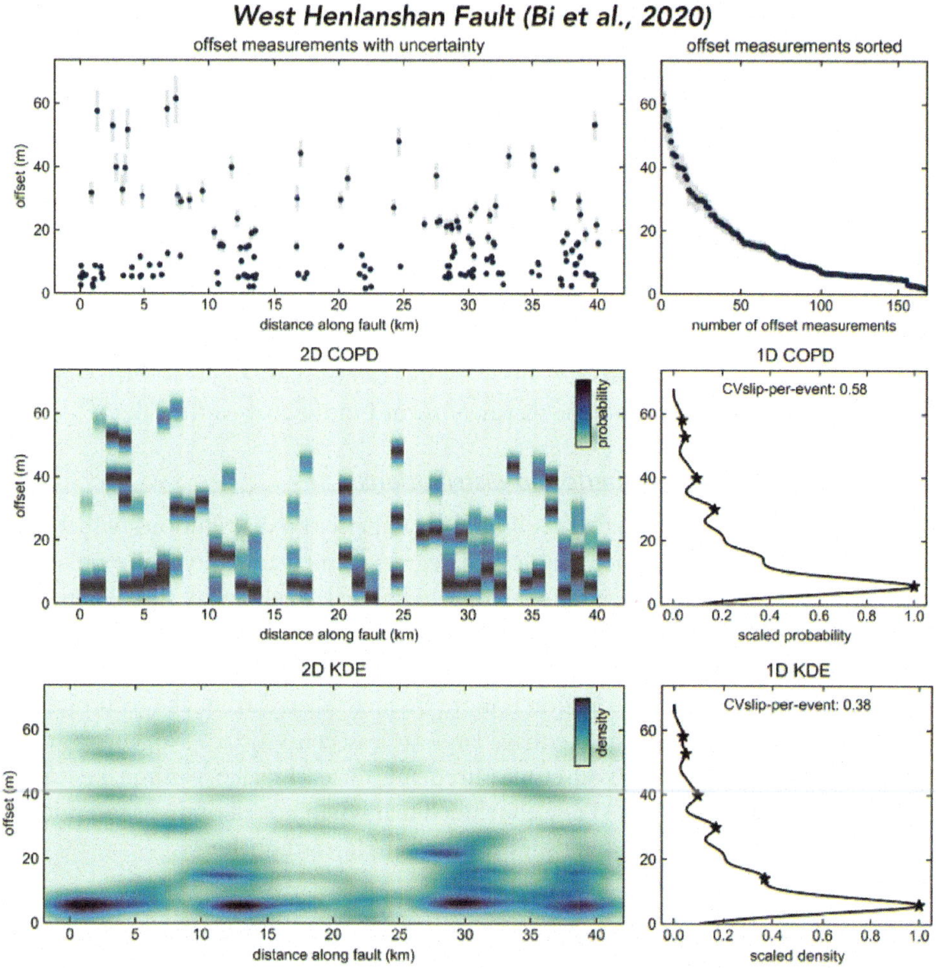

Figure 12.44 Slip distribution along strike, along with two-dimensional and one-dimensional cumulative offset probability density (COPD) plot and kernal density estimate (KDE) plots for the West Henlanshan fault. *From Reitman, N.G., Klinger, Y., Briggs, R.W., & Gold, R.D. (2024). Limited preservation of strike-slip surface displacement in the geomorphic record. Journal of Geophysical Research: Solid Earth, 129(11), p. e2024JB028692 based on data from Bi, H., Zheng, W., Zhang, P., Zeng, J., Shao, Y., Yao, Y. et al. (2022). Recovering surface slip distribution along the Sertengshan Piedmont fault (Northern China) from airborne LiDAR data. Tectonics, 41(8) https://doi.org/10.1029/2021TC007174e2021TC007174.*

as well as smaller amounts of slip from earthquake ruptures on the adjacent sections of the San Andreas fault that have propagated into the Carrizo Plain. Using stratigraphic observations from excavations at one site, Liu et al. (2004) and Liu-Zeng et al. (2006) confirm that one of the past six earthquakes on the Carrizo Plain was a small event with only 1.4 m of right-lateral slip.

Reitman et al. (2024) proposed that climatic factors and the slip rate of the fault must also be taken into account when using COPD plots to interpret the amount of slip in prehistoric earthquakes along strike-slip faults. If regional channel incision events are less frequent than earthquakes (as is the case on the Carrizo Plain section of the San Andreas fault; Grant Ludwig et al., 2010), then COPD peaks are more likely to represent more than one earthquake. On the other hand, for faults with low slip rates (and thus long intervals between earthquakes), it is more likely that new channel incision happens more frequently than do earthquakes, and new geomorphic markers are likely to form between each consecutive pair of earthquakes. In addition, Reitman et al. (2023) found that regions with higher mean annual precipitation are more likely to have large channels that record the cumulative slip in many earthquakes, whereas drier regions are more likely to have small channels that may record individual earthquakes.

12.4.2.3 Insights from synthetic slip distributions

Reitman et al. (2024) constructed COPDs for synthetic slip distributions with a variety of average slip amounts, rupture lengths and degrees of slip variability along strike. They concluded that COPD peaks only record individual earthquake slip when (1) all ruptures on the studied fault section are full-length ruptures with minimal slip variability (<30%), and (2) any smaller events have at least 50% as much slip as the larger events, and (3) geomorphic marker formation is more frequent than earthquake recurrence, and (4) all surface-rupturing earthquakes have large (at least 3 m) surface slip, and (5) there is high data density (at least ~ 40 offset measurements for each paleo-earthquake rupture), and (6) the climate is semiarid. They concluded that when not all of these conditions are met, COPD peaks often combine slips from multiple earthquakes.

12.4.3 Other methods for determining palaeo-earthquake rupture

Aside from COPD plots, other factors that may help to distinguish whether an offset marker has experienced one earthquake versus two or more earthquakes include slip gradients, relative or absolute age indicators, and multiple beheaded channels derived from the same upstream source.

12.4.3.1 Slip gradients

On the Garlock fault, California, McGill and Sieh (1991) argued that a geomorphic marker offset 5.3 ± 0.3 m and located 4 metres away from a marker offset 3.4 ± 0.5 m most likely represents the cumulative slip in two events, as must a marker offset 5.6 ± 0.7 m and located 10 m away from a marker offset 2.7 ± 0.7 m. They hypothesised that such drastic changes in lateral slip over a short distance were unlikely to occur in a single earthquake. To test the hypothesis that such drastic changes in slip along strike are unlikely to occur along a strike-slip earthquake rupture, McGill and Rubin (1999) made closely-spaced measurements of offset geomorphic and human-made markers along a 5.6-km-long portion of the 1992 Landers earthquake rupture and found that slip gradients and consequent strains were at most half as large as the one mentioned above from the Garlock fault.

In a broader study, Zielke et al. (2015) found that substantial short-wavelength (sub-km) variation in lateral slip measurements along recent earthquake ruptures was also present in many recent earthquakes, including the 1992 M_w 7.3 Landers (McGill & Rubin, 1999; Zachariasen & Sieh, 1995), the 1999 M_w 7.6 Izmit (Rockwell et al., 2002), the 2001 M7.8 Kokoxili (Klinger et al., 2005), the 2002 M_w 7.9 Denali (Haeussler et al., 2004; Schwartz et al., 2012), the 2010 M_w 7.2 El Mayor-Cucapah (e.g. Gold et al., 2013; Oskin et al., 2012), the 2010 M_w 7.1 Darfield (Elliott et al., 2012; Quigley et al., 2012), and the 2011 M_w 6.6 Fukushima–Hamadōri (Mizoguchi et al., 2012; Nissen et al., 2014; Toda & Tsutsumi, 2013) earthquakes. Large variations in dip-slip along strike have also been documented for the 1872 Owens Valley earthquake in California (Beanland & Clark, 1994) and for the most recent prehistoric rupture along the Cucamonga thrust fault in southern California (McPhillips & Scharer, 2018).

Thus when slip gradients between adjacent geomorphic offsets along a fault are extreme enough, they may potentially be used to infer that the larger offset feature represents the cumulative slip in an additional event. However, measurement uncertainties and uncertainties in preearthquake morphology of the offset maker are generally high enough that this will only rarely be a useful indicator.

12.4.3.2 Relative or absolute age indicators

A more promising approach for distinguishing markers that are offset in a single earthquake from those offset in two or more earthquakes is to obtain relative or absolute ages for the offset features. One would expect geomorphic markers that have experienced two earthquakes to be older than those that have experienced only one earthquake, although it is possible that markers of different ages have experienced the same number of earthquakes if the ages of the markers differ by an amount less than the average earthquake recurrence interval for the fault. Surface characteristics such as local roughness, channel incision, or (in arid environments) the degree of desert pavement, rock varnish formation or ventral clast rubifcation (the reddish discoloration that forms on the undersides of rocks (clasts) resting on the ground) have been used to distinguish parts of an alluvial fan that have different ages (e.g. Dühnforth et al., 2017; Frankel & Dolan, 2007; Helms et al., 2003) and may, in some case be used to infer relative ages of offset markers.

Direct dating of offset geomorphic features can be useful in distinguishing the number of earthquakes experienced by an offset marker when there is a nearby dated stratigraphic record of paleoseismic earthquakes. For example, Burns (2023) was able to

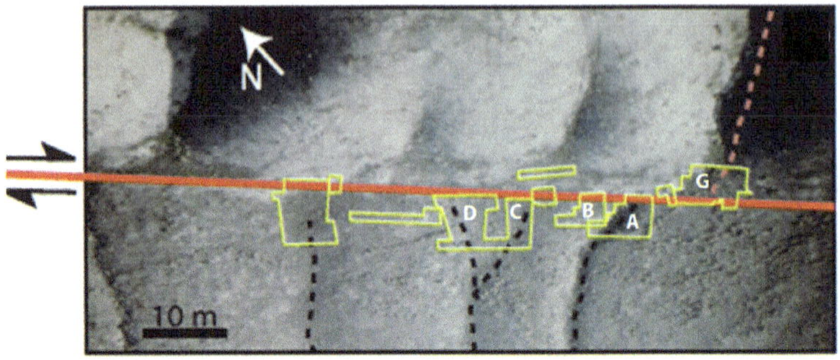

Figure 12.45 Aerial photograph along the San Andreas fault (red line) in the Carrizo plain, showing several beheaded gullies southwest of the fault. Yellow polygons mark the outlines of areas in which multiple excavations were dug by hand, parallel to the fault, to follow the buried, stratigraphic expression of the beheaded channels to the fault. Excavations north of the fault in the source gully marked with a red dashed line (area G) revealed a set of nested channels of various relative ages within that gully. Downstream from the fault, excavations A and B each revealed one buried channel, excavation C revealed two buried channels, and excavation D revealed four buried channels. Six of these eight gullies downstream from the fault could be stratigraphically matched with specific channels in the nested channel sequence in the source gully in area G. *Modified from Liu-Zeng et al. (2006).*

show that a small alluvial fan, left-laterally offset 23 (+2, -4) metres across the central Garlock fault in the El Paso Mountains (Fig. 12.38), was old enough to have experienced the four most recent earthquakes that had been dated in a paleoseismic trench 1.8 km away (Dawson et al., 2003), but was not old enough to have experienced the fifth earthquake documented in that trench. This enabled Burns (2023) to estimate an average slip of 5.75 m per earthquake for the past four events.

12.4.3.3 Multiple beheaded channels from the same source

McGill and Rubin (1999) noted that where two (or more) downstream channel segments have emanated from the same source channel on the upstream side of the fault, the beheaded channel that has been offset farther from the source channel clearly has experienced more than one earthquake. Liu et al. (2004) and Liu-Zeng et al. (2006) used this principle in their stratigraphic investigation of six beheaded channels offset from a single source across the San Andreas fault (Fig. 12.45).

12.5 Role of geomorphic indicators of past earthquakes in seismic hazard analysis and in understanding fault system behaviour

Geomorphic indicators of past earthquakes have important applications in seismic hazard analysis and mitigation as well as in understanding the physics of complex fault systems. The following two sections briefly summarise these applications.

12.5.1 Seismic hazard analysis

Fault slip-rates derived from measurements of offset geomorphic features (in conjunction with data on the age of the offset feature) provide an important input to seismic hazard analysis because faults with faster slip rates typically produce larger and/or more frequent earthquakes. Recent studies have, however, pointed to the possibility of variable slip along strike and cautioned against extrapolating slip rates measured at a single point along the fault to the entire fault section, especially for sites that are near the end of a potential rupture segment (e.g. McPhillips & Scharer, 2023).

Estimates of the amount of slip in individual past earthquakes (whether historic or prehistoric) along a fault also play an important role in seismic hazard analysis and mitigation in several ways:

1. Knowledge of the typical amount of slip in earthquakes along a fault can aid governments and businesses in designing infrastructure (e.g. pipelines, roads, tunnels) that must cross the fault, and to develop plans to quickly repair offset infrastructure after an earthquake (Fuis & Wald, 2003; Metropolitan Water District of Southern California, 2025).
2. Knowledge of the seismic moment or magnitude of past earthquakes on a fault can be used to infer the level of shaking that this fault has produced in the past, at various distances from the fault, as a guide to the distribution of shaking that may be expected from future earthquakes on the fault (Jones et al., 2008). The seismic moment of past earthquakes can be estimated from the amount of slip in past earthquakes (in conjunction with an inferred rupture length and depth), using the equation $M_o = \mu As$ (Brune, 1968), where M_o is the seismic moment of the earthquake, μ the shear modulus of the rock surrounding the fault (typically assumed to be 3×10^{10} N/m^2 in Earth's crust), A is the area of the fault plane (length times down-dip width) and s is the amount of slip on the fault during the earthquake. If significant off-fault deformation occurred during the earthquake, then the measured offsets of geomorphic features at the surface may underestimate the seismic moment, unless the off-fault deformation can be measured and included in the slip estimate (see Section 12.3.2.1.3). The seismic moment, M_w, can be converted to a moment magnitude using the equation $M_w = 2/3 \log M_o - 6.0$, where M_o is in N-m (Hanks & Kanamori, 1979).
   Alternatively, empirical scaling relationships between fault slip and earthquake magnitude can be used to estimate the magnitudes of prehistoric earthquakes (Hanks & Bakun, 2008; Shaw, 2009, 2013; Stirling et al., 2024; Wells & Coppersmith, 1994; Working Group on California Earthquake Probabilities, 2003). Most of the field studies on which these scaling relationships are based either did not include off-fault deformation at all or only included what was easily visible and measurable. Thus some amount of off-fault deformation is built into the relationships, although the amount of off-fault deformation varies between ruptures (Gold et al., 2015).

3. Knowledge of the *average* amount of slip, S, in past surface-rupturing earthquakes can be used in conjunction with a measured or inferred slip rate, R, and an appropriate recurrence model to estimate the *average* recurrence interval, I, between surface-rupturing earthquakes, using the equation $I = S/R$. (e.g. Madden et al., 2011; McGill & Sieh, 1991; Sieh & Jahns, 1984). For this reason, recent seismic hazard models have included estimates of the slip in the most recent earthquake, where available, in addition to slip rates (e.g. Madden et al., 2011).

12.5.2 Fault systems and earthquake physics

Estimates of slip rate and slip per earthquake measured from geomorphic features have also been used to investigate the physics of various earthquake processes. Some examples are given below.

1. Estimates of slip per earthquake on a fault have been used to develop and test hypotheses regarding whether a given fault section typically produces a uniform amount of slip each time it ruptures or whether the amount of slip on a fault section varies from one earthquake to the next (Schwartz & Coppersmith, 1984; Sieh, 1978, 1981, 1996; Zielke et al., 2015; Zielke, Arrowsmith, Grant Ludwig, et al., 2012).
2. Measurements of the variability in lateral offsets along strike during historical earthquakes have been used to investigate how earthquake ruptures behave in surficial sediments, the width of the zone of coseismic deformation and the distribution of deformation within it, and how surficial offsets differ in magnitude from coseismic slip at seismogenic depths as modelled from seismograms (e.g. Fialko et al., 2005; Gold et al., 2013).
3. Comparison of slip rates estimated from dated offset geomorphic features with deformation rates estimated geodetically has been used to make inferences about the amount of off-fault deformation in earthquakes (e.g. Dolan & Haravitch, 2014; Herbert et al., 2014; McCaffrey, 2005), to propose temporal oscillations in the locus of deformation along a plate boundary (e.g. Peltzer et al., 2001), and to infer rheologies and physical processes in the mantle and lower crust that are involved in the earthquake cycle (e.g. Chuang & Johnson, 2011).
4. Estimates of the amount of slip in the most recent several earthquakes, as well as slip rates calculated over a variety of different time periods, have been used to construct time histories of offset along a fault to provide insight into nonconstant slip rates and irregular paleoseismic recurrence intervals (e.g. Fougere et al., 2024; Liu-Zeng et al., 2006; Zinke et al., 2017; Zinke et al., 2019). From studies like these, it is becoming apparent that the slip rates of some faults have fluctuated over millennial time scales (Dawson et al., 2003; Dolan et al., 2016; Fougere et al., 2024; Rittase et al., 2014; Zinke et al., 2017; Zinke et al., 2019). Geomorphic studies indicate that fluctuating slip rates are more common on complex fault systems, whereas faults with simpler

geometries tend to have slip rates that are constant over millennial time scales (Gauriau et al., 2023). Dolan et al. (2024) have also observed that variable slip rates on subparallel strands of the Marlborough fault system, New Zealand, add up to a total plate boundary slip rate that is less variable over time. In this fault system, each individual fault strand has periods of higher-than-average slip-rate, lasting for several earthquake cycles totalling 20–25 m of slip before slowing down and letting another fault strand take over for a while. Cawood and Dolan (2024) proposed a model for fault strength evolution that could explain this behaviour.

12.6 Conclusions

Geomorphic indicators of prehistoric earthquakes have played a pivotal role in our understanding of the physical behaviour of fault systems as well as in mapping the distribution of seismic hazard. These have been based on an understanding of the interplay between the tectonic and geomorphic processes that create and modify landforms. The importance of geomorphology in understanding fault behaviour and seismic hazard is demonstrated by the continually growing literature on how best to interpret and measure geomorphic indicators of slip in previous earthquakes. New technologies, such as airborne, tripod-based, and hand-held Lidar, have made collection of geomorphic data easier as well as more accurate and precise, and increasingly sophisticated methods of analysis have both improved our understanding of offset geomorphic markers, while continuing to highlight the uncertainties that can plague the interpretation of some geomorphic markers.

References

Akçiz, S. O., Ludwig, L. G., Arrowsmith, J. R., & Zielke, O. (2010). Century-long average time intervals between earthquake ruptures of the San Andreas fault in the Carrizo Plain, California. *Geology, 38*(9), 787–790.

Anderson, E. M. (1905). The dynamics of faulting. *Transactions of the Edinburgh Geological Society, 8*(3), 387–402.

Antoine, S. L., Klinger, Y., Wang, K., & Bürgmann, R. (2024). Coseismic shallow slip deficit accounted for by diffuse off-fault deformation. *Geophysical Research Letters, 51*(24) e2024GL110798.

Antoine, S. L., & Liu, Z. (2025). Impact of optical imagery and topography data resolution on the measurement of surface fault displacement using sub-pixel image correlation. *Earth and Space Science, 12*(4) p. e2024EA003660.

Arrowsmith, Ramon J., & Rhodes, D. D. (1994). Original forms and initial modifications of the Galway Lake Road scarp formed along the Emerson fault during the 28 June 1992 Landers, California, earthquake. *Bulletin of the Seismological Society of America, 84*(3), 511–527.

Avouac, J. P. (1993). Analysis of scarp profiles: Evaluation of errors in morphologic dating. *Journal of Geophysical Research: Solid Earth, 98*(B4), 6745–6754.

Axen, G. J., Fletcher, J. M., Cowgill, E., Murphy, M., Kapp, P., MacMillan, I., Ramos-Velázquez, E., & Aranda-Gómez, J. (1999). Range-front fault scarps of the Sierra El Mayor, Baja California: Formed above an active low-angle normal fault? *Geology, 27*(3), 247–250.

Beanland, S., Clark, M.M. (1994). The Owens Valley fault zone, eastern California, and surface faulting associated with the 1872 earthquake, U.S. Geological Survey Bulletin, no. 1982, US Government Printing Office, 27 pp.

Bi, H., Zheng, W., Zhang, P., Zeng, J., Shao, Y., Yao, Y., et al. (2022). Recovering surface slip distribution along the Sertengshan Piedmont fault (Northern China) from airborne LiDAR data. *Tectonics, 41*(8). https://doi.org/10.1029/2021TC007174. e2021TC007174.

Brune, J. N. (1968). Seismic moment, seismicity, and rate of slip along major fault zones. *Journal of Geophysical Research, 73*(2), 777–784.

Buckley, W. C., Williams, P. L., Rockwell, T. K., & Gontz, A. M. (2025). Estimating slip distribution along the southernmost approximately 80 km of the San Andreas Fault by examining tectonically offset features. *Geomorphology, 487*. https://doi.org/10.1016/j.geomorph.2025.109937.

Burns, J. E. (2023). *Late holocene slip history of the central garlock fault, Mojave, Desert, California*. San Bernardino: California State University, 137. M.S. thesis.

Carretier, S., Lucazeau, F., Ritz, J. F., & Philip, H. (2002). Comparison of morphological dating models for cumulative reverse fault scarps. *Journal of Geophysical Research, 107*(B10). https://doi.org/10.1029/2000jb000028. ETG 4-1.

Caskey, S. J. (1995). Geometric relations of dip slip to a faulted ground surface: New nomograms for estimating components of fault displacement. *Journal of Structural Geology, 17*(8), 1197–1202.

Cawood, T., & Dolan, J. (2024). An exploration of potentially reversible controls on millennial-scale variations in the slip rate of seismogenic faults: Linking structural observations with variable earthquake recurrence patterns. *Seismica, 3*(2).

Chen, T., Liu-Zeng, J., Shao, Y., Zhang, P., Oskin, M. E., Lei, Q., & Li, Z. (2018). Geomorphic offsets along the creeping Laohu Shan section of the Haiyuan fault, northern Tibetan Plateau. *Geosphere, 14*(3), 1165–1186. https://doi.org/10.1130/GES01561.1.

Chuang, R. Y., & Johnson, K. M. (2011). Reconciling geologic and geodetic model fault slip-rate discrepancies in Southern California: Consideration of nonsteady mantle flow and lower crustal fault creep. *Geology, 39*(7), 627–630.

Clark, M. M. (1970). Some characteristics of the most recently active traces of the Garlock fault. *Geological Society of America, Abstracts with Programs, 2*, 82.

Cloos, H. (1928). *Experimente zur inneren Tektonik: Centralbl. f. Mineral, u. Pal. 1928*(B), 609–621.

Cole, D. A., Jr, & Lade, P. V. (1984). Influence zones in alluvium over dip-slip faults. *Journal of Geotechnical Engineering, 110*(5), 599–615.

Dawson, T. E., McGill, S. F., & Rockwell, T. K. (2003). Irregular recurrence of paleoearthquakes along the central Garlock fault near El Paso Peaks, California. *Journal of Geophysical Research: Solid Earth, 108*(B7).

Dawson, T. E., Weldon, R. J., & Field, E. H. (2013). Appendix B: Geologic slip-rate data and geologic deformation model. US Geol. Surv. Open-File Rept. 2013-1165-B, and California Geol. Surv. Special Rept. 228-B.

Densmore, A. L., Ellis, M. A., & Anderson, R. S. (1998). Landsliding and the evolution of normal-fault-bounded mountains. *Journal of Geophysical Research: Solid Earth, 103*(B7), 15203–15219.

Dolan, J. F., & Haravitch, B. D. (2014). How well do surface slip measurements track slip at depth in large strike-slip earthquakes? The importance of fault structural maturity in controlling on-fault slip versus off-fault surface deformation. *Earth and Planetary Science Letters, 388*, 38–47. https://doi.org/10.1016/j.epsl.2013.11.043.

Dolan, J. F., McAuliffe, L. J., Rhodes, E. J., McGill, S. F., & Zinke, R. (2016). Extreme multi-millennial slip rate variations on the Garlock fault, California: Strain super-cycles, potentially time-variable fault strength, and implications for system-level earthquake occurrence. *Earth and Planetary Science Letters, 446*, 123–136.

Dolan, J. F., Van Dissen, R. J., Rhodes, E. J., Zinke, R., Hatem, A. E., McGuire, C., Langridge, R. M., & Grenader, J. R. (2024). One tune, many tempos: Faults trade off slip in time and space to accommodate relative plate motions. *Earth and Planetary Science Letters, 625*, 118484. 10.1016/j.epsl.2023.118484.

Dühnforth, M., Densmore, A. L., Ivy-Ochs, S., Allen, P., & Kubik, P. W. (2017). Early to Late Pleistocene history of debris-flow fan evolution in western Death Valley (California) using cosmogenic ^{10}Be and ^{26}Al. *Geomorphology, 281*, 53–65. https://doi.org/10.1016/j.geomorph.2016.12.020.

DuRoss, C. B., Gold, R. D., Dawson, T. E., Scharer, K. M., Kendrick, K. J., Akciz, S. O., Angster, S. J., Bachhuber, J., Bacon, S., Bennett, S. E., & Blair, L. (2020). Surface displacement distributions for the July 2019 Ridgecrest, California, earthquake ruptures. *Bulletin of the Seismological Society of America, 110*(4), 1400–1418.

Elliott, J. R., Nissen, E. K., England, P. C., Jackson, J. A., Lamb, S., Li, Z., Oehlers, M., & Parsons, B. (2012). Slip in the 2010-2011 Canterbury earthquakes, New Zealand. *Journal of Geophysical Research, 117*(B3). https://doi.org/10.1029/2011JB008868.

Ellis, M. A., & Dunlap, W. J. (1988). Displacement variation along thrust faults: Implications for the development of large faults. *Journal of Structural Geology, 10*(2), 183–192.

von Engeln, O. D. (1942). *Geomorphology*. New York: The Macmillan Co, 655.

Fenton, C. H., Charusiri, P., & Wood, S. H. (2003). Recent paleoseismic investigations in Northern and Western Thailand. *Annals of Geophysics, 6*(5), 957–981.

Fialko, Y., Sandwell, D., Simons, M., & Rosen, P. (2005). Three-dimensional deformation caused by the Bam, Iran, earthquake and the origin of shallow slip deficit. *Nature, 435*(7040), 295–299.

Field, E. H., Arrowsmith, R. J., Biasi, G. P., Bird, P., Dawson, T. E., Felzer, K. R., Jackson, D. D., Johnson, K. M., Jordan, T. H., Madden, C., & Michael, A. J. (2014). Uniform California earthquake rupture forecast, version 3 (UCERF3)—The time-independent model. *Bulletin of the Seismological Society of America, 104*(3), 1122–1180.

Fletcher, J. M., Teran, O. J., Rockwell, T. K., Oskin, M. E., Hudnut, K. W., Mueller, K. J., Spelz, R. M., Akciz, S. O., Masana, E., Faneros, G., & Fielding, E. J. (2014). Assembly of a large earthquake from a complex fault system: Surface rupture kinematics of the 4 April 2010 El Mayor–Cucapah (Mexico) Mw 7.2 earthquake. *Geosphere, 10*(4), 797–827.

Fougere, D., Dolan, J., Rhodes, E., & McGill, S. (2024). Refined Holocene slip rate for the western and central segments of the Garlock fault: Record of alternating millennial-scale periods of fast and flow fault slip. *Seismica, 3*(2).

Frankel, K. L., & Dolan, J. F. (2007). Characterizing arid region alluvial fan surface roughness with airborne laser swath mapping digital topographic data. *Journal of Geophysical Research: Earth Surface, 112*(F2).

Fuis, G., & Wald, L. (2003). Rupture in South-Central Alaska—The Denali Fault Earthquake of 2002. *U.S. Geological Survey, Fact Sheet 014-03,* 4.

Gauriau, J., Barbot, S., & Dolan, J. F. (2023). Islands of chaos in a sea of periodic earthquakes. *Earth and Planetary Science Letters, 618*, 118274. https://doi.org/10.1016/j.epsl.2023.118274.

Gilbert, G. K. (1890). Lake Bonneville, U.S. Geological Survey Monograph, 1. US Government Printing Office, 438 pp.

Giletycz, S. J. (2015). Landscape dynamics of the Hengchun Peninsula, Southern Taiwan, Ph.D. thesis, National Central University (Taiwan), 184 p.

Gold, P. O., Behr, W. M., Rood, D., Sharp, W. D., Rockwell, T. K., Kendrick, K., & Salin, A. (2015). Holocene geologic slip rate for the Banning strand of the southern San Andreas Fault, southern California. *Journal of Geophysical Research: Solid Earth, 120*(8), 5639–5663.

Gold, P. O., Oskin, M. E., Elliott, A. J., Hinojosa-Corona, A., Taylor, M. H., Kreylos, O., & Cowgill, E. (2013). Coseismic slip variation assessed from terrestrial lidar scans of the El Mayor-Cucupah Surface Rupture. *Earth and Planetary Science Letters, 366*, 151–162. https://doi.org/10.1016/j.epsl.2013.01.040.

Gold, R. D., DuRoss, C. B., & Barnhart, W. D. (2021). Coseismic surface displacement in the 2019 Ridgecrest earthquakes: Comparison of field measurements and optical image correlation results. *Geochemistry, Geophysics, Geosystems, 22*(3), 1–22. https://doi.org/10.1029/2020GC009326.

Gold, R. D., Reitman, N. G., Briggs, R. W., Barnhart, W. D., Hayes, G. P., & Wilson, E. (2015). On- and off-fault deformation associated with the September 2013Mw 7.7 Balochistan earthquake: Implications for geologic slip rate measurements. *Tectonophysics, 660*(September 2013), 65–78. https://doi.org/10.1016/j.tecto.2015.08.019.

The Meckering and Calingiri earthquakes, October 1968 and March 1970. In F. R. Gordon, & J. D. Lewis (Vol. Eds.), *Bulletin - Geological survey of Western Australia: 126*, (pp. 229). Perth, West. Aust., Australia: Geological Survey of Western Australia. http://www.dmp.wa.gov.au/8481_162.aspx.

Grant Ludwig, L., Akciz, S. O., Noriega, G. R., Zielke, O., & Arrowsmith, J. R. (2010). Climate-modulated channel incision and rupture history of the San Andreas fault in the Carrizo Plain. *Science, 327*(5969), 1117–1119. https://doi.org/10.1126/science.1182837.

Haddon, E., Amos, C., Zielke, O., Jayko, A., & Burgmann, R. (2016). Surface slip during large Owens Valley earthquakes. *Geochemistry, Geophysics, Geosystems, 17*(6), 2239–2269. https://doi.org/10.1002/2015GC006033.

Haeussler, P. J., Schwartz, D. P., Dawson, T. E., Stenner, H. D., Lienkaemper, J. J., Sherrod, B. L., & Personius, S. F. (2004). Surface rupture and slip distribution of the Denali and Totschunda faults in the 3 November 2002M7.9 earthquake, Alaska. *Bulletin of the Seismological Society of America, 94*(6B), S23–S52. https://doi.org/10.1785/0120040626.

Hanks, T. C. (2000). The age of scarplike landforms from diffusion-equation analysis. *Quaternary Geochronology: Methods and Applications, AGU Reference Shelf, 4*(4), 313–338.

Hanks, T. C., & Bakun, W. H. (2008). M-log A observations for recent large earthquakes. *Bulletin of the Seismological Society of America, 98*(1), 490–494.

Hanks, T. C., Bucknam, R. C., Lajoie, K. R., & Wallace, R. E. (1984). Modification of wave-cut and faulting-controlled landforms. *Journal of Geophysical Research: Solid Earth, 89*(B7), 5771–5790.

Hanks, T. C., & Kanamori, H. (1979). A moment magnitude scale. *Journal of Geophysical Research: Solid Earth, 84*(B5), 2348–2350.

Helms, J. G., McGill, S. F., & Rockwell, T. K. (2003). Calibrated, late Quaternary age indices using clast rubification and soil development on alluvial surfaces in Pilot Knob Valley, Mojave Desert, southeastern California. *Quaternary Research, 60*(3), 377–393.

Herbert, J. W., Cooke, M. L., Oskin, M., & Difo, O. (2014). How much can off-fault deformation contribute to the slip rate discrepancy within the eastern California shear zone? *Geology, 42*(1), 71–75.

Hergarten, S. (2024). A simple model for faceted topographies at normal faults based on an extended stream-power law. *Earth Surface Dynamics, 12*(6), 1315–1327.

Hilley, G. E., & Arrowsmith, J. R. (2008). Geomorphic response to uplift along the Dragon's Back pressure ridge, Carrizo Plain, California. *Geology, 36*(5), 367–370.

Hilley, G. E., DeLong, S., Prentice, C., Blisniuk, K., & Arrowsmith, J. R. (2010). Morphologic dating of fault scarps using airborne laser swath mapping (ALSM) data. *Geophysical Research Letters, 37*(4).

Jones, L. M., Bernknopf, R. L., Cox, D. A., Goltz, J., Hudnut, K. W., Mileti, D. S., Perry, S., Ponti, D. J., Porter, K. A., Reichle, M. S., & Seligson, H. A. (2008). *The ShakeOut scenario: Effects of a potential M7. 8 Earthquake on the San Andreas fault in southern California.* US Geological Survey. https://doi.org/10.3133/ofr20081150. Open-File Report 2008-1150.

Klinger, Y., Avouac, J. P., Dorbath, L., Karaki, N. A., & Tisnerat, N. (2000). Seismic behaviour of the Dead Sea fault along Araba valley, Jordan. *Geophysical Journal International, 142*(3), 769–782.

Klinger, Y., Etchebes, M., Tapponnier, P., & Narteau, C. (2011). Characteristic slip for five great earthquakes along the Fuyun fault in China. *Nature Geoscience, 4*(6), 389–392.

Klinger, Y., Xu, X., Tapponnier, P., Van der Woerd, J., Lasserre, C., & King, G. (2005). High-resolution satellite imagery mapping of the surface rupture and slip distribution of the M w~ 7.8, 14 November 2001 Kokoxili earthquake, Kunlun fault, northern Tibet, China. *Bulletin of the Seismological Society of America, 95*(5), 1970–1987.

Lienkaemper, J. J. (2001). 1857 slip on the San Andreas fault southeast of Cholame, California. *Bulletin of the Seismological Society of America, 91*(6), 1659–1672.

Lin, Z., Liu-Zeng, J., Weldon, R. J., Tian, J., Ding, C., & Du, Y. (2020). Modeling repeated coseismic slip to identify and characterize individual earthquakes from geomorphic offsets on strike-slip faults. *Earth and Planetary Science Letters, 545*, 11613. https://doi.org/10.1016/j.epsl.2020.116313.

Lindvall, S. C., Rockwell, T. K., & Hudnut, K. W. (1989). Evidence for prehistoric earthquakes on the Superstition Hills fault from offset geomorphic features. *Bulletin of the Seismological Society of America, 79*(2), 342–361.

Liu, J., Klinger, Y., Sieh, K., & Rubin, C. (2004). Six similar sequential ruptures of the San Andreas fault, Carrizo Plain, California. *Geology, 32*(8), 649–652.

Liu-Zeng, J., Klinger, Y., Sieh, K., Rubin, C., & Seitz, G. (2006). Serial ruptures of the San Andreas fault, Carrizo Plain, California, revealed by three-dimensional excavations. *Journal of Geophysical Research: Solid Earth, 111*(B2).

Madden, C., Haddad, D. E., Salisbury, J. B., Zielke, O., Arrowsmith, J. R., Weldon, R., II, & Colunga, J. (2011). Appendix R—Compilation of slip-in-the-last-event data and analysis of last event, repeated slip, and average displacement for recent and prehistoric ruptures. In E. H. Field, G. P. Biasi, P. Bird, T. E. Dawson, K. R. Felzer, D. D. Jackson, K. M. Johnson, T. H. Jordan, C. Madden, A. J. Michael, K. R. Milner, M. T. Page, T. Parsons, P. M. Powers, B. E. Shaw, W. R. Thatcher, R. J. Weldon, II, & Y. Zeng (Eds.), *Uniform California earthquake rupture forecast, version 3 (UCERF3)—The time-independent model* (U.S. Geological Survey Open-File Report 2013–1165; California Geological Survey Special Report 228; Southern California Earthquake Center Publication 1792). U.S. Geological Survey. https://pubs.usgs.gov/of/2013/1165/.

Manighetti, I., Mercier, A., & De Barros, L. (2021). Fault trace corrugation and segmentation as a measure of fault structural maturity. *Geophysical Research Letters, 48*(20), e2021GL095372.

Mann, P. (2007). Global catalogue, classification and tectonic origins of restraining-and releasing bends on active and ancient strike-slip fault systems. *Geological Society, London, Special Publications, 290*(1), 13–142. (emailed librarian to see if this is accessible or whether I should order on ILL) available from Google Books (as an e-book).

McCaffrey, R. (2005). Block kinematics of the Pacific–North America plate boundary in the southwestern United States from inversion of GPS, seismological, and geologic data. *Journal of Geophysical Research: Solid Earth, 110*(B7).

McGill, S., & Sieh, K. (1993). Holocene slip rate of the central Garlock fault in southeastern Searles Valley, California. *Journal of Geophysical Research: Solid Earth, 98*(B8), 14217–14231.

McGill, S. F., & Rubin, C. M. (1999). Surficial slip distribution on the central Emerson fault during the June 28, 1992, Landers earthquake, California. *Journal of Geophysical Research: Solid Earth, 104*(B3), 4811–4833.

McGill, S. F., & Sieh, K. (1991). Surficial offsets on the central and eastern Garlock fault associated with prehistoric earthquakes. *Journal of Geophysical Research: Solid.*

McPhillips, D., & Scharer, K. (2023). Spatial and temporal analysis of geologic slip rates, Cucamonga fault, California, USA: Implications for along-strike applications and multi-fault rupture. *Journal of Geophysical Research: Solid Earth, 128*(3). p. e2022JB024942.

McPhillips, D., & Scharer, K. M. (2018). Quantifying uncertainty in cumulative surface slip along the Cucamonga Fault, a crustal thrust fault in southern California. *Journal of Geophysical Research: Solid Earth, 123*(10), 9063–9083.

Metropolitan Water District of Southern California. (2025). Metropolitan earns national award for new earthquake-resistant pipeline. https://www.mwdh2o.com/press-releases/metropolitan-earns-national-award-for-new-earthquake-resistant-pipeline/.

Milliner, C. W., Dolan, J. F., Hollingsworth, J., Leprince, S., Ayoub, F., & Sammis, C. G. (2015). Quantifying near-field and off-fault deformation patterns of the 1992 Mw 7.3 Landers earthquake. *Geochemistry, Geophysics, Geosystems, 16*(5), 1577–1598.

Mizoguchi, K., Uehara, S., & Ueta, K. (2012). Surface fault ruptures and slip distributions of the M_w 6.6 11 April Hamadori, Fukushima Prefecture, Northeast Japan, Earthquake. *Bull. Seismol. Soc. Am. 102*(1), 949–1,956.

Nash, D. B. (1986). Morphologic dating and degradation of fault scarps. In *Active tectonics, panel on active tectonics*. National Academy of Science, 181–194.

Nash, D. B., & Beaujon, J. S. (2006). Modeling degradation of terrace scarps in Grand Teton National Park, USA. *Geomorphology, 75*(3-4), 400–407.

Nissen, E., Maruyama, T., Arrowsmith, J. R., Elliott, J. R., Krishnan, A. K., Oskin, M. E., & Saripalli, S. (2014). Coseismic fault zone deformation revealed with differential lidar: Examples from Japanese Mw ~7 intraplate earthquakes. *Earth and Planetary Science Letters, 405*, 244–256. https://doi.org/10.1016/j.epsl.2014.08.031.

Oskin, M. E., Arrowsmith, J. R., Corona, A. H., Elliott, A. J., Fletcher, J. M., Fielding, E. J., Gold, P. O., Garcia, J. J. G., Hudnut, K. W., Liu-Zeng, J., & Teran, O. J. (2012). Near-field deformation from the El Mayor–Cucapah earthquake revealed by differential LIDAR. *Science, 335*(6069), 702–705.

Ouchi, S. (2005). Development of offset channels across the San Andreas Fault. *Geomorphology, 70*(1–2), 112–128. https://doi.org/10.1016/j.geomorph.2005.04.004, https://doi-org.libproxy.lib.csusb.edu/.

Palumbo, L., Benedetti, L., Bourlès, D., Cinque, A., & Finkel, R. (2004). Slip history of the Magnola fault (Apennines, Central Italy) from 36Cl surface exposure dating: evidence for strong earthquakes over the Holocene. *Earth and Planetary Science Letters, 225*(1-2), 163–176.

Peltzer, G., Crampé, F., Hensley, S., & Rosen, P. (2001). Transient strain accumulation and fault interaction in the Eastern California shear zone. *Geology, 29*(11), 975–978.

Petit, C., Gunnell, Y., Gonga-Saholiariliva, N., Meyer, B., & Seguinot, J. (2009). Faceted spurs at normal fault scarps: Insights from numerical modeling. *Journal of Geophysical Research, 114*. https://doi.org/10.1029/2008JB005955. B05403.

Philip, H., Rogozhin, E., Cisternas, A., Bousquet, J. C., Borisov, B., & Karakhanian, A. (1992). The Armenian earthquake of 1988 December 7; faulting and folding, neotectonics and palaeoseismicity. *Geophysical Journal International, 110*(1), 141–158.

Quigley, M., Van Dissen, R., Litchfield, N., Villamor, P., Duffy, B., Barrell, D., Furlong, K., Stahl, T., Bilderback, E., & Noble, D. (2012). Surface rupture during the 2010 Mw 7.1 Darfield (Canterbury) earthquake: Implications for fault rupture dynamics and seismic-hazard analysis. *Geology, 40*(1), 55–58.

Reitman, N. G., Klinger, Y., Briggs, R. W., & Gold, R. D. (2023). Climatic influence on the expression of strike-slip faulting. *Geology, 51*(1), 18–22.

Reitman, N. G., Klinger, Y., Briggs, R. W., & Gold, R. D. (2024). Limited preservation of strike-slip surface displacement in the geomorphic record. *Journal of Geophysical Research: Solid Earth, 129*(11). p. e2024JB028692.

Reitman, N. G., Mueller, K. J., & Tucker, G. E. (2022). Surface slip variability on strike-slip faults. *Earth Surface Processes and Landforms, 47*(4), 908–935.

Reitman, N. G., Mueller, K. J., Tucker, G. E., Gold, R. D., Briggs, R. W., & Barnhart, K. R. (2019). Offset channels may not accurately record strike-slip fault displacement: Evidence from landscape evolution models. *Journal of Geophysical Research: Solid Earth, 124*(12), 13427–13451.

Ren, Z., Zhang, Z., Chen, T., Yan, S., Yin, J., Zhang, P., et al. (2016). *Clustering of offsets on the Haiyuan.*

Riedel, W. (1929). *Zur Mechanik geologischer Brucherscheinungen: Centralbl. f. Mineral. Geol. u. Pal. 1929*(B), 354–368.

Rittase, W. M., Kirby, E., McDonald, E., Walker, J. D., Gosse, J., Spencer, J. Q., & Herrs, A. J. (2014). Temporal variations in Holocene slip rate along the central Garlock fault, Pilot Knob Valley, California. *Lithosphere, 6*(1), 48–58.

Rizza, M., Ritz, J. F., Braucher, R., Vassallo, R., Prentice, C., Mahan, S., McGill, S., Chauvet, A., Marco, S., Todbileg, M., & Demberel, S. (2011). Slip rate and slip magnitudes of past earthquakes along the Bogd left-lateral strike-slip fault (Mongolia). *Geophysical Journal International, 186*(3), 897–927.

Rockwell, T. K., & Klinger, Y. (2013). Surface rupture and slip distribution of the 1940 Imperial Valley earthquake, Imperial fault, southern California: Implications for rupture segmentation and dynamics. *Bulletin of the Seismological Society of America, 103*(2A), 629–640.

Rockwell, T. K., Lindvall, S., Dawson, T., Langridge, R., Lettis, W., & Klinger, Y. (2002). Lateral offsets on surveyed cultural features resulting from the 1999 Izmit and Duzce earthquakes, Turkey. *Bulletin of the Seismological Society of America, 92*(1), 79–94.

Rockwell, T. K., Pinault, C. T., & Ehlig, P. (1986). Holocene slip events on the southern Elsinore fault, Coyote Mountains, southern California. *Guidebook and Volume on Neotectonics and Faulting in Southern California,* 193–196.

Salisbury, J. B., Arrowsmith, J. R., Brown, N., Rockwell, T., Akciz, S., & Ludwig, L. G. (2018). The age and origin of small offsets at Van Matre Ranch along the San Andreas fault in the Carrizo Plain, California. *Bulletin of the Seismological Society of America, 108*(2), 639–653.

Salisbury, J. B., Rockwell, T. K., Middleton, T. J., & Hudnut, K. W. (2012). LiDAR and field observations of slip distribution for the most recent surface ruptures along the central San Jacinto fault. *Bulletin of the Seismological Society of America, 102*(2), 598–619.

Salyards, S. L. (1985). Patterns of offset associated with the 1983 Borah Peak, Idaho, earthquake and previous events. In Proceedings of Workshop XXVIII on the Borah Peak, Idaho, earthquake: US Geological Survey Open-File Report (pp. 85-290).

Schwartz, D. P., & Coppersmith, K. J. (1984). Fault behavior and characteristic earthquakes: Examples from the Wasatch and San Andreas fault zones. *Journal of Geophysical Research: Solid Earth, 89*(B7), 5681–5698.

Schwartz, D. P., Haeussler, P. J., Seitz, G. G., & Dawson, T. E. (2012). Why the 2002 Denali fault rupture propagated onto the Totschunda fault: implications for fault branching and seismic hazards. *Journal of Geophysical Research, 117*. https://doi.org/10.1029/2011JB008918. B11304.

Shaw, B. E. (2009). Constant stress drop from small to great earthquakes in magnitude-area scaling. *Bulletin of the Seismological Society of America, 99*, 871. https://doi.org/10.1785/0120080006.

Shaw, B. E. (2013). Earthquake surface slip-length data is fit by constant stress drop and is useful for seismic hazard analysis. *Bulletin of the Seismological Society of America, 103*, 876. https://doi.org/10.1785/0120110258.

Sieh, K. E. (1978). Slip along the San Andreas fault associated with the great 1857 earthquake. *Bulletin of the Seismological Society of America, 68*(5), 1421–1448.

Sieh, K. E. (1996). The repetition of large-earthquake ruptures. In: *Proceedings of the National Academy of Sciences, 93*(3), 764–3,771.

Sieh, K. E., & Jahns, R. H. (1984). Holocene activity of the San Andreas fault at Wallace creek, California. *Geological Society of America bulletin, 95*(8), 883–896.

Sieh, K. E. (1981). In D. W. Simpson, & P. G. Richards (Vol. Eds.), *Earthquake prediction, an international review, Maurice Ewing Set: vol. 4*, (pp. 181–207). Washington, D.C.: AGU.

Stirling, M., Fitzgerald, M., Shaw, B., & Ross, C. (2024). New magnitude–Area scaling relations for the New Zealand national seismic hazard model 2022. *Bulletin of the Seismological Society of America, 114*(1), 137–149.

Tchalenko, J. S. (1970). Similarities between shear zones of different magnitudes. *Geological Society of America Bulletin, 81*(6), 1625–1640.

Toda, S., & Tsutsumi, H. (2013). Spontaneous reactivation of two, subparallel, inland normal faults during the Mw 6.6 11 April 2011 Iwaki Earthquake triggered by the Mw 9.0 Tohoku-oki, Japan, Earthquake. *Bulletin of the Seismological Society of America, 103*(2B), 1,584–1,602.

Tortorici, L., Monaco, C., Tansi, C., & Cocina, O. (1995). Recent and active tectonics in the Calabrian arc (Southern Italy). *Tectonophysics, 243*(1-2), 37–55.

Treiman, J. A., Kendrick, K. J., Bryant, W. A., Rockwell, T. K., & McGill, S. F. (2002). Primary surface rupture associated with the M w 7.1 16 October 1999 Hector mine earthquake, San Bernardino County, California. *Bulletin of the Seismological Society of America, 92*(4), 1171–1191.

Tucker, G. E., Hobley, D. E., McCoy, S. W., & Struble, W. T. (2020). Modeling the shape and evolution of Normal-fault facets. *Journal of Geophysical Research: Earth Surface, 125*(3), p. e2019JF005305.

Tucker, G. E., McCoy, S. W., Whittaker, A. C., Roberts, G. P., Lancaster, S. T., & Phillips, R. (2011). Geomorphic significance of postglacial bedrock scarps on normal-fault footwalls. *Journal of Geophysical Research: Earth Surface, 116*(F1).

Vallage, A., Klinger, Y., Grandin, R., Bhat, H. S., & Pierrot-Deseilligny, M. (2015). Inelastic surface deformation during the 2013 Mw 7.7 Balochistan, Pakistan, earthquake. *Geology, 43*(12), 1079–1082. https://doi.org/10.1130/G37290.1.

Wallace, R. E. (1977). Profiles and ages of young fault scarps, north-central Nevada. *Geological Society of America bulletin, 88*(9), 1267–1281. https://doi.org/10.1130/0016-7606(1977)88<1267:PAAOYF>2.0.CO;2.

Wallace, R. E. (1978). Geometry and rates of change of fault-generated range fronts, north-central Nevada. *Journal of Research of the U.S. Geological Survey, 6*(5), 637–650.

Wallace, R. E., Dickinson, W. R., & Grantz, A. (1968). Notes on stream channels offset by the San Andreas fault, southern Coast Ranges, California. In *Conference on geologic problems of the San Andreas fault system, Vol. 11, Stanford University Publication in Geological Sciences*, 6–21.

Wells, D. L., & Coppersmith, K. J. (1994). New empirical relationships among magnitude, rupture length, rupture width, rupture area, and surface displacement. *Bulletin of the Seismological Society of America, 84*(4), 974–1002.

Wesnousky, S. G. (2006). Predicting the endpoints of earthquake ruptures. *Nature, 444*(7117), 358–360.

Working Group on California Earthquake Probabilities. (2003). Earthquake probabilities in the San Francisco Bay region; 2002–2031: U.S. Geological Survey Open-File Report 2003–214.

Yang, H., Huang, X., Cunningham, D., Zhang, H., Yang, X., & Hu, Z. (2022). Lithological and structural anatomy of an unusually well exposed strike-slip fault pressure ridge in the Nanjieshan, western China. *Journal of Structural Geology, 165*, 104757.

Yu, G., Xu, X., Klinger, Y., Diao, G., Chen, G., Feng, X., Li, C., Zhu, A., Yuan, R., Guo, T., & Sun, X. (2010). Fault-scarp features and cascading-rupture model for the M w 7.9 Wenchuan earthquake, eastern Tibetan plateau, China. *Bulletin of the Seismological Society of America, 100*(5B), 2590–2614.

Zachariasen, J., & Sieh, K. (1995). The transfer of slip between two en echelon strike-slip faults: A case study from the 1992 Landers earthquake, southern California. *Journal of Geophysical Research: Solid Earth, 100*(B8), 15281–15301.

Zielke, O., & Arrowsmith, J. R. (2012). LaDiCaoz and LiDARimager—MATLAB GUIs for LiDAR data handling and lateral displacement measurement. *Geosphere, 8*(1), 206–221.

Zielke, O., Arrowsmith, J. R., Grant Ludwig, L., & Akciz, S. O. (2012). High-resolution topography-derived offsets along the 1857 Fort Tejon earthquake rupture trace, San Andreas fault. *Bulletin of the Seismological Society of America, 102*(3), 1135–1154.

Zielke, O., Arrowsmith, J. R., Grant Ludwig, L., & Akçiz, S. O. (2010). Slip in the 1857 and earlier large earthquakes along the Carrizo Plain, San Andreas fault. *Science, 327*(5969), 1119–1122.

Zielke, O., Klinger, Y., & Arrowsmith, J. R. (2015). Fault slip and earthquake recurrence along strike-slip faults—Contributions of high-resolution geomorphic data. *Tectonophysics, 638*, 43–62.

Zinke, R., Dolan, J. F., Rhodes, E. J., Van Dissen, R., & McGuire, C. P. (2017). Highly variable latest Pleistocene-Holocene incremental slip rates on the Awatere fault at Saxton River, South Island, New Zealand, revealed by Lidar mapping and luminescence dating. *Geophysical Research Letters, 44*(22), 11–301.

Zinke, R., Hollingsworth, J., & Dolan, J. F. (2014). Surface slip and off-fault deformation patterns in the 2013 MW 7.7 Balochistan, Pakistan earthquake: Implications for controls on the distribution of near-surface coseismic slip. *Geochemistry, Geophysics, Geosystems, 15*(12), 5034–5050.

Zinke, R., Hollingsworth, J., Dolan, J. F., & Van Dissen, R. (2019). Three-dimensional surface deformation in the 2016 MW 7.8 Kaikōura, New Zealand, earthquake from optical image correlation: Implications for strain localization and long-term evolution of the Pacific-Australian plate boundary. *Geochemistry, Geophysics, Geosystems, 20*(3), 1609–1628.

Zhu, M., Guangzeng, W., Sanzhong, L., Yongjiang, L., Pengcheng, W., Lingli, G., Li, Z., Xingpeng, C., & Taihai, S. (2024). An overview of structures associated with bends of strike-slip faults: Focus on analogue and numerical models. *Marine and Petroleum Geology, 167*, 106983. https://doi.org/10.1016/j.marpetgeo.2024.106983. ISSN 0264-8172.

CHAPTER 13

Geomorphic impacts of the 1999 Chi-Chi earthquake, Taiwan

Kristen L. Cook
IRD, ISTerre, Universite Grenoble Alpes, Grenoble, France

13.1 Introduction

The M_w 7.6 Chi-Chi earthquake, also known in Taiwan as the 921 earthquake, struck western Taiwan on 21 September 1999. The earthquake resulted in 2415 fatalities and 300 billion NTD of economic damage, making it one of the most significant earthquakes in Taiwan's history and the largest of the 20th century in Taiwan. At the time, the Chi-Chi earthquake was also one of the best seismically documented major earthquakes in the world, due to the dense seismic and GPS networks operating in Taiwan. The geomorphic impacts of the earthquake were also extensively documented and the findings from this event helped drive the future study of earthquake impacts on surface processes. As a result, the Chi-Chi earthquake played a key role in the development of our understanding of major earthquakes and their impacts.

The Chi-Chi earthquake was notable for the widespread and varied geomorphic impacts that it triggered. These included extensive coseismic and postseismic landsliding, debris flows, high sediment fluxes and fluvial aggradation, hydrological changes, surface deformation and fault and fold scarp formation and fluvial uplift and incision. The study of the geomorphic impacts was facilitated by improved satellite imagery capabilities, as well as the regular collection of aerial photographs by the Agricultural and Forestry Aerial Survey Institute of Taiwan, which provided extensive coverage of very high-resolution pre- and postearthquake images. Taiwan also had an extensive hydrometric network operated by the Water Resources Agency (WRA), which provided data on river discharge, suspended-sediment concentration, precipitation and groundwater levels.

While many of the geomorphic phenomena observed during and after the Chi-Chi earthquake had been noted during previous earthquakes, the Chi-Chi earthquake provided an unprecedented dataset that allowed linkages between the different impacts to be observed and quantified, providing a uniquely comprehensive view of the geomorphic impacts of a large earthquake and a model for the study of future events.

13.1.1 Introduction to Taiwan

Taiwan is a tectonically highly active region located at a complicated subduction plate boundary, where the Philippine Sea plate is being subducted underneath the Eurasian plate

Geomorphic Impacts of Earthquakes
DOI: https://doi.org/10.1016/B978-0-443-40375-0.00013-X

to the north, while the Eurasian plate is being subducted underneath the Philippine Sea plate to the east (Fig. 13.1). The uplift and formation of the island of Taiwan have resulted from the oblique collision between the Eurasian continental crust and the Luzon Volcanic Arc, which sits on the Philippine Sea plate (Suppe, 1981). This ongoing collision, with a convergence rate of 8–9 cm/year (Sella et al., 2002; Yu et al., 1997), results in very high uplift rates of up to 4–14 mm/year in the central part of Taiwan, with geodetic rates substantially higher than geologic rates (Hsu et al., 2018; Lee et al., 2006). Taiwan has very high seismic activity, with earthquakes both on the subduction margins and on shallow structures that accommodate the ongoing shortening and deformation (Shin & Teng, 2001). Active faults are found on both sides of the island, with strike-slip faulting east of the Central Range accommodating oblique convergence and thrust faulting to the west of the Central and Hsueshan ranges accommodating ~45%–50% of the total convergence rate (Simoes & Avouac, 2006; Yu et al., 1997). In the northeast of the island, normal faults are accommodating backarc extension related to the northward subduction of the Philippine Sea plate (Suppe, 1981; Yu et al., 1997).

High uplift rates in Taiwan are balanced by high erosion rates, with cosmogenic basin-wide erosion rates of up to 2–4 mm/year reported in the Central and Hsueshan ranges (Derrieux et al., 2014) and suspended-sediment–derived erosion rates of 3–6 mm/year (Dadson et al., 2003). Erosion rates are generally higher on the eastern side of the Central Range and lower on the western side of the Central and Hsueshan ranges (Dadson et al., 2003; Derrieux et al., 2014).

Taiwan has a high mean annual precipitation throughout the island, ranging from 2154 mm in central Taiwan to 2932 mm in northern Taiwan over the period 1949–2009 (Water Resources Agency, 2025). On average, 78% of the total precipitation falls from May to October. Much of this rain falls during typhoon storms, which frequently hit Taiwan and can bring extremely high amounts of precipitation (up to metres per day). In addition to typhoons, Taiwan experiences convective precipitation events during the summer and *mei-yu* (plum rains, or spring monsoon) in the spring. Individual storm events in Taiwan are strongly influenced by topography, with the Central Range forming a major orographic barrier. As a result, the spatial distribution of typhoon rainfall depends strongly on the typhoon track and rainfall amounts can vary substantially over short distances.

Many place and feature names in Taiwan have multiple potential spellings arising from the usage of different romanisation systems to render the Chinese names. For names that appear in the literature under different variants, I provide alternative names in parentheses.

13.1.2 Introduction to the Chi-Chi earthquake

The Chi-Chi earthquake occurred on the Chelungpu Fault, one of the frontal faults of the Western foothills. The Chelungpu Fault is a shallow thrust fault rooted in the

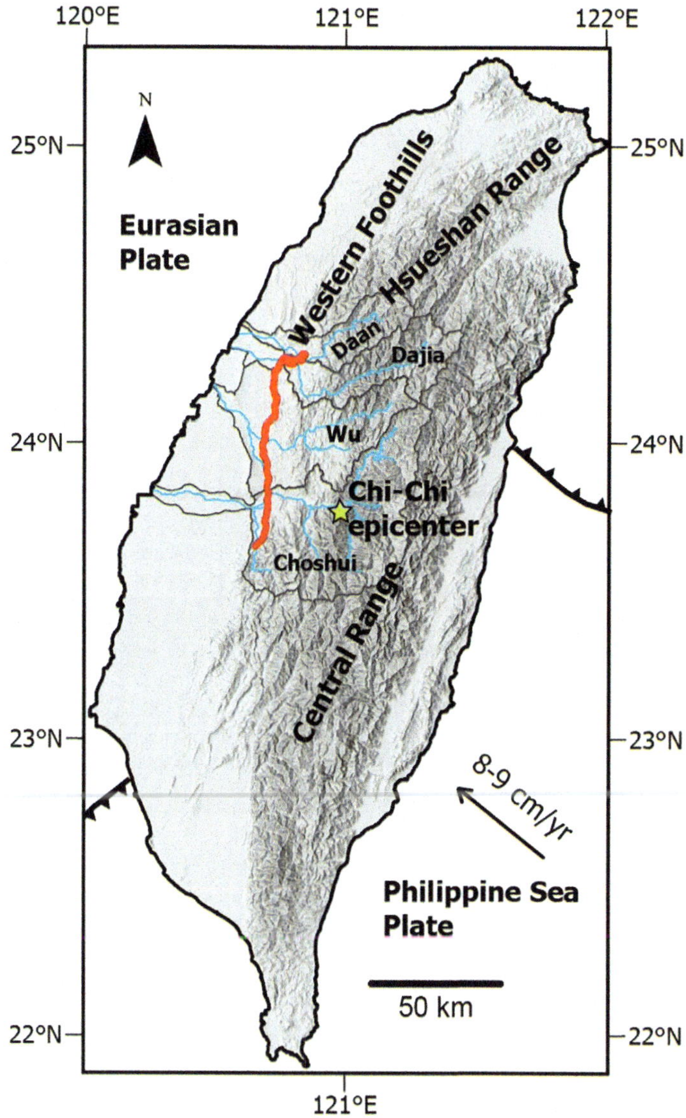

Figure 13.1 Overview of Taiwan. Black lines with triangles show the subduction zones, red line shows the Chi-Chi surface rupture, yellow star shows the earthquake epicentre, blue lines and grey outlines show the major rivers and their catchments in the earthquake area. The black arrow shows the velocity vector of the Philippine Sea plate relative to the Eurasian plate. Map lines delineate study areas and do not necessarily depict accepted national boundaries.

Pliocene Chinshui Shale (Fig. 13.2) and has accumulated ~14 km of total slip since 1 Ma (Yue et al., 2005). The fault has a ramp and flat geometry, with the base of the ramp at 5–6 km depth and a ramp angle that is generally ~40–55 degrees near the surface and

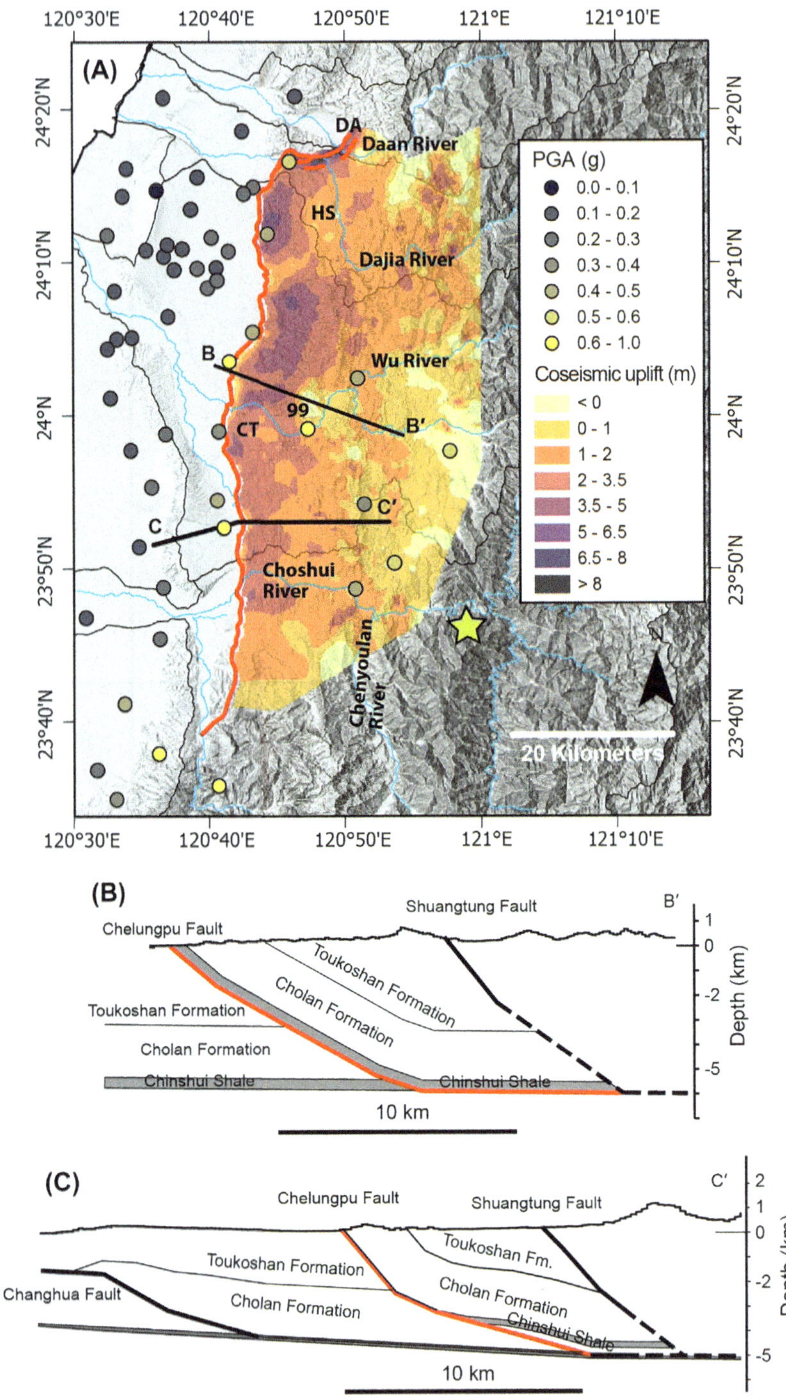
(A)
120°30'E
120°40'E
120°50'E
121°E
121°10'E
24°20'N
24°10'N
24°N
23°50'N
23°40'N
DA
Daan River
HS
Dajia River
B
Wu River
99
B'
CT
C
C'
Choshui River
Chenyoulan River
20 Kilometers
N
PGA (g)
0.0 - 0.1
0.1 - 0.2
0.2 - 0.3
0.3 - 0.4
0.4 - 0.5
0.5 - 0.6
0.6 - 1.0
Coseismic uplift (m)
< 0
0 - 1
1 - 2
2 - 3.5
3.5 - 5
5 - 6.5
6.5 - 8
> 8
(B)
Shuangtung Fault
B'
Chelungpu Fault
Toukoshan Formation
Cholan Formation
Toukoshan Formation
Cholan Formation
Chinshui Shale
Chinshui Shale
10 km
Depth (km)
1
0
-2
-5
(C)
Chelungpu Fault
Shuangtung Fault
C'
Toukoshan Fm.
Toukoshan Formation
Changhua Fault
Cholan Formation
Cholan Formation
Chinshui Shale
10 km
Depth (km)
2
1
0
-2
-5

15–25 degrees from ~3 to 5–6 km depth (Yue et al., 2005). The fault structure becomes significantly more complex at the northern end of the Chi-Chi rupture (Lee et al., 2005; Yue et al., 2005), which is reflected in the surface deformation (see below).

The earthquake had an epicentre at 8 km depth, near the town of Chi-Chi in Nantou County and the rupture propagated to the north in a complicated fashion with irregular ruptures of asperities (Chi et al., 2001; Shin & Teng, 2001). The shallow depth and large magnitude of the earthquake resulted in high-intensity ground shaking. The earthquake was recorded by a network of 650 strong-motion seismic stations comprising the Taiwan Strong-Motion Instrumentation Program, which was installed by the Central Weather Bureau (CWB) in 1997, in addition to the 75 6-component (short-period and broadband) stations of the CWB seismic network (Shin & Teng, 2001), allowing detailed reconstructions of the earthquake dynamics. The maximum values of peak ground acceleration (PGA) were greater than 1 g (980 gal), observed at the southern end of the rupture on the hanging wall east of the fault (Fig. 13.2). The PGA decreased to the north, with acceleration at the northern end of the rupture zone reaching nearly 0.5 g (490 gal) (Shin & Teng, 2001). In contrast, the coseismic peak ground velocity (PGV) and peak seismic ground displacement increased from south to north, up to maximum values of 2.92 m/s and 8.7 m, respectively (Shin & Teng, 2001; Wang et al., 2002). PGA decreased rapidly with distance from the fault, falling below 0.1 g (98 gal) at a distance of 55 km from the fault (Wang et al., 2002). PGA values were systematically higher in the east-west direction than in the north-south direction, particularly near the fault, indicating increased shaking normal to the fault strike (Wang et al., 2002).

13.2 Coseismic uplift and surface ruptures

The ~100 km long Chi-Chi surface rupture was a significant geomorphic feature (Fig. 13.3), with vertical displacements ranging from 2 m at the southern end to more

Figure 13.2 (A) Map of the earthquake-affected area. Red line shows the surface rupture, yellow star shows the epicentre, colours in the hanging wall show coseismic surface uplift based on InSAR and GPS from Chuang et al. (2020) and circles show maximum peak ground acceleration (PGA) values. *DA*, Dongshi Anticline; *HS*, Hsinshe fold scarps; *CT*, Caotun fold scarp. (B) and (C) Cross sections showing the ramp-flat structure of the Chelungpu Fault, modified from Yue et al. (2005). Red lines show the Chelungpu Fault. Map lines delineate study areas and do not necessarily depict accepted national boundaries. Source: *A, Obtained from Lee, S. J., Chen, H. W., & Ma, K. F. (2007). Strong ground motion simulation of the 1999 Chi-Chi, Taiwan earthquake from a realistic three-dimensional source and crustal structure.* Journal of Geophysical Research: Solid Earth, 112*(B6); B and C, simplified from Fig. 13.5 of Yue, L.-F., Suppe, J., & Hung, J.-H. (2005). Structural geology of a classic thrust belt earthquake: The 1999 Chi-Chi earthquake Taiwan (Mw = 7.6),* Journal of Structural Geology, 27*(11), 2058–2083.*

than 10 m at the northern end (Chen et al., 2001; Yu et al., 2001). At the northern end of the rupture zone, the Chelungpu Fault did not reach the surface, but slip at a depth of 550–600 m caused coseismic folding and substantial surface deformation, with vertical deformation of up to 10 m in the Dongshi (Tungshih) anticline (Fig. 13.2A; Lee et al., 2005). Horizontal surface displacement, obtained from subpixel correlation of SPOT images, also increased from south to north, with 5–6 m of displacement in the epicentral area and 10–11 m at the northern end of the rupture zone (Dominguez et al., 2003), matching the 2.4–10 m offsets obtained from GPS data (Yu et al., 2001). The surface rupture passed through densely populated areas, greatly contributing to the human and economic impacts of the earthquake.

One of the most notable sites of the surface rupture was the deformation of the running track at the Kuang-Fu Junior High School in Taichung County (Fig. 13.3). The track and the destroyed school buildings have been preserved as part of the 921 Earthquake Museum. Another dramatic image of surface deformation was from near the northern end of the rupture, where the earthquake created a 6 m high waterfall in the Dajia River and destroyed the Beifeng (Pifung) bridge (Fig. 13.3).

In addition to the uplift along the surface rupture, surface deformation occurred above the fault ramp, in the hanging wall of the Chelungpu Fault (Chen et al., 2007; Lai et al., 2006). Fold scarps formed by the Chi-Chi earthquake have been described from Hsinshe (Xinshe) and Caotun (Tsaotun), located about 6 and 2 km east of the surface rupture, respectively. Both of these sites contained preexisting scarps that had been previously interpreted as fault scarps (Chen et al., 2007; Lai et al., 2006; Lee et al., 2005).

At Caotun, folding over multiple earthquakes has created a broad zone of tilting ~40 m across and the Chi-Chi fold scarp was less than 1 m high (Lee et al., 2005; Lee et al., 2010). At Hsinshe, the coseismic fold scarp accommodated ~2 m of differential uplift, based on GPS data from both sides of the fold (Lai et al., 2006). The maximum accumulated fold scarp height at Hsinshe is 193 m, on a terrace dated to 30 ka (Le Béon et al., 2014). Analysis of the preserved fold scarps on Hsinshe terraces ranging from 9 to 30 ka allowed the estimation of the magnitude and direction of the cumulative slip on the fault ramp; cumulative slip over the past 30 ky was 267 ± 58 m with an azimuth of 338 ± 6 degrees, indistinguishable from the azimuth of the coseismic slip vectors obtained from local GPS stations (Le Béon et al., 2014). This suggests that the kinematics of the fault have remained consistent over that time period and that the Chi-Chi earthquake is representative of previous earthquakes on the Chelungpu Fault (Le Béon et al., 2014).

13.3 Coseismic landslides

The Chi-Chi earthquake was a major landslide-triggering event due to the combination of high PGA and PGV, steep topography, fault geometry, and relatively weak

Figure 13.3 Chi-Chi earthquake surface ruptures. (A) Photograph of the deformed running track of the Kuang-Fu Junior High School, now preserved as part of the 921 Earthquake Museum (B) Aerial photograph of the Dajia River fault scarp (marked with the solid line). (C) Photograph of the surface rupture in a pineapple field near Wufeng (photograph taken in 2008). (D) Coseismic offset of the Shigang Dam. The right (northern) section of the dam was uplifted by 1.99m relative to its preearthquake elevation, while the left (southern) part of the dam was uplifted by 12.32m.

lithologies. More than 20,000 landslides were triggered coseismically (Lin et al., 2004; Wang, 2000). A comprehensive landslide inventory was created by Taiwan's National Science and Technology Center for Disaster Prevention using 20 m resolution SPOT satellite imagery (Lin, Liao, et al., 2000), as well as by the Bureau of Soil and Water Conservation (Liao & Lee, 2000; Wang et al., 2002) (Fig. 13.4). Additional coseismic landslides were mapped using SPOT imagery combined with aerial photographs from the Agricultural and Forestry Aerial Survey Institute of Taiwan (Wang et al., 2002).

Coseismic landslides covered a total area of 117.6 km^2 (Liao & Lee, 2000); however, despite the comprehensive landslide mapping, it is difficult to confidently estimate the total volume of landslides triggered by the Chi-Chi earthquake. This is related to the

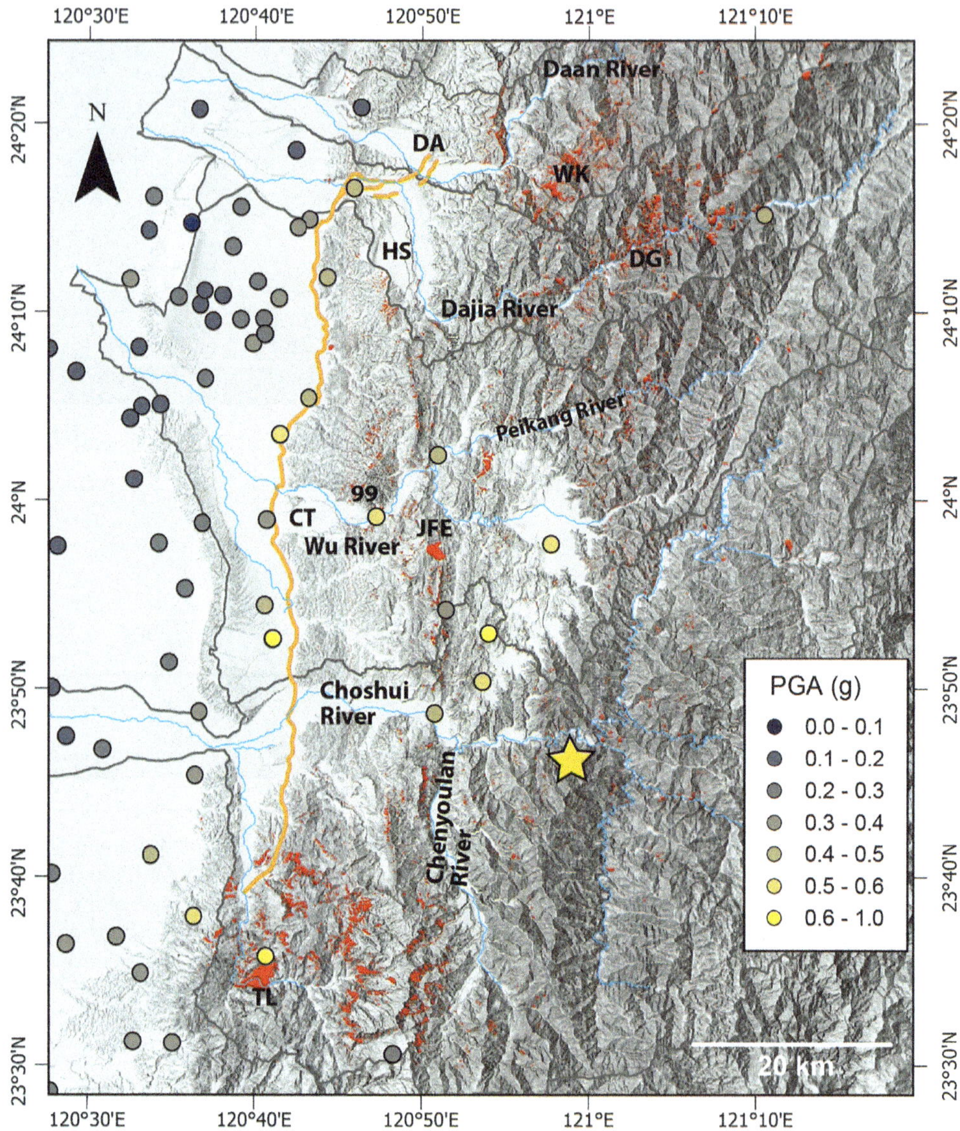

Figure 13.4 Map of coseismic landslides (shown in red) from the landslide inventory of Liao and Lee (2000), with peak ground acceleration (PGA) values obtained from Lee et al. (2007). *DG*, Dajia Gorge region; *WK*, Wushikeng; *99*, 99 Peaks area; *JFE*, Jiufengershan landslide; *TL*, Tsaoling landslide. The yellow star shows the epicentre, while the orange line shows the surface rupture. Catchment boundaries of the major rivers (labelled) are shown in dark grey. Map lines delineate study areas and do not necessarily depict accepted national boundaries.

problem of landslide polygon amalgamation noted by Marc and Hovius (2015), where the application of landslide area-volume scaling relationships can produce large errors if multiple failures are mapped as a single polygon. For inventories such as Chi-Chi, with extensive areas of small shallow slides, this is a particular concern (Marc & Hovius, 2015). Following correction of the largest 173 landslide polygons for amalgamation, Marc and Hovius (2015) proposed a volume of 0.45 km^3 for the inventory of Liao and Lee (2000).

The coseismic landslides show a clear spatial pattern, with landslide numbers and area density peaking in the hanging wall 10–15 km from the fault and decreasing to the east (Wang et al., 2002) (Figs. 13.4 and 13.5). This spatial distribution of landslides can be directly linked to the spatial pattern of ground acceleration during the earthquake (see Chapter 4). The detailed landslide inventories, combined with the geophysical observations of the Chi-Chi earthquake, enabled insights into the relationship between coseismic shaking and landslide triggering. Wang et al. (2002) looked at the relationship between PGA during the earthquake and landslide occurrence and found that the landslide numbers and area density peaked between 300 and 600 gal (0.3–0.6 g). They found that lithology also played a role, with Pliocene and Plio-Pleistocene sedimentary units having a higher landslide area density than older and more indurated units at similar distances to the fault (Wang et al., 2002) (Fig. 13.4). Meunier et al. (2007) found a close relationship between landslide density and PGA and identified critical values of PGA for landslide triggering of 100 gal for vertical acceleration and 170 gal for horizontal acceleration. This tight link between PGA and landsliding allowed them to develop a landslide density attenuation equation that reflects the regional attenuation of seismic waves (Meunier et al., 2007).

A number of studies have noted differences between the spatial characteristics of coseismic compared to rainfall-triggered landslides in the Chi-Chi earthquake rupture area, including their occurrence across a broader range of gradients and basin sizes, their tendency to originate higher in the landscape (Lin et al., 2004; Sepúlveda et al., 2005), and to vary with slope aspect, with more failures on south-facing slopes (Lin & Tung, 2004) (Fig. 13.5). Sepúlveda et al. (2005) suggested that topographic amplification of seismic waves was required to explain a cluster of rockslides in the steep Dajia gorge region (Fig. 13.4), based on a stability analysis and on the ratio between the height of hillslopes and the dominant wavelength of the propagating seismic waves, which showed that the failures occurred preferentially on slopes in a narrow range of this ratio. Meunier et al. (2008) took the question of topographic effects further and demonstrated that the earthquake-triggered landslides occupied a different portion of the landscape from rainfall-triggered landslides in Taiwan, with the coseismic landslides more likely to occur higher on the hillslopes near ridge crests, while typhoon-triggered landslides

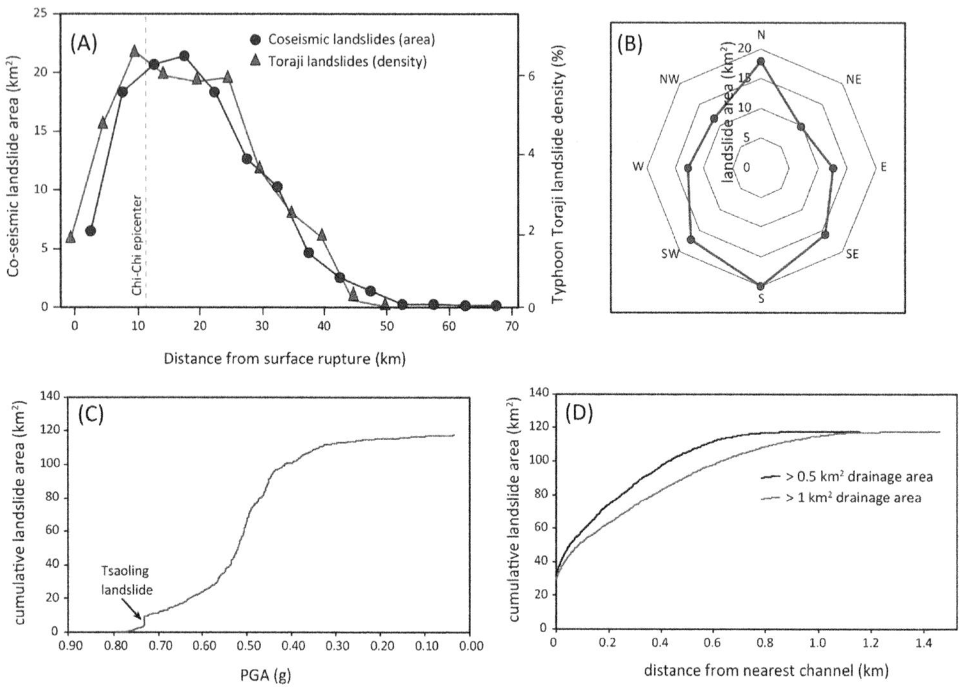

Figure 13.5 Landslide statistics, based on the landslide inventory of Liao and Lee (2000). (A) Coseismic landslide area (left y-axis) in 5 km bins with distance from the surface rupture and landslide density (right y-axis) from postearthquake Typhoon Toraji in a 50 km swath through the Choshui catchment that is perpendicular to the strike of the surface rupture (Hovius et al., 2011). (B) Coseismic landslide area by slope aspect. (C) Cumulative coseismic landslide area versus peak ground acceleration (PGA); note that the x-axis increases to the left. (D) Coseismic landslide connectivity to the fluvial network, measured as cumulative landslide area plotted as a function of distance from the nearest channel for drainage areas of more than 0.5 km^2 (black) and 1 km^2 (grey).

clustered near the base of the hillslopes. This can be attributed, on one hand, to the amplification of seismic waves by the topography, and on the other to the hydrological factors controlling the precipitation triggering of landslides (Meunier et al., 2008). The landslides also showed a strong asymmetry, with landslide densities five times higher on slopes dipping away from the epicentre than on slopes facing towards the epicentre, which is consistent with topographic amplification resulting in the highest shaking on ridge sides facing away from the incoming waves (Meunier et al., 2008).

The Chi-Chi earthquake-triggered landslides were predominantly shallow failures, rockfalls, and debris failures (see Chapter 7 for definitions). However, there were some notable large coseismic landslides which are described below (see Fig. 13.4 for locations).

Tsaoling landslide: The largest landslide triggered by the Chi-Chi earthquake was the Tsaoling (Caoling) landslide, a bedding plane dip-slope failure that involved a release of 124–125 x 10^6 m^3 of material and a deposit of 150 x 10^6 m^3 (with the volume increase inferred to be due to fragmentation) (Chen, & Chang, Angelier, et al., 2006; Wang et al., 2002). This event impacted a local village, resulting in 36 fatalities (Hung, 2000). The landslide dammed the Jingshui (Chingsui) River, creating a lake with a volume of ~45 x 10^6 m^3 (Hung, 2000). In response, an emergency spillway was constructed through the landslide deposit, and overflow of the dam began in December 1999 with no catastrophic breach. Erosion of the landslide deposit proceeded rapidly in the years following the earthquake, with estimates of 9.8 x 10^6 m^3 removed between 1999 and July 2000 and a further 30.7 x 10^6 m^3 removed between 2000 and 2003 (Chen, & Chang, Angelier, et al., 2006). This erosion was concentrated in a gorge up to 60 m deep, cut by the Jingshui River.

The Tsaoling region had experienced repeated previous large dip-slope failures, including a large river-blocking failure thought to be earthquake-triggered in 1862, a 100 x 10^6 m^3 rockslide triggered by an earthquake in 1941, and a larger postseismic rainfall-triggered failure in 1942. The 1862 and 1942 failures created landslide dams that broke 36 and 9 years later, respectively (Hung, 2000); the latter dam break failure killed 137 people. In 1979, two rainfall-triggered failures totalling 66 x 10^6 m^3 in volume created a short-lived landslide dam that burst after 9 days (Hung, 2000).

Jiufengershan landslide: The second-largest landslide during the Chi-Chi earthquake was another bedding plane failure that occurred at Jiufengershan (Chiu-fen-erh-shan, Juo-Feng-Err-shan, Chiufenershan), about 12 km from the earthquake epicentre. The failure had a volume of ~42 x 10^6 m^3, estimated by comparing airborne lidar data from 2002 to a preevent DEM from 1986 (Chang et al., 2005). Observations of seepage and springs on the sliding surface several days after the event point to the possible role of high pore-water pressure in triggering the slide (Chang et al., 2005). Along the sliding surface, Lin et al. (2001) found pseudotachylyte and inferred that frictional melting occurred during sliding. The deposit dammed two small rivers and buried a village, resulting in 39 fatalities. Unlike the Tsaoling landslide, the Jiufengershan slope had not experienced previous historical failures in the ~100 years prior to the earthquake (Shou & Wang, 2003).

99 Peak failures: Another notable area of landsliding was the 99 Peaks region, or Jiujiufeng. In contrast to the Tsaoling and Jiufengershan sites, the 99 Peaks experienced widespread shallow failures that stripped the soil and vegetation over an area of 8.6 km^2, for a landslide area density of 48% (Chen & Wu, 2006). These slides were very shallow, with depths generally less than 1 m and an average depth estimated to be ~0.2 m (Chen & Wu, 2006). The underlying lithology is highly permeable conglomerates of the Pliocene–Quaternary Toukoshan Formation, topped with thin, highly weathered soil and extensive forest cover. The 99 Peaks area is extremely steep, with more than 50% of the area having slopes of 60–70 degrees (Chen & Wan, 2004). This combination of very steep slopes and weak lithology accounts for the very high density of failures in the area.

13.4 Postseismic landslides

In addition to coseismic landslides, the Chi-Chi earthquake was notable for high rates of postseismic landsliding. Precipitation-triggered landslides in Taiwan are primarily related to typhoon storms, so the occurrence of postseismic landslides was dominated by the first major typhoons following the earthquake. In July 2001, Typhoon Toraji brought a mean of 364 mm of precipitation to the Choshui (Zhuoshui) catchment in central western Taiwan. Three years later, in June 2004, Typhoon Mindulle brought 650 mm of precipitation. These postearthquake typhoon events can be compared to Typhoon Herb, which struck in July 1996, with a mean precipitation of 486 mm (Lin et al., 2006). The postearthquake typhoons triggered significantly more landslides than the preearthquake Typhoon Herb and this increase in landslides showed a clear spatial pattern, with the 2001 and 2004 landslides decreasing with decreasing coseismic PGA and increasing distance from the Chelungpu Fault (Fig. 13.5; Dadson et al., 2004; Lin et al., 2006; Lin, Chen, Chen, et al., 2008; Lin, Chen, Hovius, et al., 2008).

In addition to the map-view relationship between the postearthquake landsides and PGA, Meunier et al. (2008) found that the topographic clustering of landslides triggered by Typhoon Toraji resembled that of the coseismic landslides, with more landslides triggered near ridge crests and was very different from the pattern during Typhoon Herb.

The large number of landslides triggered by Typhoon Mindulle significantly interrupted postearthquake recovery efforts, such as the reconstruction of the Central Cross-Island Highway. Landslides and debris flows triggered by Typhoon Mindulle, most of which involved reactivation of coseismic landslides, destroyed 45 km of the highway, which had been recently reconstructed after the Chi-Chi earthquake at an estimated cost of about 53 million USD (Chen & Petley, 2005). As of 2025, the road had not been reopened to public traffic. Typhoon Mindulle also extensively reactivated the 99 Peaks area, triggering failures covering ~4.2 km^2, approximately half of the area affected by coseismic landslides (Chen & Wu, 2006).

These associations between the earthquake PGA and postearthquake landslides continued for at least 5 years following the earthquake and in addition to the increased landslide rates, provide strong evidence of a causal link between the earthquake and postseismic landsliding, suggesting that the coseismic shaking reduced the stability of the hillslopes for an extended period of time.

A comprehensive analysis of post-Chi-Chi landsliding was undertaken by Marc, Hovius, Meunier, et al. (2015), who mapped new landslides (excluding reactivation of preexisting landslides) from 1994 to 2016 in a portion of the Choshui (Zhuoshui) catchment with high temporal resolution (for a total of 39 epochs) (Fig. 13.6). When normalised for precipitation, the landslide rate rose following the earthquake and then had a gradual recovery over a period of 5 years, with a postearthquake trend that can be fit by an exponential function. Marc, Hovius, Meunier, et al. (2015) suggested that this

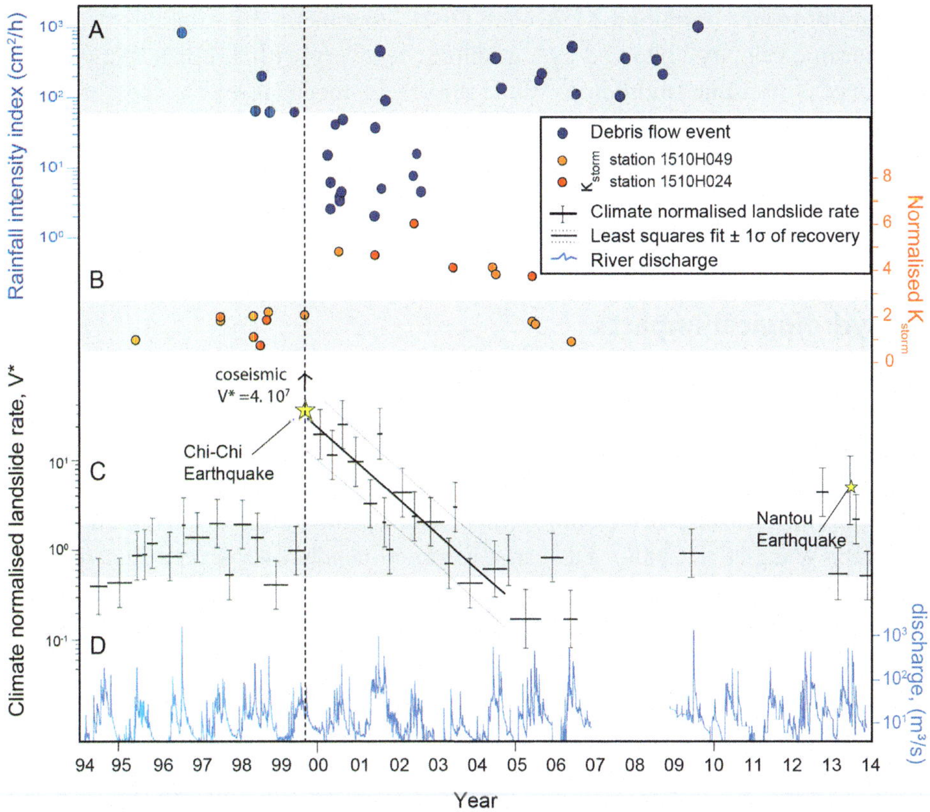

Figure 13.6 Postseismic perturbations to debris-flow triggering, suspended-sediment concentration and landslide rate. The vertical dashed line shows the time of the Chi-Chi earthquake. (A) Debris-flow events from the Chenyoulan catchment (blue circles) plotted by triggering rainfall intensity index (the product of maximum daily rainfall and maximum hourly intensity). (B) Unit storm suspended-sediment concentrations from two stations on the Chenyoulan River (orange and red circles). (C) Climate-normalised landslide rate (black bars), representing landslide volume per area corrected for cumulative precipitation over the mapping interval, black line shows the least-squares fit to landslide rates from the earthquake to the end of 2005, with fine grey lines showing ± 1 standard deviation. (D) Chenyoulan River discharge (blue line). Grey bars represent background preearthquake conditions for each dataset. Source: *A, Data are taken from Fig. 13.3 of Chen et al. (2013); B, data from Fig. 13.4 of Hovius et al. (2011); C and D, Modified from Fig. 13.2 of Marc, Hovius, Meunier, et al. (2015).*

could be related to damage to the shallow subsurface by shaking in the earthquake and subsequent healing. Marc et al. (2021) took the next step to investigate changes to the shallow subsurface (the upper 10 seconds of metres) using passive seismic interferometry, which can measure small changes in the relative velocity of seismic waves through time. Because seismic velocity depends on material properties, relative velocity changes, after correction for factors such as temperature and groundwater levels, reflect changing subsurface conditions. Marc et al. (2021) documented a coseismic drop in the relative seismic velocity, followed by a nonlinear recovery with a similar recovery time as the rate of postseismic landsliding. While the precise mechanisms for a coseismic drop in shallow seismic velocity and the subsequent recovery are unclear, the coincidence between the seismic and landslide recovery patterns suggests that both relate to damage and a reduction of strength of the upper few 10 seconds of metres of the subsurface (Marc et al., 2021).

13.5 Hydrological impacts

Further indication of coseismic impacts on the subsurface is the hydrological changes observed following the Chi-Chi earthquake. An examination of river gauge data from 1999 showed an increase in discharge (recorded daily) following the earthquake at 23 out of the 32 gauges in the Daan, Dajia, Wu and Choshui catchments; at seven other gauges, changes due to reservoir releases and formation of a landslide dam were observed (Liu et al., 2018; Wang et al., 2004). Discharge increased rapidly over several days following the earthquake, generally reaching a peak after 2–3 days. The magnitude of the change varied from 58% to 833% increase in daily discharge, with the largest increases generally occurring on the tributaries with the smallest discharges (Liu et al., 2018). For gauge sites that did not experience other postearthquake perturbations (such as rainfall events or small landslide dam failures), the discharge gradually returned to preearthquake levels over ~1.5–2 months after the earthquake (Liu et al., 2018). These increases in discharge result from an increase in permeability of the subsurface, which allows more groundwater to drain to the river. At one gauge station, 4 km from the earthquake epicentre on Shueili Creek in the Choshui catchment, a decrease in discharge was observed after the earthquake and comparison between this gauge and one 3 km upstream on the same channel suggested that a section of the creek was losing discharge for approximately 8 months following the earthquake (Liu et al., 2018). This gauge is located just downstream of the Shueilikung fault. Although this fault was not active during the Chi-Chi earthquake, it is likely associated with extensive local fractures that could have been affected by the earthquake, providing conduits for water to drain to the subsurface. Hydrological changes were also recorded in well data in the Choshui River alluvial fan, west of the fault. The pattern of coseismic changes in the water levels of monitoring wells depended on the type of aquifer (confined vs unconfined), but could be related to the distribution of coseismic strain (Lee et al., 2002).

13.6 Postseismic debris flows

Landslide debris produced by both co- and postseismic rockfalls and landslides was delivered to the fluvial system, although the connectivity between the coseismic failures and the river system was initially limited. Dadson et al. (2004) reported that only 8% of the coseismic landslides in the Chenyoulan catchment were connected directly to the fluvial system. An analysis of the landslide inventory of Liao and Lee (2000), corrected for amalgamation by Marc and Hovius (2015), shows that for all coseismic landslides, 8.7% of landslides and 25% of landslide area were directly connected to a channel with at least 1 km^2 upstream drainage area, and 11% of landslides and 27% of landslide area were connected to a channel with at least 0.5 km^2 drainage area (Fig. 13.5D). The higher connectivity in the landslide area is dominated by the large Tsaoling and Jiufengershan landslides described above.

The large amounts of landslide debris stored on hillslopes and in small channels after the earthquake primed the landscape for debris-flow activity and led to an increase in debris-flow occurrence. A number of studies have investigated changes in debris-flow triggering thresholds following the earthquake (Lin et al., 2004; Liu et al., 2008; Shieh et al., 2009; Yu et al., 2014). All studies found increased debris-flow activity in the years after the earthquake, corresponding to a decrease in rainfall duration-intensity thresholds for debris-flow triggering. Several studies focused on the Chenyoulan catchment, where preearthquake debris flows were reported only during four typhoons from 1963 to 1992 and during and following Typhoon Herb in 1996 (Chen et al., 2013; Lin et al., 2004). Following the Chi-Chi earthquake, debris-flow activity increased dramatically, with numerous events in 2000 and 2001, many triggered by relatively minor rain events (Chen et al., 2013; Chen, 2011; Lin et al., 2004). Debris flows in 2000–02 were triggered at lower values of rainfall intensity and total accumulation than at any other time from 1985 to 2010, with estimates of triggering rainfall intensity approximately five times lower than before the earthquake, for rainfall of a given duration (Chen et al., 2007) (Fig. 13.6).

Debris flows after the earthquake occurred in a wider range of catchments and channels than previously and several studies noted that debris flows were observed in steeper and smaller catchments (Lin et al., 2004; Liu et al., 2008; Chen, 2011), reflecting the distribution of coseismic landslides as opposed to rainfall-triggered landslides.

The evolution of debris-flow occurrence following the earthquake was influenced by the postearthquake Typhoons Toraji and Mindulle, which both triggered large numbers of debris flows throughout the affected region. However, these typhoons also triggered large numbers of new landslides, as described above, so their effect on sediment storage in potential debris-flow catchments depended on the relative volumes of sediment delivered by landslides and sediment evacuated by debris flows. Dong et al. (2009) constructed a sediment budget model to investigate the impacts of the earthquake-typhoon-typhoon sequence in the Dongshi region, at the northern end of the Chi-Chi earthquake-affected area. In this region, they suggested that the typhoon-triggered landslides and debris flows were largely in balance, with net sediment storage increasing slightly after Typhoon Toraji and decreasing slightly after Typhoon

Mindulle (Dong et al., 2009). This suggests that debris-flow susceptibility remained high in this region through 2004. In contrast, Shieh et al. (2009) found that in the Wushikeng and 99 Peaks regions, sediment storage in debris-flow channels increased after the earthquake and then declined, with relatively little new landslide input after the earthquake. Similarly, the debris-flow triggering threshold decreased substantially after the earthquake and then steadily increased; however, in 2006, the threshold was still significantly below the preearthquake level. In contrast, in the Chenyoulan catchment, Chen (2011) and Chen et al. (2013) suggested a recovery time of 5 years for debris-flow triggering thresholds. This variability in the response time and magnitude likely reflects the varying susceptibility to postseismic sediment loading and the variations in the local magnitude of the postearthquake rain events. In 2009, Taiwan was hit by Typhoon Morakot, which brought up to 3 m of rain over three days in areas of southern Taiwan. This was a major landslide-triggering event and a significant perturbation to the landscape, potentially obscuring continued recovery from the Chi-Chi earthquake.

13.7 Postseismic sediment flux

A network of river gauging stations operated by the WRA of Taiwan that included regular suspended-sediment measurements has allowed detailed analyses of the impacts of the Chi-Chi earthquake on suspended-sediment dynamics. Dadson et al. (2004) used suspended sediment and discharge data from the WRA to calculate unit sediment concentration (a measure of sediment concentration independent of discharge) for both pre- (1970–99) and postearthquake (until the end of 2001) periods. They found that the unit sediment concentration increased by a factor of four in the epicentral area of the earthquake, and related this to co- and postseismic landslide density (Dadson et al., 2004).

Extending this analysis later, Hovius et al. (2011) calculated unit sediment concentrations for individual typhoon events from 1984 to 2007 at four gauging stations on the Choshui (Zhuoshui) River. They found that this metric, termed K_{storm}, remained consistent over all of the recorded storms prior to the earthquake, and then increased by a factor of three in the first storm event following the earthquake. The storm unit sediment concentration reached a maximum in July 2001 with Typhoon Toraji and then began to decline, returning to the preearthquake levels 6 ± 0.8 years after the earthquake (Hovius et al., 2011) (Fig. 13.6).

This timescale of sediment-flux recovery corresponds to the timescale of elevated landslide rates in the Chenyoulan catchment, as well as the recovery of debris-flow triggering thresholds following the earthquake, suggesting that the suspended sediment, which is supply-limited in this setting, was continuously supplied by postseismic landslides and debris flows (Hovius et al., 2011).

While most of the analyses of postearthquake sediment fluxes have focused on suspended sediment due to the availability of data, the landslides also delivered significant amounts of coarse sediment to the fluvial system. However, the response of the fluvial system to this influx of coarse sediment has not been systematically evaluated. Yanites et al. (2010) documented aggradation in the Peikang River (a tributary of the Wu River) and estimated that it will take decades for this river to evacuate the coarse sediment produced by the Chi-Chi earthquake. In many rivers, the fluvial response to increased sediment delivery from co- and postseismic landslides was complicated by the presence of dams and reservoirs. Huang and Montgomery (2012) examined reservoir sedimentation rates through 2010 for six reservoirs on the Dajia (Tachia) River. They found that, on average, the rates increased by a factor of three in the decade following the Chi-Chi earthquake, but with complicated spatiotemporal patterns involving increases in sedimentation in different reservoirs following the earthquake, Typhoon Toraji and Typhoon Mindulle. Overall, Typhoon Mindulle resulted in the greatest increase in sedimentation rates and led to aggradation in the river channel (Huang & Montgomery, 2012). Due to the difficulty in measuring bedload transport, particularly in large Taiwanese rivers during the floods that move most of the material, there is no direct information about bedload flux changes after the earthquake.

13.8 Fluvial uplift and erosion

One of the most notable impacts of the Chi-Chi earthquake was the dramatic fluvial incision in several of the rivers affected by the earthquake. The high magnitude of coseismic uplift, particularly at the northern end of the Chi-Chi surface rupture, had a particularly strong impact on several of the fault-crossing rivers. Over its 100 km length, the rupture zone crossed four major rivers draining the Central Range of Taiwan, creating scarps that extended across the river beds.

The lithologies located in the vicinity of the Chelungpu Fault are Pliocene and Pleistocene sedimentary units, which are poorly lithified and have very high erodibility. Erosion of these units is highly sensitive to the presence or absence of coarse sediment (Stock et al., 2005). A reduction in sediment supply may lead to bedrock erosion if the gravel cover is stripped from the river bed and the bedrock is exposed; however, if coarse sediment supply reduces too far, erosion will slow due to a lack of bedload tools for driving abrasion (Sklar & Dietrich, 2004). The response of each river crossing the Chelungpu Fault was controlled by the geometry of the local deformation, the local conditions controlling bedload flux and engineering works in the river, as outlined below.

In the southern portion of the uplift zone, the scarps that formed in the Choshui (Zhuoshui) and Wu Rivers were relatively small, at 1.2 and 1 m high, respectively (Huang et al., 2015). Yanites et al. (2010) suggested that the uplift on the Zhuoshui River resulted in an upstream propagating knickpoint and channel narrowing. However, Chen et al. (2024) argued that the incision of the Zhuoshui River is the result of sediment trapping by the Jiji Dam from 2001 onwards, with incision propagating downstream from the dam as the sediment cover was removed, rather than upstream from the surface rupture. Thus the earthquake did not generate significant incision in the Zhuoshui River, although the scarp may have slightly exacerbated the incision due to exposure of the bedrock downstream of the dam. The Wu River had a similarly small coseismic scarp and there it was clear that the magnitude of surface uplift was smaller than the thickness of coarse sediment in the river bed, so the scarp quickly diffused in the sedimentary cover and no bedrock incision was observed (Huang et al., 2015).

The most dramatic fluvial impacts, however, were observed at the northern end of the rupture zone, where the surface uplift was the highest.

13.8.1 Daan (Taan) River

At the northernmost end of the rupture zone, the Daan River crosses the Dongshih anticline, which was activated by slip on the Chelungpu Fault at depth. As described above, the coseismic folding resulted in surface uplift of up to 10 m, bounded by upstream and downstream fold scarps located about 1 km apart in the Daan River channel (Lee et al., 2005). This extreme perturbation of the river channel led to rapid fluvial erosion and the formation of the Daan River gorge following the earthquake. The postearthquake evolution of the Daan River has been studied extensively (Chen & Chen, 2025; Chen et al., 2024; Cook et al., 2013, 2014, 2020; Huang et al., 2013) and highlights the controls of structure and coarse sediment supply on fluvial incision in response to sudden uplift.

Before the Chi-Chi earthquake, the Daan River was a typical Taiwanese gravel-covered braided river with an active channel ~500 m wide. The coseismic uplift was asymmetric, with a broad scarp ~400 m long at the downstream (western) end of the uplift zone and a steep, near-vertical ~6.5 m high fold scarp at the upstream end that formed a dam across the Daan River. This upstream scarp played a key role in the response of the river to coseismic uplift, as it acted as a barrier to bedload transport (Cook et al., 2013). In order to reduce flooding in the surrounding areas upstream of the anticline, a small channel was excavated across the upstream scarp within days of the earthquake; however, the scarp continued to pond the river upstream and to block bedload transport into the uplifted area. Over the first 4 years after the earthquake, relatively little erosion took place in the uplifted zone (Fig. 13.7), but the gravel cover was gradually removed from the channel bed. At the downstream scarp, a steep knickpoint developed on a layer of more resistant sandstone, but no knickpoint propagation took place due to a lack of bedload tools to drive erosion (Cook et al., 2013).

Knickpoint propagation and incision began in 2004, after Typhoons Mindulle and Aere deposited enough bedload to fill the channel upstream of the scarp and reestablish bedload transport across the uplifted area (Cook et al., 2013). Knickpoint retreat then

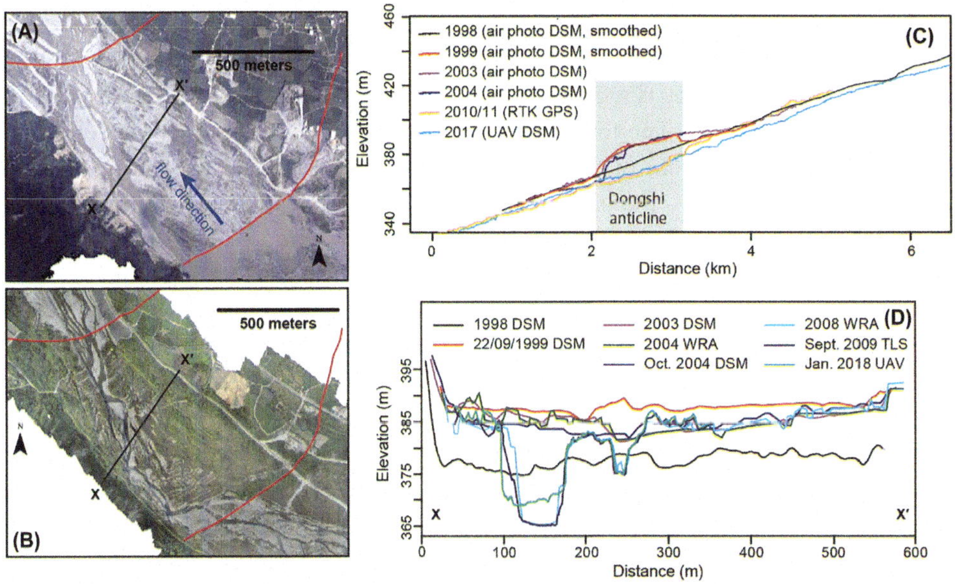

Figure 13.7 Daan River incision. (A) Aerial photograph of the Dongshi anticline and Daan River on 22 September, 1999. Red lines show the fold scarps and the black line shows the location of the cross sections in D. (B) UAV-derived orthoimage of the Daan Gorge in January 2019, showing the same area as in panel A. (C) Longitudinal profiles of the channel from 1998 to 2017. (D) Cross sections through the river channel from 1998 to 2018.

proceeded rapidly and by 2009 had propagated ~1 km and reached the upstream scarp. In the wake of the knickpoint, the river cut a gorge up to 20 m deep and less than 10 m wide at its narrowest (Cook et al., 2014). After 2009, vertical incision of the gorge ceased and by 2012, approximately 5 m of aggradation had occurred throughout the gorge (Cook et al., 2020) (Fig. 13.7), due to the increase in bedload flux into the gorge after the knickpoint crossed the upstream scarp and tapped into the sediment ponded upstream.

Lateral erosion and gorge widening followed knickpoint propagation, with average widening rates of 5.1 ± 0.1 m/year from 2005 to 2008, slowing to 1.49 ± 0.05 m/year from 2008 to 2013. The most rapid rates of lateral erosion took place at the upstream scarp, which was eroded at an average rate of 17 ± 0.1 m/year between 2008 and 2013 (Cook et al., 2014). Cook et al. (2014) proposed that focused erosion of the upstream scarp, which they termed downstream sweep erosion, could remove the uplifted topography of the Dongshi anticline in as little as 50 years after the earthquake. However, Chen and Chen (2025) noted that the pace of this erosion has slowed considerably due to aggradation against the scarp, so their proposed timescale of resetting of the valley floor was about four times longer than the estimate of Cook et al. (2014), although still shorter than the return time of a Chi-Chi-sized earthquake.

The impact of the earthquake on the Daan River was not confined to the area that experienced coseismic deformation. The trapping of sediment behind the Dongshi anticline starved the river downstream of the anticline of sediment, resulting in the removal of the gravel cover for at least 3 km downstream of the gorge. As in the Zhoshui River downstream of the Jiji Dam, this resulted in fluvial incision. By 2001, bedrock was exposed in the river beneath the Zhuolan (Cholan) Bridge, 1.9 km downstream of the gorge exit. Significant incision began here in 2004 and by 2006, the river had cut an inner channel, which damaged one of the piers of the bridge and caused a partial bridge collapse in 2006. The river downstream of the bridge had incised ~10 m by 2008, including up to ~6.5 m of bedrock incision.

In response to 10 m of coseismic uplift, the Daan River incised a gorge up to 20 m deep (Fig. 13.7), resulting in a river bed significantly below its preearthquake level. This overincision was dominantly due to the reduction in bedload flux across the upstream scarp from 1999 to 2009, with some additional contribution from the channel narrowing. The downstream sediment removal and subsequent erosion lowered the base level at the outlet of the gorge by ~7 m, accounting for most of the excess incision. The excess incision had further consequences upstream of the uplift zone, as the base level for the upstream channel was also lowered by ~10 m. This led to the upstream propagation of bedrock incision far beyond the initial uplifted area.

The Daan River illustrates the potential for far-reaching effects from a spatially limited coseismic perturbation. The Chi-Chi earthquake uplifted about 1 km of river bed, but the effects of changes in coarse sediment flux and channel width on fluvial bedrock erosion led to an extremely complex channel response that resulted in incision propagating both upstream and downstream of the initial perturbation. In the end, at least 5 km of the Daan River experienced bedrock incision as a result of the Chi-Chi earthquake (Fig. 13.7). Unfortunately, due to extensive anthropogenic activities within the river, particularly gravel extraction upstream and engineering works downstream in the later stages of the response, the complete evolution of the fluvial response to the earthquake cannot be evaluated.

13.8.2 Dajia (Dachia, Tachia) River

As described above, the Dajia River was cut by a coseismic fault scarp ~6 m high, which created a steep waterfall in the river bed (Fig. 13.3B). Similarly to the response of the Daan River, the Dajia River responded to coseismic uplift by rapidly incising into its bedrock. However, the response of the Dajia River was strongly influenced by the presence of the Shigang Dam 1 km upstream of the fault scarp and by the local bedrock structure.

Although the evolution of the Dajia scarp is often described as a knickpoint retreat (Chen et al., 2024; Huang et al., 2015), a detailed analysis of yearly aerial photographs and DEMs from 1999 to 2007 shows a more complex pattern of fluvial erosion (Fig. 13.8). The steep coseismic scarp quickly relaxed and began to retreat following the earthquake. After this initial phase of relaxation, a fixed knickpoint developed ~25 m upstream of the original scarp position within a year of the earthquake. From 2000 to 2008, this knickpoint gradually lowered in elevation, but did not translate upstream. Instead, incision upstream of the knickpoint occurred rapidly in a ~300 m stretch of the channel, creating a reach with a very low gradient that pooled at low flow conditions (Fig. 13.8). From this zone upstream to the dam, the channel morphology was very heavily influenced by the underlying bedrock structure. As in the Daan River, the bedrock in the Dajia River consists of alternating layers with different erodibilities; however, in the Dajia River, the bedding is steeply dipping and strikes obliquely across the channel. Water that was released by the Shigang Dam on the north side of the

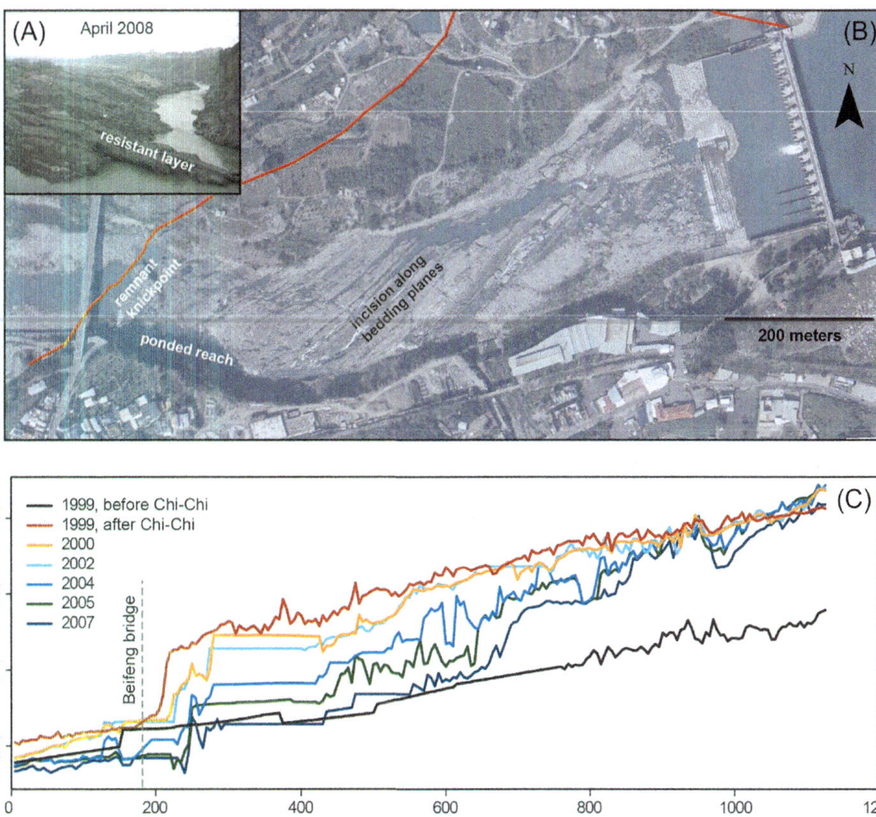

Figure 13.8 Dajia River incision. (A) Photograph of the scarp remnant in April 2008, looking upstream from the Beifeng Bridge, (B) aerial photograph of the incised reach in 2007, red line shows the coseismic scarp, (C) longitudinal profiles of the scarp and river channel.

channel flowed obliquely across the channel along the more erodible layers, creating a series of deeply incised but very narrow erosional features.

The bedrock layer where the knickpoint propagation stalled was visible as a bedrock step in the river prior to the Chi-Chi earthquake and qualitative field observations suggest that these rocks are more resistant than those upstream (Fig. 13.8A), suggesting a preexisting lithological control on the style of incision. The strong control of bedrock properties on the incision of the Dajia River is likely related to the very limited amount of bedload available in the uplifted reach. Bedload trapping by the Shigang Dam had stripped the active channel bed of coarse sediment prior to the earthquake and deprived the river of tools for erosion, as in the Daan River prior to 2004. This meant that only the weakest layers were able to incise quickly and the stronger layers exerted more control.

Channel engineering work was undertaken in the Dajia River in order to protect the south wall of the channel from erosion. Intensive work began in 2009, including large-scale excavation of a new channel through the bedrock and multiple attempts at channel regulation followed in later years. Therefore the continued adjustment of the river cannot be observed.

13.9 Long-term legacy of the Chi-Chi earthquake

The return period for the Chi-Chi earthquake has been investigated in several trenches excavated across the fault, with estimates ranging from 150–700 years; the penultimate earthquake occurred 300–450 years before present (Chen et al., 2007; Lee et al., 2001; Streig et al., 2007). Studies of folded and deformed terraces and sedimentary units provide evidence that the Chi-Chi earthquake was characteristic of the Chelungpu Fault and that at least 20 previous Chi-Chi events have occurred in the late Quaternary (Graveleau et al., 2014; Le Béon et al., 2014). The Chi-Chi earthquake is therefore the latest in a series of similar events and has contributed another increment to the continued topographic development at the front of the Western Foothills fold and thrust belt. As outlined above, most of the earthquake impacts on sediment production and delivery had a recovery period of about 6 years, although the evacuation of coarse material produced by the earthquake, particularly the deposits of the largest landslides, is expected to take significantly longer (Yanites et al., 2010; Chen et al., 2022). The fluvial erosion caused by the earthquake, particularly in the Daan River, is expected to take up to several 100 years to reset the river channel; however, the lack of any signature of previous earthquakes in the channels of all four major fault-crossing rivers (Fig. 13.1) suggests that the coseismic uplift and subsequent incision will not leave a lasting geomorphic mark on the rivers.

Hovius et al. (2011) estimated the mass balance of the Chi-Chi earthquake in the Choshui catchment, based on time series of suspended-sediment flux and landsliding balanced against the mass added based on geodetic data. They estimated that within 7 years following the earthquake, the mass wasting and suspended-sediment flux had recovered to preearthquake levels and that during this recovery time ~30% of the coseismically added mass had been removed from the catchment (Hovius et al., 2011). If these findings are representative of the rest of the affected region, it suggests that the short-term geomorphic impacts of the earthquake were not sufficient to balance the coseismic uplift and that continued evacuation of coseismic landslide debris or erosion of coseismically added mass may constitute part of the 'background' geomorphic evolution in Taiwan (Hovius et al., 2011).

While the geomorphic impacts of the earthquake may be relatively short-lived, the Chi-Chi earthquake has left a significant and long-lasting scientific and disaster management legacy in Taiwan. The Chi-Chi earthquake stimulated a generation of Earth scientists in Taiwan and helped to establish Taiwan as a centre of geophysical and geomorphological research. In addition to driving research about earthquake processes and impacts specifically, the aftermath of the earthquake motivated an increased focus on landslide and debris-flow triggering and processes more generally.

The Chi-Chi earthquake also highlighted the need for improved disaster management policies and procedures in Taiwan and led to the passing of the Disaster Prevention and Protection Act of 2000. The earthquake drove efforts to reduce seismic risk through modifications to building codes and retrofitting and the establishment of the national earthquake insurance programme, TREIF. On the scientific side, the National Science and Technology Center for Disaster Reduction was established in 2003 to conduct and coordinate research on disaster reduction and related applications (https://www.ncdr.nat.gov.tw/). These efforts to improve our understanding of large earthquakes and their impacts, combined with the efforts to improve disaster management and response, ensure that Taiwan is now substantially more resilient to the next earthquake like Chi-Chi.

13.10 Conclusions

The Chi-Chi earthquake had an intense, but transient, geomorphic impact due to the high magnitudes of ground acceleration and coseismic deformation, combined with a landscape that was primed for erosion and landsliding due to steep slopes, highly erodible lithologies, and extreme rainfall events. The earthquake's effects on the surface and subsurface were linked to coseismic and postseismic landsliding, which were in turn linked to debris-flow triggering and suspended-sediment export, all of which had a recovery period of ~4–5 years. The postearthquake processes were strongly influenced by the sequencing of typhoon storms after the earthquake, which extended the hazard related to the earthquake in both space and time and hampered reconstruction and recovery efforts. The coseismic uplift generated a complex pattern of fluvial erosion, strongly controlled by impacts on coarse sediment availability, resulting in incision both deeper and more widespread than the uplift that drove it. The Chi-Chi earthquake was the first large event with extensive geomorphic observations both immediately after the earthquake and throughout the postseismic response of the landscape, allowing connections to be made between the different components of the impact and response and enabling the estimation of recovery periods for multiple processes. This represented a significant step forward in our understanding of how large shallow earthquakes perturb geomorphic systems.

References

Chang, K. J., Taboada, A., & Chan, Y. C. (2005). Geological and morphological study of the Jiufengershan landslide triggered by the Chi-Chi Taiwan earthquake. *Geomorphology, 71*(3-4), 293–309.

Chen, C.-M., Shyu, J. B. H., Tsui, H.-K., & Hsieh, Y.-C. (2022). Preservation and transportation of large landslide deposits under decadal and millennial timescales in the Taiwan orogenic belt. *Geomorphology, 415*, 108402. https://doi.org/10.1016/j.geomorph.2022.108402.

Chen, H., & Petley, D. N. (2005). The impact of landslides and debris flows triggered by Typhoon Mindulle in Taiwan. *Quarterly Journal of Engineering Geology and Hydrogeology, 38*(3), 301–304.

Chen, H., & Wan, J. P. (2004). The effect of orientation and shape distribution of gravel on slope angles in central Taiwan. *Engineering Geology, 72*(1-2), 19–31.

Chen, H. E., Chiu, Y. Y., Cheng, C. Y., & Chen, S. C. (2024). Knickpoints and fixed points: The evolution of fluvial morphology under the combined effect of fault uplift and dam obstruction on a soft bedrock river. *Earth Surface Dynamics, 12*, 1329–1346.

Chen, J. C. (2011). Variability of impact of earthquake on debris-flow triggering conditions: Case study of Chen-Yu-Lan watershed, Taiwan. *Environmental Earth Sciences, 64*(7), 1787–1794.

Chen, J. C., Jan, C. D., & Huang, W. S. (2013). Characteristics of rainfall triggering of debris flows in the Chenyulan watershed, Taiwan. *Natural Hazards and Earth System Sciences, 13*(4), 1015–1023.

Chen, R. F., Chang, K. J., Angelier, J., Chan, Y. C., Deffontaines, B., Lee, C. T., & Lin, M. L. (2006). Topographical changes revealed by high-resolution airborne LiDAR data: The 1999 Tsaoling landslide induced by the Chi–Chi earthquake. *Engineering Geology, 88*(3-4), 160–172.

Chen, S. C., & Wu, C. H. (2006). Slope stabilization and landslide size on Mt. 99 Peaks after Chichi Earthquake in Taiwan. *Environmental Geology, 50*(5), 623–636.

Chen, W. S., Lee, K. J., Lee, L. S., Streig, A. R., Rubin, C. M., Chen, Y. G., Yang, H. C., Chang, H. C., & Lin, C. W. (2007). Paleoseismic evidence for coseismic growth-fold in the 1999 Chichi earthquake and earlier earthquakes, central Taiwan. *Journal of Asian Earth Sciences, 31*(3), 204–213.

Chen, W. T., & Chen, Y. C. (2025). Channel aggradation decelerates downstream sweep erosion in Daan River Gorge, Taiwan. *Geoscience Letters, 12*(1), 62.

Chen, Y. G., Chen, W. S., Lee, J. C., Lee, Y. H., Lee, C. T., Chang, H. C., & Lo, C. H. (2001). Surface rupture of 1999 Chi-Chi earthquake yields insights on active tectonics of central Taiwan. *Bulletin of the Seismological Society of America, 91*(5), 977–985.

Chen, Y. G., Lai, K. Y., Lee, Y. H., Suppe, J., Chen, W. S., Lin, Y. N. N., Wang, Y., Hung, J. H., & Kuo, Y. T. (2007). Coseismic fold scarps and their kinematic behavior in the 1999 Chi-Chi earthquake Taiwan. *Journal of Geophysical Research: Solid Earth, 112*(B3).

Chi, W. C., Dreger, D., & Kaverina, A. (2001). Finite-source modeling of the 1999 Taiwan (Chi-Chi) earthquake derived from a dense strong-motion network. *Bulletin of the Seismological Society of America, 91*(5), 1144–1157.

Chuang, R. Y., Lu, C. H., Yang, C. J., Lin, Y. S., & Lee, T. Y. (2020). Coseismic uplift of the 1999 Mw7. 6 Chi-Chi earthquake and implication to topographic change in frontal mountain belts. *Geophysical Research Letters, 47*(15), e2020GL088947.

Cook, K. L., Turowski, J. M., & Hovius, N. (2013). A demonstration of the importance of bedload transport for fluvial bedrock erosion and knickpoint propagation. *Earth Surface Processes and Landforms, 38*(7), 683–695.

Cook, K. L., Turowski, J. M., & Hovius, N. (2014). River gorge eradication by downstream sweep erosion. *Nature Geoscience,* 7(9), 682–686.

Cook, K. L., Turowski, J. M., & Hovius, N. (2020). Width control on event-scale deposition and evacuation of sediment in bedrock-confined channels. *Earth Surface Processes and Landforms, 45*(14), 3702–3713.

Dadson, S. J., Hovius, N., Chen, H., Dade, W. B., Hsieh, M. L., Willett, S. D., Hu, J. C., Horng, M. J., Chen, M. C., Stark, C. P., & Lague, D. (2003). Links between erosion, runoff variability and seismicity in the Taiwan orogen. *Nature, 426*(6967), 648–651.

Dadson, S. J., Hovius, N., Chen, H., Dade, W. B., Lin, J. C., Hsu, M. L., Lin, C. W., Horng, M. J., Chen, T. C., Milliman, J., & Stark, C. P. (2004). Earthquake-triggered increase in sediment delivery from an active mountain belt. *Geology, 32*(8), 733–736.

Derrieux, F., Siame, L. L., Bourlès, D. L., Chen, R. F., Braucher, R., Léanni, L., Lee, J. C., Chu, H. T., & Byrne, T. B. (2014). How fast is the denudation of the Taiwan mountain belt? Perspectives from in situ cosmogenic 10Be. *Journal of Asian Earth Sciences, 88*, 230–245.

Dominguez, S., Avouac, J. P., & Michel, R. (2003). Horizontal coseismic deformation of the 1999 Chi-Chi earthquake measured from SPOT satellite images: Implications for the seismic cycle along the western foothills of central Taiwan. *Journal of Geophysical Research: Solid Earth, 108*(B2).

Dong, J.-J., Lee, C.-T., Tung, Y.-H., Liu, C.-N., Lin, K.-P., & Lee, J.-F. (2009). The role of the sediment budget in understanding debris flow susceptibility. *Earth Surface Processes and Landforms, 34*(12), 1612–1624. https://doi.org/10.1002/esp.1850.

Graveleau, F., Suppe, J., Chang, K. J., Dominguez, S., Ustaszewski, M., & Huang, M. H. (2014). Incremental vs. geological growth of folds: Examples of Tungshih anticline and Neiwan syncline during Taiwan Mw7. 6 1999 ChiChi earthquake. *Geophysical Research Abstracts, Vol. 16*.

Hovius, N., Meunier, P., Lin, C. W., Chen, H., Chen, Y. G., Dadson, S., Horng, M. J., & Lines, M. (2011). Prolonged seismically induced erosion and the mass balance of a large earthquake. *Earth and Planetary Science Letters, 304*(3-4), 347–355.

Hsu, Y. J., Lai, Y. R., You, R. J., Chen, H. Y., Teng, L. S., Tsai, Y. C., Tang, C. H., & Su, H. H. (2018). Detecting rock uplift across southern Taiwan mountain belt by integrated GPS and leveling data. *Tectonophysics, 744*, 275–284.

Huang, M. W., Pan, Y. W., & Liao, J. J. (2013). A case of rapid rock riverbed incision in a coseismic uplift reach and its implications. *Geomorphology, 184*, 98–110.

Huang, M. W., Pan, Y. W., & Liao, J. J. (2015). The processes and rates of knickpoint migration in western Taiwan. In *10th IAEG Asian Regional Conference*, Kyoto, Japan.

Huang, M. Y. F., & Montgomery, D. R. (2012). Fluvial response to rapid episodic erosion by earthquake and typhoons, Tachia River, central Taiwan. *Geomorphology, 175*, 126–138.

Hung, J. J. (2000). Chi-Chi earthquake induced landslides in Taiwan. *Earthquake Engineering and Engineering Seismology, 2*(2), 25–33.

Lai, K. Y., Chen, Y. G., Hung, J. H., Suppe, J., Yue, L. F., & Chen, Y. W. (2006). Surface deformation related to kink-folding above an active fault: Evidence from geomorphic features and co-seismic slips. *Quaternary International, 147*(1), 44–54.

Le Béon, M., Suppe, J., Jaiswal, M. K., Chen, Y. G., & Ustaszewski, M. E. (2014). Deciphering cumulative fault slip vectors from fold scarps: Relationships between long-term and coseismic deformations in central Western Taiwan. *Journal of Geophysical Research: Solid Earth, 119*(7), 5943–5978.

Lee, Y.-H., Lu, S.-T., Shih, T.-S., Hsieh, M.-L., & Wu, W.-Y. (2005). Structures associated with the northern end of the 1999 Chi-Chi Earthquake rupture, central Taiwan: implications for seismic-hazard assessment. *Bulletin of the Seismological Society of America, 95*, 471–485. https://doi.org/10.1785/0120020170.

Lee, J. C., Chen, Y. G., Sieh, K., Mueller, K., Chen, W. S., Chu, H. T., Chan, Y. C., Rubin, C., & Yeats, R. (2001). A vertical exposure of the 1999 surface rupture of the Chelungpu fault at Wufeng, western

Taiwan: Structural and paleoseismic implications for an active thrust fault. *Bulletin of the Seismological Society of America, 91*(5), 914–929.

Lee, M., Liu, T. K., Ma, K. F., & Chang, Y. M. (2002). Coseismic hydrological changes associated with dislocation of the September 21, 1999 Chichi earthquake, Taiwan. *Geophysical Research Letters, 29*(17), 5. 1.

Lee, S. J., Chen, H. W., & Ma, K. F. (2007). Strong ground motion simulation of the 1999 Chi-Chi, Taiwan earthquake from a realistic three-dimensional source and crustal structure. *Journal of Geophysical Research: Solid Earth, 112*(B6).

Lee, Y. H., Chen, C. C., Liu, T. K., Ho, H. C., Lu, H. Y., & Lo, W. (2006). Mountain building mechanisms in the Southern Central Range of the Taiwan Orogenic Belt—From accretionary wedge deformation to arc–continental collision. *Earth and Planetary Science Letters, 252*(3-4), 413–422.

Lee, Y. H., Wu, K. C., Rau, R. J., Chen, H. C., Lo, W., & Cheng, K. C. (2010). Revealing coseismic displacements and the deformation zones of the 1999 Chi-Chi earthquake in the Tsaotung area, central Taiwan, using digital cadastral data. *Journal of Geophysical Research: Solid Earth, 115*(B3).

Liao, H.-W., & Lee, C. (2000). Landslides triggered by Chi-Chi earthquake. In *Proceedings of the 21st Asian Conference on Remote Sensing* (Vol. 1, pp. 383–388).

Lin, A., Chen, A., Liau, C. F., Lee, C. T., Lin, C. C., Lin, P. S., Wen, S. C., & Ouchi, T. (2001). Frictional fusion due to coseismic landsliding during the 1999 Chi-Chi (Taiwan) ML 7.3 Earthquake. *Geophysical Research Letters, 28*(20), 4011–4014.

Lin, C. W., Liu, S. H., Lee, S. Y., & Liu, C. C. (2006). Impacts of the Chi-Chi earthquake on subsequent rainfall-induced landslides in central Taiwan. *Engineering Geology, 86*(2-3), 87–101.

Lin, C. W., Shieh, C. L., Yuan, B. D., Shieh, Y. C., Liu, S. H., & Lee, S. Y. (2004). Impact of Chi-Chi earthquake on the occurrence of landslides and debris flows: example from the Chenyulan River watershed, Nantou, Taiwan. *Engineering Geology, 71*(1-2), 49–61.

Lin, G. W., Chen, H., Chen, Y. H., & Horng, M. J. (2008). Influence of typhoons and earthquakes on rainfall-induced landslides and suspended sediments discharge. *Engineering Geology, 97*(1-2), 32–41.

Lin, G. W., Chen, H., Hovius, N., Horng, M. J., Dadson, S., Meunier, P., & Lines, M. (2008). Effects of earthquake and cyclone sequencing on landsliding and fluvial sediment transfer in a mountain catchment. *Earth Surface Processes and Landforms: The Journal of the British Geomorphological Research Group, 33*(9), 1354–1373.

Lin, M. L., & Tung, C. C. (2004). A GIS-based potential analysis of the landslides induced by the Chi-Chi earthquake. *Engineering Geology, 71*(1-2), 63–77.

Lin, M. L., Liao, H. J., & Ueng, Z. S. (2000). The geotechnical hazard caused by Chi-Chi earthquake. In *Proceedings of the International Workshop on the September 21, 1999 Chi-Chi Earthquake* (pp. 113–123).

Liu, C.-N., Huang, H.-F., & Dong, J.-J. (2008). Impacts of September 21, 1999 Chi-Chi earthquake on the characteristics of gully-type debris flows in central Taiwan. *Natural Hazards, 47*(3), 349–368. https://doi.org/10.1007/s11069-008-9223-9.

Liu, C. Y., Chia, Y., Chuang, P. Y., Wang, C. Y., Ge, S., & Teng, M. H. (2018). Streamflow changes in the vicinity of seismogenic fault after the 1999 Chi–Chi earthquake. *Pure and Applied Geophysics, 175*(7), 2425–2434.

Marc, O., & Hovius, N. (2015). Amalgamation in landslide maps: Effects and automatic detection. *Natural Hazards and Earth System Science, 15*(4), 723–733.

Marc, O., Hovius, N., Meunier, P., Uchida, T., & Hayashi, S. I. (2015). Transient changes of landslide rates after earthquakes. *Geology, 43*(10), 883–886.

Marc, O., Sens-Schönfelder, C., Illien, L., Meunier, P., Hobiger, M., Sawazaki, K., Rault, C., & Hovius, N. (2021). Toward using seismic interferometry to quantify landscape mechanical variations after earthquakes. *Bulletin of the Seismological Society of America, 111*(3), 1631–1649.

Meunier, P., Hovius, N., & Haines, A. J. (2007). Regional patterns of earthquake-triggered landslides and their relation to ground motion. *Geophysical Research Letters, 34*(20).

Meunier, P., Hovius, N., & Haines, J. A. (2008). Topographic site effects and the location of earthquake induced landslides. *Earth and Planetary Science Letters, 275*(3-4), 221–232.

Sella, G. F., Dixon, T. H., & Mao, A. (2002). REVEL: A model for recent plate velocities from space geodesy. *Journal of Geophysical Research: Solid Earth, 107*(B4), ETG–11.

Sepúlveda, S. A., Murphy, W., & Petley, D. N. (2005). Topographic controls on coseismic rock slides during the 1999 Chi-Chi earthquake, Taiwan. *Quarterly Journal of Engineering Geology and Hydrogeology, 38*(2), 189–196.

Shieh, C. L., Chen, Y. S., Tsai, Y. J., & Wu, J. H. (2009). Variability in rainfall threshold for debris flow after the Chi-Chi earthquake in central Taiwan, China. *International Journal of Sediment Research, 24*(2), 177–188. https://doi.org/10.1016/s1001-6279(09)60025-1.

Shin, T. C., & Teng, T. L. (2001). An overview of the 1999 Chi-Chi, Taiwan, earthquake. *Bulletin of the Seismological Society of America, 91*(5), 895–913.

Shou, K. J., & Wang, C. F. (2003). Analysis of the Chiufengershan landslide triggered by the 1999 Chi-Chi earthquake in Taiwan. *Engineering Geology, 68*(3-4), 237–250.

Simoes, M., & Avouac, J. P. (2006). Investigating the kinematics of mountain building in Taiwan from the spatiotemporal evolution of the foreland basin and western foothills. *Journal of Geophysical Research: Solid Earth, 111*(B10).

Sklar, L. S., & Dietrich, W. E. (2004). A mechanistic model for river incision into bedrock by saltating bed load. *Water Resources Research, 40*(6).

Stock, J. D., Montgomery, D. R., Collins, B. D., Dietrich, W. E., & Sklar, L. (2005). Field measurements of incision rates following bedrock exposure: Implications for process controls on the long profiles of valleys cut by rivers and debris flows. *Geological Society of America Bulletin, 117*(1-2), 174–194.

Streig, A. R., Rubin, C. M., Chen, W. S., Chen, Y. G., Lee, L. S., Thompson, S. C., Madden, C., & Lu, S. T. (2007). Evidence for prehistoric coseismic folding along the Tsaotun segment of the Chelungpu fault near Nan-Tou, Taiwan. *Journal of Geophysical Research: Solid Earth, 112*(B3).

Suppe, J. (1981). Mechanics of mountain building and metamorphism in Taiwan. *Memoirs of the Geological Society of China, 4*, 67–89.

Wang, C. Y., Wang, C. H., & Manga, M. (2004). Co-seismic release of water from mountains: Evidence from the 1999 (Mw = 7.5) Chi–Chi, Taiwan, earthquake. *Geology, 32*(9), 769–772.

Wang, G. Q., Zhou, X. Y., Zhang, P. Z., & Igel, H. (2002). Characteristics of amplitude and duration for near fault strong ground motion from the 1999 Chi-Chi, Taiwan earthquake. *Soil Dynamics and Earthquake Engineering, 22*(1), 73–96.

Wang, W. N. (2000). A photogeologic study on the earthquake-induced slope failure in central Taiwan on September 21, 1999. In *Proceedings of the 39th Conference of the Japan Landslide Society* (2000). The Japan Landslide Society.

Wang, W. N., Nakamura, H., Tsuchiya, S., & CHEN, C. C. (2002). Distributions of landslides triggered by the Chi-chi Earthquake in Central Taiwan on September 21, 1999. *Landslides, 38*(4), 318–326.

Water Resources Agency. *Precipitation in Taiwan*. Retrieved August 20, 2025, from https://www.wra.gov.tw/en/cp.aspx?n=5121.

Yanites, B. J., Tucker, G. E., Mueller, K. J., & Chen, Y. G. (2010). How rivers react to large earthquakes: Evidence from central Taiwan. *Geology, 38*(7), 639–642.

Yu, S. B., Chen, H. Y., & Kuo, L. C. (1997). Velocity field of GPS stations in the Taiwan area. *Tectonophysics, 274*(1-3), 41–59.

Yu, S. B., Kuo, L. C., Hsu, Y. J., Su, H. H., Liu, C. C., Hou, C. S., Lee, J. F., Lai, T. C., Liu, C. C., Liu, C. L., & Tseng, T. F. (2001). Preseismic deformation and coseismic displacements associated with the 1999 Chi-Chi, Taiwan, earthquake. *Bulletin of the Seismological Society of America, 91*(5), 995–1012.

Yu, B., Wu, Y., & Chu, S. (2014). Preliminary study of the effect of earthquakes on the rainfall threshold of debris flows. *Engineering Geology, 182*, 130–135. https://doi.org/10.1016/j.enggeo.2014.04.007.

Yue, L.-F., Suppe, J., & Hung, J.-H. (2005). Structural geology of a classic thrust belt earthquake: The 1999 Chi-Chi earthquake Taiwan (Mw = 7.6). *Journal of Structural Geology, 27*(11), 2058–2083.

CHAPTER 14

Case study of the 2008 Wenchuan earthquake, China

Tristram C. Hales[1], Erin L. Harvey[2], Lanxin Dai[3], and Xuanmei Fan[3]
[1]School of Earth and Environmental Sciences, Cardiff University, Cardiff, United Kingdom,
[2]Department of Geography, Durham University, Durham, United Kingdom,
[3]State Key Laboratory of Geohazard Prevention and Geoenvironment Protection, Chengdu University of Technology, Chengdu, Sichuan, P.R. China

14.1 Introduction

The 2008 M_w 7.9 Wenchuan earthquake was a large continental earthquake that formed along the range front of the Longmen Shan. The earthquake generated a highly active and complex surface process cascade (Fan, Scaringi, Korup, et al., 2019; Yanites et al., 2025). The cascading effects of this earthquake were caused by the epicentral location in steep, high-relief topography combined with the ~3 km^3 volume of coseismic landslide debris (Li et al., 2014). When combined with the accessibility of the fault trace, the earthquake's occurrence at the advent of widespread, high-resolution and frequently acquired satellite imagery enabled a level of analysis that was not possible for previous earthquakes of comparable magnitude (Chen, Tang, Xiong et al., 2024; Fan, Scaringi, Domènech, et al., 2019) (Fig. 14.1). This chapter summarises the Wenchuan earthquake, its characteristics, and the resulting geomorphic cascade. We consider four major advances that have been precipitated by observations of this earthquake. (1) The Wenchuan earthquake has changed our understanding of how large earthquakes contribute to the uplift and erosion of orogens (Li et al., 2014; Marc et al., 2016; Parker et al., 2011). (2) The earthquake has changed our perception of the nature and length of the geomorphic and hazard cascade associated with large continental earthquakes (Domènech et al., 2019; Fan et al., 2018; Huang & Fan, 2013; Huang et al., 2016; Wang et al., 2024). (3) Enhanced rates of earthquake-related large and very large debris flows (sensu Major et al., 2005) continue well beyond time scales observed in other earthquakes (Fan, Tang, et al., 2020; Yang et al., 2021). These large and very large debris flows were undocumented in this region prior to the earthquake, but continue to occur within the epicentral region of the earthquake (Harvey, 2022). Finally, (4) the nature of the postearthquake geomorphic cascade affects the mass balance of terrestrial carbon. Preliminary evidence suggests that coseismic landslide debris may act as a carbon sink over multiple earthquake cycles (Liu, 2024).

Geomorphic Impacts of Earthquakes
DOI: https://doi.org/10.1016/B978-0-443-40375-0.00002-5

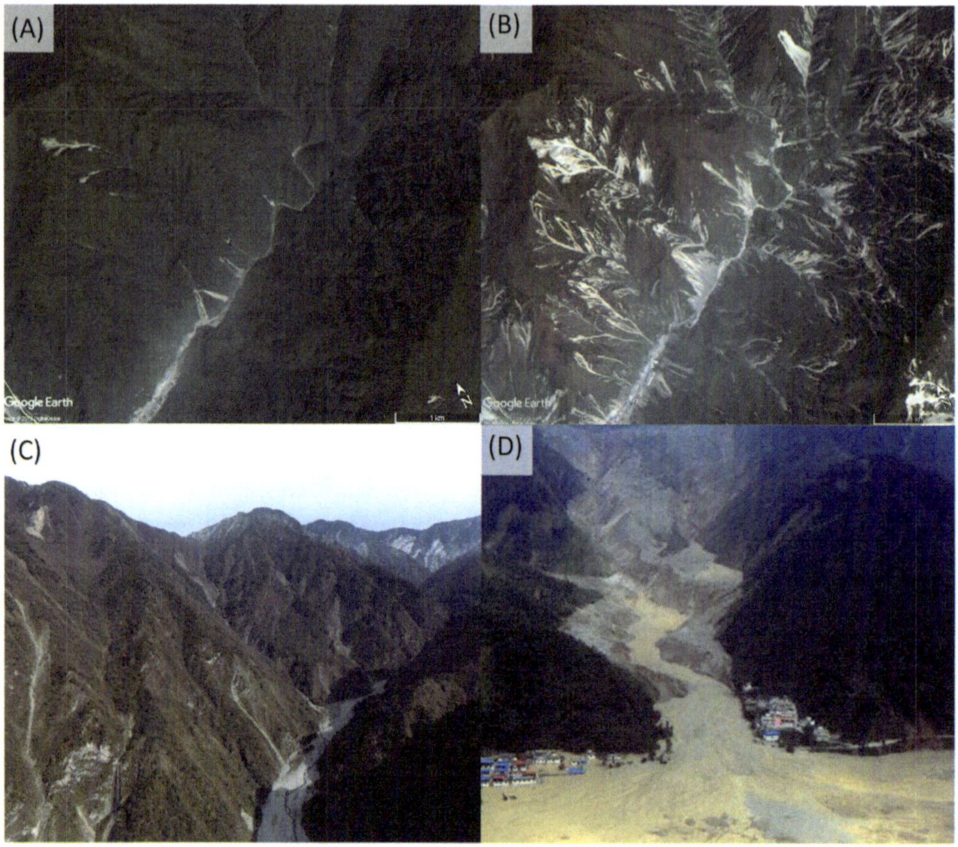

Figure 14.1 Geomorphic consequences of the Wenchuan earthquake. Google Earth imagery of the Taoguan catchment (A) before (2005) and (B) after (2011) the Wenchuan earthquake. (C) Photo of the Luoquan catchment in 2019. The photo shows the revegetation of landslide scars, with a small number of coseismic landslides still actively generating debris flows. The channel has been filled with sediment that has been remobilised by multiple large debris flows, with the most recent triggered 2 months prior to the photograph. Approximately 2 km of the channel can be seen in this figure. (D) The consequences of the 2010 large debris flow from the Wenjia Gully (Tang et al., 2011). The width of the channel is approximately 100 m. This flow buried many houses built during the recovery period from the Wenchuan earthquake. Source: *Maps data: Google, © 2019 Digital Globe.*

14.2 Wenchuan earthquake geomorphic setting and dynamics

The Longmen Shan is one of the steepest mountain ranges in the world, with topographic relief exceeding 5000 m over 50 km in length (Densmore et al., 2007). The mountain range divides the western edge of the Sichuan Basin (average elevation of 500 m) from the eastern rim of the Tibetan Plateau (average elevation 4000 m) (Liu-zeng et al., 2009). The steep topography is dissected by several tributaries of the Yangtze River, including the Dadu, Qingyi, Min, Fu, Tuo and Jialing Rivers

(Kirby et al., 2003). The Longmen Shan has a complex tectonic history as part of the Mesozoic closure of the Paleo-Tethys to form the Indosinian orogeny and subsequent orogenic events in the early Cretaceous, and is associated with the current Indian–Asian collision (Yong et al., 2003). Hence, the mountains contain varied lithologies, including metaigneous and metasedimentary Proterozoic basement, such as the Pengguan Massif, as well as a wide range of Palaeozoic and younger sedimentary and metasedimentary units (Burchfiel et al., 1995). Prior to the Wenchuan earthquake, a number of active faults were identified (Burchfiel et al., 1995), but geodetic measurements recorded limited shortening, suggesting a possible role for lower crustal flow in driving surface uplift (Royden et al., 1997). However, significant shortening associated with the Wenchuan earthquake supports an important, possibly exclusively coseismic, surface uplift element (Hubbard & Shaw, 2009).

Climate in the Longmen Shan is controlled by the East Asian and Indian monsoons, which are characterised by warm, wet summers and an orographic rainfall effect due to the steep topography (Liu-Zeng et al., 2011; Wang et al., 2015). The orogen has an average annual precipitation of ~1200 mm at its eastern edge, adjacent to the Sichuan Basin. The majority (70%–80%) of rainfall occurs between May and September (Liu-Zeng et al., 2011). Heavy monsoonal rainfall corresponds to high suspended sediment concentrations, with 97% of all suspended sediment transported between May and October in both pre- and postearthquake settings (Wang et al., 2015). Hillslopes are also susceptible to rainfall-triggered landsliding during the monsoon (Ouimet et al., 2007). The high precipitation intensity during summer has considerably shaped the post-earthquake sediment cascade, with annual reactivations of coseismic landslides, often in the form of debris flows, for several years after the earthquake (Fan et al., 2018; Francis et al., 2022).

The M_w 7.9 Wenchuan earthquake occurred on 12 May 2008 along the middle segment of the Longmen Shan thrust belt, with the epicentre close to the town of Yingxiu and a focal depth of 14–19 km, generating shaking intensities up to Modified Mercalli Intensity XI (Xu, Wen, et al., 2009). Two main fault segments ruptured along the Beichuan and Pengguan faults (Fig. 14.2). Both faults had a history of surface-rupturing earthquakes in the late Pleistocene to Holocene (Densmore et al., 2007). The Beichuan fault created a 240 km surface rupture characterised by right lateral strike-slip and thrusting with up to 10 m of vertical displacement (Liu-zeng et al., 2009; Xu, Wen, et al., 2009). The Pengguan fault ruptured along ~72 km with up to 3.5 m of dominantly thrust offset. The offset along the surface ruptures varied along strike, with the towns of Yingxiu and Beichuan impacted by particularly large slip and moment release. GPS recordings and measurements along the faults conducted prior to the earthquake indicated that preearthquake motion on these faults was primarily strike-slip (Densmore et al., 2007; Zhang et al., 2004). Over 1300 aftershocks of M_w>3.5 occurred, with a greater number concentrated in the southern portion of the fault

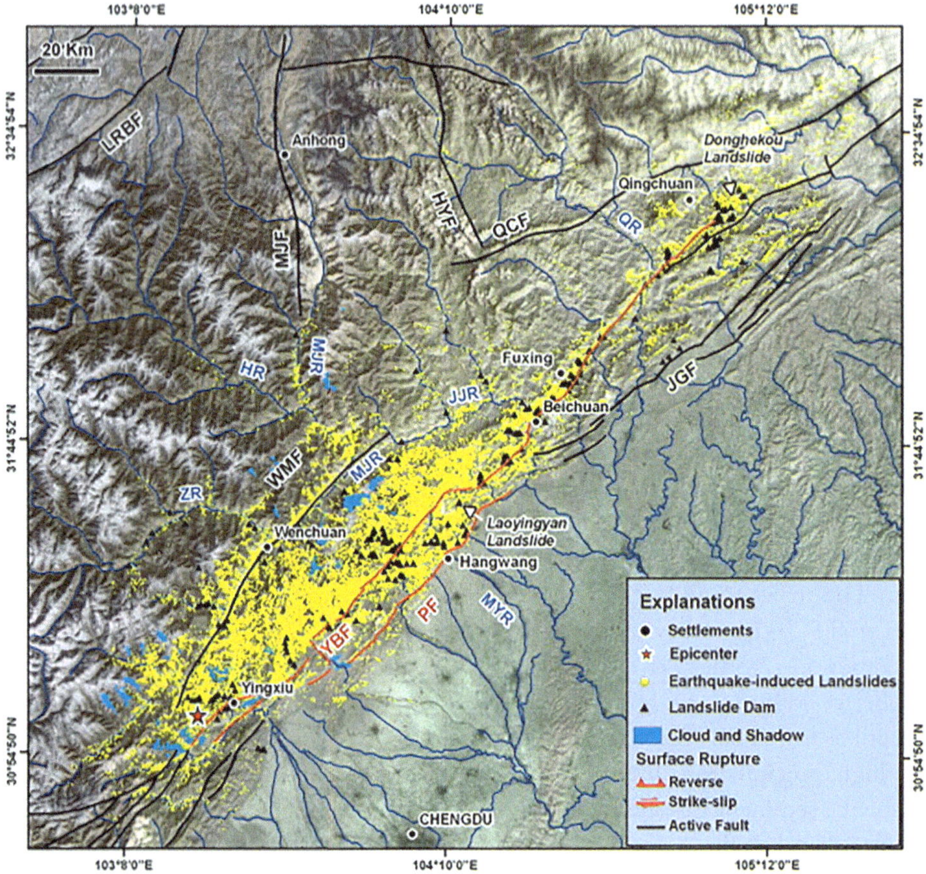

Figure 14.2 Map of the Wenchuan earthquake-affected area with the ruptured faults (*YBF*, Yingxiu-Beichuan fault; *PF*, Pengguan fault) in red and other known active faults in black (*JGF*, Jiangyou–Guanxian fault; *QCF*, Qingchuan fault; *HYF*, Huya fault; *MJF*, Min Jiang fault). Landslides and landslide dams mapped by Gorum et al. (2011) are shown. The following rivers are labelled: *MJR*, Min River; *ZR*, Zagunao River; *HR*, Heishuehe River; *MYR*, Mianyuan River; *JJR*, Jian River; and *QR*, Qingzhu River. Map lines delineate study areas and do not necessarily depict accepted national boundaries. Source: *Figure taken from Gorum, T., Fan, X., van Westen, C. J., Huang, R. Q., Xu, Q., Tang, C., & Wang, G. (2011). Distribution pattern of earthquake-induced landslides triggered by the 12 May 2008 Wenchuan earthquake.* Geomorphology, 133, *152–167. https://doi.org/10.1016/j.geomorph.2010.12.030.*

system to depths of approximately 30 km (Cai et al., 2011). The 2008 Wenchuan earthquake added $2.6 \pm 1.2\,km^3$ of crustal volume to the mountain range (Li et al., 2014; Parker et al., 2011; Xu, Wen, et al., 2009).

The 2008 earthquake led to almost 70,000 fatalities, with a further 18,000 people missing (Cui et al., 2011). About one-third of the fatalities (~20,000) were attributed to coseismic landslides (Tang, Zhu, Qi, 2011). Landslides also destroyed homes and blocked highways and bridges, isolating the city of Wenchuan and many other towns and villages (Tang, Zhu, Qi, 2011). Thirty-four large barrier lakes were formed from landslide dams in the area, which led to secondary hazards such as flash flooding and debris flows (Tang, Zhu, Qi, 2011). A series of catastrophic debris-flow events in the years that followed the earthquake have resulted in further fatalities and infrastructural damage (Tang et al., 2011; Yang et al., 2021).

14.2.1 Remote sensing of the geomorphic cascade of the Wenchuan earthquake

The timing of the Wenchuan earthquake coincided with advances in satellite data availability, and the earthquake occurred in a relatively accessible area of China that could support abundant fieldwork. The resulting sediment cascade was observed and recorded in detail using both remotely sensed and field data, providing some of the most detailed observations of an earthquake sediment cascade at that point (Fan, Scaringi, Korup, et al., 2019; Huang & Li, 2014). The result was an abundance of data about landslides, landslide dams and other geomorphic hazards that were available within weeks of the main shock. Subsequent high-resolution imagery has allowed for process-level interpretation of the geomorphic cascade (Fan, Scaringi, Domènech, et al., 2019).

A wide range of remotely sensed imagery allowed for rapid characterisation of coseismic landslides using semiautomated (Gorum et al., 2011; Li et al., 2014; Parker et al., 2011) and manual (Dai et al., 2011; Xu et al., 2014) mapping from satellite imagery, and field-based (Huang & Li, 2009) methods. Coseismic landslide datasets were developed from preearthquake satellite images by compiling data from Advanced Spaceborne Thermal Emission and Reflection Radiometer (ASTER; 15 m resolution, image dates between 2000 and 2006), Advanced Land Observing Satellite (ALOS; 10 m resolution in 2007, 2.5 m resolution in 2006, 2008), CARTOSAT-1 imagery (2.5 m resolution in 2007) (Gorum et al., 2011), Satellite Pour l'Observation de la Terre 5 (SPOT-5; 2.5 m resolution), Resourcesat-1 (IRS-P6; 5.8 m resolution) and China–Brazil Earth Resources Satellite (CBERS; 19.5 m resolution) (Dai et al., 2011). Postearthquake images from the ASTER, ALOS, SPOT-5, IKONOS (2.5 + 4 m resolution), QuickBird (0.6 and 2.4 m resolutions), DigitalGlobe Worldview-1 (0.5 m resolution) and CARTOSAT-1 platforms were collected in different parts of the region in the months after the earthquake (Dai et al., 2011; Gorum et al., 2011; Li et al., 2014; Parker et al., 2011; Xu et al., 2014). This information was supplemented by aerial orthophotos flown by Chinese governmental agencies (Dai et al., 2011). The frequency of satellite imagery allowed for accurate estimation of the landslides triggered by the main shock and aftershocks, but was not sufficient to discriminate between main shock landslides and subsequent coseismic landslides.

A small number of multitemporal landslide datasets have been collected since the Wenchuan earthquake (Chen, Tang, Xiong et al., 2024; Fan et al., 2018; Fan, Scaringi, Domènech, et al., 2019; Tang, Zhu, Qi & Ding, 2011; Wang et al., 2022; Yang, Tang, et al., 2023). Imagery from very-high-resolution satellites (e.g. WorldView) was used to assess how coseismic landslide deposits in Wenchuan evolved through time and identify specific reactivation processes (e.g. rilling and debris-flow triggering). Publicly accessible geospatial platforms, such as Google Earth Pro and Bing Maps, supported landslide detection and inventory development through 3D terrain visualisation and historical imagery archives (Dahlquist & West, 2019). From these multitemporal inventories, the spatiotemporal evolution and activity of landslides following an earthquake could be addressed for the first time.

14.3 The geomorphic and sediment cascade of the Wenchuan earthquake

The geomorphic and sediment cascade associated with the Wenchuan earthquake allowed the science community to explore the complex interactions and stochastic events controlling the movement of earthquake-generated landslide debris through hillslope and channel networks (Fig. 14.3) (Chen, Tang, Xiong et al., 2024; Dai et al., 2021; Fan et al., 2018; Fan, Scaringi, Domènech, et al., 2019; Francis et al., 2022; Yang, Tang, et al., 2023). The specific geomorphic processes and extent of the Wenchuan sediment cascade are both dictated by the volume, composition and spatial (and temporal) distribution of coseismic landslide debris. Most landslide sediment was deposited on hillslopes above the channel network or in low-order streams with low stream power (Li et al., 2014). However, notable large landslides ran out directly into the channel network, creating significant landslide dam reservoirs, some with potentially catastrophic hazard consequences (Xu, Fan, et al., 2009). While many of these reservoirs have been modified for hazard-reduction purposes, some remain as long-term stores of upstream sediment. Hillslope sediment was mobile during and immediately after the earthquake, generating numerous postearthquake landslides and debris flows, with some cutting off transport networks and destroying infrastructure (Tang et al., 2009; Tang et al., 2011; Hu & Huang, 2017; Xu et al., 2012). Mobilisation of sediment from hillslopes has inundated lower-order channels, which have progressively become the source of large debris flow and flood events such as in 2019 (Fig. 14.1C) (Xiong et al., 2021; Yang et al., 2021). However, much of the coseismic sediment has revegetated in the 17 years after the earthquake. Currently, the epicentral area shows little evidence of the vast quantities of earthquake sediment that are stored on hillslopes and in the upper part of the channel network, raising questions about the future fate of this sediment.

The generation of coseismic landslides is often seen as the starting point for a series of events that have somewhat overlapping terminology in the literature, reflecting the

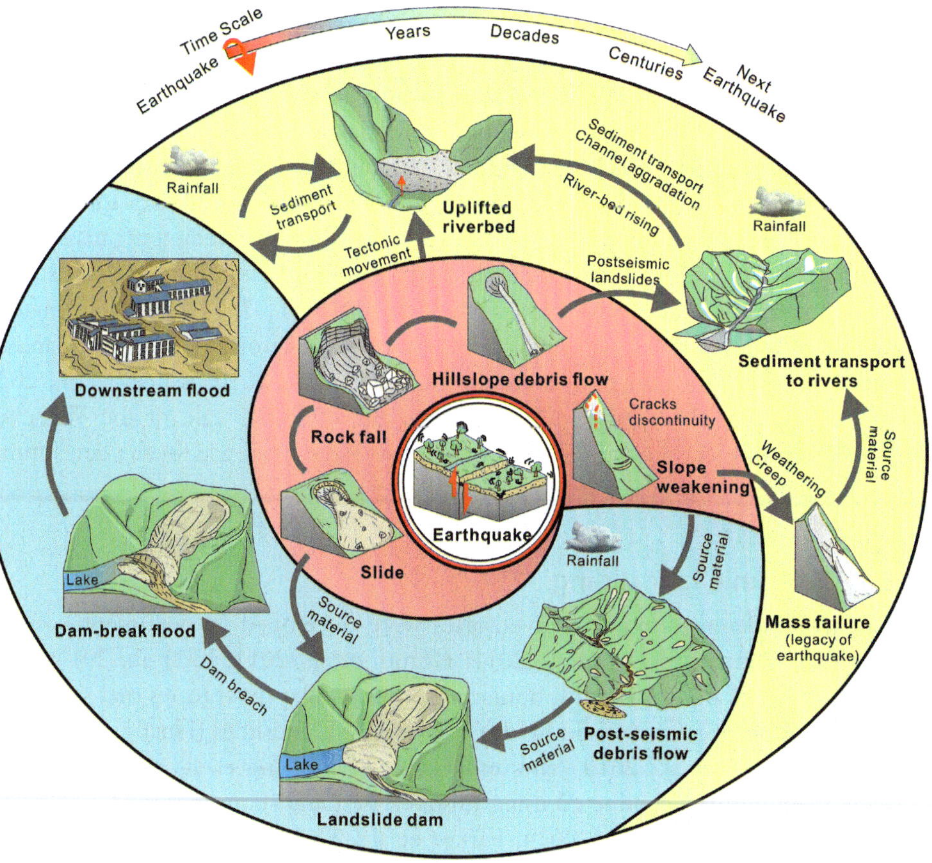

Figure 14.3 The Wenchuan earthquake sediment cascade. The complexity of the sediment cascade is related to the myriad of different processes that can deliver sediment between the major hillslope and channel stores of sediment and the time scales over which these processes occur (Fan, Scaringi, Korup, et al., 2019).

relationship between the cascading movement of sediment from hillslopes to rivers via sediment reservoirs (a sediment cascade; Davies & Korup, 2010) and a suite of geomorphic processes (a geomorphic cascade). We separate the geomorphic cascade from the sediment cascade in this contribution, since the movement of sediment (and water) between different reservoirs in the Wenchuan region can occur via multiple chains of geomorphic processes (Francis et al., 2022). Hence, we use the term geomorphic cascade when we have sufficient resolution to describe the individual processes, and sediment cascade when we are referring more generally to sediment movements without necessarily knowing the specific processes. Finally, the range of cascading geomorphic processes can also exacerbate hazardous processes, particularly when combined with other geophysical (e.g. aftershocks) and anthropogenic (e.g. dam collapse and industrial waste spills) hazards, to create a hazard cascade (Fan, Scaringi, Korup, et al., 2019; Gill & Malamud, 2014; Yanites et al., 2025).

The nature and scale of the Wenchuan earthquake sediment cascade contrasts with sediment cascades in other large continental earthquakes with respect to the dominance of the remobilisation of coseismic landslide sediment and the presence of extremely large, highly fluidised landslides. In contrast, the 1999 M_w 7.7 Chi-Chi earthquake in Taiwan (see Chapter 13) generated more new landslide sediment mass during postearthquake Typhoon Toraji than through coseismic landsliding (Hovius et al., 2011). More recent large earthquakes, such as the 2015 M_w 7.8 Gorkha earthquake in Nepal (see Chapter 15), have involved both remobilisation of coseismic landslide sediment and generation of new landslides (Khattak et al., 2010; Kincey et al., 2021), but none have generated large debris flows at the rate and scale of the Wenchuan earthquake. In this section, we will review the nature and evidence for the development of the Wenchuan earthquake sediment cascade.

14.3.1 Coseismic landsliding

At least five different, complete inventories were developed to document coseismic landslides in Wenchuan (Dai et al., 2011; Gorum et al., 2011; Li et al., 2014; Parker et al., 2011; Xu et al., 2014), with numerous other smaller inventories that subsampled particular regions or are restricted by size to the largest landslides (Huang and Li, 2009; Ouimet, 2010; Qi et al., 2010; Sato and Harp, 2009; Tang et al., 2016). Coseismic landslide mapping was conducted using unsupervised and semisupervised classification (Gorum et al., 2011; Li et al., 2014; Parker et al., 2011) as well as manual mapping (Dai et al., 2011; Xu et al., 2014). Many of the differences in the number of observed landslides likely relate to methodological differences between studies and differences in the types and resolution of the imagery used. This methodological variation between studies yielded four estimates of 56,000 to 60,000 coseismic landslides and one estimate of 196,000 landslides triggered by the earthquake (Dai et al., 2011; Gorum et al., 2011; Li et al., 2014). Comparisons of different methods suggest that no single method is necessarily more accurate or precise than another (Milledge et al., 2022; Rault et al., 2019), making it difficult to compare the accuracy of different methods to establish the total volume of the event or the magnitude-frequency relationships of the coseismic landslides. DEMs of difference may provide a more accurate view of the total coseismic landslide volume (Jones et al., 2025), but are not available for this earthquake. While no specific accuracy comparison has been made between the inventories, Tanyaş et al. (2017) compared the five Wenchuan coseismic inventories against eight accuracy criteria, with three inventories fulfilling four of these criteria (Dai et al., 2011; Gorum et al., 2011; Li et al., 2014), one dataset fulfilling five of the criteria (Xu et al., 2014) and one dataset fulfilling six criteria (Tang et al., 2016). Visual comparison of these landslide datasets against high-resolution imagery demonstrates inaccuracies in all the landslide inventories, both between the different datasets and in the location of specific landslides relative to the imagery. The methodological differences also affect estimates of the total landslide area associated with the earthquake. Therefore when volumes are calculated from area-volume scaling relationships they range from $2.8 \pm 0.9\,km^3$ to as high as $12\,km^3$ (Xu et al., 2016).

The distribution of coseismic landslides triggered by the Wenchuan earthquake was primarily governed by seismic parameters, whereas lithologic and topographic factors exerted comparatively less influence. Coseismic landslides occurred at higher frequencies along the hanging walls of the faults, decaying rapidly away from the trace of the main Beichuan fault (Fig. 14.2) (Chen et al., 2012; Dai et al., 2011; Gorum & Carranza, 2015; Gorum et al., 2011; Huang & Li, 2009; Yuan et al., 2013). The hanging wall of the southwestern segment of the Beichuan fault alone contained more than 70% of the total landslides (Dai et al., 2011; Gorum et al., 2011). The nature of slip along the Beichuan fault rupture varied from dominantly thrust motion in the southwest, close to the epicentre, to right lateral strike-slip motion at its north-easterly extent (Hao et al., 2009; Shen et al., 2009). Landslide frequency was higher where motion was dominantly thrust, and less where it was dominantly strike-slip (Chen et al., 2012; Gorum & Carranza, 2015; Yuan et al., 2013). Frequency-density mapping of landslides along the fault strike demonstrated that changes in fault motion along strike also controlled the width of the area affected by landslides (Gorum & Carranza, 2015) and the frequency of large landslides, such that large landslides were only found within 5 km of the dominantly strike-slip northeastern section of the fault but were found up to 15 km from the thrust-dominated southwestern section of the fault (Chen et al., 2012). Distance from the hanging wall was a more significant control on landslide triggering than other common metrics such as distance from the epicentre and Mercalli Intensity (Dai et al., 2011).

Topographically, coseismic landslides cooccurred with areas of steeper and more incised topography and were particularly noticeable along deeply incised rivers such as the Min, Jian and Zagunao. Landslide densities correlated (although not strongly) with slope angle and elevation but did not correlate in a simple way with slope aspect or curvature (Xu et al., 2014). There were no clear controls on the topographic characteristics of the largest ($>50{,}000\,m^2$) landslides (Chen et al., 2012). While there was some spatial clustering of landslides towards ridges, there was no consistent pattern across the landscape, with estimates of the proportion of landslides triggered close to the ridge crest depending strongly on the coseismic landslide dataset used (Rault et al., 2019). Coseismic landslide material remained concentrated high in the drainage network, with between 20% and 90% of the landslides considered to be 'connected' to the stream network (Li et al., 2016). These combined observations demonstrated that coseismic landslides were more likely to occur in the upper parts of catchments and on steeper slopes. However, steep slopes occur frequently across the whole of the epicentral region, and once differences in peak ground acceleration and distance from fault were accounted for, there were no statistically significant topographic factors that specifically increased the probability of more or larger coseismic landslides.

Lithology also acted as a secondary control on the distribution of Wenchuan earthquake coseismic landslides. Coseismic landslide densities were higher in carbonate rocks (Gorum et al., 2011) and slightly lower in igneous lithologies (Xu et al., 2014), but these observations were not statistically significant. The impact of lithology was quantified by calculating the optimal strength parameter by inverting the Newmark equation based on peak ground acceleration and slope. Rock strength increased with distance from the fault, suggesting that rock fracture associated with seismic activity acted as a strong control on the lithologic effect on coseismic landslides (Gallen et al., 2015). As a result, despite a wide range of different geologies, the role of lithology remained secondary to the seismic parameters in explaining the distribution of coseismic landslides.

14.3.2 Landslide dams

Landslides that blocked rivers and formed lakes were a major and immediate hazard because of their high potential for generating destructive floods once overtopped. These dams required urgent risk management measures, including engineered stabilisation and precautionary evacuations (Fan et al., 2012; Peng & Zhang, 2012). Remote sensing imagery interpretation identified 828 landslide dams triggered by the Wenchuan earthquake, accounting for ~1.4% of the >60,000 mapped coseismic landslides (Fan et al., 2012; Fig. 14.2; Gorum et al., 2011). Spatial clustering analysis within a 1 km^2 moving circular window revealed maximum values of 4.5 landslide dams per km^2, where their density reached 8.3% of 54.1 coseismic landslides per km^2. Of the 828 landslide dams, 501 (61%) caused complete river blockage, while the rest resulted in partial damming or channel diversion. Completely river-blocking landslides covered areas ranging from 1249 to 7.1×10^6 m^2, slightly larger than partially river-blocking landslides (768 to 1.3×10^6 m^2) and produced lakes with areas spanning from 217 to 6.5×10^6 m^2. Collectively, the area covered by landslide dams equated to 54 km^2, which is ~7% of the total coseismic landslide area. Additionally, Fan et al. (2012) estimated that these landslide dams temporarily stored ~0.6 km^3 of water and sediment, highlighting their significant role in altering local hydro-sedimentary dynamics. To reduce the risk of catastrophic outburst floods, 32 dams were artificially drained using explosives and heavy machinery (Xu, Fan, et al., 2009).

The large number of landslide dams associated with the Wenchuan earthquake allowed detailed evaluation of complex and multifaceted uncertainties associated with assessing their stability (Fan, Dufresne, et al., 2020). Two years after the Wenchuan

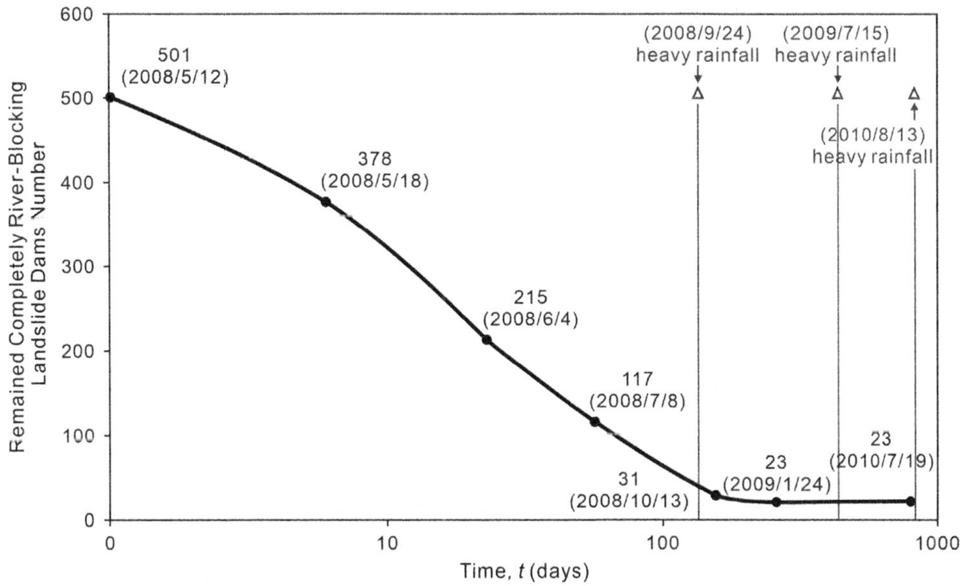

Figure 14.4 Decay of the number of remaining completely river-blocking landslide dams with time after the Wenchuan earthquake. Triangles show the dates of heavy rainfall in the summers of 2008, 2009 and 2010 (Fan et al., 2012). Source: *Taken from Fan, X., van Westen, C. J., Xu, Q., Gorum, T., & Dai, F. (2012b). Analysis of landslide dams induced by the 2008 Wenchuan earthquake.* Journal of Asian Earth Sciences, 57. *https://doi.org/10.1016/j.jseaes.2012.06.002.*

earthquake, only 23 of the 501 completely river-blocking landslide dams remained in place, and those were predominantly formed by complex deep-seated, large-volume slope failures (Fig. 14.4; Fan et al., 2012). During the immediate postseismic period (within 1 month), 60% of landslide dams (covering ~43% of the total dam area) failed. Although the earthquake region experienced three heavy rainstorms within 3 years postearthquake (triangle symbols in Fig. 14.5), there was no strong correlation between these rainstorms and the increased frequency of dam breaches. Landslide reactivation can cause postearthquake landslide dam formation driven by aftershocks and/or river erosion that undercuts hillslopes or remobilises loose materials deposited by prior landslides (Tacconi Stefanelli et al., 2015). In the Beichuan area, the Tangjiawan landslide triggered by the 2008 Wenchuan earthquake initially dammed the Duba River, forming a barrier lake. This landslide reactivated 8 years later on 5 September 2016, redamming the river at the same site with a 20-m high barrier and creating a new 0.6 million m^3 lake, demonstrating the persistent long-term hazard of such earthquake-induced landslide dams (Fan et al., 2018).

The primary geomorphic and hazard questions after the Wenchuan earthquake involved the stability and longevity of landslide dams. The stability of landslide dams is governed by morphometric, hydrogeological and geotechnical parameters

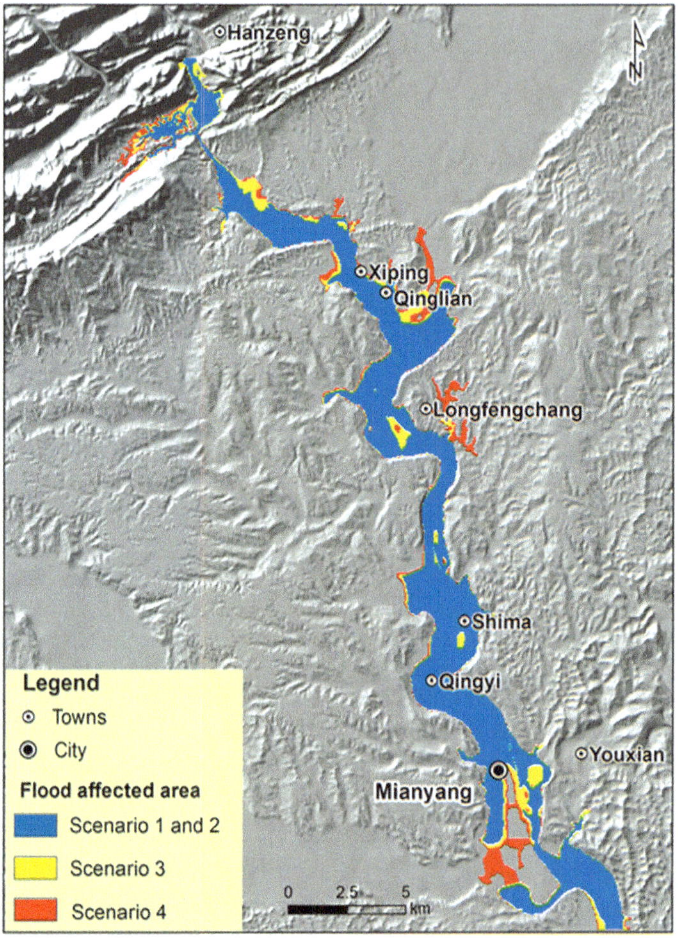

Figure 14.5 Modelled flood areas for different scenarios of the breach of the Tangjiashan landslide dam, located just upstream of Beichuan town. See Fig. 14.2 for location. The yellow and red areas represent the increase in flooded areas for scenarios 3 and 4, respectively. Map lines delineate study areas and do not necessarily depict accepted national boundaries. Source: *Taken from Fan, X., Tang, C. X., Van Westen, C. J., & Alkema, D. (2012). Simulating dam-breach flood scenarios of the Tangjiashan landslide dam induced by the Wenchuan Earthquake.* Natural Hazards and Earth System Science, 12, *3031–3044. https://doi.org/10.5194/nhess-12-3031-2012.*

(Casagli & Ermini, 1999; Fan, Dufresne, et al., 2020). Key volumetric and dimensional attributes (e.g. dam volume, height, width, length, lake volume and upstream catchment area) informed rapid field assessment of landslide dam stability (Dong et al., 2009). Blockage index (BI), the ratio of landslide dam volume to catchment area (Casagli & Ermini, 1999; Ermini & Casagli, 2003), has been effectively applied to explain the distribution of dammed lakes following the 2008 Wenchuan earthquake

(Cui et al., 2009; Xu, Fan, et al., 2009). Where BI<2, earthquake-related landslide dams did not form in Wenchuan, while dams with BI>5 maintained stability after 1 year (Fan et al., 2012). Such rapid stability and hazard assessment constitutes a critical priority within landslide dam emergency management protocols.

Concurrent with geomorphologic methods for identifying potential breaching events, considerable work was conducted to calculate the scale of potential breaching floods. Flood modelling is often conducted using peak discharges estimated with independent dam-breach models as input conditions. The most important flood modelling effort simulated a potential breach of the Tangjiashan landslide dam, just upstream of Beichuan town, which impounded approximately 300 million m^3 of water. This lake threatened catastrophic outburst flooding, endangering over 2.5 million residents across downstream settlements and Mianyang City, situated 85 km from the dam site (Fig. 14.5; Xu, Fan, et al., 2009). The worst-case geotechnical collapse fully drained the lake, and that Scenario 4 yielded the highest modelled peak discharge (75,500 m^3/s) with the shortest duration of flow at a discharge >1000 m^3/s (Fan et al., 2012). Developing a unified numerical model replicating the sequential stages from landslide dam emplacement to dam-break flooding remains an outstanding objective for future simulations.

14.3.3 Debris flows

The frequency and magnitude of debris-flow events increased in the epicentral region after the Wenchuan earthquake. As of 2024, this region was still experiencing elevated rates of debris flow activity when compared to an admittedly short baseline period (Chen, Tang, Xiong et al., 2024; Wang et al., 2024). Debris flows were first observed within days of the earthquake; these debris flows hampered recovery from the earthquake, resulting in dozens of fatalities and further displacing many thousands of people (Xie et al., 2009). Many major debris-flow events occurred during the peak monsoon period in July and August of 2008, especially in the Beichuan area (Fan, Scaringi, Domènech, et al., 2019; Wang et al., 2022). Although debris flows have been recorded in most years since 2008, major events in 2010, 2013 and 2019 were associated with fatalities and substantial damage to infrastructure (Fan et al., 2018; Xu et al., 2012; Yang et al., 2021). Particularly noteworthy are the 37 large debris flows with volumes greater than 10^6 m^3, which caused a disproportionate amount of damage (these are often called catastrophic debris flows in the literature; e.g. Tang et al., 2011).

The 2008 Wenchuan earthquake drastically changed the distribution of debris-flow magnitudes when compared to before the earthquake, while also increasing the frequency of debris flows at all magnitudes (Harvey, 2022). The increase in postearthquake debris-flow frequency has been attributed to coseismic landslide deposits providing a readily available source of debris-flow material (Tang et al., 2009).

Remobilisation of landslide material promotes debris-flow development when the mixed grain size and high porosity deposit becomes saturated and collapses to create a flowing mass of water and sediment (Iverson et al., 1997). Increased landslide frequencies coincided with a decrease in the volume of rainfall required to trigger debris flows, as measured using rainfall intensity–duration thresholds (Caine, 1980; Tang et al., 2009). Rainfall intensity–duration thresholds are correlations between the intensity (measured as the maximum rainfall rate in an event) and duration (the total time over which rain falls in an event) that are sufficient to trigger debris flows within a geographical area (usually a catchment) (Caine, 1980). Immediately after the earthquake, rainfall intensity–duration thresholds for debris-flow triggering dropped when compared to the preearthquake curves (Guo et al., 2016). These rainfall intensity–duration values recovered to ~50% of preearthquake levels within 8 years after the earthquake (Guo et al., 2016). Current estimates suggest that thresholds will remain below preearthquake values for between 10 and 20 years following the earthquake (Fan, Scaringi, Korup, et al., 2019). Whilst rainfall intensity–duration curves have limitations (Chang et al., 2021; Guzzetti et al., 2008), these systematic observations of changes to rainfall thresholds argue that debris-flow triggering conditions are coevolving in the immediate postearthquake period.

The intensity of the annual monsoon affecting the Wenchuan area, as measured by total monsoon precipitation, varied throughout the postearthquake period. No strong correlation was observed between overall monsoon intensity and debris-flow triggering, suggesting a systematic hydrological explanation for changing rainfall intensity–duration thresholds is unlikely. Instead, large debris-flow events are typically associated with localised, high-intensity precipitation cells (Tang et al., 2011; Hu & Huang, 2017; Xu et al., 2012; Yang et al., 2021). Some authors have suggested that material properties may instead control the changing rainfall intensity–duration threshold of debris flows, with coseismic debris thought to evolve after an earthquake (Domènech et al., 2019; Francis et al., 2022). Decreases in debris-flow frequency have been attributed to multiple changing conditions, including decreases in sediment supply (Zhang et al., 2013), increased vegetation cover (Yunus et al., 2020) and a change in the grain size distribution of debris-flow source material (Chen, Zhang, et al., 2014; Domènech et al., 2019). Decreased debris-flow frequency is unlikely to relate to the exhaustion of sediment on hillslopes, as a large proportion of coarse sediment remains on hillslopes. Therefore processes responsible for stabilising hillslope deposits are more likely (Fan et al., 2021). The coarsening of debris-flow deposits as fines are winnowed via previous overland flow could partly explain the decrease in debris-flow frequency, since fine grains are considered critical for sustaining debris flows (De Haas et al., 2015). Field evidence supports this hypothesis because rapid erosion of fine soil fractions was observed within years through consecutive debris flows in identical areas, coinciding with observed coarsening of postearthquake debris flows (Chen, Chen et al., 2014;

Zhang et al., 2014). Artificial slope experiments confirm that hydrologically driven grain coarsening through fine-particle erosion increases hydraulic conductivity and interparticle contact forces (Chen, Chen, et al., 2014; Hu et al., 2017; Zhang et al., 2014). However, grain size is exceptionally difficult to measure at the field scale, particularly for these large debris flows with wide grain size distributions, meaning that accurate constraints on changes in grain size are limited (Harvey, Hales, et al., 2022). There has also been evidence of coarse deposits remobilising as large debris flows, suggesting that deposit winnowing is not the only explanation for the reduction in debris-flow frequency (Harvey et al., 2025; Yang et al., 2021). The revegetation of landslide deposits provides an alternative explanation for decreased debris-flow frequency, with vegetation known to increase the shear strength of deposits and increase the triggering threshold for remobilisation (Shen et al., 2017). However, the main colonising plants on deposits in the Longmen Shan were grasses and shrubs, which are thought to have weak root structures, suggesting they are unlikely to be the sole mechanism for the decreased frequency of debris flows (Francis et al., 2022; Hales, 2018; Shen et al., 2020). It is probable that the general decrease in debris-flow frequency can be attributed to a combination of these and other processes. Nonetheless, the abundance of literature exploring the evolution in debris-flow activity after the Wenchuan earthquake has provided us with a greater understanding of how long debris-flow activity may persist after an earthquake and the complex controls on this persistence.

14.3.4 The fluvial system

The sedimentary response of the major river systems that drain the earthquake-affected area was muted when compared to previous events such as the 1999 Chi-Chi earthquake in Taiwan. Immediately after the Wenchuan earthquake, suspended sediment discharges increased by 3–7 times preearthquake levels (Wang et al., 2015). Elevated sediment yields returned to pre-earthquake levels within 5 years, resulting in an additional 92.5 (±9.2) Mt of 1 suspended sediment export (Wang et al., 2015). Increases in suspended load have led to measurable increases in sedimentation within the Three Gorges Reservoir (Liu et al., 2025). Increases in bedload have been observed in tributary channels, but rarely in the major river systems such as the Min. Sedimentation rates measured in the Zipingpu Reservoir (10 km downstream of the earthquake epicentre) have not increased dramatically, although changes in magnetic susceptibility identify key postearthquake rainfall events (Zhang et al., 2019). There is evidence of more significant postearthquake sedimentation at the inlet to the reservoir, where repeat surveys of sediment volumes suggest deposition of around 10%–20% of the total sediment generated upstream of that point (Li et al., 2025). The sedimentation rate at the inlet to the Zipingpu Reservoir is consistent with rates of sediment export from lower-order catchments (Francis et al., 2022), suggesting efficient sediment transport within the large trunk river systems such as the Min River.

Looking forward, the impact of the remaining sediment on the fluvial system remains largely unknown. Channels in the Sichuan Basin downstream of the mountain front incised immediately after the earthquake, which was argued to be due to a reduction in sediment supply from the mountain front after the earthquake (Fan et al., 2016). The complex nature of the sediment cascade associated with the Wenchuan earthquake limits the volume, calibre and rate of sediment input to the fluvial network, which is in turn limited by the connectivity of the channel network to coseismic landslide debris (Li et al., 2016) and the rate of sediment delivery by debris flows and other hillslope processes (Francis et al., 2022; Xiong et al., 2022).

14.4 Advances in our understanding

14.4.1 Earthquake mass balance

The sediment generated by the Wenchuan earthquake (Parker et al., 2011) represents the largest volume of coseismic landslide material ever mapped by a landslide inventory and is surpassed only by the 1950 Assam earthquake in terms of overall landslide volume in the historical record (Kingdon-Ward, 1955; Marc et al., 2016). The volume of coseismic landslide material approached or exceeded the volume of crustal material that was uplifted during the earthquake, highlighting that the earthquake may be erosional in nature across the interseismic period (Fig. 14.6) (Li et al., 2014; Parker et al., 2011). This observation led to important questions about the role of earthquakes and other mechanisms of uplift, such as isostasy, in controlling orogenic uplift across different time scales (Densmore et al., 2012; Francis et al., 2020; Li et al., 2014; Marc et al., 2016; Molnar, 2012). Since the original work of Parker et al. (2011), estimates of the total landslide volume have been revised, with most subsequent studies, though not all (e.g. Xu et al., 2014), suggesting a substantially smaller landsliding event (e.g. Gorum et al., 2011). However, even with smaller estimated landsliding events, this still remains an important erosional event (Fig. 14.6).

In addition to the total mass of landsliding, the spatial concentration of landsliding near the range front may control the long-term incision of the orogen as a whole (Li et al., 2017). The Longmen Shan maintains considerable relief along its range front that is cut by large and deep river systems such as the Min, Tuo and Fu. Deeply incised rivers reduce normal stresses on the fault systems that act as a fundamental driver of slip activity on the central segment of the Longmen Shan Thrust Belt (Tan et al., 2018). These river systems were also areas of concentrated coseismic landsliding (Gorum et al., 2011; Ouimet, 2010) and had consistently high rates of pre and postearthquake suspended sediment discharge (Li et al., 2017; Liu-Zeng et al., 2011). The instantaneous

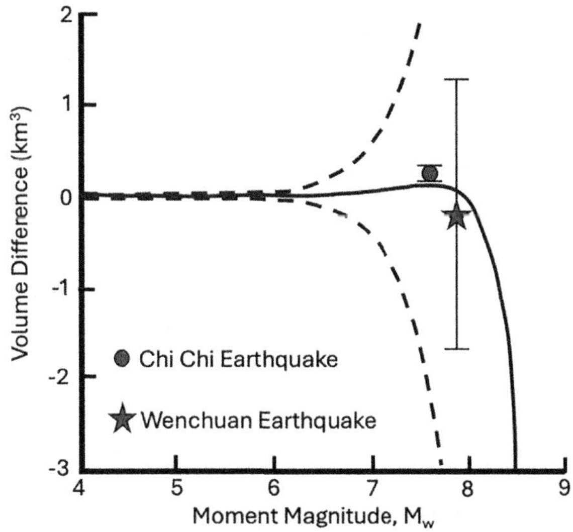

Figure 14.6 Plot of the modelled relationship of the net gain or loss of volume (volume difference) as a function of different earthquake magnitudes. The estimated volume differences for the 1999 Chi-Chi earthquake and 2008 Wenchuan earthquakes fall within this range. Source: *Modified from Li, G., West, A. J., Densmore, A. L., Jin, Z., Parker, R. N., Hilton, & R. G. (2014). Seismic mountain building: Landslides associated with the 2008 Wenchuan earthquake in the context of a generalized model for earthquake volume balance.* Geochemistry, Geophysics, Geosystems, 15, *833–844. https://doi.org/10.1002/2013GC005067.*

volume of coseismic landslide-derived debris averaged across the return period of large earthquakes (called the 'seismic erosion rate' by Li et al., 2017) may therefore account for most or all of the exhumation measured at the range front across multiple time scales (Li et al., 2017).

14.4.2 The nature and length of a geomorphic (hazard) cascade

The rapidly evolving rates and complex interactions between hillslope and fluvial processes in the post-Wenchuan earthquake landscape have shed light on how mountain hazard cascades develop and over what time scales. The addition of large volumes of coseismic landslide material to both the hillslopes and the fluvial network has increased flooding frequency and debris-flow generation (Fan, Scaringi, Korup, et al., 2019), consistent with enhanced activity of these processes following other earthquakes (Marc et al., 2015). However, the Wenchuan earthquake geomorphic cascade differs from others in two interesting ways; first, the frequency of large (catastrophic) debris flows has remained high since the earthquake, despite the reduced frequency of smaller debris flows (Tang et al., 2011; Hu & Huang, 2017; Xu et al., 2012; Yang et al., 2021). Enhanced rates of large debris flows may relate to the presence of considerable stored

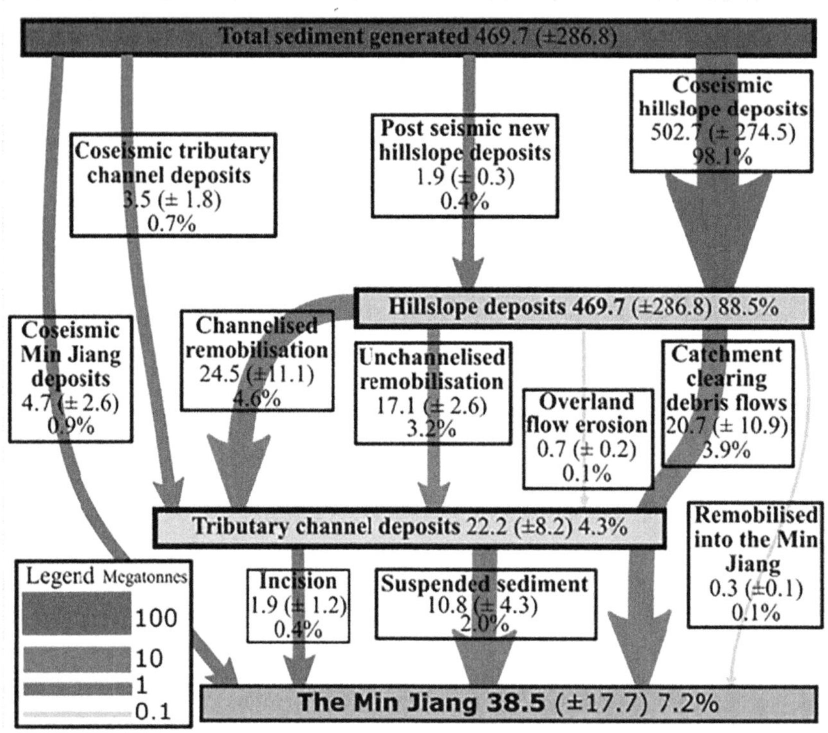

Figure 14.7 The sediment cascade for the Wenchuan earthquake, as quantified in detail for catchments draining the Min River (Min Jiang). Each value represents the mass of sediment in megatonnes (and proportion of total generated sediment, represented as a %) that has either been deposited or transported since the Wenchuan earthquake. Most of the mass of sediment generated by the earthquake remains stored as hillslope deposits. Sediment delivered to the channel network primarily occurs by debris-flow processes. Source: *Taken from Francis, O., Fan, X., Hales, T., Hobley, D., Xu, Q., & Huang, R. (2022). The fate of sediment after a large earthquake.* Journal of Geophysical Research: Earth Surface, 127. *https://doi.org/10.1029/2021JF006352.*

sediment high in steep tributary channels (Harvey et al., 2025), something that has not been observed in other earthquake events. Second, the hazard cascade is dominated by remobilisation of coseismic sediment deposited on hillslopes, rather than from new landslides preconditioned by the earthquake, as observed in other earthquakes such as the 1999 Chi-Chi earthquake in Taiwan (Hovius et al., 2011). The Longmen Shan is subjected to regular, intense and long-duration monsoonal precipitation events. Hence, the reduction in landsliding rate is more likely to relate to changes in material properties of coseismic landslide sediment, including coarsening and winnowing and changes in surface cohesion due to revegetation (Domènech et al., 2019), although strong evidence for the exact nature of how material properties evolve in this landscape remains elusive.

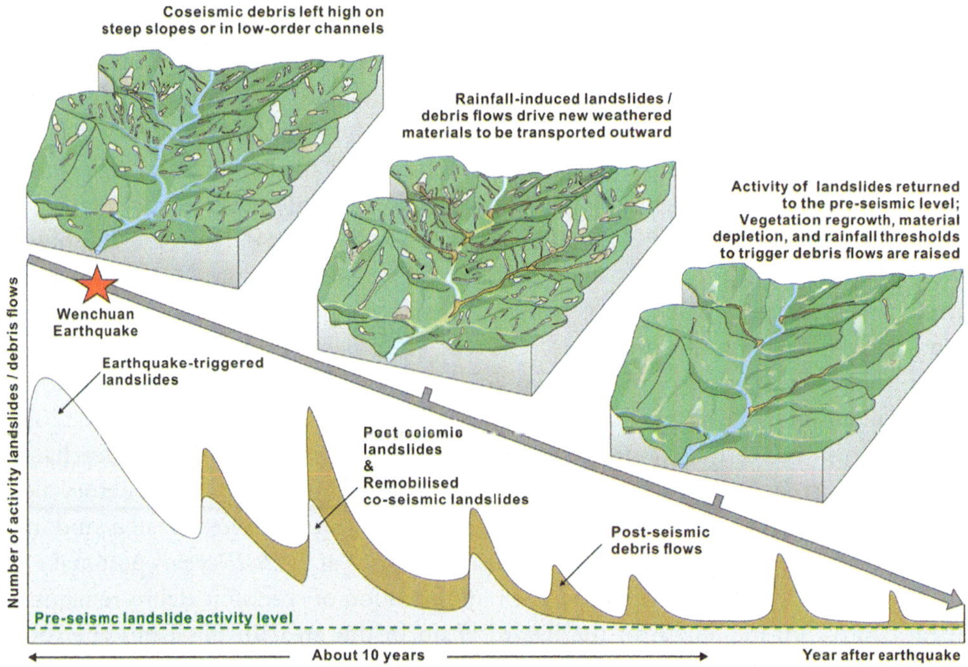

Figure 14.8 Schematic figure highlighting the geomorphic cascade for the Wenchuan earthquake. Immediately after the earthquake, new landslide material sits on hillslopes (grey curve), with the frequency of new landsliding decaying quickly as aftershocks subside. During subsequent monsoons, landslide material is transferred along hillslopes and into channels, creating material that feeds postseismic debris flows (brown curve). Within a decade, the generation of new landslides subsides, but the geomorphic cascade is maintained by remobilised debris.

A detailed sediment budget of the tributary catchments of the Min River which were affected by Wenchuan earthquake-triggered landsliding provides insight into the nature of the interacting hillslope and fluvial processes in the area (Francis et al., 2022). Sediment remobilised by the Min River may have been eroded or transported by up to four different geomorphic processes (Fig. 14.7). The most efficient pathways for the remobilisation of coseismic sediment were by large debris flows (catchment-clearing debris flows in Fig. 14.7) or by remobilisation of fine-grained landslide material into tributary channels, where it can enter the wash load during storm events (Francis et al., 2022). These postearthquake landslide activities are stochastic and rely on intense rainfall events as the primary trigger (Domènech et al., 2019; Tang, Zhu, Qi & Ding, 2011) (Fig. 14.8).

At least six long-term postearthquake landslide datasets were annually catalogued (2008–18) for severely impacted catchments following the Wenchuan earthquake (Chen et al., 2020; Fan et al., 2018; Li et al., 2018; Tang et al., 2009; Yang et al., 2017; Zhang et al., 2016). In regional-scale assessments, the landslide recovery time is defined as the period required for a seismically affected area to return to its preearthquake landsliding rates (Marc et al., 2015). Landslides that remobilise previous landslide debris reduced by >70% by volume within 3 years postearthquake, with near-complete return to preearthquake rates achieved after 7 years (Fig. 14.8; Fan, Scaringi, Korup, et al., 2019). Analyses of postseismic landslide remobilisation further indicate that smaller coseismic landslides stabilised rapidly, whereas larger landslide deposits remained active for longer (Chen et al., 2020; Fan et al., 2018). Regrettably, establishing a comprehensive postearthquake landslide database across the entire Wenchuan earthquake zone remains highly challenging, as automated identification of reactivated deposits is still technically unfeasible. Current assessments primarily rely on indirect quantification of vegetation recovery to achieve an integrated evaluation of landslide activity intensity in seismically disturbed regions (Wang et al., 2024; Yunus et al., 2020).

Mechanisms governing postearthquake landslide recovery time involve interchangeable or concurrent factors. Analysis of Fan et al. (2018)'s multitemporal inventory along the Min River reveals significant activity reduction in both reactivated and new landslides. Critically, reactivated-to-new landslide ratios at 3-/5-/7-years postquake are 7:1, 4:1 and 36:1, demonstrating how the remobilisation of coseismic debris remains the major control on the length of the hazard chain in this area. In this landscape today, debris flows still occur at elevated rates, and considerable coseismic sediment remains on hillslopes. Process-based understanding of how coseismic debris is 'stabilised', combined with greater recognition of the interactions between processes within the geomorphic cascade, is required for the reduction of risks by surface processes in this area.

14.4.3 Catastrophic debris flows

The most impressive element of the geomorphic cascade of the Wenchuan earthquake was the initiation of a series of large debris flows ($>10^6$ m^3; Major et al., 2005). These debris flows are characterised as turbulent, highly liquefied flows that traverse several stream orders and deposit only when reaching a major river, such as the Min River (Fig. 14.1C) (Harvey et al., 2024; Yang et al., 2021). These large debris flows have been the subject of much research following the Wenchuan earthquake and are often referred to as catastrophic debris flows due to the substantial impact they have had on populations and infrastructure (Fig. 14.1D) (Chen, Tang, Chang, et al., 2024; Guo et al., 2016; Tang et al., 2012; Utley et al., 2024; Yang et al., 2021). The Wenchuan earthquake region represents one of only four other locations globally, and the only earthquake, to have seen a similar number (>10) of large debris flows in short succession spatially and temporally. Numerical modelling has demonstrated that for these flows to reach such large volumes, they must entrain large volumes of sediment (Horton et al., 2019). This entrainment requires high bed porosity and large volumes of sediment stored on hillslopes or in channels, conditions possible in Wenchuan due to the abundance of coseismic material and high monsoon rainfall (Francis et al., 2022; Horton et al., 2019). Modelling has also demonstrated that differences in entrainment rate could lead to the development of two distinct populations of debris flows by volume: typical hillslope debris flows, and large, fast-flowing debris flows that entrain large amounts of sediment to bulk up in size (Harvey, 2022; Horton et al., 2019).

High-resolution datasets generated from the Wenchuan earthquake present a unique opportunity to explore how the magnitude-frequency distribution of catastrophic debris flows fits with the distribution of all postearthquake debris flows. Through this comparison, it is apparent that catastrophic debris flows follow a different magnitude-frequency relationship, with larger flows occurring more frequently than may be

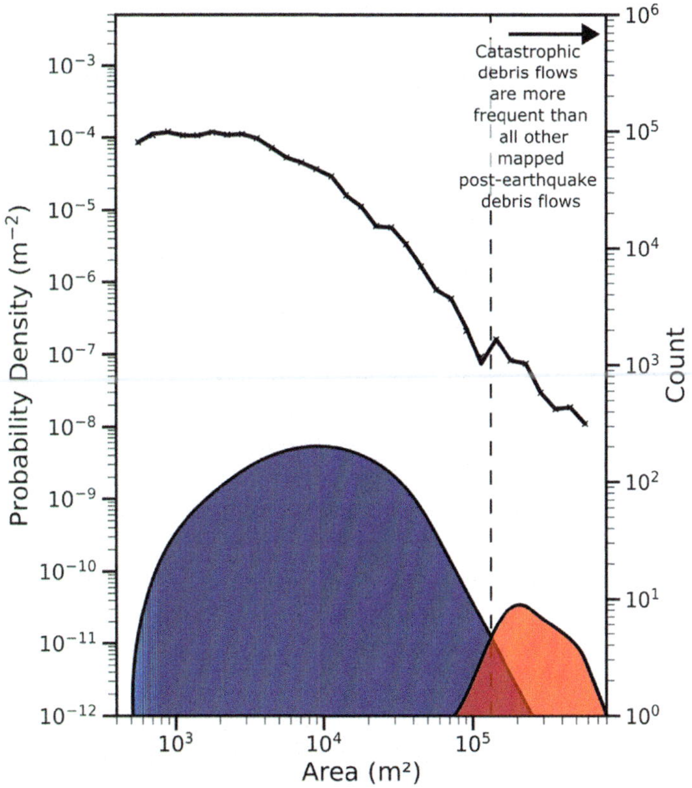

Figure 14.9 Controls on postearthquake and catastrophic debris-flow magnitude-frequency relationships. Probability density function for all debris flows (>500 m^2) triggered in catchments along the Min River between 2008 and 2019 (Fan et al., 2018). The secondary axis shows the number of debris flows classed as catastrophic debris flows (red) and all other postearthquake debris flows (blue). At an area of ~10^5 m^2, the number of catastrophic debris flows exceeds the number of all other debris flows, which corresponds with a shift in the probability density function for the dataset. These trends are further explained in Harvey (2022).

expected based on all debris flows in the inventory (Fig. 14.9). This relationship has been attributed to the convergence of conditions in Wenchuan that promote the evolution of these flows, including large volumes of sediment in channels available for entrainment, monsoon rainfall that can saturate channel sediment, as well as hillslope sediment stores from which debris flows can be initiated (Harvey, 2022; Marino et al., 2022). Previously, postearthquake hazard assessments considered debris flows along a single magnitude–frequency curve (Huang et al., 2020). However, these findings, made possible due to the unique conditions in Wenchuan and the unprecedented number of large debris flows, suggest that catastrophic debris flows can introduce epistemic uncertainty into existing hazard and prediction models.

14.4.4 Impact on the terrestrial carbon cycle

Earthquakes have the potential to disrupt the terrestrial carbon cycle by burying carbon in landslide deposits over hundreds to thousands of years before it is eroded and transported through the fluvial system. The 2008 Wenchuan earthquake appears to have induced abrupt and extensive shifts in terrestrial carbon stocks consistent with these observations. Field studies in the epicentral region reported that landslides and debris flows reduced vegetation and soil carbon by 89.6%–97.8% within landslide deposits, giving rise to a net loss of 0.118×10^6 Mg of carbon in affected catchments (Liu et al., 2022; Liu et al., 2026). Forest inventories documented a loss of roughly 8 Mg of carbon per square kilometre per year across the earthquake-affected area, and simulations estimated a transfer of about 235 Tg of carbon to dead pools in landslide-prone areas (Qiu et al., 2015). Meanwhile, studies of riverine systems indicated that both biospheric carbon discharge and particulate organic carbon export doubled postevent, with one report noting debris flows that expelled a year's worth of lignin in a single day (Wang et al., 2016).

Remote sensing and other observations revealed slow, uneven recovery across ecosystems (Yang, Fan, et al., 2023). In landslide scars, only 45.36% of net primary productivity was restored over 13 years, while other locations showed improvements in vegetation cover within 2–9 years (Yang, Fan, et al., 2023). Spatial analyses have shown that losses were concentrated near the epicentre and along fault zones, with damage diminishing sharply beyond approximately 75 km from the rupture trace (Liu, 2024). In one assessment, forest biomass loss in this event alone offset 0.23% of China's living carbon stock, and the modelled loss equalled roughly 2% of annual global fossil fuel emissions. Collectively, these data indicate that the earthquake caused immediate, quantifiable carbon losses that have changed carbon cycling across multiple terrestrial pools (Qiu et al., 2015). Overall, the total loss of carbon from the orogen (Wang et al., 2016) was considerably lower than the total mass of carbon mobilised by the earthquake. While revegetation of landslide debris adds carbon to the orogen, it is likely that the Wenchuan earthquake was a net sink of carbon in the short term (Liu et al., 2026; Liu, 2024).

14.5 Conclusions

The 2008 M_w 7.9 Wenchuan earthquake has provided geomorphologists with a rich and complex set of cascading processes that have generated insight into how earthquakes affect mountains. The earthquake generated one of the largest coseismic landslide events, whose volume was high enough for us to question whether all earthquakes are constructive events. The sediment generated by coseismic landslides has generated a long and complex geomorphic cascade that is dominated by debris-flow transport of sediment from hillslopes to the fluvial network. The combination of hillslope and fluvial processes in the decade after the earthquake has eroded around ~10% of the coseismic landslide volume and transported it out of the orogen. The remaining ~90% of that sediment continues to generate debris flows, particularly hazardous large debris flows, and work continues to better understand the geomorphic pathways of that sediment in the future. Work also continues to characterise this earthquake and explore its impact on the geomorphic, sedimentary and carbon cycles in the orogen.

References

Burchfiel, B. C., Zhiliang, C., Yuping, L., & Royden, L. H. (1995). Tectonics of the Longmen Shan and adjacent regions, Central China. *International Geology Review*, 661–735.

Cai, C., Yu, C., Tao, K., Hu, X., Tian, Y., Zhang, H., Cui, X., & Ning, J. (2011). Spatial distribution and focal mechanism solutions of the Wenchuan earthquake series: Results and implications. *Earthquake Science, 24*, 115–125. https://doi.org/10.1007/s11589-011-0775-5.

Caine, N. (1980). The rainfall intensity-duration control of shallow landslides and debris flows. *Geografiska Annaler. Series A, Physical Geography, 62*, 23–27. https://doi.org/10.2307/520449.

Casagli, N., & Ermini, L. (1999). Geomorphic analysis of landslide dams in the Northern Apennine. *Transactions of the Japanese Geomorphological Union, 20*, 219–249.

Chang, M., Dou, X., Hales, T. C., & Yu, B. (2021). *Patterns of rainfall-threshold for debris-flow occurrence in the Wenchuan seismic region*. Southwest China. https://doi.org/10.1007/s10064-020-02080-7/Published.

Chen, H. X., Zhang, L. M., & Zhang, S. (2014). Evolution of debris flow properties and physical interactions in debris- flow mixtures in the Wenchuan Earthquake zone. *Engineering Geology, 182*, 136–147. https://doi.org/10.1016/j.enggeo.2014.08.004.

Chen, J. G., Chen, X. Q., Wang, T., Zou, Y. H., & Zhong, W. (2014). Types and causes of debris flow damage to drainage channels in the Wenchuan earthquake area. *Journal of Mountain Science, 11*, 1406–1419. https://doi.org/10.1007/s11629-014-3045-x.

Chen, M., Tang, C., Chang, M., & Xiong, J. (2024). Seismically induced hillslope disturbance, sediment connectivity and mass wasting: Insights from the 2008 Wenchuan earthquake. *Geomorphology, 449*. https://doi.org/10.1016/j.geomorph.2024.109064.

Chen, M., Tang, C., Xiong, J., Chang, M., & Li, N. (2024). Spatio-temporal mapping and long-term evolution of debris flow activity after a high magnitude earthquake. *Catena (Amst), 236*. https://doi.org/10.1016/j.catena.2023.107716.

Chen, M., Tang, C., Xiong, J., Shi, Q. Y., Li, N., Gong, L. F., Wang, X. D., & Tie, Y. (2020). The long-term evolution of landslide activity near the epicentral area of the 2008 Wenchuan earthquake in China. *Geomorphology, 367*. https://doi.org/10.1016/j.geomorph.2020.107317.

Chen, X. L., Ran, H. L., & Yang, W. T. (2012). Evaluation of factors controlling large earthquake-induced landslides by the Wenchuan earthquake. *Natural Hazards and Earth System Science, 12*, 3645–3657. https://doi.org/10.5194/nhess-12-3645-2012.

Cui, P., Chen, X.-Q., Zhu, Y.-Y., Su, F.-H., Wei, F.-Q., Han, Y.-S., Liu, H.-J., & Zhuang, J.-Q. (2011). The Wenchuan Earthquake (May 12, 2008), Sichuan Province, China, and resulting geohazards. *Natural Hazards*, 19–36. https://doi.org/10.1007/s11069-009-9392-1.

Cui, P., Zhu, Y.yan, Han, Y.shun, Chen, X.qing, & Zhuang, J.qi (2009). The 12 May Wenchuan earthquake-induced landslide lakes: Distribution and preliminary risk evaluation. *Landslides, 6*. https://doi.org/10.1007/s10346-009-0160-9.

Dahlquist, M. P., & West, A. J. (2019). Initiation and runout of post-seismic debris flows: Insights from the 2015 Gorkha earthquake. *Geophysical Research Letters, 46*, 9658–9668. https://doi.org/10.1029/2019GL083548.

Dai, F. C., Xu, C., Yao, X., Xu, L., Tu, X. B., & Gong, Q. M. (2011). Spatial distribution of landslides triggered by the 2008 Ms 8.0 Wenchuan earthquake, China. *Journal of Asian Earth Sciences, 40*, 883–895. https://doi.org/10.1016/j.jseaes.2010.04.010.

Dai, L., Scaringi, G., Fan, X., Yunus, A. P., Liu-Zeng, J., Xu, Q., & Huang, R. (2021). Coseismic debris remains in the orogen despite a decade of enhanced landsliding. *Geophysical Research Letters, 48*. https://doi.org/10.1029/2021GL095850.

Davies, T. R. H., & Korup, O. (2010). Sediment cascades in active landscapes. In T. Burt, & R. Allison (Eds.). *Sediment cascades: An integrated approach*. John Wiley and Sons.

De Haas, T., Braat, L., Leuven, J. R. F. W., Lokhorst, I. R., & Kleinhans, M. G. (2015). Effects of debris flow composition on runout, depositional mechanisms, and deposit morphology in laboratory experiments. *Journal of Geophysical Research: Earth Surface, 120*. https://doi.org/10.1002/2015JF003525.

Densmore, A. L., Ellis, M. A., Li, Y., Zhou, R., Hancock, G. S., & Richardson, N. (2007). Active tectonics of the Beichuan and Pengguan faults at the eastern margin of the Tibetan Plateau. *Tectonics, 26*. https://doi.org/10.1029/2006TC001987.

Densmore, A. L., Parker, R. N., Rosser, N. J., de Michele, M., Yong, L., Runqiu, H., Whadcoat, S., & Petley, D. N. (2012). Reply to 'Isostasy can't be ignored' *Nature Geoscience, 5*, 83. https://doi.org/10.1038/ngeo1385.

Domènech, G., Fan, X., Scaringi, G., Asch, T. W. J., Van, Xu, Q., Huang, R., & Hales, T. C. (2019). Modelling the role of material depletion, grain coarsening and revegetation in debris flow occurrences after the 2008 Wenchuan earthquake. *Engineering Geology, 250*, 34–44. https://doi.org/10.1016/j.enggeo.2019.01.010.

Dong, J. J., Tung, Y. H., Chen, C. C., Liao, J. J., & Pan, Y. W. (2009). Discriminant analysis of the geomorphic characteristics and stability of landslide dams. *Geomorphology, 110*, 162–171. https://doi.org/10.1016/j.geomorph.2009.04.004.

Ermini, L., & Casagli, N. (2003). Prediction of the behaviour of landslide dams using a geomorphological dimensionless index. *Earth Surface Processes and Landforms, 28*. https://doi.org/10.1002/esp.424.

Fan, N., Nie, R., Wang, Q., & Liu, X. (2016). Dramatic undercutting of piedmont rivers after the 2008 Wenchuan Ms 8.0 earthquake. *Scientific Reports, 6*, 37–108. https://doi.org/10.1038/srep37108.

Fan, X., Domènech, G., Scaringi, G., Huang, R., Xu, Q., Hales, T. C., Dai, L., Yang, Q., & Francis, O. (2018). Spatio-temporal evolution of mass wasting after the 2008 M w 7.9 Wenchuan earthquake revealed by a detailed multi-temporal inventory. *Landslides, 15*, 2325–2341. https://doi.org/10.1007/s10346-018-1054-5.

Fan, X., Scaringi, G., Domènech, G., Yang, F., Guo, X., & Dai, L. (2019). Two multi-temporal datasets that track the enhanced landsliding after the 2008 Wenchuan earthquake. *Earth System Science Data, 11*, 35–55. https://doi.org/10.5194/essd-11-35-2019.

Fan, X., Scaringi, G., Korup, O., West, A. J., van Westen, C. J., Tanyas, H., Hovius, N., Hales, T. C., Jibson, R. W., Allstadt, K. E., Zhang, L., Evans, S. G., Xu, C., Li, G., Pei, X., Xu, Q., & Huang, R. (2019). Earthquake-induced chains of geologic hazards: Patterns, mechanisms, and impacts. *Reviews of Geophysics*. https://doi.org/10.1029/2018RG000626.

Fan, X., Tang, C. X., Van Westen, C. J., & Alkema, D. (2012). Simulating dam-breach flood scenarios of the Tangjiashan landslide dam induced by the Wenchuan Earthquake. *Natural Hazards and Earth System Science, 12*, 3031–3044. https://doi.org/10.5194/nhess-12-3031-2012.

Fan, X., Yunus, A. P., Scaringi, G., Catani, F., Siva Subramanian, S., Xu, Q., & Huang, R. (2021). Rapidly evolving controls of landslides after a strong earthquake and implications for hazard assessments. *Geophysical Research Letters, 48*. https://doi.org/10.1029/2020GL090509.

Fan, X., Zhan, W., Dong, X., van Westen, C., Xu, Q., Dai, L., Yang, Q., Huang, R., & Havenith, H. B. (2018). Analyzing successive landslide dam formation by different triggering mechanisms: The case of the Tangjiawan landslide, Sichuan, China. *Engineering Geology, 243*. https://doi.org/10.1016/j.enggeo.2018.06.016.

Fan, X., Tang, J., Tian, S., & Jiang, Y. (2020). Rainfall-induced rapid and long-runout catastrophic landslide on July 23, 2019 in Shuicheng, Guizhou, China. *Landslides, 17*, 2161–2171. https://doi.org/10.1007/s10346-020-01454-y.

Fan, X., Dufresne, A., Siva Subramanian, S., Strom, A., Hermanns, R., Tacconi Stefanelli, C., Hewitt, K., Yunus, A. P., Dunning, S., Capra, L., Geertsema, M., Miller, B., Casagli, N., Jansen, J. D., & Xu, Q. (2020). The formation and impact of landslide dams – State of the art. *Earth-Science Reviews*. https://doi.org/10.1016/j.earscirev.2020.103116.

Fan, X., van Westen, C. J., Korup, O., Gorum, T., Xu, Q., Dai, F., Huang, R., & Wang, G. (2012). Transient water and sediment storage of the decaying landslide dams induced by the 2008 Wenchuan earthquake, China. *Geomorphology*, 171–172. https://doi.org/10.1016/j.geomorph.2012.05.003.

Fan, X., van Westen, C. J., Xu, Q., Gorum, T., & Dai, F. (2012). Analysis of landslide dams induced by the 2008 Wenchuan earthquake. *Journal of Asian Earth Sciences, 57*. https://doi.org/10.1016/j.jseaes.2012.06.002.

Francis, O., Fan, X., Hales, T., Hobley, D., Xu, Q., & Huang, R. (2022). The fate of sediment after a large earthquake. *Journal of Geophysical Research: Earth Surface, 127*. https://doi.org/10.1029/2021JF006352.

Francis, O. R., Hales, T., Hobley, D. E. J., Fan, X., Horton, A., Scaringi, G., & Huang, R. (2020). The impact of earthquakes on orogen-scale exhumation. *Earth Surface Dynamics, 8*, 579–593. https://doi.org/10.5194/esurf-8-579-2020.

Gallen, S. F., Clark, M. J., & Godt, J. W. (2015). Coseismic landslides reveal near-surface rock strength in a high-relief, tectonically active setting. *Geology, 43*, 11–14. https://doi.org/10.1130/G36080.1.

Gill, J. C., & Malamud, B. D. (2014). Reviewing and visualizing the interactions of natural hazards. *Reviews of Geophysics, 52*, 680–722. https://doi.org/10.1002/2013RG000445.

Gorum, T., & Carranza, E. J. M. (2015). Control of style-of-faulting on spatial pattern of earthquake-triggered landslides. *International Journal of Environmental Science and Technology, 12*, 3189–3212. https://doi.org/10.1007/s13762-015-0752-y.

Gorum, T., Fan, X., van Westen, C. J., Huang, R. Q., Xu, Q., Tang, C., & Wang, G. (2011). Distribution pattern of earthquake-induced landslides triggered by the 12 May 2008 Wenchuan earthquake. *Geomorphology, 133*, 152–167. https://doi.org/10.1016/j.geomorph.2010.12.030.

Guo, X., Cui, P., Li, Y., Ma, L., Ge, Y., & Mahoney, W. B. (2016). *Geomorphology intensity – duration threshold of rainfall-triggered debris flows in the Wenchuan Earthquake affected area, China, 253*, 208–216. https://doi.org/10.1016/j.geomorph.2015.10.009.

Guo, X., Cui, P., Li, Y., Zhang, J., Ma, L., & Mahoney, W. B. (2016). Spatial features of debris flows and their rainfall thresholds in the Wenchuan earthquake-affected area. *Landslides, 13*, 1215–1229. https://doi.org/10.1007/s10346-015-0608-z.

Guzzetti, F., Peruccacci, S., Rossi, M., & Stark, C. P. (2008). The rainfall intensity-duration control of shallow landslides and debris flows: An update. *Landslides, 5*. https://doi.org/10.1007/s10346-007-0112-1.

Hales, T. C. (2018). Modelling biome-scale root reinforcement and slope stability. *Earth Surface Processes and Landforms, 43*, 2157–2166. https://doi.org/10.1002/esp.4381.

Hao, K. X., Si, H., Fujiwara, H., & Ozawa, T. (2009). Coseismic surface-ruptures and crustal deformations of the 2008 Wenchuan earthquake Mw7.9, China. *Geophysical Research Letters, 36*. https://doi.org/10.1029/2009GL037971.

Harvey, E. (2022). *Grain size and magnitude of post-earthquake debris flows*. Cardiff University.

Harvey, E. L., Hales, T. C., Hobley, D. E. J., Liu, J., & Fan, X. (2022). Measuring the grain-size distributions of mass movement deposits. *Earth Surface Processes and Landforms, 47*, 1599–1614. https://doi.org/10.1002/esp.5337.

Harvey, E. L., Hales, T. C., Horton, A. J., Francis, O. R., Yang, F., Liu, J., & Fan, X. (2025). The hazard of large debris flows. *Science Advances, 11*, eadz4625. https://doi.org/10.1126/sciadv.adz4625.

Harvey, E. L., Hales, T. C., Liu, J., Hobley, D. E. J., Yang, F., Xia, B., & Fan, X. (2024). Grain-size variability in debris flows of different runout lengths, Wenchuan, China. *Bulletin of the Geological Society of America, 136*, 2676–2688. https://doi.org/10.1130/B37027.1.

Horton, A. J., Hales, T. C., Ouyang, C., & Fan, X. (2019). Identifying post-earthquake debris flow hazard using Massflow. *Engineering Geology, 258*. https://doi.org/10.1016/j.enggeo.2019.05.011.

Hovius, N., Meunier, P., Lin, C. W., Chen, H., Chen, Y. G., Dadson, S., Horng, M. J., & Lines, M. (2011). Prolonged seismically induced erosion and the mass balance of a large earthquake. *Earth and Planetary Science Letters, 304*, 347–355. https://doi.org/10.1016/j.epsl.2011.02.005.

Hu, T., & Huang, R. Q. (2017). A catastrophic debris flow in the Wenchuan earthquake area, July 2013: Characteristics, formation, and risk reduction. *Journal of Mountain Science, 14*, 15–30. https://doi.org/10.1007/s11629-016-3965-8https://doi.org/10.1007/s11629-016-3965-8.

Hu, W., Scaringi, G., Xu, Q., Pei, Z., Van Asch, T. W. J. J., & Hicher, P.-Y. Y. (2017). Sensitivity of the initiation and runout of flowslides in loose granular deposits to the content of small particles: An insight from flume tests. *Engineering Geology, 231*, 34–44. https://doi.org/10.1016/j.enggeo.2017.10.001.

Huang, J., Hales, T. C., Huang, R., Ju, N., Li, Q., & Huang, Y. (2020). A hybrid machine-learning model to estimate potential debris-flow volumes. *Geomorphology, 367*, 107333. https://doi.org/10.1016/j.geomorph.2020.107333.

Huang, R., & Fan, X. (2013). The landslide story. *Nature Geoscience*. https://doi.org/10.1038/ngeo1806.

Huang, R., & Li, W. (2014). Post-earthquake landsliding and long-term impacts in the Wenchuan earthquake area, China. *Engineering Geology, 182*, 111–120. https://doi.org/10.1016/j.enggeo.2014.07.008.

Huang, R., Yu, B., Hu, W., Li, W., Fan, X., Tang, C., & Xu, Q. (2016). The long-term geologic hazards and consequent risk after the Wenchuan earthquake. *Landslides and Engineered Slopes. Experience, Theory and Practice*, 233–258. https://doi.org/10.1201/b21520-21.

Huang, R. Q., & Li, W. L. (2009). Analysis of the geo-hazards triggered by the 12 May 2008 Wenchuan Earthquake, China. *Bulletin of Engineering Geology and the Environment, 68*, 363–371. https://doi.org/10.1007/s10064-009-0207-0.

Hubbard, J., & Shaw, J. H. (2009). Uplift of the Longmen Shan and Tibetan plateau, and the 2008 Wenchuan (M=7.9) earthquake. *Nature, 458*, 194. https://doi.org/10.1038/nature07837, https://www.nature.com/articles/nature07837#supplementary-information.

Iverson, R. M., Reid, M. E., & Lahusen, R. G. (1997). Debris-flow mobilization from landslides. *Annual Review of Earth and Planetary Sciences, 25*, 85–138. https://doi.org/10.1146/annurev.earth.25.1.85.

Jones, K. E., Howarth, J. D., Massey, C. I., Luković, B., Sirguey, P., Singeisen, C., Gasston, C., Morgenstern, R., & Ries, W. (2025). An alternative to landslide volume-area scaling relationships: An ensemble approach adopting a difference model to estimate the total volume of landsliding triggered by the 2016 Kaikōura earthquake, New Zealand. *Landslides, 22*, 2219–2236. https://doi.org/10.1007/s10346-025-02479-x.

Khattak, G. A., Owen, L. A., Kamp, U., & Harp, E. L. (2010). Evolution of earthquake-triggered landslides in the Kashmir Himalaya, northern Pakistan. *Geomorphology, 115*, 102–108. https://doi.org/10.1016/j.geomorph.2009.09.035.

Kincey, M. E., Rosser, N. J., Robinson, T. R., Densmore, A. L., Shrestha, R., Pujara, D. S., Oven, K. J., Williams, J. G., & Swirad, Z. M. (2021). Evolution of coseismic and post-seismic landsliding after the 2015 Mw 7.8 Gorkha earthquake, Nepal. *Journal of Geophysical Research: Earth Surface, 126*. https://doi.org/10.1029/2020JF005803.

Kingdon-Ward, F. (1955). Aftermath of the great Assam earthquake of 1950. *The Geographical Journal, 121*, 290–303.

Kirby, E., Whipple, K. X., Tang, W., & Chen, Z. (2003). Distribution of active rock uplift along the eastern margin of the Tibetan Plateau: Inferences from bedrock channel longitudinal profiles. *Journal of Geophysical Research: Solid Earth, 108*. https://doi.org/10.1029/2001jb000861.

Li, C., Wang, M., & Liu, K. (2018). A decadal evolution of landslides and debris flows after the Wenchuan earthquake. *Geomorphology, 323*, 1–12. https://doi.org/10.1016/j.geomorph.2018.09.010.

Li, G., West, A. J., Densmore, A. L., Hammond, D. E., Jin, Z., Zhang, F., Wang, J., Hilton, R. G., Gen, L., Joshua, W. A., L, D. A., E, H. D., Zhangdong, J., Fei, Z., Jin, W., & G, H. R. (2016). Connectivity of earthquake-triggered landslides with the fluvial network: Implications for landslide sediment transport after the 2008 Wenchuan earthquake. *Journal of Geophysical Research: Earth Surface, 121*, 703–724. https://doi.org/10.1002/2015JF003718.

Li, G., West, A. J., Densmore, A. L., Jin, Z., Parker, R. N., & Hilton, R. G. (2014). Seismic mountain building: Landslides associated with the 2008 Wenchuan earthquake in the context of a generalized model for earthquake volume balance. *Geochemistry, Geophysics, Geosystems, 15*, 833–844. https://doi.org/10.1002/2013GC005067.

Li, G., West, A. J., Densmore, A. L., Jin, Z., Zhang, F., Wang, J., Clark, M., & Hilton, R. G. (2017). Earthquakes drive focused denudation along a tectonically active mountain front. *Earth and Planetary Science Letters, 472*, 253–265. https://doi.org/10.1016/j.epsl.2017.04.040.

Li, G. K., West, A. J., Jin, Z., Qiu, H., Zhang, F., Wang, J., Hammond, D. E., Densmore, A. L., Hilton, R. G., Dong, S., Atwood, A., Fischer, W. W., & Lamb, M. P. (2025). Large riverbed sediment flux sustained for a decade after an earthquake. *Nature, 644*, 398–403. https://doi.org/10.1038/s41586-025-09354-8.

Liu, J. (2024). *The impact of earthquakes on the terrestrial ecosystem carbon cycle*. Cardiff.

Liu, J., Fan, X., Hales, T., Harvey, E. L., West, A. J., Jansen, J. D., Tang, X., & Xu, Q. (2026). Earthquakes act as a capacitor for terrestrial organic carbon. *Nature Communications*. https://doi.org/10.1038/s41467-026-68341-3.

Liu, J., Fan, X., Tang, X., Xu, Q., Harvey, E. L., Hales, T. C., & Jin, Z. (2022). Ecosystem carbon stock loss after a mega earthquake. *Catena (Amst), 216*. https://doi.org/10.1016/j.catena.2022.106393.

Liu, S., Wang, Z., Wang, D., Zhang, P., Yang, S., Hu, J., & Xu, Z. (2025). The 2008 Wenchuan earthquake and intense post-earthquake precipitation: Their combined impact on sedimentation in the Three Gorges Reservoir, China. *Hydrological Processes, 39*. https://doi.org/10.1002/hyp.70125.

Liu-Zeng, J., Wen, L., Oskin, M., & Zeng, L. (2011). Focused modern denudation of the Longmen Shan margin, eastern Tibetan Plateau. *Geochemistry, Geophysics, Geosystems, 12*. https://doi.org/10.1029/2011GC003652.

Liu-zeng, J., Zhang, Z., Wen, L., Tapponnier, P., Sun, J., Xing, X., Hu, G., Xu, Q., Zeng, L., Ding, L., Ji, C., Hudnut, K. W., & Woerd, J. V. D. (2009). Co-seismic ruptures of the 12 May 2008, M s 8. 0 Wenchuan earthquake, Sichuan: East – west crustal shortening on oblique, parallel thrusts along the eastern edge of Tibet. *Earth and Planetary Science Letters, 286*, 355–370. https://doi.org/10.1016/j.epsl.2009.07.017.

Major, J. J., Pierson, T. C., & Scott, K. M. (2005). Debris flows at Mount St. Helens, Washington, USA. In M. Jakob, & O. Hungr (Eds.). *Debris-flow hazards and related phenomena* (pp. 685–731). Springer Berlin Heidelberg. https://doi.org/10.1007/3-540-27129-5_27.

Marc, O., Hovius, N., & Meunier, P. (2016). The mass balance of earthquakes and earthquake sequences. *Geophysical Research Letters, 43*, 3708–3716. https://doi.org/10.1002/2016GL068333.

Marc, O., Hovius, N., Meunier, P., Uchida, T., & Hayashi, S. (2015). Transient changes of landslide rates after earthquakes. *Geology, 43*, 883–886. https://doi.org/10.1130/G36961.1.

Marc, O., Hovius, N., Meunier, P., Gorum, T., & Uchida, T. (2016). A seismologically consistent expression for the total area and volume of earthquake-triggered landsliding. *Journal of Geophysical Research: Earth Surface, 121*, 640–663. https://doi.org/10.1002/2015JF003732.

Marino, P., Siva Subramanian, S., Fan, X., & Greco, R. (2022). Changes in debris-flow susceptibility after the Wenchuan earthquake revealed by meteorological and hydro-meteorological thresholds. *Catena (Amst), 210*. https://doi.org/10.1016/j.catena.2021.105929.

Milledge, D. G., Bellugi, D. G., Watt, J., & Densmore, A. L. (2022). Automated determination of landslide locations after large trigger events: Advantages and disadvantages compared to manual mapping. *Natural Hazards and Earth System Sciences, 22*, 481–508. https://doi.org/10.5194/nhess-22-481-2022.

Molnar, P. (2012). Isostasy can't be ignored. *Nature Geoscience, 5*, 83. https://doi.org/10.1038/ngeo1383.

Ouimet, W. B. (2010). Landslides associated with the May 12, 2008 Wenchuan earthquake: Implications for the erosion and tectonic evolution of the Longmen Shan. *Tectonophysics, 491*, 244–252. https://doi.org/10.1016/j.tecto.2009.09.012.

Ouimet, W. B., Whipple, K. X., Royden, L. H., Sun, Z., & Chen, Z. (2007). The influence of large landslides on river incision in a transient landscape: Eastern margin of the Tibetan Plateau (Sichuan, China). *Bulletin of the Geological Society of America, 119*, 1462–1476. https://doi.org/10.1130/B26136.1.

Parker, R. N., Densmore, A. L., Rosser, N. J., de Michele, M., Li, Y., Huang, R., Whadcoat, S., & Petley, D. N. (2011). Mass wasting triggered by the 2008 Wenchuan earthquake is greater than orogenic growth. *Nature Geosci, 4*, 449–452. https://www.nature.com/ngeo/journal/v4/n7/abs/ngeo1154.html#supplementary-information.

Peng, M., & Zhang, L. M. (2012). Breaching parameters of landslide dams. *Landslides, 9*, 13–31. https://doi.org/10.1007/s10346-011-0271-y.

Qi, S., Xu, Q., Lan, H., Zhang, B., & Liu, J. (2010). Spatial distribution analysis of landslides triggered by 2008.5.12 Wenchuan Earthquake, China. *Engineering Geology, 116*, 95–108. https://doi.org/10.1016/j.enggeo.2010.07.011.

Qiu, S., Xu, M., Zheng, Y., Li, R., Wong, M. H. G., Zhang, L., Liu, L., Lai, C., & Zhang, W. (2015). Impacts of the Wenchuan earthquake on tree mortality and biomass carbon stock. *Natural Hazards, 77*, 1261–1274. https://doi.org/10.1007/s11069-015-1653-6.

Rault, C., Robert, A., Marc, O., Hovius, N., & Meunier, P. (2019). Seismic and geologic controls on spatial clustering of landslides in three large earthquakes. *Earth Surface Dynamics, 7*, 829–839. https://doi.org/10.5194/esurf-7-829-2019.

Royden, L. H., Burchfiel, B. C., King, R. W., Wang, E., Chen, Z., Shen, F., & Liu, Y. (1997). Surface deformation and lower crustal flow in eastern Tibet. *Science, 276*(1979), 788–790. https://doi.org/10.1126/science.276.5313.788.

Sato, H. P., & Harp, E. L. (2009). Interpretation of earthquake-induced landslides triggered by the 12 May 2008, M7.9 Wenchuan earthquake in the Beichuan area, Sichuan Province, China using satellite imagery and Google Earth. *Landslides, 6*, 153–159. https://doi.org/10.1007/s10346-009-0147-6.

Shen, P., Zhang, L. M., Chen, H. X.; & Gao, L. (2017). Role of vegetation restoration in mitigating hillslope erosion and debris flows. *Engineering Geology, 216*. https://doi.org/10.1016/j.enggeo.2016.11.019.

Shen, P., Zhang, L. M., Fan, R. L., Zhu, H., & Zhang, S. (2020). Declining geohazard activity with vegetation recovery during first ten years after the 2008 Wenchuan earthquake. *Geomorphology, 352*. https://doi.org/10.1016/j.geomorph.2019.106989.

Shen, Z. K., Sun, J., Zhang, P., Wan, Y., Wang, M., Bürgmann, R., Zeng, Y., Gan, W., Liao, H., & Wang, Q. (2009). Slip maxima at fault junctions and rupturing of barriers during the 2008 Wenchuan earthquake. *Nature Geoscience, 2*, 718–724. https://doi.org/10.1038/ngeo636.

Tacconi Stefanelli, C., Catani, F., & Casagli, N. (2015). Geomorphological investigations on landslide dams. *Geoenvironmental Disasters, 2*. https://doi.org/10.1186/s40677-015-0030-9.

Tan, X., Yue, H., Liu, Y., Xu, X., Shi, F., Xu, C., Ren, Z., Shyu, J. B. H., Lu, R., & Hao, H. (2018). Topographic loads modified by fluvial incision impact fault activity in the Longmenshan thrust belt, eastern margin of the Tibetan Plateau. *Tectonics, 37*, 3001–3017. https://doi.org/10.1029/2017TC004864.

Tang, C., Van Westen, C. J., Tanyaş, H., & Jetten, V. G. (2016). Analysing post-earthquake landslide activity using multi- temporal landslide inventories near the epicentral area of the 2008 Wenchuan earthquake. *Natural Hazards and Earth System Science, 16*, 2641–2655. https://doi.org/10.5194/nhess-2016-193https://doi.org/10.5194/nhess-16-2641-2016.

Tang, C., Zhu, J., Chang, M., Ding, J., & Qi, X. (2012). An empirical and statistical model for predicting debris-flow runout zones in the Wenchuan earthquake area. *Quaternary International, 250*, 63–73. https://doi.org/10.1016/j.quaint.2010.11.020.

Tang, C., Zhu, J., Ding, J., Cui, X., Chen, L., & Zhang, J. (2011). Catastrophic debris flows triggered by a 14 August 2010 rainfall at the epicenter of the Wenchuan earthquake. *Landslides, 8*, 485–497. https://doi.org/10.1007/s10346-011-0269-5.

Tang, C., Zhu, J., Li, W. L., & Liang, J. T. (2009). Rainfall-triggered debris flows following the Wenchuan earthquake. *Bulletin of Engineering Geology and the Environment, 68*, 187–194. https://doi.org/10.1007/s10064-009-0201-6.

Tang, C., Zhu, J., & Qi, X. (2011). Landslide hazard assessment of the 2008 Wenchuan earthquake: A case study in Beichuan area. *Canadian Geotechnical Journal, 48*. https://doi.org/10.1139/T10-059.

Tang, C., Zhu, J., Qi, X., & Ding, J. (2011). Landslides induced by the Wenchuan earthquake and the subsequent strong rainfall event: A case study in the Beichuan area of China. *Engineering Geology, 122*, 22–33. https://doi.org/10.1016/j.enggeo.2011.03.013.

Tanyaş, H., van Westen, C. J., Allstadt, K. E., Anna Nowicki Jessee, M., Görüm, T., Jibson, R. W., Godt, J. W., Sato, H. P., Schmitt, R. G., Marc, O., & Hovius, N. (2017). Presentation and analysis of a worldwide database of earthquake-induced landslide inventories. *Journal of Geophysical Research: Earth Surface, 122*, 1991–2015. https://doi.org/10.1002/2017JF004236.

Utley, I., Hales, T., Hussain, E., & Fan, X. (2024). Transformations in exposure to debris flows in post-earthquake Sichuan, China. *Natural Hazards and Earth System Science, 25*, 2699–2716. https://doi.org/10.5194/egusphere-2024-2277 https://doi.org/10.5194/nhess-25-2699-2025 China.

Wang, J., Jin, Z., Hilton, R. G., Zhang, F., Densmore, A. L., Li, G., Joshua West, A., & West, A. J. (2015). Controls on fluvial evacuation of sediment from earthquake-triggered landslides. *Geology, 43*, 115–118. https://doi.org/10.1130/g36157.1.

Wang, J., Jin, Z., Hilton, R. G., Zhang, F., Li, G., Densmore, A. L., Gröcke, D. R., Xu, X., & Joshua West, A. (2016). Earthquake-triggered increase in biospheric carbon export from a mountain belt. *Geology, 44*, 471–474. https://doi.org/10.1130/G37533.1.

Wang, L., Chang, M., Le, J., Xiang, L., & Ni, Z. (2022). Two multi-temporal datasets to track debris flow after the 2008 Wenchuan earthquake. *Scientific Data, 9*. https://doi.org/10.1038/s41597-022-01658-y.

Wang, X., Fan, X., Fang, C., Dai, L., Zhang, W., Zheng, H., & Xu, Q. (2024). Long-term landslide evolution and restoration after the Wenchuan earthquake revealed by time-series remote sensing images. *Geophysical Research Letters, 51*. https://doi.org/10.1029/2023GL106422.

Xie, H., Zhong, D., Jiao, Z., & Zhang, J. (2009). Debris flow in Wenchuan quake-hit area in 2008. *Journal of Mountain Science, 27*, 501–509.

Xiong, J., Tang, C., Chen, M., Zhang, X., Shi, Q., & Gong, L. (2021). Activity characteristics and enlightenment of the debris flow triggered by the rainstorm on 20 August 2019 in Wenchuan County, China. *Bulletin of Engineering Geology and the Environment, 80*, 873–888. https://doi.org/10.1007/s10064-020-01981-x/Published.

Xiong, J., Tang, C., Gong, L., Chen, M., Li, N., Shi, Q., Zhang, X., Chang, M., & Li, M. (2022). How landslide sediments are transferred out of an alpine basin: Evidence from the epicentre of the Wenchuan earthquake. *Catena (Amst), 208*. https://doi.org/10.1016/j.catena.2021.105781.

Xu, C., Xu, X., Shen, L., Yao, Q., Tan, X., Kang, W., Ma, S., Wu, X., Cai, J., Gao, M., & Li, K. (2016). Optimized volume models of earthquake-triggered landslides. *Scientific Reports, 6*. https://doi.org/10.1038/srep29797.

Xu, C., Xu, X., Yao, X., & Dai, F. (2014). Three (nearly) complete inventories of landslides triggered by the May 12, 2008 Wenchuan Mw 7.9 earthquake of China and their spatial distribution statistical analysis. *Landslides, 11*, 441–461. https://doi.org/10.1007/s10346-013-0404-6.

Xu, Q., Fan, X. M., Huang, R. Q., & Westen, C. V. (2009). Landslide dams triggered by the Wenchuan Earthquake, Sichuan Province, south west China. *Bulletin of Engineering Geology and the Environment, 68*, 373–386. https://doi.org/10.1007/s10064-009-0214-1.

Xu, Q., Zhang, S., Li, W. L., & Asch, T. W. J. V. (2012). *The 13 August 2010 catastrophic debris flows after the 2008 Wenchuan earthquake, China Hongchun gully*, 201–216. https://doi.org/10.5194/nhess-12-201-2012.

Xu, X., Wen, X., Yu, G., Chen, G., Klinger, Y., Hubbard, J., & Shaw, J. (2009). Coseismic reverse- and oblique-slip surface faulting generated by the 2008 Mw 7.9 Wenchuan earthquake, China. *Geology, 37*, 515–518. https://doi.org/10.1130/G25462A.1.

Yang, F., Fan, X., Siva Subramanian, S., Dou, X., Xiong, J., Xia, B., Yu, Z., & Xu, Q. (2021). Catastrophic debris flows triggered by the 20 August 2019 rainfall, a decade since the Wenchuan earthquake, China. *Landslides, 18*, 3197–3212. https://doi.org/10.1007/s10346-021-01713-6.

Yang, W., Qi, W., Wang, M., Zhang, J., & Zhang, Y. (2017). Spatial and temporal analyses of post-seismic landslide changes near the epicentre of the Wenchuan earthquake. *Geomorphology, 276*. https://doi.org/10.1016/j.geomorph.2016.10.010.

Yang, Y., Fan, X., Wang, X., Lv, L., Zou, C., & Feng, Z. (2023). Net primary productivity changes associated with landslides induced by the 2008 Wenchuan Earthquake. *Land Degradation & Development, 34*. https://doi.org/10.1002/ldr.4514.

Yang, Y., Tang, C., Tang, C., Chen, M., Cai, Y., Bu, X., & Liu, C. (2023). Spatial and temporal evolution of long-term debris flow activity and the dynamic influence of condition factors in the Wenchuan earthquake-affected area, Sichuan, China. *Geomorphology, 435*. https://doi.org/10.1016/j.geomorph.2023.108755.

Yanites, B. J., Clark, M. K., Roering, J. J., West, A. J., Zekkos, D., Baldwin, J. W., Cerovski-Darriau, C., Gallen, S. F., Horton, D. E., Kirby, E., Leshchinsky, B. A., Mason, H. B., Moon, S., Barnhart, K. R., Booth, A., Czuba, J. A., McCoy, S., McGuire, L., Pfeiffer, A., & Pierce, J. (2025). Cascading land surface hazards as a nexus in the Earth system. *Science (1979), 388*. https://doi.org/10.1126/science.adp9559.

Yong, L., Allen, P. A., Densmore, A. L., & Qiang, X. (2003). Evolution of the Longmen Shan foreland basin (western Sichuan, China) during the Late Triassic Indosinian orogeny. *Basin Research, 15*, 117–138. https://doi.org/10.1046/j.1365-2117.2003.00197.x.

Yuan, R. M., Deng, Q. H., Cunningham, D., Xu, C., Xu, X. W., & Chang, C. P. (2013). Density distribution of landslides triggered by the 2008 Wenchuan earthquake and their relationships to peak ground acceleration. *Bulletin of the Seismological Society of America, 103*, 2344–2355. https://doi.org/10.1785/0120110233.

Yunus, A. P., Fan, X., Tang, X., Jie, D., Xu, Q., & Huang, R. (2020). Remote sensing of environment decadal vegetation succession from MODIS reveals the spatio-temporal evolution of post-seismic landsliding after the 2008 Wenchuan earthquake. *Remote Sensing of Environment, 236*, 111476. https://doi.org/10.1016/j.rse.2019.111476.

Zhang, F., Jin, Z., West, A. J., An, Z., Hilton, R. G., Wang, J., Li, G., Densmore, A. L., Yu, J., Qiang, X., Sun, Y., Li, L., Gou, L., Xu, Y., Xu, X., Liu, X., Pan, Y., & You, C.-F. (2019). Monsoonal control on a delayed response of sedimentation to the 2008 Wenchuan earthquake. *Science Advances, 5*, eaav7110. https://doi.org/10.1126/sciadv.aav7110.

Zhang, P. Z., Shen, Z., Wang, M., Gan, W., Bürgmann, R., Molnar, P., Wang, Q., Niu, Z., Sun, J., Wu, J., Hanrong, S., & Xinzhao, Y. (2004). Continuous deformation of the Tibetan Plateau from global positioning system data. *Geology, 32*, 809–812. https://doi.org/10.1130/G20554.1.

Zhang, S., Zhang, L., Lacasse, S., & Nadim, F. (2016). Evolution of mass movements near epicentre of Wenchuan earthquake, the first eight years. *Scientific Reports, 6*. https://doi.org/10.1038/srep36154.

Zhang, S., Zhang, L. M., & Chen, H. X. (2014). Relationships among three repeated large-scale debris flows at Pubugou Ravine in the Wenchuan earthquake zone. *Canadian Geotechnical Journal, 51*, 951–965. https://doi.org/10.1139/cgj-2013-0368.

Zhang, S., Zhang, L. M., Chen, H. X., Yuan, Q., & Pan, H. (2013). Changes in runout distances of debris flows over time in the Wenchuan earthquake zone. *Journal of Mountain Science, 10*. https://doi.org/10.1007/s11629-012-2506-y.

CHAPTER 15

Case study: 2015 Gorkha earthquake, Nepal

Mark Kincey[1] **and Nick J. Rosser**[2]
[1]School of Geography, Politics and Sociology, Newcastle University, Newcastle upon Tyne, United Kingdom,
[2]Department of Geography, Institute of Hazard Risk and Resilience, Durham University, Durham, United Kingdom

15.1 Introduction

The M_w 7.8 Gorkha earthquake of 25 April 2015 represented one of the most significant geomorphic perturbations to the central Himalayan region in recent history, impacting an area of >28,000 km^2 and initiating a series of cascading geomorphic processes that are still being experienced over a decade later. The earthquake triggered at least 25,000 landslides (Roback et al., 2018), resulting in an extensive and immediate coseismic impact but also an ongoing postseismic legacy of hillslope damage and enhanced supply of sediment potentially available to the fluvial network (Kincey et al., 2021, 2023; Tian, Owen, Xu, Ma, et al., 2020; Tian, Owen, Xu, Shen, et al., 2020). Although the heightened rate of postseismic landsliding and debris flows reduced markedly within 1–2 years of the earthquake (Dahlquist & West, 2019), the geomorphic consequences are still playing out, especially following intense postseismic monsoon seasons such as that of 2020 (Rosser et al., 2021).

This chapter initially outlines the geomorphic context of the 2015 Gorkha earthquake, considering the seismic and topographic setting and the key broad-scale geomorphic consequences that occurred at the time of the earthquake and in the years since. We then focus on the extent to which preexisting landslides responded to the earthquake shaking, including evidence for weakening of hillslopes and associated enhanced susceptibility to further failure, as well as direct examples of reactivation of previously failed slopes during the earthquake itself. We then summarise the growing body of research on the postseismic evolution of hillslopes and landslides within the earthquake-affected area, focusing on spatial and temporal patterns of landslide and debris-flow activity in the years following the earthquake. Using two new forms of data around landslide cluster evolution and long-term landslide time series, we explore how the landscape perturbation generated by the 2015 earthquake fits into the longer-term geomorphic context of this region of the Himalayas, drawing upon the recent compilation of long-term landslide histories from archive satellite imagery. Finally, we provide a reflection on how the geomorphic effects of the earthquake have impacted

Geomorphic Impacts of Earthquakes
DOI: https://doi.org/10.1016/B978-0-443-40375-0.00018-9

people and infrastructure in the region, fundamentally altering patterns of geohazard risk both in the immediate aftermath of the event and for many years afterwards.

15.2 Geomorphic context of the 2015 Gorkha earthquake

Nepal lies within the Himalayan orogenic belt, formed by convergence between the Indian and Eurasian continental plates, whereby the Indian plate is being underthrust northwards beneath Eurasia at a rate of approximately 40–50 mm/year (Bilham et al., 1997). Previous large historical seismic events in the region have included a M_w 7.7 earthquake north of Kathmandu in 1833, the devastating M_w 8.1 Nepal–Bihar earthquake in 1934, and a M_w 6.8 earthquake in 1988 that occurred 160 km southeast of Kathmandu and led to approximately 1000 fatalities (Bilham, 2004; Collins & Jibson, 2015). Palaeoseismic evidence suggests that multiple other high-magnitude events ($M_w > 8$) have occurred across Nepal since CE 1100 (Bilham, 2004; Lavé et al., 2005; Schwanghart et al., 2016).

The steep terrain and annual influence of the Indian monsoon also make Nepal a global hotspot for a wide range of mountain hazards. Nepal is one of the most landslide-prone countries in the world, experiencing >10% of all rainfall-triggered fatal landslides globally, despite having <0.4% of the global population (Dilley et al., 2005; Froude & Petley, 2018). Longer-term studies have suggested that both landslide occurrence and the resulting fatalities have increased over recent decades (Adhikari & Tian, 2021; Petley et al., 2007). The seasonal landslide peak, coincident with the timing of the South Asia Summer Monsoon, combines with other episodic and cascading hazards, such as the persistent and potentially increasing occurrence of glacial lake outburst floods (GLOFs) (Allen et al., 2022; Cook et al., 2018) and river floods (Pink et al., 2025) within the major transboundary catchments draining the Himalaya. Following high-magnitude earthquakes, such as the 2015 Gorkha event, these multiple processes interact in complex ways to influence the temporal trajectories and spatial variability of postseismic landscape change and recovery (Mani et al., 2023; Yanites et al., 2025).

On 25 April 2015, a M_w 7.8 earthquake ruptured the Main Himalaya Thrust fault with its epicentre in Gorkha District of Nepal (28.147°N, 84.708°E), approximately 76 km northwest of the capital Kathmandu (Avouac et al., 2015; Gualandi et al., 2017) (Fig. 15.1). The mainshock occurred at a hypocentral depth of ~15 km, centred on a section of fault along the subduction boundary between the Indian and Eurasian plates with a dip angle of 7–12 degrees (Galetzka et al., 2015; Hayes et al., 2017). The earthquake propagated eastwards along a 140-km-long segment of fault without rupturing the ground surface and was followed by >550 aftershocks, at least 30 of which were $M_w > 5.0$ (Adhikari et al., 2015; Parameswaran et al., 2015). A M_w 7.3 aftershock occurred on 12 May 2015, approximately 150 km east-northeast of the mainshock epicentre, extending the rupture area further east but also leaving a ~20 km

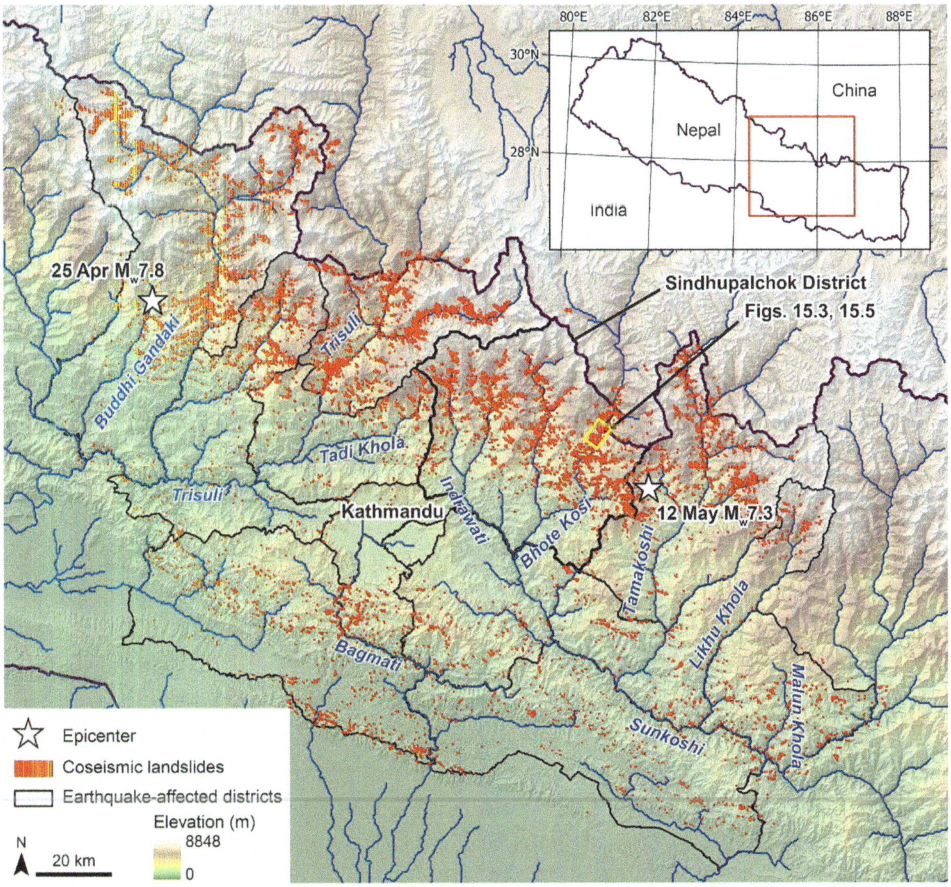

Figure 15.1 Map of the area affected by the 2015 Gorkha earthquake in Nepal, showing the locations of the mainshock (25 April, M_w 7.8) and the largest aftershock (12 May, M_w 7.3). Landslides triggered by the earthquake mapped by Kincey et al. (2021) are shown in red, along with the 14 administrative districts that were most intensively impacted by the earthquake (outlined in black). Note the position of Sindhupalchok District, northeast of Kathmandu, where coseismic and postseismic landslide impacts were especially pronounced. The yellow box shows the approximate location of Figs. 15.3 and 15.5. Map lines delineate study areas and do not necessarily depict accepted national boundaries. Source: *Background elevation data are derived from the ASTER Global Digital Elevation Model V003 (NASA/METI/AIST/Japan Spacesystems and U.S./Japan ASTER Science Team).*

gap where no coseismic slip occurred (Lindsey et al., 2015). The total earthquake-affected area spanned >28,000 km^2 of central Nepal, extending from the High Himalaya in the north through the Middle Hill region that forms the central west–east physiographic belt through the country (Roback et al., 2018).

Multiple coseismic landslide inventories were produced following the earthquake (e.g. Kargel et al., 2016; Martha et al., 2017; Meena & Tavakkoli Piralilou, 2019;

Roback et al., 2018; Tiwari et al., 2017; Williams et al., 2018; Xu et al., 2016; Xu, 2018), resulting in estimates of coseismic landslide numbers ranging from <5000 to ~47,000. This diversity reflects in part the variety of mapping and modelling techniques that were applied (Milledge et al., 2022), in addition to the range of motivations, from rapid response (e.g. Williams et al., 2018; Gallen et al., 2017) to multiinventory time-series monitoring (e.g. Kincey et al., 2021). The most complete coseismic inventory documented at least 25,000 landslides with a total area of ~ 87 km^2 (Roback et al., 2018). For comparison, this represents around 4–6 years of monsoon-induced landsliding within this region occurring within a single year (Marc et al., 2019). The majority of the coseismic landslides in the Gorkha earthquake sequence were triggered by the M_w 7.8 mainshock, with only ~10% thought to have been triggered by the M_w 7.3 aftershock (Tiwari et al., 2017), although considerable uncertainty arises due to the limited imagery available between these events.

Although widespread, the number of large coseismic landslides (those with areas of ≥10,000 m^2) triggered by the Gorkha earthquake was only ~2000, which was notably lower than the ~25,500 landslides of this size that resulted from the M_w 7.9 2008 Wenchuan earthquake (Xu et al., 2016; see Chapter 14). Estimates of total landslide volume released by the Gorkha earthquake range between ~251 and 1502 Mm^3, which was considerably less than the ~2134 Mm^3 of overall volume added to the Himalaya through vertical rock uplift in the earthquake (Valagussa et al., 2021). The spatial density of landslides was controlled by the broad-field seismic deformation, with landslides concentrated on the downthrown block to the north of the main fault rupture (Kargel et al., 2016). At a more local scale, the coseismic landslide distribution was influenced by a complex set of seismo-tectonic, lithological and topographic controls which have been explored in detail by a range of studies (e.g. Dahal et al., 2024; Dunham et al., 2022; Kargel et al., 2016; Martha et al., 2017; Roback et al., 2018; Tian, Owen, Xu, Ma, et al., 2020; Tian, Owen, Xu, Shen, et al., 2020).

Approximately 15% (c.3750) of the coseismic landslides mapped by Roback et al. (2018) ran out directly into the fluvial network, representing 33% of the total landslide area and 50% of the total estimated volume of mobilised sediment. Initial postseismic surveys identified that at least 74 (c.2%) of these landslides formed valley-blocking dams within the earthquake-affected area (Collins & Jibson, 2015), with significant, albeit short-term, impoundment forming along the Marsyangdi River in the Annapurna region and the lower Tom River in the Manaslu Conservation Area (Kargel et al., 2016). Most landslide-dam-related impoundments were relatively small and only 50% of landslide dams had lakes impounded, either because rivers were dry or more often due to almost instant breaching. Lake surface areas were small (mean 1700 m^2); the largest was at Baisari which resulted in a c.3 km lake backed up along the Kali Gandaki River (Collins & Jibson, 2015). This event generated perhaps the most significant outburst flood, but no fatalities were recorded.

Similarly, although postseismic cracking was identified on moraine dams impounding some glacial lakes (e.g. Tsho Rolpa, Rolwaling Valley), and landslide densities proximal to glacial lakes increased (Byers III et al., 2017), no major GLOFs occurred in the weeks after the Gorkha earthquake (Kargel et al., 2016). There has since been speculation around the role of shaking-induced weakening of hillslope materials from the Gorkha earthquake as a contributing factor to the occurrence of large-scale flood events sourced in the High Himalaya within the earthquake-affected area (e.g. Talchabhadel et al., 2023). The potential exacerbation of GLOF impacts, where floods propagate through earthquake-damaged landscapes, has also been demonstrated by Chen et al. (2023) in a study of flood impact amplification along the Bhote Kosi River in Sindhupalchok District in July 2016. Using passive seismic monitoring of the same 2016 flood, Cook et al. (2018) proposed that longer-term landscape erosion rates may well be driven by GLOF magnitude and frequency, rather than by precipitation, where such events occur.

The 2015 Gorkha earthquake has led to an ongoing geomorphic legacy for central Nepal, notably in terms of enhanced rates of hillslope activity. Although the rate of new landsliding and debris flows reduced within the first 1–2 years after the earthquake (Dahlquist & West, 2019; Marc et al., 2019), multitemporal landslide inventory mapping across the entire earthquake-affected area indicates that the overall number and area of landslides visible in satellite imagery continued to increase in the 5 years following the earthquake (Kincey et al., 2021; Rosser et al., 2021). Many coseismic landslides have remained active in the years immediately following the earthquake (Tian, Owen, Xu, Ma, et al., 2020; Tian, Owen, Xu, Shen, et al., 2020), with systematic upslope and northward shifts in the spatial footprint of landsliding through time (Kincey et al., 2021), coupled with downslope routing of mobilised sediment that becomes increasingly connected to the fluvial network (Kincey et al., 2023). Despite the widespread co- and postseismic landsliding, the impact on fluvial sediment yields in the 6 years following the Gorkha earthquake remained relatively localised and limited (Graf et al., 2024) in comparison to other recent mountain region earthquakes, such as the 2008 Wenchuan earthquake (e.g. Fan et al., 2018; Chapter 14). Since landslide-channel connectivity for the Gorkha earthquake was relatively high (Roback et al., 2018) and landslides on the hillslopes in many instances did not rapidly revegetate (Kincey et al., 2021), this relatively limited influence on fluvial sediment dynamics is likely the result of a combination of valley morphology and the location and timing of specific postseismic meteorological triggering events (Graf et al., 2024).

Alongside the geomorphic effects, the 2015 Gorkha earthquake had a profound impact on the people of Nepal. Almost 9000 people died as an immediate result of the earthquake, and a further 22,300 were injured, with approximately 8 million people, one-third of Nepal's population, being directly impacted (Government of Nepal, 2015). In total, around 300,000 buildings were entirely destroyed, and a further 270,000 were partially damaged, with a direct economic cost to Nepal of over US$5 billion

(Zhao, 2016). Many immediate losses likely resulted from landslides and rockfalls triggered by the earthquake, including the catastrophic debris avalanche that buried multiple settlements within the Langtang Valley, killing at least 350 people (Lacroix, 2016), and extensive impacts on densely populated valley-bottom highway corridors in the affected area (Collins & Jibson, 2015). As with many mountain region earthquakes, the overall attribution of losses from secondary geohazards, as compared with other earthquake impacts such as building collapse, remains difficult to ascertain (Budimir, 2015). Following recognition of the severe landslide impacts triggered by the 2015 Gorkha earthquake, and the well-established dominance of the annual monsoon in driving landsliding and consequent fatalities (Petley et al., 2007), the Government of Nepal initiated the National Geohazards Assessment to evaluate landslide hazards and associated risks, with the aim of informing emergency response and reconstruction planning (Rosser et al., 2021). The assessment involved rapid field investigations, followed by a more extensive phase undertaken under the National Reconstruction Authority, which by July 2020 had identified 1053 settlements as potentially at risk from geohazards. Of these, 40% were classified as safe, 30% as requiring mitigation and 30% as at an unacceptably high risk that was unfeasible to manage; this resulted in recommendations that over 4000 households be supported to relocate to safer positions in the landscape (Rosser et al., 2021). As the geomorphic effects have evolved in the decade following the earthquake, so have the risks to people, with notable shifts in patterns of exposure to potential secondary hazards (Arrell et al., 2024; Kincey et al., 2023).

15.3 Reactivation of preexisting landslides

The Nepal Himalaya affected by the 2015 Gorkha earthquake contains numerous palaeo-landslides, many of which are dormant or exhibit only minor episodic movement. These features are commonly mantled by poorly consolidated colluvial soils that are repeatedly disturbed by the construction and maintenance of agricultural terracing (Tsou et al., 2018), rendering their identification from surface morphology or deformation patterns alone difficult. Despite their apparent quiescence, such slopes remain unstable and are frequently occupied by densely populated communities, where even low-magnitude deformation can be highly disruptive (Ferrer et al., 2024). Previous studies have demonstrated that palaeo-landslides may respond in a variety of ways to seismic shaking, exhibiting delayed reactivation, transient acceleration, or even a negligible response (Lacroix et al., 2022).

Bhandari (2024) documented widespread postseismic reactivation of ancient landslide deposits during monsoon seasons following the Gorkha earthquake, particularly within the heavily affected districts of Sindhupalchok, Nuwakot and Rasuwa. Field investigations identified 114 palaeo-landslides that reactivated between 2019 and 2023 in response to intense monsoon rainfall. Geotechnical testing of disturbed soil samples revealed low cohesion (3–10.5 kPa) and moderate plasticity, indicating materials highly susceptible to strength loss upon saturation. While monsoon rainfall was identified as the immediate trigger, Bhandari (2024) proposed that the 2015 earthquake acted as a critical preconditioning mechanism, weakening slopes through fracturing and deformation and thereby lowering resistance to subsequent hydrometeorological forcing.

In contrast, evidence for systematic postseismic acceleration of slow-moving landslides following the Gorkha earthquake remains equivocal. Using InSAR-derived deformation rates spanning the preevent period and up to 2.5 years postevent, Bekaert et al. (2020) observed negligible accelerations and in some cases decelerations, of active slow-moving landslides over annual timescales. Similarly, Tsou et al. (2018), focusing on 'coherent' slow-moving landslides in the Trishuli River valley of central Nepal, reported limited sensitivity to seismic shaking, although they noted clustering of shallow, rapid coseismic landslides initiated within the larger landslide bodies. Comparable clustering of new landslides on slow-moving landslide deposits was observed along the Arniko and Pasang Lhamu highways (Tian, Owen, Xu, Ma, et al., 2020; Tian, Owen, Xu, Shen, et al., 2020).

Delayed seismic triggering has nevertheless been documented in several cases. At Baisari, a rockslide failed 29 days after the mainshock, damming the Kali Gandaki River. Eyewitness accounts describe crack initiation following the April M_w 7.8 mainshock, progressive widening after the 12 May M_w 7.3 aftershock, rock spalling approximately 10 days later and ultimate failure on 24 May. Postevent investigations have suggested that the hillslope was already near failure prior to the Gorkha earthquakes, with seismic shaking promoting progressive deformation along incipient failure surfaces, consistent with limited occurrence of any other coseismic landslides nearby (Collins & Jibson, 2015).

A detailed site-specific study by Wu et al. (2017) examined seismic impacts on an ancient landslide at the Tatopani Dry Port near the Nepal–China border in the Upper Bhote Koshi Valley in Sindhupalchok District (Figs. 15.2 and 15.3). The site comprises a large amphitheatre-shaped palaeo-landslide with evidence of multiple prior failures and ongoing deformation, previously attributed to river undercutting and to undercutting of the lower part of the landslide mass through road construction and terracing. Field observations documented earthquake-induced failures and the development of tension cracks near the crest and along the western flank of the landslide mass following the mainshock and aftershocks. The landslide body experienced approximately 3 m of vertical displacement within 6 months of the earthquake (Fig. 15.2), and previously stable slope sections developed fresh cracks and localised failures due to intense coseismic ground shaking. Numerical stability modelling indicated that the landslide would remain stable under nonseismic conditions but could be remobilised if peak ground acceleration during a future earthquake exceeds ~0.35 g (Wu et al., 2017). Sequential satellite imagery of the site (Fig. 15.3) shows destabilisation of the toe of the

Figure 15.2 Extensive lateral scarp at Purano gumba, Duguna Gadhadi, Sindhupalchok (whatthreewords location: ///climate.ministers.mildest), showing c.3 m of vertical displacement during the 2015 Gorkha earthquake. Photo taken in November 2015. Deformation of the same landslide is shown in Fig. 15.3c of Lacroix et al. (2022), and the site was studied in detail by Wu et al. (2017).

landslide mass in 2017, but not immediately after the earthquake in 2015, in addition to a rapidly evolving network of incised debris-flow channels over the top of the main landslide, routing coseismically released debris from the surrounding rockslopes.

Lacroix et al. (2022) investigated the response of slow-moving landslides to the 2015 Gorkha earthquake using time-series synthetic aperture radar (Sentinel-1) and high-resolution optical (Pléiades) imagery. Within a ~750 km^2 portion of the earthquake-affected area, they identified 11 slow-moving landslides and quantified displacement before, during and after the earthquake using time-series image correlation. The results show that postseismic displacements in the 2 months following the mainshock substantially exceeded coseismic motion, locally reaching ~34 m. Notably, at least two landslides exhibited pronounced acceleration only several days (>4 days) after the earthquake, indicating a delayed response not readily attributable to rainfall or aftershocks. The authors attributed this behaviour to postseismic groundwater diffusion and progressive saturation driven by coseismic contraction and enhanced permeability, highlighting the critical role of hydrological processes in controlling slow-moving landslide deformation following strong seismic forcing (see also Chapter 13). Complementary observations from passive seismic monitoring installed shortly after the earthquake suggest short-lived (<1 year) postseismic modifications of near-surface permeability that may alter pore-pressure distributions and frictional weakening (Illien et al., 2022), consistent with the timescales of transient postseismic landslide responses

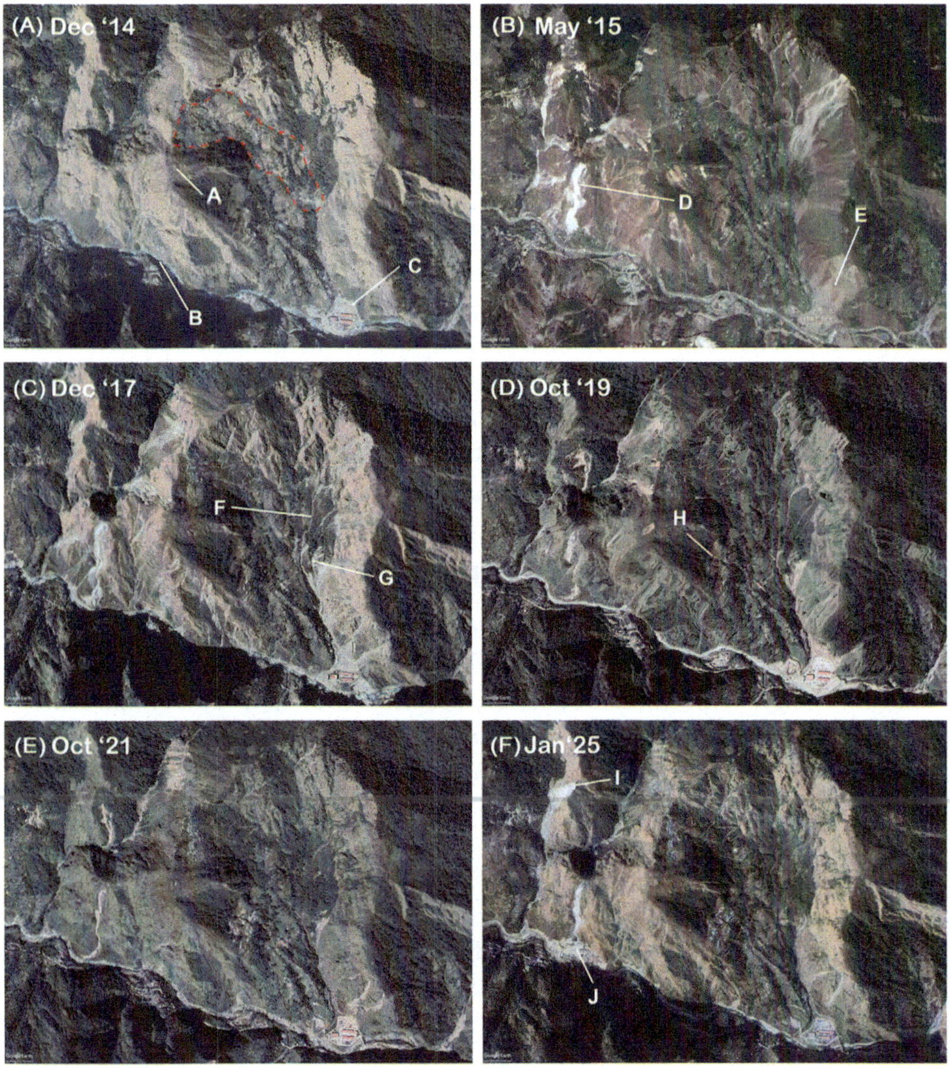

Figure 15.3 Satellite image time-series showing pre- to postearthquake landsliding along a 3.9 km section of the Upper Bhote Kosi River, from Hindi (left) to Larcha (right). See Fig. 15.1 for location. Features referred to in the text are annotated: (A) Duguna Gadhi Fort; (B) Upper Bhote Kosi River; (C) Tatopani Dry Port; (D) Tushari rockslide; (E) debris avalanche onto the dry port; (F) postearthquake debris flows incised into slow-moving landslide mass; (G) evolving landslides at the toe of a slow-moving landslide; (H) resettlement area; (I) large rockslope failure; (J) new hydropower development beneath Tushari rockslide. The slow-moving landslide shown in Fig. 15.2 and discussed by Wu et al. (2017) is outlined by the red dashed line in (A). Source: *Image © 2015 Maxar Technologies; Image © 2014/17/19/21 CNES/Airbus; Image © 2025 Airbus.*

reported elsewhere (Marc, Hovius, Meunier, et al., 2015). In the absence of in situ monitoring during the Gorkha earthquake, these studies provide rare quantitative evidence linking postseismic landslide kinematics to seismically induced hydrological perturbations.

Ao et al. (2020) examined the long-term deformation of the catastrophic Jure landslide on the Sunkoshi River in Sindhupalchok District, which severely impacted the Arniko Highway and adjacent valley. Although the initial failure occurred in August 2014, prior to the Gorkha earthquake, the site provided an important opportunity to assess pre-, co- and postseismic landslide behaviour due to its continued activity. Using distributed scatterer and conventional InSAR applied to multimission SAR datasets (ALOS PALSAR, TerraSAR-X, Sentinel-1), the authors showed that the landslide exhibited limited sensitivity to rainfall but experienced a clear acceleration following the 2015 Gorkha earthquake, followed by a progressive deceleration towards quasilinear motion from early 2016 onward.

15.4 Postseismic evolution of landslides

Consistent with other similar high-magnitude earthquakes, the coseismic geomorphic impacts of the Gorkha event have been followed by a complex set of evolving postseismic longer-term effects. Rates of occurrence of new postseismic landslides remained elevated for several years after the earthquake (KC et al., 2024). Estimates of landslide recovery timescales, when considered in terms of the time over which susceptibility to new landsliding is elevated postevent, suggest that hillslopes within the footprint of the 2015 Gorkha earthquake took between 1.5 and 3.5 years to recover (Tanyaş et al., 2021). This moderately long recovery timescale reflects the high topographic relief and monsoonal climate of Nepal but is notably shorter in duration than recovery timescales for other recent high-magnitude earthquakes (e.g. 2008 M_w 7.9 Wenchuan earthquake, 7–25 years recovery; Tanyaş et al., 2021).

Hillslope 'recovery' timescales are, however, highly dependent on the methodology and metrics employed in their quantification (Kincey et al., 2021), with entirely new landslides representing only one component of the overall postearthquake–landslide hazard cascade (Rosser et al., 2021). When hillslope recovery is considered to include both new and existing landslides visible in the landscape within the Gorkha earthquake-affected region, longer-term multitemporal studies have shown that the total number and area of landslides were both still higher in 2020 than they were on the day of the mainshock itself (Kincey et al., 2021; Kincey et al., 2023). This partially reflects the relatively long timescale required for hillslopes to revegetate following landsliding in this region (Fig. 15.3), but also a notable increase in landsliding from 2017 onwards has been tentatively linked to widespread road building and infrastructure development associated with local elections (Rosser et al., 2021), with poorly engineered road construction known to be an important factor in landslide density in mountain areas (Sudmeier-Rieux et al., 2019; Tanyaş et al., 2022). Across the entire 2015 Gorkha earthquake footprint, areas of high landslide density shifted to higher elevations in the north of the region following the earthquake, with a corresponding reduction in landslide density on the lower slopes of the major north–south river valleys through Gorkha, Rasuwa, Dhading, Sindhupalchok and Dolakha districts. These spatial differences in recovery and new landsliding likely reflect the coincidence of seismic hillslope damage and local-scale topographic controls, such as elevation, slope angle and slope aspect (Kincey et al., 2021).

15.4.1 Postseismic evolution of individual landslides

Understanding how and when individual landslides evolve is a fundamental requirement of postseismic hazard assessments, but it is challenging to quantify when landslide inventories have been mapped independently as separate epochs and where the complexity of the landslide history and topography is combined (Kincey et al., 2021). Recent work has aimed to resolve this through the development of a semiautomated method for temporally linking landslide polygons from separate inventory time steps (Shrestha et al., 2026). This approach applies a series of structured spatial overlay steps with amalgamation-checking routines (Marc & Hovius, 2015) and geomorphological slope unit attribution (Alvioli et al., 2016), in order to group individual landslide polygons into unique clusters from which an evolutionary history can be extracted. Each cluster contains multiple polygons representing the temporal evolution of the same landslide, allowing metrics such as changes in shape, size, head and toe position and degree of branching to be quantified.

This method has been applied to >200,000 independently mapped landslide polygons spanning 1 year before the 2015 Gorkha earthquake and 5 years after the event (2014–20), providing insights into postseismic landslide dynamics and the controls on landslide evolution (Shrestha et al., 2026). Of the >32,300 unique landslide clusters identified across the earthquake-affected region in this period, ~6500 were newly triggered by the 2015 Gorkha earthquake. Analysis of the temporal evolution of these coseismic landslide clusters showed that they doubled in area between 2015 and 2020. New landslide clusters triggered in the subsequent 2015–17 monsoon seasons after the earthquake also increased by a similar proportion, but more recent new landslide clusters (2018–19) increased in size at a lower rate. Overall, the number of existing landslide clusters in 2020 remained at least three times that of the preearthquake baseline number.

Individual coseismic landslides display complex patterns of postseismic evolution, with the degree of upslope retrogression and downslope runout being at least partially controlled by the size of the landslide, the time since the initial failure and the local topographic context. For example, landslides that experienced the highest amount (95th percentile) of change to their toe position, typically representing postseismic debris flows, were more commonly found on slopes between 35–45 degrees. In contrast, landslides that recorded the largest amount of headward expansion (retrogression) were more typically on steeper slopes between 40–45 degrees. This increased headward expansion has the potential to supply additional sediment to lower-slope positions, in turn increasing the volume of sediment that is available to runout through future debris flows.

Landslide cluster size was also an important control on subsequent hillslope evolution, with the average change in head and toe position for the largest landslides being an order of magnitude greater than for the overall landslide population. Larger coseismic landslides from other mountain-region earthquakes have also been shown to remain visible and/or active longer (Fan et al., 2018), relating in part to the more rapid revegetation and sediment exhaustion that occurs with smaller landslides (Fan et al., 2018; Shen et al., 2020).

Landslides with the largest changes in head and toe position preferentially occurred within mid-slope positions, reflecting the elevated potential for additional erosional processes at both ridge-top and channel-bank positions, but also the availability of hillslope space into which the initial landslide could develop. Across the entire earthquake-affected area, we also note a grouping of highly active landslide clusters within the same or neighbouring hillslopes, highlighting spatial variability in recovery rates and a recognised persistence of postseismic activity in areas more densely impacted by landslides (Kincey et al., 2021; Yang et al., 2018).

In terms of timing, the landslides that were identified as most likely to expand through upslope regression or downslope runout were those triggered within a year before and after the earthquake, potentially relating to the occurrence of the earthquake immediately before the 2015 monsoon (Jones et al., 2021; Marc, Hovius, Meunier, et al., 2015). We also find that the rate of recovery of individual landslides was related to the initial timing and size of the failures, with new postseismic landslides triggered during the first 2 years after the earthquake revegetating more rapidly than the total coseismic landslide population (Fig. 15.4). However, this trend was disrupted in later years, with new landslides triggered more than 2 years after the earthquake taking longer to revegetate, indicating that other factors, such as the intensity of monsoonal rains, also influence patterns of postseismic recovery.

The 2020 monsoon had a significant impact on geomorphic recovery following the 2015 Gorkha earthquake, with both the rate of new landsliding and the reactivation of preexisting landslide clusters being higher in 2020 than in earlier postseismic monsoon seasons. The intense 2020 monsoon triggered extensive landsliding and debris flows across central Nepal (Rosser et al., 2021), and our results show that landslide clusters initially triggered during the earthquake and the subsequent 2015 monsoon were those which were most active during 2020. This precipitation-driven reactivation of co- and immediately postseismic landslides has been noted elsewhere (e.g. after the 2016–17

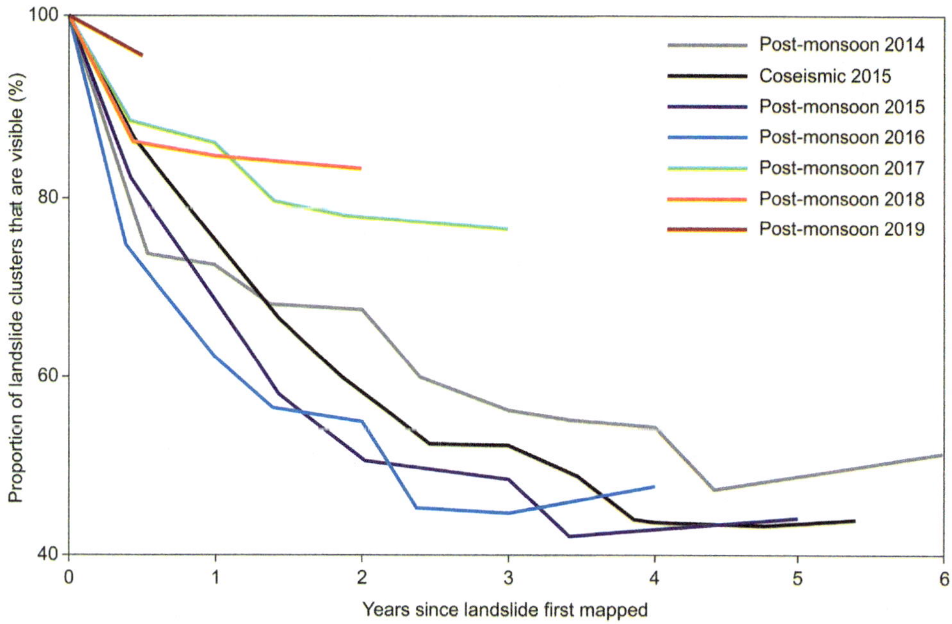

Figure 15.4 Time history of unique landslide clusters, based upon the initial date of cluster formation. Each coloured line relates to the initial epoch in which the landslide clusters first appeared, indicating the proportion (in percent) of landslides that were visible in each subsequent timestep. Landslides initially triggered in the years immediately following the earthquake (postmonsoon 2015 and 2016, blue lines) revegetated more quickly than those first triggered towards the end of the time series (2017–19, green to red lines), with a higher proportion of these later landslides persisting for the first 1–3 years after failure.

central Italy earthquake sequence, Song et al., 2022) and implies a degree of path dependency in landslide susceptibility, and specifically where new landslides occur in relation to previous landslides (Roberts et al., 2021; Samia et al., 2017). The extensive impact of the 2020 monsoon demonstrates the longer-term potential for large-scale postseismic disruption by intense precipitation events several years after the initial earthquake occurred, and fits with observations of decadal-scale recovery timescales from other high-magnitude earthquakes (Huang & Li, 2014; Li et al., 2018; Tang et al., 2016).

Overall, ~50% of all preearthquake, coseismic and immediately postseismic (i.e. triggered by the 2015 monsoon) landslide clusters were still visible in the landscape 5 years after the earthquake. In contrast, postseismic landslides initially triggered 1–2 years after the earthquake were much quicker to revegetate, reducing to only 40% of the overall population still visible after just 2 years. Although the presence of a visible landslide scar does not necessarily mean the landslide is active, these results highlight the potential for preexisting clusters to change or reactivate years after their initial triggering. Understanding the spatial location and topographic context for each landslide, as well as when it was first triggered, therefore provides important insights into how we might expect a given hillslope to evolve in the years following a high-magnitude earthquake, which in turn provides invaluable information for managing post-seismic hazard and risk.

15.4.2 Potential for future landslide hazard

The potential for remobilisation of sediment generated by coseismic landslides through rapid, fluidised debris flows has been widely demonstrated for other high-magnitude events, most notably the 2008 Wenchuan earthquake (Fan et al., 2019; Fan et al., 2018). Modelling of landslide runout pathways across the entire region impacted by the 2015 Gorkha earthquake demonstrated that, while an area of 150 km^2 was directly impacted by coseismic landsliding, a further 614 km^2 was at potential risk of debris-flow runout from those landslide source areas (Kincey et al., 2023). Importantly, the total runout-susceptible area had increased to 777 km^2 after the 2019 monsoon, 4.5 years after the initial earthquake, but had also evolved spatially as a result of changes to the postseismic landslide distribution. As this landslide distribution evolved, the threat to local populations also changed, becoming more pronounced at mid- and lower-hillslope positions (Arrell et al., 2024). In total, 14% of the overall area susceptible to runout from coseismic landslides experienced a landslide at some point within the 4.5-year time series, increasing to 32% when the model output only considered areas classified as higher susceptibility (Kincey et al., 2023). The rate at which coseismic modelled runout areas were realised as landslides did decrease considerably, though, with the majority occurring within the first 1–2 years and so corresponding well with observed findings from the analysis of postseismic debris-flow inventories (Dahlquist & West, 2019).

Overall, this indicates that in the case of the 2015 Gorkha earthquake, there was a relatively short-lived increase in postseismic debris-flow activity, followed by a reduction. However, despite this reduction in activity, postseismic debris flows linked to coseismic landslides are still occurring in the earthquake-affected area, years after the initial event. For example, analysis of landslides along the Upper Bhote Kosi (Sindhupalchok District) and Trishuli (Rasuwa District) river valleys recorded persistent activity still occurring from over half of the coseismic landslides under study three years after the earthquake, with debris flows in particular becoming more frequent and damaging within certain localised areas (Tian, Owen, Xu, Ma, et al., 2020; Tian, Owen, Xu, Shen, et al., 2020). Comparison of modelled runout pathways from coseismic landslides with postseismic multitemporal landslide inventories (Fig. 15.5) also shows persistent activity across hillslopes that were badly impacted during the earthquake, such as areas with extensive ground cracking (e.g. Fig. 15.6). Other damaging debris flows have since occurred within the earthquake-affected area during more

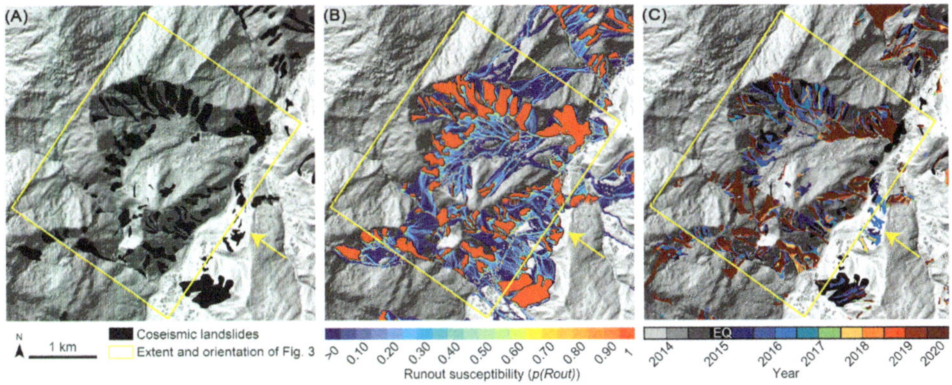

Figure 15.5 (A) Coseismic landslide distribution for an area of the Upper Bhote Kosi between Hindi (southwest) and Larcha (northeast). (B) Modelled landslide runout susceptibility, showing the predicted debris-flow propagation under extreme rainfall, using the 2015 coseismic landslides as source areas. Colours show the runout susceptibility or likelihood of eventual inundation by landslide debris from source areas located upslope; note that coseismic landslide areas have a susceptibility of 1 by definition. (C) Multitemporal landslide inventory, showing the mapped evolution of landslide hazard between 2014 and 2020. The yellow box and arrow show the approximate extent and orientation of the images in Fig. 15.3. Source: *Landslide inventory data are from Kincey et al. (2021); modelled runout data are from Kincey et al. (2023). Shaded relief elevation map derived from a 5-m ALOS DEM of the study area (credit: © NTT DATA, RESTEC/©JAXA)..*

Figure 15.6 Ground cracking above Yarmasing, Duguna Gadhadi, Sindhupalchok District (whatthreewords position: \\\whiten.classy.backsliding). Photo taken in November 2015.

recent intense monsoon seasons, such as the three fatal events at Jambu, Lidhi and Ghumtang that all occurred in Bahrabise Municipality in 2020 (Rosser et al., 2021; Shrestha et al., 2024). In the case of the Jambu event, coseismic landslides had been identified in the headwaters, with postevent modelling of runout from these landslides suggesting that the valley-floor settlement could be at risk of subsequent debris flows. However, determining exactly which of the many potentially at-risk settlements will be impacted by future debris flows, and the timescales over which this might occur, remains highly challenging.

15.5 Putting Gorkha into a longer-term context

Harvey et al. (2024) showed that the area impacted by the 2015 Gorkha earthquake, and Nepal more widely, has only limited inventory data describing long-term patterns in landsliding. To address this gap, several studies have created landslide time histories from archived satellite imagery to investigate controls on interannual variability in landsliding. Focussing on landslides with areas >1000 m^2, Jones et al. (2021) identified a link between years with extreme events, including cloudbursts (e.g. 1993, 2002; Dhital, 2003), flooding (e.g. 2017) and the 2015 Gorkha earthquake, and changes to the mapped landslide distribution. The coincidence of the footprint of extreme events with areas of high infrastructure and urban development, such as the rainfall in central Nepal in 2002, also demonstrated the role of rural road construction in increasing landslide susceptibility. For data from 2015 specifically, Jones et al. (2021) noted a shift to landsliding on steeper (35–60 degrees shifting to 45–70 degrees) and higher (250–800 m to 500–1200 m) parts of the landscape, and further from channels, as is typical for coseismic landslide distributions (e.g. Meunier et al., 2008).

Later work building a 30-year time series of landsliding from satellite image archives across the Hindu Kush Himalaya at 30 m resolution places the 2015 Gorkha earthquake impacts into a longer-term context (Chen et al., 2024; Fig. 15.7). An analysis of landsliding in Sindhupalchok, one of the districts most intensively impacted by landsliding in the 2015 Gorkha earthquake, shows a complex trend. The spatial footprint of landsliding shows pervasive small-scale failures across the landscape present in all years, with the impact of the 2015 Gorkha earthquake clearly superimposed from 2016 onwards. Years with documented high monsoon rainfall and consequent flooding (e.g. 1998, 2002, 2008: Yogacharya & Gautam, 2008) display heightened valley-bottom concentrations of landslides which persist through the remainder of the time series, occasionally linked to large individual landslide scars. The transient landslide response after the Gorkha earthquake is also visible at this scale, showing both a gradual reduction in landsliding in the landscape from 2016 onwards, and the persistence of concentrated landsliding in major north–south oriented valleys, including in the tributaries of the Bhote Kosi River (Fig. 15.7).

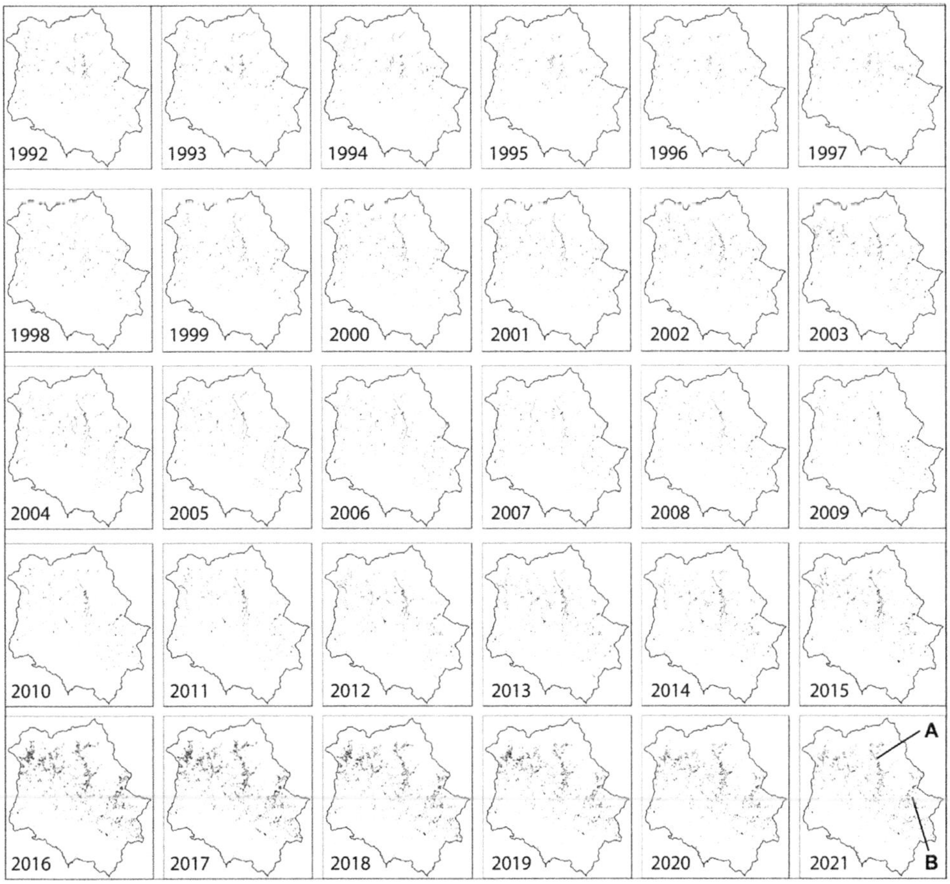

Figure 15.7 Time series of annual landslide mapping from 1992 to 2021, based upon the analysis of Chen et al. (2024), showing the extent of mapped landsliding at 30 m cell resolution for all areas under 4000 masl within Sindhupalchok District. For scale, Sindhupalchok is c.60 km E–W and c.66 km N–S. Labels (A) and (B) in the 2021 panel show areas of high landslide density in the upper tributaries of the Bhote Kosi River. Note that landsliding mapped in any given calendar year is displayed here in the following year, so the impact of the 2015 Gorkha earthquake is shown in 2016. See Chen et al. (2024) for explanation of how data were derived. Map lines delineate study areas and do not necessarily depict accepted national boundaries.

Notably, the proportion of land-surface area subject to landsliding (Fig. 15.8, red line) varied between c.0.5% to c.1% in the period between 1992 and immediately prior to the 2015 earthquake. This fluctuation appeared to systematically increase and then decrease around a peak in 2003, which corresponds to landsliding in 2002. This peak coincides with extensive documented landsliding in central Nepal associated with cloudburst rainfall, as reported by Dhital (2003) and also observed by Jones et al. (2021). Interestingly, based on this long-term record, the immediate preearthquake landslide

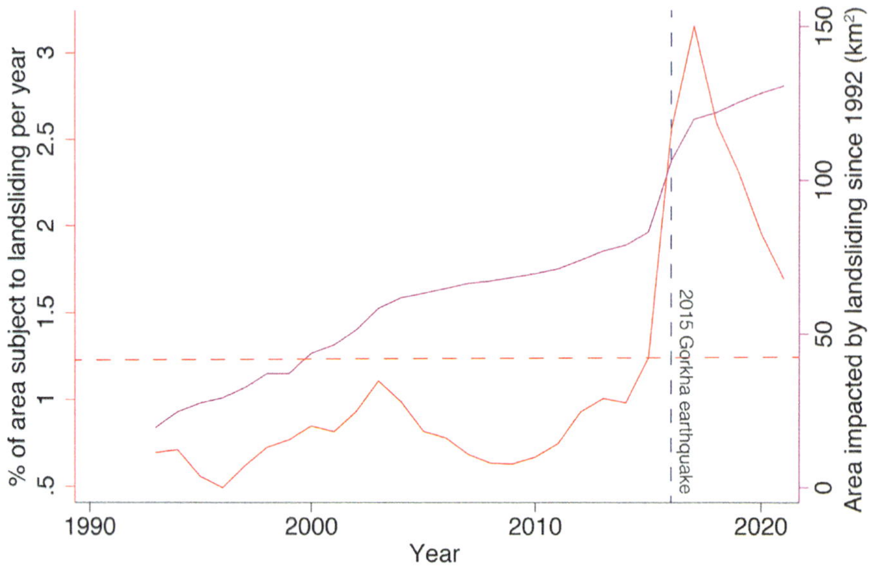

Figure 15.8 Time series of annual landslide areas from Sindhupalchok District over the period 1992 to 2021. Pink line (right y-axis) shows the cumulative area impacted by landsliding through time, in km^2; red line (left y-axis) shows the proportion of the total district area experiencing landsliding per year, in percent. The horizontal red dashed line shows the preearthquake area subject to landsliding (1.24%). Source: *Data are based on Chen et al. (2024).*

area (1.24%, here shown as '2015') was the highest on record since 1992, suggesting that the 2015 earthquake occurred at a time when landsliding was at its highest in – at least – the 18 preceding years. The area in Sindhupalchok impacted by landsliding (Fig. 15.8, pink line) is a cumulative measure of the landscape area that experiences a landslide at least once in this time-series dataset. These data show a near-steady increase through time, both pre- and postearthquake, implying that over the 30-year timescale of this study and despite the widespread landslide impacts of the 2015 Gorkha earthquake, new locations able to host landslides in the landscape have not been exhausted. Finally, these data also enable landslide age to be tracked, showing that new landslide areas in each year on average represent ~1% of all landslide areas mapped in the region, with no notable increase in the rate of new landsliding through time. This mirrors observations that whilst landslide impacts may be increasing in this region (e.g. Froude & Petley, 2018), landslide occurrence itself has remained more stable, whereby the rate of landslide occurrence per year has remained roughly constant (Chen et al., 2024).

Compiling a consistent 30-year record of landslide data enables mapping of the frequency of landslide *activations*, areas with *persistence* in landsliding and the *recurrence* of discrete periods of landsliding across Sindhupalchok (Fig. 15.9A, C, E). *Occurrence*, the number of years during which a pixel is mapped as a landslide through the time series, is

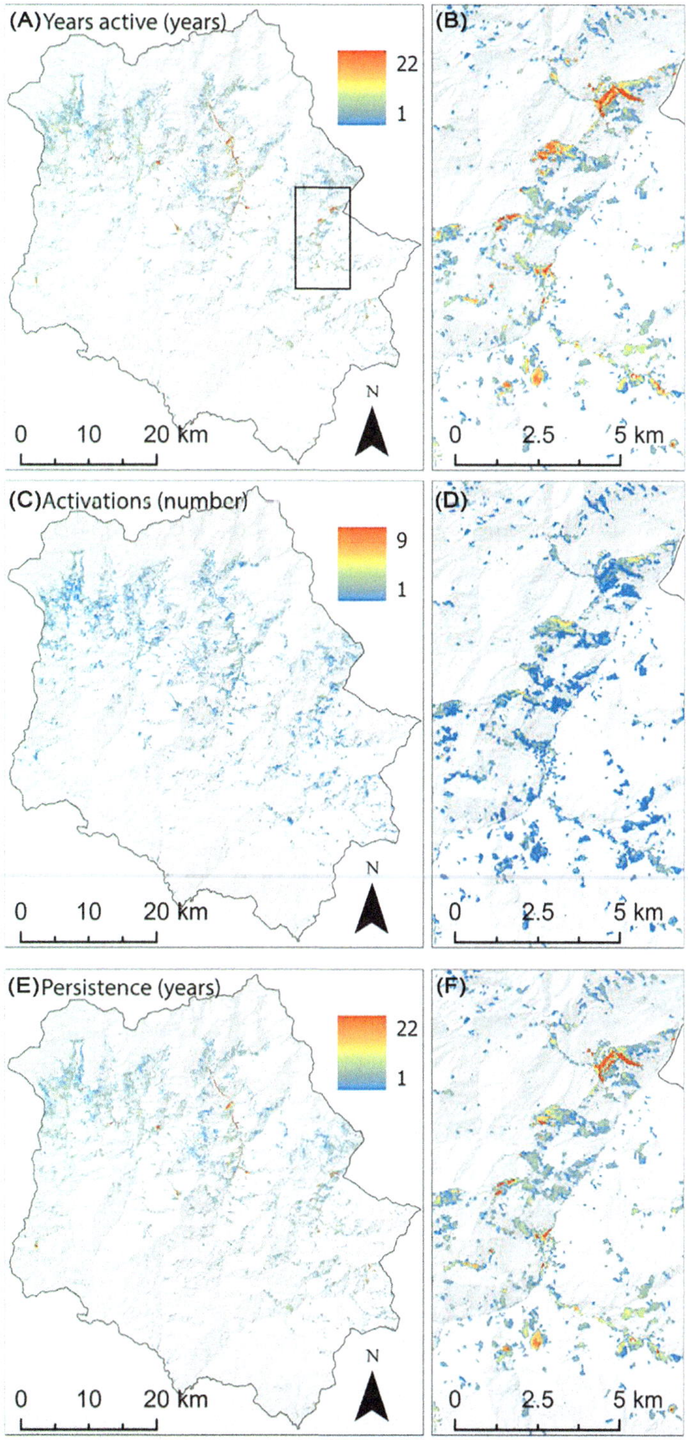

Figure 15.9 Summary statistics of long-term landslide mapping from 1992 to 2021 for Sindhupalchok District (A, C, E) and for a section of the Upper Bhote Kosi river valley (B, D, F), defined by the inset black rectangle in (A). Map lines delineate study areas and do not necessarily depict accepted national boundaries. Source: *Data are based on Chen et al. (2024). Background elevation data are derived from the ASTER Global Digital Elevation Model V003 (NASA/METI/AIST/Japan Spacesystems and U.S./Japan ASTER Science Team).*

dominated by a number of relatively large landslide scars across the district, many of which link directly into the fluvial network, in addition to debris that has runout into the channel network. In the inset example in Fig. 15.9 (panels B, D, F) covering the Upper Bhote Kosi, areas of established long-term slope failure are highlighted, dominated by source zones that continuously release fresh material. *Activations*, the number of unique landslide periods in the data, are dominated by coseismic landslides triggered by the 2015 Gorkha earthquake, reflecting failures in locations not previously exposed to landsliding in the time series. The *persistence* measure is again dominated by two behaviours: the short-term presence of coseismic landslides, and the long-term continuous presence of continually unstable hillslopes. A critical value in these metrics is the identification of commonalities between locations with similar time-dependent landslide characteristics. The ability to predict measures such as persistence and recurrence across the landscape holds particular value for anticipating future landslide hazard and risk.

15.6 Implications for longer-term geohazard risk

The Gorkha earthquake had far-reaching immediate consequences for the people of Nepal, resulting in around 9000 fatalities, >22,000 injuries and more than half a million collapsed or damaged houses (Government of Nepal, 2015). Damage to and blockages along Nepal's road infrastructure were widespread, leading to a short-term loss of access and communication during the postdisaster response and recovery phases, and the longer-term need for extensive investment in road repair and renewal (Government of Nepal, 2015). This disruption impacted a transport system that was already experiencing the dual pressures of rapid development and annual exposure to damage by intense monsoonal rains (McAdoo et al., 2018). Nepal's energy network was also severely impacted, with an immediate loss of ~15%–20% of the national hydropower supply capacity, largely due to landslide and rockfall damage to plant infrastructure (Pehlivan et al., 2017; Underwood et al., 2020). In more rural areas, between 50% and 90% of all electrical infrastructure was destroyed or damaged, with an estimated 600,000 people entirely losing their electricity supply as a direct result of the earthquake (Underwood et al., 2020). Livelihoods were also impacted through damage to agricultural land and terraces and changes to the reliability of water sources and springs due to alteration to groundwater pathways during the earthquake (Sargeant et al., 2020).

The impact of the 2015 earthquake was not just a short-lived experience, however; local populations were still feeling the effects of the event as of early 2026. In particular, the evolution of the geomorphic context in the form of hillslope weakening and elevated postseismic landsliding has meant changes to the nature and timing of associated population exposure and risk. The evolving nature of coseismic and postseismic landslides has resulted in altered patterns of downslope runout susceptibility – the potential of being impacted by landslides in the landscape above (Kincey et al., 2023) – which in turn has led to changing threats to local people and infrastructure (Arrell et al., 2024). The total number of individual buildings under threat from landsliding continually increased in the 5-year period after the earthquake, with an increasing proportion of these exposed to the presence of new landslides or to postseismic changes to the footprint of coseismic landslides (Fig. 15.10). A progressive spatial shift in threat has also been observed through time, with a notable increase in threat to buildings at mid- and lower-slope positions as landslides progressively developed downslope relative to the coseismic situation. Where housing is nonuniformly distributed, with clusters

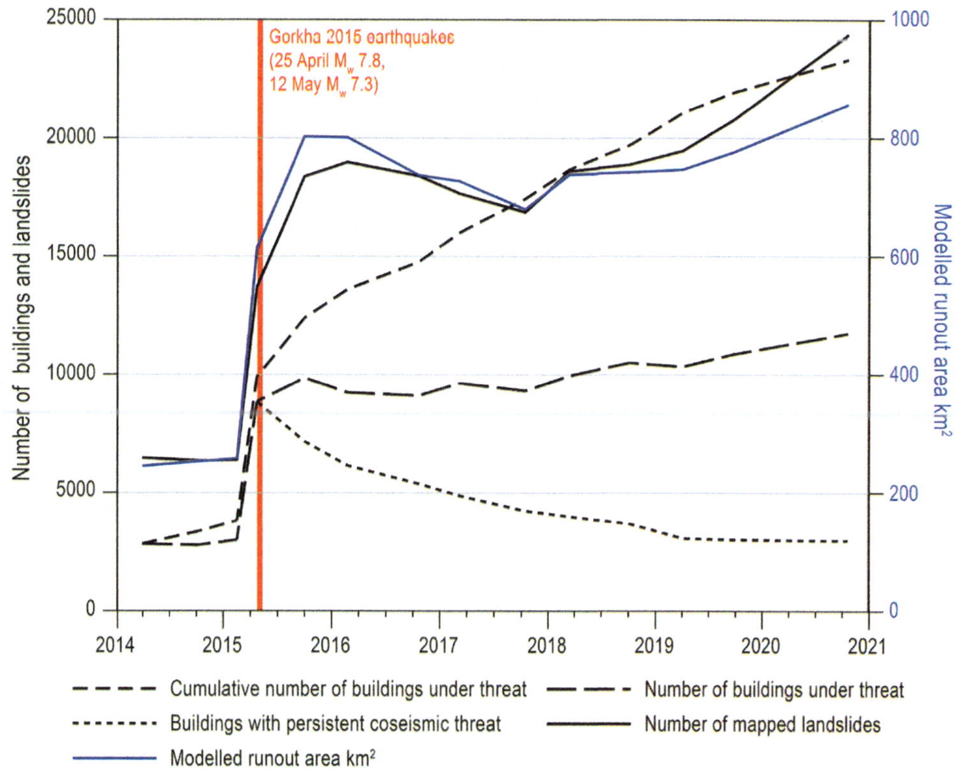

Figure 15.10 Temporal trends in mapped landslide hazard and the threat from modelled landslide runout. The number of buildings that fall within the modelled runout area through time, and hence are considered under threat, is shown as the cumulative count of individual buildings under threat (medium-dashed line), the total per-epoch number of buildings under threat (long-dashed line), and the number of buildings persistently under threat since the earthquake (short-dashed line). In addition, the number of mapped landslides (black solid line) and the associated modelled runout area (blue solid line) are displayed. Source: *Modified from Arrell, K., Rosser, N. J., Kincey, M. E., Robinson, T. R., Horton, P., Densmore, A. L., Oven, K. J., Shrestha, R., & Pujara, D. S. (2024). The dynamic threat from landslides following large continental earthquakes.* PLOS One, 19*(8), p.e0308444.*

commonly in valley bottoms adjacent to linear infrastructure, this represents a growing degree of exposure as the landslide footprint evolves. Importantly, these spatial patterns of runout hazard are often highly localised over distances of as little as a few tens of metres. Whilst this necessitates high-resolution assessments of risk, conversely, the threat to buildings can be potentially reduced through local-scale relocation over similar length scales (Arrell et al., 2024). Understanding both the geomorphic impacts of a high-magnitude earthquake and the spatio-temporal evolution of the postseismic hazard is therefore of paramount importance for reducing associated risk to local populations.

15.7 Conclusions

The 2015 M_w 7.8 Gorkha earthquake was a defining geomorphic event for central Nepal, with impacts that have extended far beyond the immediate destruction associated with seismic shaking. This chapter has shown that the earthquake acted not only as a trigger of widespread landsliding but as a long-term perturbation to the Himalayan landscape, initiating a cascade of geomorphic processes that continue to influence hillslope behaviour, sediment transfer, and geohazard risk more than a decade later. Coseismic landsliding was extensive, with at least 25,000 failures mapped across more than 28,000 km^2, equivalent to several years of background monsoon-driven landsliding, but here compressed into a single event. Yet, compared to some other large mountain region earthquakes, such as the 2008 Wenchuan earthquake in China, the 2015 Gorkha earthquake generated fewer very large landslides and a more modest total landslide volume relative to the amount of tectonic rock uplift. This highlights the importance of regional controls, including topography, lithology and seismic deformation patterns, in shaping earthquake–landslide responses. Despite this, the earthquake fundamentally altered hillslope stability, creating a legacy of weakened hillslopes, disrupted hydrology and enhanced sediment availability.

Postseismic landscape evolution has proven to be complex and spatially heterogeneous. While the rate of new landsliding declined within one to 2 years, multi-temporal inventories demonstrate that the number, area and activity of landslides remained elevated for at least the following decade, with significant reactivation during intense monsoon seasons, particularly in 2020. The persistence and growth of coseismically initiated landslides, coupled with shifting spatial patterns of activity towards higher elevations and evolving downslope connectivity, underline that 'recovery' cannot be defined simply by a return to preearthquake landslide rates, which themselves are shown to be highly variable through time and space. Instead, the earthquake reset hillslope trajectories, with long-lived impacts on sediment routing and hazard distribution.

Crucially, this chapter demonstrates that preexisting landslides play a central role in postseismic landscape dynamics. Evidence from field studies, InSAR and optical imagery shows a spectrum of responses, from negligible change to delayed acceleration and reactivation, often mediated by hydrological processes rather than direct seismic forcing alone. These findings emphasise that earthquakes may act as preconditioners, lowering stability thresholds and amplifying the geomorphic effectiveness of subsequent triggers such as monsoon rainfall and floods. Placed within a longer-term context, the Gorkha earthquake emerges as one extreme event superimposed on a landscape already characterised by frequent landsliding driven by climate variability, human activity and no doubt previous earthquakes. Longer-term satellite-based records show that while the earthquake produced a clear transient signal, it did not exhaust the landscape's capacity for future landsliding, nor fundamentally alter long-term rates of new landslide occurrence.

For communities and infrastructure, these evolving geomorphic processes translate directly into changing patterns of exposure. The National Geohazards Assessment and subsequent relocation efforts since 2015 have highlighted both the necessity and the difficulty of managing risk in a dynamic postseismic landscape. Understanding these long-term, coupled geomorphic and societal effects is essential for improving disaster impact reduction, land-use planning and resilience in seismically active mountain environments.

Acknowledgements

The original research in this chapter has been supported by the Global Challenges Research Fund Multi-Hazard and Systemic Risk programme through the 'Sajag-Nepal' project (NE/T01038X/1), NASA's LCLUC grant NNX17AH98G, the Gaylord Donnelley Environmental Postdoctoral Fellowship at Yale and the UKRI-DFID SHEAR programme (201844-112). This work has also benefited from discussions with the wider Sajag-Nepal project team.

References

Ao, M., Zhang, L., Dong, Y., Su, L., Shi, X., Balz, T., & Liao, M. (2020). Characterizing the evolution life cycle of the Sunkoshi landslide in Nepal with multi-source SAR data. *Scientific Reports, 10*(1). https://doi.org/10.1038/s41598-020-75002-y.

Adhikari, B. R., & Tian, B. (2021). Spatiotemporal distribution of landslides in Nepal. In S. Eslamian, & F. Eslamian (Eds.). *Handbook of disaster risk reduction for resilience*. Springer, https://doi.org/10.1007/978-3-030-61278-8_20.

Adhikari, L. B., Gautam, U. P., Koirala, B. P., Bhattarai, M., Kandel, T., Gupta, R. M., Timsina, C., Maharjan, N., Maharjan, K., Dahal, T., Hoste-Colomer, R., Cano, Y., Dandine, M., Guilhem, A., Merrer, S., Roudil, P., & Bollinger, L. (2015). The aftershock sequence of the 2015 April 25 Gorkha–Nepal earthquake. *Geophysical Journal International, 203*(3), 2119–2124.

Allen, S. K., Sattar, A., King, O., Zhang, G., Bhattacharya, A., Yao, T., & Bolch, T. (2022). Glacial lake outburst flood hazard under current and future conditions: Worst-case scenarios in a transboundary Himalayan basin. *Natural Hazards and Earth System Sciences, 22*(11), 3765–3785.

Alvioli, M., Marchesini, I., Reichenbach, P., Rossi, M., Ardizzone, F., Fiorucci, F., & Guzzetti, F. (2016). Automatic delineation of geomorphological slope units with r. slopeunits v1. 0 and their optimization for landslide susceptibility modeling. *Geoscientific Model Development, 9*(11), 3975–3991.

Arrell, K., Rosser, N. J., Kincey, M. E., Robinson, T. R., Horton, P., Densmore, A. L., Oven, K. J., Shrestha, R., & Pujara, D. S. (2024). The dynamic threat from landslides following large continental earthquakes. *PLoS One, 19*(8), e0308444.

Avouac, J. P., Meng, L., Wei, S., Wang, T., & Ampuero, J. P. (2015). Lower edge of locked main Himalayan thrust unzipped by the 2015 Gorkha earthquake. *Nature Geoscience, 8*(9), 708–711.

Bekaert, D. P. S., Handwerger, A. L., Agram, P., & Kirschbaum, D. B. (2020). InSAR-based detection method for mapping and monitoring slow-moving landslides in remote regions with steep and mountainous terrain: An application to Nepal. *Remote Sensing of Environment, 249*, 111983. https://doi.org/10.1016/j.rse.2020.111983.

Bhandari, B. P. (2024). Paleo landslide reactivation and threat to the people in the Nepal Himalaya. *Asian Journal of Engineering Geology, 1*(Special Issue), 11–12.

Bilham, R. (2004). Earthquakes in India and the Himalaya: Tectonics, geodesy and history. *Annals of Geophysics, 47*(2-3).

Bilham, R., Larson, K., & Freymueller, J. (1997). GPS measurements of present-day convergence across the Nepal Himalaya. *Nature, 386*(6620), 61–64.

Budimir, M. (2015). *Cascading natural hazards: Probability and loss modelling for earthquakes and earthquake-triggered landslides* (Doctoral thesis, University of Southampton, School of Geography). 315 pp.

Byers III, A. C., Byers, E. A., McKinney, D. C., & Rounce, D. R. (2017). A field-based study of impacts of the 2015 earthquake on potentially dangerous glacial lakes in Nepal. *HIMALAYA, the Journal of the Association for Nepal and Himalayan Studies, 37*(2), 7.

Chen, N., Liu, M., Allen, S., Deng, M., Khanal, N. R., Peng, T., Tian, S., Huggel, C., Wu, K., Rahman, M., & Somos-Valenzuela, M. (2023). Small outbursts into big disasters: Earthquakes exacerbate climate-driven cascade processes of the glacial lakes failure in the Himalayas. *Geomorphology, 422*, 108539.

Chen, T.-H. K., Kincey, M. E., Rosser, N. J., & Seto, K. C. (2024). Identifying recurrent and persistent landslides using satellite imagery and deep learning: A 30-year analysis of the Himalaya. *Science of The Total Environment, 922*, 171161. https://doi.org/10.1016/j.scitotenv.2024.171161.

Collins, B. D., & Jibson, R. W. (2015). *Assessment of existing and potential landslide hazards resulting from the April 25, 2015 Gorkha, Nepal earthquake sequence* (Version 1.1, August 2015). United States Geological Survey Open-File Report 2015–1142, p. 50. https://doi.org/10.3133/ofr20151142

Cook, K. L., Andermann, C., Gimbert, F., Adhikari, B. R., & Hovius, N. (2018). Glacial lake outburst floods as drivers of fluvial erosion in the Himalaya. *Science (New York, N.Y.), 362*(6410), 53–57.

Dahal, A., Tanyas, H., Mai, P. M., van der Meijde, M., van Westen, C., & Lombardo, L. (2024). Quantifying the influence of topographic amplification on the landslides triggered by the 2015 Gorkha earthquake. *Communications Earth & Environment, 5*(1), 678.

Dahlquist, M. P., & West, A. J. (2019). Initiation and runout of post-seismic debris flows: Insights from the 2015 Gorkha earthquake. *Geophysical Research Letters, 46*(16), 9658–9668.

Dhital, M. R. (2003). Causes and consequences of the 1993 debris flows and landslides in the Kulekhani watershed, central Nepal. In D. Rickenmann, & C. L. Chen (Eds.). *Proceedings of 3rd International Conference on Debris-Flow Hazards Mitigation: Mechanics, Prediction and Assessment* (pp. 931–942). Millpress.

Dilley, M., Chen, R. S., Deichmann, U., Lerner-Lam, A. L., & Arnold, M. (2005). *Natural disaster hotspots: A global risk analysis*. The World Bank Retrieved from https://openknowledge.worldbank.org/handle/1098.

Dunham, A. M., Kiser, E., Kargel, J. S., Haritashya, U. K., Watson, C. S., Shugar, D. H., Hughes, A., & DeCelles, P. G. (2022). Topographic control on ground motions and landslides from the 2015 Gorkha earthquake. *Geophysical Research Letters, 49*(10) p.e2022GL098582.

Fan, X., Domènech, G., Scaringi, G., Huang, R., Xu, Q., Hales, T. C., Dai, L., Yang, Q., & Francis, O. (2018). Spatio-temporal evolution of mass wasting after the 2008 Mw 7.9 Wenchuan earthquake revealed by a detailed multi-temporal inventory. *Landslides, 15*(12), 2325–2341.

Fan, X., Scaringi, G., Korup, O., West, A. J., van Westen, C. J., Tanyas, H., Hovius, N., Hales, T. C., Jibson, R. W., Allstadt, K. E., Zhang, L., Evans, S. G., Xu, C., Li, G., Pei, X., Xu, Q., & Huang, R. (2019). Earthquake-induced chains of geologic hazards: patterns, mechanisms, and impacts. *Reviews of Geophysics, 57*(2), 421–503. https://doi.org/10.1029/2018rg000626.

Ferrer, J. V., Samprogna Mohor, G., Dewitte, O., Pánek, T., Reyes-Carmona, C., Handwerger, A. L., Hürlimann, M., Köhler, L., Teshebaeva, K., Thieken, A. H., & Tsou, C. Y. (2024). Human settlement pressure drives slow-moving landslide exposure. *Earth's Future, 12*(9) p.e2024EF004830.

Froude, M. J., & Petley, D. N. (2018). Global fatal landslide occurrence from 2004 to 2016. *Natural Hazards and Earth System Sciences, 18*(8), 2161–2181.

Galetzka, J., Melgar, D., Genrich, J. F., Geng, J., Owen, S., Lindsey, E. O., Xu, X., Bock, Y., Avouac, J. P., Adhikari, L. B., & Upreti, B. N. (2015). Slip pulse and resonance of the Kathmandu basin during the 2015 Gorkha earthquake, Nepal. *Science (New York, N.Y.), 349*(6252), 1091–1095.

Gallen, S. F., Clark, M. K., Godt, J. W., Roback, K., & Niemi, N. A. (2017). Application and evaluation of a rapid response earthquake-triggered landslide model to the 25 April 2015 Mw 7.8 Gorkha earthquake, Nepal. *Tectonophysics, 714*, 173–187.

Government of Nepal. (2015). *Nepal earthquake 2015. Post disaster needs assessment: Executive summary.* Government of Nepal, National Planning Commission. www.worldbank.org/content/dam/Worldbank/document/SAR/nepal-pdna-executive-summary.pdf.

Graf, E. L., Sinclair, H. D., Attal, M., Gailleton, B., Adhikari, B. R., & Baral, B. R. (2024). Geomorphological and hydrological controls on sediment export in earthquake-affected catchments in the Nepal Himalaya. *Earth Surface Dynamics, 12*(1), 135–161.

Gualandi, A., Avouac, J. P., Galetzka, J., Genrich, J. F., Blewitt, G., Adhikari, L. B., Koirala, B. P., Gupta, R., Upreti, B. N., Pratt-Sitaula, B., & Liu-Zeng, J. (2017). Pre-and post-seismic deformation related to the 2015, Mw7. 8 Gorkha earthquake, Nepal. *Tectonophysics, 714*, 90–106.

Harvey, E. L., Kincey, M. E., Rosser, N. J., Gadtaula, A., Collins, E., Densmore, A. L., Dunant, A., Oven, K. J., Arrell, K., Basyal, G. K., Dhital, M. R., Robinson, T. R., Van Wyk de Vries, M., Paudyal, S., Pujara, D. S., & Shrestha, R. (2024). Review of landslide inventories for Nepal between 2010 and 2021 reveals data gaps in global landslide hotspot. *Natural Hazards, 121*(5), 5075–5101. https://doi.org/10.1007/s11069-024-07013-1.

Hayes, G. P., Myers, E. K., Dewey, J. W., Briggs, R. W., Earle, P. S., Benz, H. M., Smoczyk, G. M., Flamme, H. E., Barnhart, W. D., Gold, R. D., & Furlong, K. P. (2017). *Tectonic summaries of magnitude 7 and greater earthquakes from 2000 to 2015.* United States Geological Survey Open-File Report 2016–1192, p. 148. https://doi.org/10.3133/ofr20161192.

Huang, R., & Li, W. (2014). Post-earthquake landsliding and long-term impacts in the Wenchuan earthquake area, China. *Engineering Geology, 182*, 111–120.

Illien, L., Sens-Schönfelder, C., Andermann, C., Marc, O., Cook, K. L., Adhikari, L. B., & Hovius, N. (2022). Seismic velocity recovery in the subsurface: Transient damage and groundwater drainage following the 2015 Gorkha earthquake, Nepal. *Journal of Geophysical Research: Solid Earth, 127*(2) p. e2021JB023402.

Jones, J. N., Boulton, S. J., Stokes, M., Bennett, G. L., & Whitworth, M. R. (2021). 30-year record of Himalaya mass-wasting reveals landscape perturbations by extreme events. *Nature Communications, 12*(1), 6701.

Kargel, J. S., Leonard, G. J., Shugar, D. H., Haritashya, U. K., Bevington, A., Fielding, E. J., Fujita, K., Geertsema, M., Miles, E. S., Steiner, J., & Anderson, E. (2016). Geomorphic and geologic controls of geohazards induced by Nepal's 2015 Gorkha earthquake. *Science (New York, N.Y.), 351*(6269) p.aac8353.

KC, R., Sharma, K., Dahal, B. K., Aryal, M., & Subedi, M. (2024). Study of the spatial distribution and the temporal trend of landslide disasters that occurred in the Nepal Himalayas from 2011 to 2020. *Environmental Earth Sciences, 83*(1), 42.

Kincey, M. E., Rosser, N. J., Densmore, A. L., Robinson, T. R., Shrestha, R., Singh Pujara, D., Horton, P., Swirad, Z. M., Oven, K. J., & Arrell, K. (2023). Modelling post-earthquake cascading hazards: Changing patterns of landslide runout following the 2015 Gorkha earthquake, Nepal. *Earth Surface Processes and Landforms, 48*(3), 537–554.

Kincey, M. E., Rosser, N. J., Robinson, T. R., Densmore, A. L., Shrestha, R., Pujara, D. S., Oven, K. J., Williams, J. G., & Swirad, Z. M. (2021). Evolution of coseismic and post-seismic landsliding after the 2015 Mw 7.8 Gorkha earthquake, Nepal. *Journal of Geophysical Research: Earth Surface, 126*(3) p.e2020JF005803.

Lacroix, P. (2016). Landslides triggered by the Gorkha earthquake in the Langtang valley, volumes and initiation processes. *Earth, Planets and Space, 68*(1), 1–10.

Lacroix, P., Gavillon, T., Bouchant, C., Lavé, J., Mugnier, J.-L., Dhungel, S., & Vernier, F. (2022). SAR and optical images correlation illuminates post-seismic landslide motion after the Mw 7.8 Gorkha earthquake (Nepal). *Scientific Reports, 12*(1). https://doi.org/10.1038/s41598-022-10016-2.

Lavé, J., Yule, D., Sapkota, S., Basant, K., Madden, C., Attal, M., & Pandey, R. (2005). Evidence for a great medieval earthquake (~ 1100 AD) in the central Himalayas, Nepal. *Science (New York, N.Y.), 307*(5713), 1302–1305.

Li, C., Wang, M., & Liu, K. (2018). A decadal evolution of landslides and debris flows after the Wenchuan earthquake. *Geomorphology, 323*, 1–12.

Lindsey, E. O., Natsuaki, R., Xu, X., Shimada, M., Hashimoto, M., Melgar, D., & Sandwell, D. T. (2015). Line-of-sight displacement from ALOS-2 interferometry: Mw 7.8 Gorkha earthquake and Mw 7.3 aftershock. *Geophysical Research Letters, 42*(16), 6655–6661.

Mani, P., Allen, S., Evans, S. G., Kargel, J. S., Mergili, M., Petrakov, D., & Stoffel, M. (2023). Geomorphic process chains in high-mountain regions—A review and classification approach for natural hazards assessment. *Reviews of Geophysics, 61*(4).

Marc, O., Behling, R., Andermann, C., Turowski, J. M., Illien, L., Roessner, S., & Hovius, N. (2019). Long-term erosion of the Nepal Himalayas by bedrock landsliding: The role of monsoons, earthquakes and giant landslides. *Earth Surface Dynamics, 7*(1), 107–128.

Marc, O., & Hovius, N. (2015). Amalgamation in landslide maps: Effects and automatic detection. *Natural Hazards and Earth System Science, 15*(4), 723–733.

Marc, O., Hovius, N., Meunier, P., Uchida, T., & Hayashi, S. I. (2015). Transient changes of landslide rates after earthquakes. *Geology, 43*(10), 883–886.

Martha, T. R., Roy, P., Mazumdar, R., Govindharaj, K. B., & Kumar, K. V. (2017). Spatial characteristics of landslides triggered by the 2015 Mw 7.8 (Gorkha) and Mw 7.3 (Dolakha) earthquakes in Nepal. *Landslides, 14*(2), 697–704.

McAdoo, B. G., Quak, M., Gnyawali, K. R., Adhikari, B. R., Devkota, S., Rajbhandari, P. L., & Sudmeier-Rieux, K. (2018). Roads and landslides in Nepal: How development affects environmental risk. *Natural Hazards and Earth System Sciences, 18*(12), 3203–3210.

Meena, S. R., & Tavakkoli Piralilou, S. (2019). Comparison of earthquake-triggered landslide inventories: A case study of the 2015 Gorkha earthquake, Nepal. *Geosciences, 9*(10), 437. https://doi.org/10.3390/geosciences9100437.

Meunier, P., Hovius, N., & Haines, J. A. (2008). Topographic site effects and the location of earthquake induced landslides. *Earth and Planetary Science Letters, 275*(3-4), 221–232. https://doi.org/10.1016/j.epsl.2008.07.020.

Milledge, D. G., Bellugi, D. G., Watt, J., & Densmore, A. L. (2022). Automated determination of landslide locations after large trigger events: advantages and disadvantages compared to manual mapping. *Natural Hazards and Earth System Sciences, 22*(2), 481–508. https://doi.org/10.5194/nhess-22-481-2022.

Parameswaran, R. M., Natarajan, T., Rajendran, K., Rajendran, C. P., Mallick, R., Wood, M., & Lekhak, H. C. (2015). Seismotectonics of the April–May 2015 Nepal earthquakes: An assessment based on the aftershock patterns, surface effects and deformational characteristics. *Journal of Asian Earth Sciences, 111*, 161–174.

Pehlivan, M., Madugo, C. M., Macdonald, A., Rayamajhi, D., Hashash, Y. M., & Tiwari, B. (2017). Hydropower infrastructure performance after the 2015 Gorkha, Nepal, earthquake sequence. *Earthquake Spectra, 33*(1_suppl), 115–132.

Petley, D. N., Hearn, G. J., Hart, A., Rosser, N. J., Dunning, S. A., Oven, K., & Mitchell, W. A. (2007). Trends in landslide occurrence in Nepal. *Natural Hazards, 43*(1), 23–44.

Pink, I., Reaney, S. M., Hardy, R. J., & Bovolo, C. I. (2025). Increased rainfall-runoff drives flood hazard intensification in Central Himalayan river systems. *Scientific Reports, 15*(1), 42277.

Roback, K., Clark, M. K., West, A. J., Zekkos, D., Li, G., Gallen, S. F., Chamlagain, D., & Godt, J. W. (2018). The size, distribution, and mobility of landslides caused by the 2015 Mw7. 8 Gorkha earthquake, Nepal. *Geomorphology, 301*, 121–138.

Roberts, S., Jones, J. N., & Boulton, S. J. (2021). Characteristics of landslide path dependency revealed through multiple resolution landslide inventories in the Nepal Himalaya. *Geomorphology, 390*, 107868.

Rosser, N., Kincey, M., Oven, K., Densmore, A., Robinson, T., Pujara, D. S., Shrestha, R., Smutny, J., Gurung, K., Lama, S., & Dhital, M. R. (2021). Changing significance of landslide hazard and risk after the 2015 Mw 7.8 Gorkha, Nepal earthquake. *Progress in Disaster Science, 10*, 100159.

Samia, J., Temme, A., Bregt, A., Wallinga, J., Guzzetti, F., Ardizzone, F., & Rossi, M. (2017). Do landslides follow landslides? Insights in path dependency from a multi-temporal landslide inventory. *Landslides, 14*(2), 547–558.

Sargeant, S., Finlayson, A., Dijkstra, T., Flinn, B., Schofield, H., Morel, L. M., Twigg, J., Lovell, E., Stephenson, V., & Adhikari, B. R. (2020). The influence of the physical environment on self-recovery after disasters in Nepal and the Philippines. *International Journal of Disaster Risk Reduction, 50*, 101673.

Schwanghart, W., Bernhardt, A., Stolle, A., Hoelzmann, P., Adhikari, B. R., Andermann, C., Tofelde, S., Merchel, S., Rugel, G., Fort, M., & Korup, O. (2016). Repeated catastrophic valley infill following medieval earthquakes in the Nepal Himalaya. *Science (New York, N.Y.), 351*(6269), 147–150.

Shen, P., Zhang, L. M., Fan, R. L., Zhu, H., & Zhang, S. (2020). Declining geohazard activity with vegetation recovery during first ten years after the 2008 Wenchuan earthquake. *Geomorphology, 352*, 106989.

Shrestha, M., Noppradit, P., Shrestha, R. P., & Dahal, R. K. (2024). Understanding livelihood insecurity due to landslides in the mid-hill of Nepal: A case study of Bahrabise Municipality. *International Journal of Disaster Risk Reduction, 104*, 104399.

Shrestha, R., Harvey, E. L., Kincey, M. E., Rosser, N. J., Arrell, K., Basyal, G. K., Pujara, D. S., Paudyal, S., Densmore, A. L., Van Wyk de Vries, M., & Jimee, G. K. (2026). Multi-temporal clustering method to detect individual landslide response to the 2015 Gorkha earthquake. In *An International Conference Series of Nepal Geotechnical Society* (pp. 117–129). Springer Nature. https://doi.org/10.1007/978-981-96-7921-8_10

Song, C., Yu, C., Li, Z., Utili, S., Frattini, P., Crosta, G., & Peng, J. (2022). Triggering and recovery of earthquake accelerated landslides in Central Italy revealed by satellite radar observations. *Nature Communications, 13*(1), 7278.

Sudmeier-Rieux, K., McAdoo, B. G., Devkota, S., Rajbhandari, P. C. L., Howell, J., & Sharma, S. (2019). Invited perspectives: Mountain roads in Nepal at a new crossroads. *Natural Hazards and Earth System Sciences, 19*(3), 655–660.

Talchabhadel, R., Maskey, S., Gouli, M. R., Dahal, K., Thapa, A., Sharma, S., Dixit, A. M., & Kumar, S. (2023). Multimodal multiscale characterization of cascading hazard on mountain terrain. *Geomatics, Natural Hazards and Risk, 14*(1), 2162443.

Tang, C., Van Westen, C. J., Tanyas, H., & Jetten, V. G. (2016). Analysing post-earthquake landslide activity using multi-temporal landslide inventories near the epicentral area of the 2008 Wenchuan earthquake. *Natural Hazards and Earth System Sciences, 16*, 2641–2655. https://doi.org/10.5194/nhess-16-2641-2016.

Tanyaş, H., Görüm, T., Kirschbaum, D., & Lombardo, L. (2022). Could road constructions be more hazardous than an earthquake in terms of mass movement? *Natural Hazards, 112*(1), 639–663.

Tanyaş, H., Kirschbaum, D., Görüm, T., van Westen, C. J., Tang, C., & Lombardo, L. (2021). A closer look at factors governing landslide recovery time in post-seismic periods. *Geomorphology, 391*, 107912.

Tian, Y., Owen, L. A., Xu, C., Ma, S., Li, K., Xu, X., Figueiredo, P. M., Kang, W., Guo, P., Wang, S., & Liang, X. (2020). Landslide development within 3 years after the 2015 M w 7.8 Gorkha earthquake, Nepal. *Landslides, 17*(5), 1251–1267.

Tian, Y., Owen, L. A., Xu, C., Shen, L., Zhou, Q., & Figueiredo, P. M. (2020). Geomorphometry and statistical analyses of landslides triggered by the 2015 Mw 7.8 Gorkha earthquake and the Mw 7.3 aftershock, Nepal. *Frontiers in Earth Science, 8*, 572449.

Tiwari, B., Ajmera, B., & Dhital, S. (2017). Characteristics of moderate-to large-scale landslides triggered by the Mw 7.8 2015 Gorkha earthquake and its aftershocks. *Landslides, 14*(4), 1297–1318.

Tsou, C. Y., Chigira, M., Higaki, D., Sato, G., Yagi, H., Sato, H. P., Wakai, A., Dangol, V., Amatya, S. C., & Yatagai, A. (2018). Topographic and geologic controls on landslides induced by the 2015 Gorkha earthquake and its aftershocks: An example from the Trishuli Valley, central Nepal. *Landslides, 15*(5), 953–965.

Underwood, G., Hill, D., & Lamichhane, S. (2020). Earthquakes, blockades and energy crises: A conceptual framework for energy systems resilience applied to Nepal. *Energy Research & Social Science, 69*, 101609.

Valagussa, A., Frattini, P., Valbuzzi, E., & Crosta, G. B. (2021). Role of landslides on the volume balance of the Nepal 2015 earthquake sequence. *Scientific Reports, 11*(1), 3434.

Williams, J. G., Rosser, N. J., Kincey, M. E., Benjamin, J., Oven, K. J., Densmore, A. L., Milledge, D. G., Robinson, T. R., Jordan, C. A., & Dijkstra, T. A. (2018). Satellite-based emergency mapping using optical imagery: experience and reflections from the 2015 Nepal earthquakes. *Natural Hazards and Earth System Sciences, 18*(1), 185–205. https://doi.org/10.5194/nhess-18-185-2018.

Wu, G., Cunningham, D., Yuan, R.-M., Zhou, Q., Zeng, X., & Yang, X.-P. (2017). Mass-wasting effects induced by the 2015 Gorkha (Nepal) Mw 7.8 earthquake within a large paleo-landslide site adjacent to the Tatopani Border Station, Nepal: implications for future development along the critical Bhote Koshi River valley transport corridor between Nepal and China. *Landslides, 14*(3), 1147–1160. https://doi.org/10.1007/s10346-016-0792-5.

Xu, C., Xu, X., Tian, Y., Shen, L., Yao, Q., Huang, X., Ma, J., Chen, X., & Ma, S. (2016). Two comparable earthquakes produced greatly different coseismic landslides: The 2015 Gorkha, Nepal and 2008 Wenchuan, China events. *Journal of Earth Science, 27*(6), 1008–1015. https://doi.org/10.1007/s12583-016-0684-6.

Xu, C. (2018). Landslides triggered by the 2015 Gorkha, Nepal earthquake. The International Archives of the Photogrammetry. *Remote Sensing and Spatial Information Sciences, 42*, 1989–1993.

Yang, W., Qi, W., & Zhou, J. (2018). Effects of precipitation and topography on vegetation recovery at landslide sites after the 2008 Wenchuan earthquake. *Land Degradation & Development, 29*(10), 3355–3365.

Yanites, B. J., Clark, M. K., Roering, J. J., West, A. J., Zekkos, D., Baldwin, J. W., Cerovski-Darriau, C., Gallen, S. F., Horton, D. E., Kirby, E., & Leshchinsky, B. A. (2025). Cascading land surface hazards as a nexus in the Earth system. *Science (New York, N.Y.), 388*(6754).

Yogacharya, K. S., & Gautam, D. K. (2008). *Floods in Nepal: Genesis, magnitude, frequency and consequences. Society of Hydrologists and Meteorologists-Nepal.*

Zhao, B. (2016). April 2015 Nepal earthquake: Observations and reflections. *Natural Hazards, 80*(2), 1405–1410.

CHAPTER 16

Implications for coseismic landslide risk analysis: advances and future directions

Tom R. Robinson
School of Earth and Environment, University of Canterbury, Christchurch, New Zealand

16.1 Introduction

Of the range of geomorphic consequences of earthquakes, coseismic landslides represent perhaps the most significant, and often underappreciated, component. Since 2000, several major events – including the 2008 M_w 7.9 Wenchuan earthquake in China and the 2016 M_w 7.8 Kaikōura earthquake in New Zealand – have shown that coseismic landslides can cause more damage and disruption than the initial ground shaking. These events produced tens of thousands of slope failures, which, in the case of Wenchuan, resulted in thousands of fatalities, widespread and costly infrastructure disruption, as well as long-term landscape change that has resulted in chronic issues for affected populations.

The 2008 Wenchuan earthquake (Chapter 14) is especially notable for the extreme landslide impacts that it triggered. The earthquake resulted in at least 60,000 landslides (Gorum et al., 2011; some estimates suggest close to 200,000), which killed perhaps as many as 20,000 people, accounting for up to a third of the total fatalities (Yin et al., 2009). While this event was by no means the first earthquake to have caused such extensive landsliding, the sheer number of fatalities clearly demonstrated the risk faced by people living in seismically active mountain belts. The timing of the Wenchuan earthquake with respect to the increasing availability of aerial and satellite imagery kick-started a global research agenda to better inventory coseismic landslides, resulting in it arguably being one of the best and most widely mapped events on record (Tanyaş et al., 2017). This increase in the desire and ability to more accurately inventory such events resulted in an increase in the use of these inventories to train and test a wide variety of spatial coseismic landslide models, leading to an almost exponential increase in available models (Fan et al., 2019). However, whether the increase in the number of available models has led to increased predictive accuracy remains contentious.

In addition to fatalities, coseismic landslides are also responsible for widespread infrastructure damage, particularly to transport networks. Bird and Bommer (2004) first

Geomorphic Impacts of Earthquakes
DOI: https://doi.org/10.1016/B978-0-443-40375-0.00015-3

quantified this, showing that in earthquakes with 'significant' transportation damage, ground failure (liquefaction and landslides) was responsible for 50% of the total recorded damage. In events with 'moderate' transportation damage, it was closer to 75%. This is epitomised by the 2016 M_w 7.8 Kaikōura earthquake Chapter 16), which affected the northeastern part of New Zealand's South Island. Whilst the low population density meant that just two people were killed by the event, over 30,000 landslides caused extensive damage to road and rail transport links, completely isolating the popular tourist centre of Kaikōura and endangering terrestrial access to most of the South Island (Massey et al., 2020). It took over a year and NZ$2 billion to clear the landslide debris, rebuild the transport infrastructure and install suitable mitigation measures to protect road and rail users (Saul & Anderson, 2019). As a result, multiple authors have sought to identify methods to model coseismic landslides in near-real-time following an event (e.g. Allstadt et al., 2018; Robinson et al., 2018), whilst others have focused on forecasting – also in near-real-time – where landslides will impact transport routes (Lin et al., 2025; Robinson, Davies, Wilson, Orchiston, & Barth, 2016).

Yet coseismic landslides are not temporally static hazards that only present a threat during or immediately after ground shaking. The longevity and dynamic nature of landslide hazard following major earthquakes are being increasingly recognised. Marc et al. (2015) used several case studies, including the 1999 M_w 7.7 Chi-Chi earthquake in Taiwan (Chapter 13), to show that while landslide hazard peaked immediately after the earthquake, the decay back to preearthquake rates of landsliding lasted up to 4 years. In Nepal, studies in the years after the 2015 M_w 7.9 Gorkha earthquake (Chapter 15) revealed that elevated landslide rates persisted for even longer, accompanied by spatial shifts in hazard that highlight critical flaws in static recovery planning (Kincey et al., 2021; Rosser et al., 2021). Incorporating this understanding into hazard and risk analyses is critical but has so far proved challenging.

Despite the major advances in our understanding of coseismic landslides since the start of the 21st century, much of the focus has been on the physical processes and mechanisms driving landslide hazard. There has been far less focus on the exposure and vulnerability of people and assets to landsliding. As a result, there has been limited progress in understanding the risk posed by coseismic landslides. To effectively reduce the impacts caused by coseismic landslides, there is a need to increase our focus on landslide exposure and vulnerability, which requires interdisciplinary research spanning the physical and social sciences as well as engineering. This chapter aims to synthesise recent advances in coseismic landslide hazard and risk analysis, and to identify the most pressing research challenges that remain. While the focus of this chapter is on the first component of the seismic hazard cascade (Fan et al., 2019), the main outcomes apply to the whole cascade, as well as to natural hazards more generally.

16.2 Definitions and terminology

Before discussing the issues related to coseismic landslide hazard and risk, it is essential to define the core terms.

- Risk in the past has had a multitude of definitions, but the Sendai Framework for Disaster Risk Reduction 2015–30 provides the most widely accepted and used definitions, which are followed here. It defines risk as 'the potential loss of life, injury, or destroyed or damaged assets which could occur to a system, society or a community in a specific period of time, determined probabilistically as a function of hazard, exposure, vulnerability and capacity' (Fig. 16.1; UNDRR, 2017).

$$Risk = f\,(hazard, \quad exposure, \quad vulnerability, \quad capacity) \tag{16.1}$$

The use of the term 'probabilistically' implies that risk is always assessed quantitatively, and while this is common, it is not always possible to fully quantify these four components – particularly vulnerability and coping capacity. As a result, risk assessments can come in both quantitative and qualitative forms, with the relative

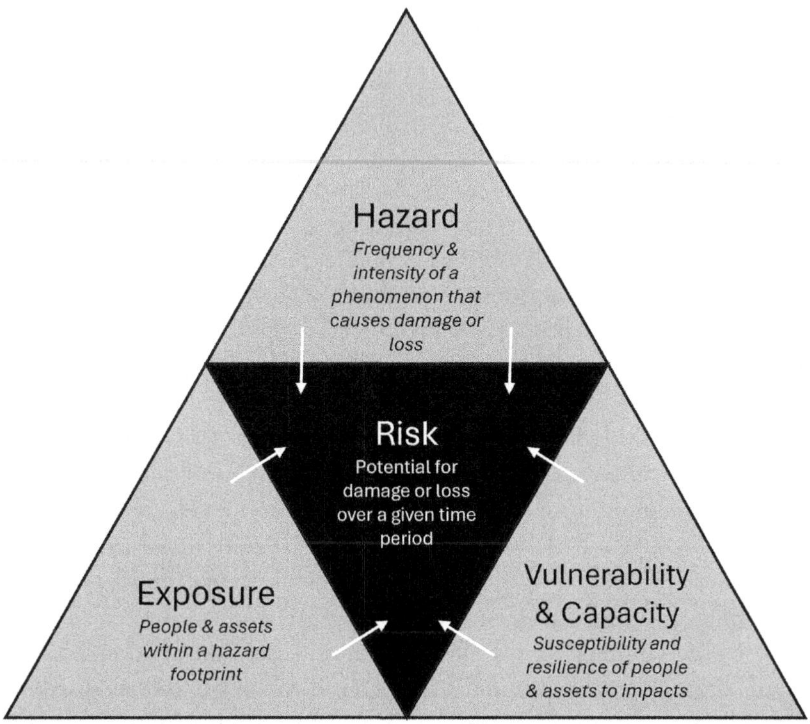

Figure 16.1 Conceptual model of risk, showing its constituent components. Risk can only occur where hazard, exposure and vulnerability are all nonzero.

strengths of either determined by the wider risk context. Nevertheless, all measures of risk are essentially probabilistic.

- Hazard is 'a process, phenomenon or human activity that may cause loss of life, injury or other health impacts, property damage, social and economic disruption or environmental degradation' (UNDRR, 2017). Typically, this is quantified in terms of the frequency of occurrence of a given magnitude or intensity of the hazard in question, within a specified area. Frequency is usually considered in terms of annual probability, giving rise, for example, to the so-called '1-in-100 yr event', equivalent to an event with a 1% chance of occurrence annually. Magnitude or intensity refers to some physical property of the hazard that describes its level of threat – for earthquakes, this is usually the moment magnitude or the peak ground acceleration or velocity (see Chapters 3 and 4). For landslides, the relevant magnitude or intensity is less clear, but most frequently it is measured in terms of landslide area or volume. Malamud et al. (2004) introduced the concept of landslide event magnitude to describe the intensity of landsliding when multiple landslides are triggered in a single event, but this is not the same as the intensity of an individual landslide.
- Exposure is 'the situation of people, infrastructure, housing, production capacities and other tangible human assets located in hazard-prone areas' (UNDRR, 2017). Measures of exposure typically refer, for example, to the total number of people, the length of road, or the economic value of assets within a hazard footprint. Exposure is perhaps the easiest component to conceptualise, but this simple definition masks complexities, particularly for landslides. Defining the hazard footprint itself can be challenging, and whilst some assets like roads and railways are spatially and temporally fixed, people are not, meaning that the timing of an event can dramatically influence the number of people exposed.
- Vulnerability is 'the conditions determined by physical, social, economic and environmental factors or processes which increase the susceptibility of an individual, a community, assets or systems to the impacts of hazards' (UNDRR, 2017). In a quantitative sense, it can be thought of as the probability of a given impact on a person or asset resulting from a given hazard magnitude. However, vulnerability also contains more qualitative elements such as social cohesion and deprivation, health and wellbeing, economic activity and status, education levels, age, gender and other individual and collective characteristics. This complexity makes it one of the more contentious components but also the most transdisciplinary, requiring a combination of engineering and social sciences to fully comprehend.
- Capacity is 'the combination of all the strengths, attributes and resources available within an organisation, community or society to manage and reduce disaster risks and strengthen resilience' (UNDRR, 2017). Often confused with vulnerability or resilience, capacity can be considered as the ability to cope with and recover from

the impacts caused by a hazardous event. It is a difficult component to conceptualise and quantify, with most quantitative risk assessments either excluding this component or incorporating it within vulnerability.

- Resilience is 'the ability of an exposed community or society to resist, absorb, accommodate, adapt to, transform and recover from the effects of a hazard in a timely and efficient manner' (UNDRR, 2017). Like capacity, resilience is a contentious term often confused with or used interchangeably with capacity. It is also a difficult component to conceptualise and quantify.

Understanding this terminology is not simply a matter of linguistics but also a critical constituent of risk assessments. Colloquially, hazard and risk are often used interchangeably, leading to the misplaced belief that quantifying the likelihood of a given hazard magnitude is akin to quantifying the risk. In reality, no matter the likelihood of a hazard, if there is no exposure or vulnerability to that hazard, then by definition there is no risk. Risk is therefore an inherently interdisciplinary concept that requires, in the broadest possible terms, physical scientists, engineers and social scientists to work together so that it can be identified, quantified and ultimately reduced.

As demonstrated throughout this book, by far the biggest recent gains in understanding the geomorphic consequences of earthquakes have come from larger investments in quantifying the physical processes and mechanisms involved. This has translated directly to a far greater appreciation of and ability to quantify and model the hazard. However, the gains in terms of understanding and ultimately reducing risk have been more limited due to a lack of similar investment in understanding exposure and vulnerability. In fact, many landslide risk assessments focus on developing highly detailed quantitative models of landslide hazard that are combined with overly simplistic qualitative models of exposure and/or vulnerability – and in some instances, vulnerability is completely ignored. Risk models are only as good as their weakest element, and both the spatial resolution and uncertainty of risk estimates are driven by the lowest common denominator. It is therefore pertinent to question what value really lies in developing highly resolved and detailed models of landslide hazard only to combine them with poorly resolved and overly simplistic exposure and vulnerability models.

This chapter will provide an overview of the current science of coseismic landslide hazard, exposure and vulnerability, and outline what the future holds for landslide risk assessment.

16.3 Advances in hazard modelling

Landslide hazard assessments can be undertaken at the scale of individual hillslopes up to national or even global scales (Khan et al., 2022). While studies of individual slopes can often involve far greater and more detailed understanding of the specific processes and mechanisms involved, studies with wider application often struggle, particularly at the regional scale. Considering that large earthquakes can affect tens of thousands of square kilometres, this chapter will focus on efforts at the regional scale to understand the extent and distribution of coseismic landslide hazard.

Landslide hazard (whether coseismically triggered or otherwise) requires understanding where in the landscape landslides are expected to occur (initiation), as well as where they will travel (runout). These two processes involve widely varying mechanisms, and thus, they are often considered separately.

16.3.1 Landslide susceptibility

Landslide susceptibility modelling is the primary approach used to identify the locations where landslides are more likely to initiate. Landslide susceptibility models are distinct from landslide hazard models in that susceptibility models do not consider the landslide trigger or its timing. Landslide hazard models are derived by combining underlying susceptibility models with some trigger mechanism, typically ground shaking for coseismic landslides. This can be applied using probabilistic estimates of ground shaking to produce probabilistic coseismic landslide hazard, or using scenarios, either of potential future events or past events, to evaluate the resulting landslide hazard for the given scenario.

Susceptibility models yield estimates of the relative or absolute probability of a landslide initiating at any given location, given a sufficient trigger. Two main approaches exist: physically based models and statistical models. Virtually all physically based models are derived from Newmark (1965), who considered a rigid block sliding on an inclined plane during shaking, with downslope displacement induced when shaking exceeds a given threshold, known as critical (or yield) acceleration (Jibson, 2007). Whilst numerous updates and refinements to the method have been developed, conceptually these models all derive from Newmark's initial model (Fan et al., 2019). The rigid block model has been critiqued, however, particularly for its lack of consideration of internal deformation and, when applied at the regional scale, the application of overly generic input data, such as rock strength or critical acceleration (Gallen et al., 2017).

Statistical methods are far more varied in their approach, although conceptually they all rely on statistical relationships between landslide occurrence and various causative factors or their proxies (Reichenbach et al., 2018). These approaches first emerged in the 1970s and 1980s (Keefer, 1984; Wilson & Keefer, 1983) and have since evolved significantly, driven by advances in computational methods and the increasing availability of spatial data. Early approaches employed bivariate statistical techniques, such as frequency ratios and weights-of-evidence methods, which were relatively simple to implement but limited in their ability to capture complex, nonlinear relationships.

The 1990s and early 2000s saw the introduction of multivariate approaches, particularly logistic regression, which allowed for the simultaneous consideration of multiple conditioning factors and their interactions (Guzzetti et al., 2005). In the mid-2010s, a key advance showed that some of these statistical relationships between landslides and causative factors appeared to be similar for earthquakes in different environments (Kritikos et al., 2015), suggesting the potential for a globally applicable model (Nowicki Jessee et al., 2018). Nevertheless, as computational power has increased, so too has the complexity of statistical methods, with more sophisticated algorithms offering the ability to handle large, high-dimensional datasets (Ye et al., 2025). More recently, deep learning techniques, such as convolutional neural networks, have been applied particularly in conjunction with high-resolution imagery and big data analytics (Dong et al., 2023).

A critical factor in the rapid evolution of these approaches has been the concomitant growth in the availability and quality of landslide inventories, which serve as the foundational input for statistical modelling. Traditionally, inventories were compiled through field surveys and manual interpretation of aerial photographs, a process that was time-consuming, labour-intensive and spatially limited (Guzzetti et al., 2012). However, the advent of high-resolution satellite imagery and the development of semiautomated and automated landslide detection techniques have expanded the spatial and temporal coverage of landslide inventories (Milledge et al., 2022; Novellino et al., 2024). Remote sensing technologies, including optical sensors (e.g. Landsat, Sentinel-2), synthetic aperture radar and lidar, have enabled the rapid and systematic mapping of landslides across large and often inaccessible areas (Ray et al., 2020). Object-based image analysis and machine learning classifiers have further enhanced the accuracy and efficiency of landslide detection from imagery (Amatya et al., 2021). These advances have led to the creation of global and regional landslide databases, such as the Global Landslide Catalogue[1] and the Globally Distributed Coseismic Landslide Dataset[2] (Fang et al., 2024), which are likely to become important resources for training and validating statistical models.

Since the 1994 M_w 6.7 Northridge earthquake in California, the cumulative number of accessible landslide inventories has grown almost exponentially (Tanyaş et al., 2017), and so too has the cumulative number of statistical models (Fan et al., 2019; Reichenbach et al., 2018). In fact, the number of approaches to statistical landslide modelling is now vast – virtually every major landslide-triggering earthquake and rainstorm is rapidly followed by publication of new or improved models. Yet the general model fit and predictive performance of these models has remained broadly similar, and Reichenbach et al. (2018) highlighted that nearly 40% of published articles

[1] https://data.nasa.gov/dataset/global-landslide-catalog-export
[2] https://zenodo.org/records/13612636

on landslide susceptibility modelling did not measure the model's predictive performance in any way. Even early simple statistical approaches were able to report high model fit and/or predictive performance, with reported area under the receiver operating curve (AUROC, one of the most common performance measures) values of >0.9, albeit often with sizeable overprediction rates (Gallen et al., 2017; Robinson et al., 2017; Robinson et al., 2018). Later and more advanced approaches have largely reported similar success rates, whilst also suffering from overprediction, suggesting diminishing returns in terms of our ability to model landslide susceptibility.

The utility of continuing to develop ever more advanced statistical approaches to model landslide occurrence with diminishing returns in terms of predictive outcomes is questionable, particularly in the context of the comparative lack of investment in exposure and vulnerability analyses. However, there is a clear benefit in improving the quality of landslide susceptibility models. Reichenbach et al. (2018) proposed a useful metric to qualitatively assess model quality, showing that, as of 2016, very few published models achieved the 'high quality' rating, with large numbers of 'low quality' assessments. Importantly, quality as defined by Reichenbach et al. (2018) is distinct from model performance, with quality considering the independence of landslide information used to quantify performance, as well as the presence of error estimates, model fit and predictive ability associated with the model. An updated analysis of works since 2016 using the approach of Reichenbach et al. (2018) would be valuable in ascertaining whether the quality of landslide susceptibility assessments has generally increased, despite no clear increase in their performance.

16.3.2 Landslide runout

Runout – the farthest distance a landslide travels from its source – is a critical parameter in hazard and risk assessments, determining the delineation of impact zones and informing potential mitigation strategies. Landslide runout modelling has undergone significant evolution since the 1980s, transitioning from simple empirical approaches to sophisticated numerical and probabilistic models.

Early efforts in runout prediction primarily relied on empirical methods, such as the reach angle (also known as the Fahrböschung angle), which relates the vertical drop to horizontal travel distance (Corominas, 1996; Heim, 1932). These models, while computationally efficient, were limited by their inability to account for site-specific topographic and material variability. Despite their simplicity, empirical models remain useful for rapid assessments and regional-scale hazard screening (Milledge et al., 2019; Peruzzetto et al., 2020).

The 1990s and early 2000s marked a shift toward physically based numerical runout models, driven by advances in computational power and the increasing availability of digital elevation models. Notably, the development of depth-averaged continuum models, such as DAN and its three-dimensional extension DAN3D (McDougall & Hungr, 2004), enabled more realistic simulations of flow-like landslides. These models incorporate rheological laws – such as Voellmy, Bingham and frictional-plastic formulations – to simulate the dynamics of mass movement over complex terrain.

Parallel developments led to the creation of rapid mass movement simulation (RAMMS), which integrates a Voellmy type rheology with high resolution topography to simulate debris flows and rock avalanches (Christen et al., 2010). RAMMS has been widely applied in alpine regions and is valued for its user-friendly interface and robust calibration framework (Kumar et al., 2024). Similarly, r.avaflow, an open-source GIS-based tool, enables the simulation of complex, multiphase mass flows, including cascading hazards such as landslide-induced floods (Mergili et al., 2017).

Recent advances have focused on machine learning approaches, which are emerging as complementary tools for estimating runout distances, particularly in data-rich environments. For example, Hibert et al. (2025) demonstrated the use of gradient boosting algorithms to predict runout from seismic signals, offering a novel pathway for near-real-time hazard assessment.

Despite these advances, challenges remain, particularly in terms of integrating runout models with susceptibility models for comprehensive hazard assessments. Susceptibility models are often trained on landslide inventories that do not distinguish between landslide source and landslide deposit, meaning elements of runout behaviour can be incorporated into susceptibility models. Coupling these models with runout simulations can therefore be difficult, and requires assumptions about landslide volume, rheology and initiation thresholds – parameters that are often poorly constrained at broad spatial scales (Mergili et al., 2017). Moreover, most physically based runout models, such as DAN3D or RAMMS, are designed for site-specific applications and require detailed input data, making them computationally intensive and difficult to scale (Christen et al., 2010; McDougall & Hungr, 2004).

To address this, simplified empirical or probabilistic approaches are sometimes used at regional scales (Arrell et al., 2024; Kincey et al., 2023), but these can introduce significant uncertainty, particularly in heterogeneous terrain (Peruzzetto et al., 2020). The lack of consistent landslide inventories and terrain data further complicates model calibration and validation across large areas (Reichenbach et al., 2018). Despite these limitations, recent efforts to develop modular frameworks and scalable algorithms – such as r.avaflow and Bayesian ensemble methods – offer promising pathways for integrating initiation and runout processes in regional hazard modelling (Aaron et al., 2019; Mergili et al., 2017). Nonetheless, achieving reliable and operationally useful outputs at regional scales remains an ongoing challenge.

16.3.3 Temporal evolution of hazard

While the advances in susceptibility and runout modelling have been significant, these have largely treated landslide hazard as temporally static. However, research over the last decade has shown that landslide hazard is highly dynamic, particularly in the years to decades following a large earthquake. Marc et al. (2015) were amongst the first to demonstrate that earthquakes don't just increase landslide hazards during the shaking, but in fact result in elevated rates of landsliding for years afterwards. Extending this work, Kincey et al. (2021) and Rosser et al. (2021) used extensive time-series inventories of landslides following the 2015 M_w 7.9 Gorkha earthquake in Nepal (Chapter 15) to show that landslide hazard had actually increased over a period of 4 years following the earthquake. This increase was temporally and spatially heterogeneous, with some areas showing marked decreases in landslide rates in the years following, while others appeared to continually increase.

Work following the 2016 M_w 7.8 Kaikōura earthquake in New Zealand has shown a similar outcome, with rates of landsliding dramatically increasing during the earthquake, and then remaining high for several years afterwards (Polwart, 2023). As with the Gorkha earthquake, this dynamic change appears to be spatially and temporally heterogeneous, with the coastal areas surrounding the affected transport networks appearing to recover faster than inland regions (Fig. 16.2). Polwart (2023) suggested postearthquake mitigation to clear debris and reinstate the transport networks as a possible explanation. However, Daysh et al. (2025) analysed rockfall rates on mitigated and nonmitigated slopes along the coastal transport network corridor and found no obvious evidence that mitigated slopes recovered faster than nonmitigated slopes. The reason for this spatial heterogeneity in postearthquake landslide rates remains unclear.

The precise duration of this elevated landslide activity remains contentious, with estimates ranging from a few years (Hovius et al., 2011; Marc et al., 2015) to a quarter of a century or more (Croissant et al., 2017). This question of landslide hazard longevity is, arguably, one of the biggest issues in landslide hazard and risk research today. Our inability to answer it leaves us drastically unprepared for the ongoing impacts from postseismic landslides. Furthermore, it limits our understanding of the role of coseismic landslides in the erosion of mountains, with controversy over whether earthquakes are the primary driver of erosion (Marc et al., 2016) or play a more minor role (Robinson, Davies, Wilson, Orchiston, & Barth, 2016). We urgently need to address this knowledge gap to transform our understanding and management of the longer-term hazards and risks that arise postearthquake.

Part of the problem with addressing this issue, though, is that virtually all landslide inventories are constructed from a single point in time, mapping landslides soon after the cessation of shaking. Multi-temporal inventories that map landslides in a consistent way through multiple pre-, co- and postseismic epochs are increasing but remain rare. Consequently, the factors that drive postseismic landslides remain controversial. To overcome this, we require a fundamental change in our approach that focuses on the creation of multitemporal landslide inventories to inform dynamic geospatial landslide hazard models that consider changing landslide hazard through time.

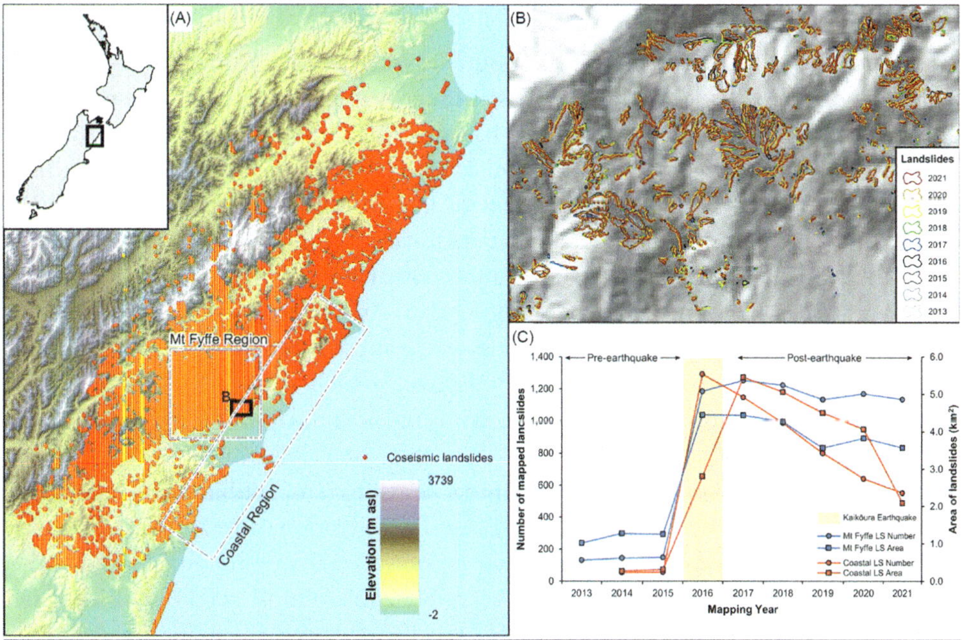

Figure 16.2 (A) Spatial distribution of coseismic landslides from the 2016 M_W 7.8 Kaikōura earthquake. (B) Mapped landslide polygons from 2013 to 2021 in a part of the Mt Fyffe region (area shown in panel A), showing persistent activity of postearthquake landslides. (C) Change in number and area of landslides in the Mt Fyffe region (blue) and in the coastal region (red) since 2013; see panel A for areal extents. Both the number and area of landslides in the coastal region show marked decreases since the earthquake, while recovery in the Mt Fyffe region is negligible, suggesting varying spatial controls on landslide behaviour. Map lines delineate study areas and do not necessarily depict accepted national boundaries. Source: *A, From Massey, C. I., Townsend, D. T., Lukovic, B., Morgenstern, R., Jones, K., Rosser, B., & de Vilder, S. (2020). Landslides triggered by the MW7. 8 14 November 2016 Kaikōura earthquake: An update.* Landslides, *17(10), 2401–2408.*

16.4 Modelling exposure

Quantifying exposure to landslides is a fundamental step in risk assessment, as it identifies the people, infrastructure and assets that are located within areas likely to be affected by landslide hazards. However, unlike earthquakes, landslides have a comparatively small hazard footprint, and distances of just a few metres can result in large changes to who or what is exposed (Arrell et al., 2024). While the increasing availability of global datasets has enabled more consistent and scalable exposure assessments, particularly in data-scarce regions (Emberson et al., 2020), accurately modelling exposure to landslides remains challenging. Continued development of global exposure frameworks is essential for improving the accuracy and utility of landslide risk assessments.

16.4.1 Spatial overlays

At their simplest, landslide exposure assessments take landslide susceptibility or hazard maps and perform a spatial overlay analysis to identify where people or infrastructure are located relative to landslide susceptibility/hazard. This can be done for individual assets or for multiple assets, with various approaches used to combine and weight the different assets into a single layer, typically in terms of monetary value (Pellicani et al., 2014). Whilst this is most typically undertaken at regional or national scales, Emberson et al. (2020) used global datasets of population and infrastructure to assess exposure at a global level. They showed – not unexpectedly – that Asia dominated global population exposure to landslides, with India and China as the most exposed nations. Such work is backed by data from Froude and Petley (2018), which show that global landslide fatalities are concentrated in rural Asia, with India and China recording the most fatalities between 2004 and 2016.

Nevertheless, a key issue with these approaches is that susceptibility and hazard maps are intended to represent the probability of landslide initiation, not the full spatial extent of potential runout, although often the processes are unintentionally combined. Consequently, overlaying assets within susceptible zones almost certainly misrepresents the actual risk (Woodard & Mirus, 2025). Challenges with integrating susceptibility models and runout models at the regional scale, however, make addressing this issue problematic. Robinson et al. (2016) proposed a simple approach that flipped the concept from understanding where a landslide might go from any given initiation location, to instead looking from an asset point of view and identifying the area from within which a landslide could reach a given asset. This approach has been built on by Robinson et al. (2018) and Lin et al. (2025), who incorporated the reach angle concept and showed that the approach performed favourably but suffered from high over-prediction rates. Further, the approach struggles to account for terrain conditions that can cause nonlinear runout paths, whilst also suffering from issues of spatial resolution. It does, however, have the advantage of identifying those areas of the landscape from which landslides pose a direct threat.

Others have attempted to apply more robust runout models to denote potential landslide footprints and thus estimate exposure. For example, de Vilder et al. (2022) used RAMMS to model potential runout paths for different-sized rockfalls, rock avalanches and flows in two glacial valleys in New Zealand to determine both the locations of highest exposure and the relative differences in exposure to different types of landslide movement. Similarly, Arrell et al. (2024) used Flow-R to model flow-type landslides from prescribed source areas defined from a susceptibility map. Both studies showed that exposure varied substantially over lateral distances on the scale of metres, highlighting the need for a high-resolution view of both landslide hazard and exposure datasets. This is rarely possible, however, and the resolution mismatch between susceptibility maps and exposure datasets – such as coarse-scale hazard layers intersected with high-resolution building footprints or vice versa – can introduce significant uncertainty (Tran et al., 2024). Furthermore, such runout models require significant assumptions in terms of how landslides are expected to move, or, in the case of Arrell et al. (2024), assume that all landslides will runout with the same style of motion, highlighting significant issues in denoting the relevant hazard footprint to use.

16.4.2 Dynamic exposure

Whilst existing spatial overlay approaches have their issues, conceptually they are well-suited for static assets like buildings and infrastructure. However, modelling population exposure presents an altogether different challenge. Humans are nonstatic, and our mobility is both spatially and temporally heterogeneous (Haraguchi et al., 2022). While the issue of human mobility has been recognised in many hazard exposure analyses, quantifying it has typically been reduced to simple diurnal patterns (Pellicani et al., 2014). The base assumption here is that people are at home in residential buildings at night, and away from those homes – probably at a place of work – during the day. While such assumptions allow exposure analyses to consider how risk varies diurnally, they do little to capture the complex reality of human mobility and its impacts on our exposure to landslides.

Little work has been done on understanding the temporal variability in population exposure to landslides. In fact, the majority of the work that has been done focuses on the temporal variability of the landslides themselves, rather than the population (Jones et al., 2021; Mudugno et al., 2022). While this variability is important for rainfall-triggered and postseismic landslides, the timing of coseismic and aseismic landslides is essentially random on human timescales, and therefore, it is human mobility that controls whether someone is exposed at the time of the event or not.

Recent work has begun to explore how landslide exposure varies over years to decades as a result of urbanisation and other land-use changes, as well as societal changes. Kühnl et al. (2023) showed that population growth in Medellín, Colombia, since 1994 has translated to dramatic changes in population exposure to landslides, especially in highly vulnerable informal settlements. Others have looked at how climate change, combined with population growth and land use change, might influence landslide exposure in the future, suggesting that in some places, exposure may increase (Ge et al., 2024). However, very few studies have looked at exposure at finer time scales that are more appropriate for day-to-day human mobility.

One recent attempt at this has sought to understand how changes in human mobility on the scale of minutes can influence population exposure. Darling et al. (2025) used WiFi probes to count the number of mobile phones – as a proxy for people – within a defined area every 5 minutes over a period of nearly 3 years. By applying this method to a popular tourist site in New Zealand, they showed that population exposure could change by up to an order of magnitude over 5 minutes, showing that the precise timing of an event was crucial in determining how many people were exposed.

Combined with findings from Arrell et al. (2024), this result highlights the fine margins and minute details that influence population exposure to landslides. A change in location of just a few metres, or a difference in timing of just a few minutes, can lead to dramatically different numbers of people exposed. Currently, there are no methods to combine exposure modelling approaches at these fine spatial and temporal resolutions, and as a result, our understanding of population exposure to landslides is limited. Compared to recent advances in landslide hazard modelling, exposure modelling lags far behind, with most methods relying on simple spatial overlays with little to no consideration of temporal variations. There are gains to be made in developing more advanced methods to assess exposure at the high spatial and temporal resolutions required, while also accounting for the known temporal variability in landslide hazard.

16.5 Vulnerability and capacity

Landslide vulnerability assessments are one of the most understudied components of landslide risk analysis (Singh et al., 2017). Much of the challenge comes from the sheer range of landslide types and the various possible measures of magnitude or intensity (e.g. volume, velocity, depth, or impact pressure). This is further complicated by the fact that vulnerability can be broken down into four different categories: physical vulnerability (to infrastructure or people), social vulnerability (in terms of coping capacity, poverty, or education), environmental vulnerability (such as natural resources or land use), and economic vulnerability (in terms of direct and indirect losses, insurance, or business interruption). Despite these complications, vulnerability assessments are critical for understanding landslide risk, and the limited focus on this core component currently severely hampers our ability to quantify and reduce that risk.

16.5.1 Physical and economic vulnerability

Physical and economic vulnerability are most typically addressed from the physical science and engineering perspective, with a focus on the level of damage from landslides to buildings, infrastructure and people. The scale of damage or loss expected for a given asset exposed to a landslide is often the main point of a risk assessment. Yet there are no universally accepted approaches to quantifying the vulnerability of people and assets to landslides. Luo et al. (2023) found that since 1987, <200 articles have been published on physical landslide vulnerability, compared to ~600 articles over a similar period on landslide susceptibility modelling alone (Reichenbach et al., 2018).

What limited research there has been on physical vulnerability has predominantly focused on the vulnerability of buildings. This has spanned methods involving static analysis, field investigations, experimental tests and numerical simulations (Luo et al., 2023). However, the challenge that remains is accounting for the extremely high levels of variability in damage and the often location-specific factors that can influence these (Fig. 16.3). For instance, a building present in the source zone of a landslide is likely to suffer from foundational collapse or slippage and become part of the landslide itself, with resulting damage correlating with the level of displacement. Yet buildings immediately below the landslide in the runout path are likely to sustain damage from direct impact, and so damage is more likely to correlate with debris depth, velocity, or impact pressure. But in dense clusters of buildings, the buildings themselves will change how the landslide moves, the degree of which will depend on the style of motion – slow, wet flow-type events may be impeded and deflected by buildings, while high-velocity rock or debris avalanches are less likely to be affected. In addition, the style of building – timber frame, masonry or concrete – influences the physical response of the building itself, resulting in different outcomes despite experiencing the same hazard intensity. While some of these factors can be accounted for with local-scale and numeric modelling, at a regional level, it is almost impossible with current approaches to fully capture such details.

When it comes to human vulnerability to assess casualties, many studies take a simple binary approach, assuming that anyone directly exposed to a landslide is killed while

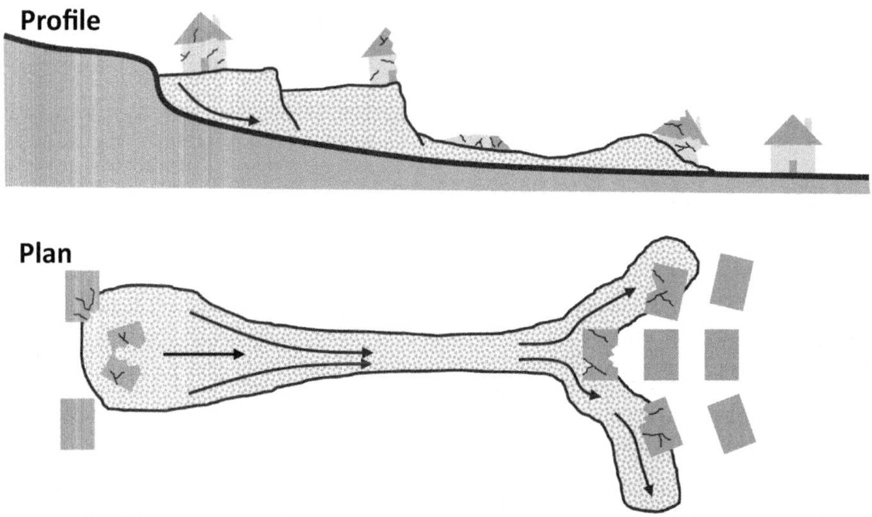

Figure 16.3 Conceptual plan and profile views of landslide runout impacting buildings, showing examples of different modes of damage (slippage near the landslide source versus impact in the runout zone) and the shielding influence that buildings can have in directing landslide runout.

those not exposed survive uninjured. Such simplistic assumptions are often based on subjective expert judgement since the underlying causes of landslide mortality are rarely explored. One of the few studies to attempt to derive nonbinary, data-driven human vulnerability models showed that binary outcomes are overly simplistic and that, in fact, the probability of death can be linked to landslide debris depth (Pollock & Wartman, 2020). In fact, people are more likely to survive than not (probability of death <50%) at debris depths of less than ~2 m. For depths between ~1 and 6 m, it is the actions people take during the landslide, such as attempting to move out of the path or shield themselves behind rigid structures, that influence whether they are killed, not the landslide intensity. Yet this highlights one of the greatest uncertainties in modelling potential landslide casualties – effectively accounting for human behaviour during a landslide. While it's arguable that this problem is simplified somewhat for coseismic landslides, when strong ground shaking means that most exposed people will be unable to move out of a landslide's path, even this assumes that landslides are triggered immediately during the strongest phase of ground shaking.

As with exposure modelling, our current understanding of physical landslide vulnerability is hampered by a lack of data and research. Improving our understanding of the link between damage and hazard intensity in conjunction with our ability to model local-scale interactions that modify hazard intensity is clearly a critical need. While Pollock & Wartman (2020) have shown that human vulnerability to landslides is far from binary, it's important to start to consider human behaviour during a landslide to understand the outcomes. In short, further investment is desperately needed to bring our understanding of physical vulnerability to landslides to match our current understanding of landslide hazard processes.

16.5.2 Social vulnerability and capacity

In comparison to physical and economic vulnerability, social vulnerability and capacity are mostly studied from the social science perspective, because they require an understanding of the underlying factors that lead to people and communities being unable to withstand the impacts of landslides (Nam et al., 2023). Social vulnerability here refers to the socio-economic stressors that influence the level of harm a community might experience, such as poverty, inequality, access to resources and demographics. As a result, social vulnerability and capacity have been assumed to be inversely related to community resilience (Bergstrand et al., 2015), although this remains contentious, with others highlighting a more complex relationship (Cutter et al., 2008; Manyena, 2006). Nevertheless, directly quantifying social vulnerability and capacity is challenging, as many of the factors that influence the extent to which a community can absorb landslide impacts cannot be directly measured.

To address the challenge of quantifying social vulnerability, most studies have opted to use either index-based or proxy measures (Siders, 2019). Index-based methods identify and measure (quantitatively or qualitatively) a set of determinants before aggregating these into an overall 'score'. In comparison, proxy-based approaches compare values for individual determinants to a selected proxy variable to identify which determinants correlate significantly.

In the landslide context, the most widely used approach is based on the Social Vulnerability Index (SoVI) first proposed by Cutter et al. (2003). This approach combines metrics of social vulnerability, such as education levels, life expectancy, poverty and demographics to quantify the level of social vulnerability within a community through a semiquantitative index. It was initially developed for the United States and therefore has limited direct transferability to other contexts (Guillard-Gonçalves et al., 2014). One of the strengths of the SoVI, however, is its adaptability – for instance, several key variables include the percentage of ethnic and racial minorities specific to the USA that can be adapted to other contexts, assuming sufficient data exist. Consequently, there are a number of SoVI studies for landslides for a range of different contexts, including local scale (Xiao et al., 2022), regional scale (Sohel et al., 2022) and national scale (Nor Diana et al., 2021).

The challenge here, however, is one of resolution. By their nature, index- and proxy-based approaches like the SoVI are often applied at census block levels, which, even at the finest resolutions, are significantly larger than the typical spatial footprint of an individual landslide. Consequently, analyses that use such approaches can only be assessed at the coarsest spatial resolution, with risk often being uniformly applied across a census block. This poses the obvious question of the utility of applying landslide susceptibility, hazard and exposure models at fine spatial resolutions if critical concepts of social vulnerability cannot be resolved at similar spatial resolutions. Furthermore, the use of census blocks creates boundary issues, whereby vulnerability can appear to change significantly across a census boundary despite that boundary often having no obvious physical definition. In such cases, two or more people could be just metres apart within a single landslide footprint, but be assessed as having very different risks as a result of vulnerability scores attributed from data that considers populations at far greater distances.

Vulnerability to landslides – whether physical or social – is extremely complex and, unlike other aspects of landslide risk modelling, requires transdisciplinary research across the physical, social and engineering sciences. Beyond the issues related to spatial scale and resolution, approaches that can easily combine physical and social vulnerability are limited, meaning most studies that do consider vulnerability only consider one aspect. Our current understanding of landslide risk is therefore equally limited.

16.6 Future of coseismic landslide risk assessment

Despite challenges in unequal progress across hazard, exposure and vulnerability, we have made tremendous progress in understanding coseismic landslide risk since the 1970s, both conceptually and practically. However, this does not appear to have translated into reduced landslide impact – while global annual landslide fatalities have not obviously increased, neither have they substantially decreased (Froude & Petley, 2018). While increasing global population might suggest that annual fatality rates have decreased, major earthquakes in the Himalaya alone since 2000 have shown the potential for tens or hundreds of thousands of people to be killed by coseismic landslides, essentially at any moment. With that spectre, we must consider the key research challenges that remain for coseismic landslide risk assessment.

16.6.1 Do we need more hazard analysis?

Clearly, the majority of our progress in coseismic landslide risk has come from our efforts to understand the underlying processes and hazards. Equally clearly, there remains much uncertainty in many aspects of coseismic landslide hazard that requires attention. Hence, at face value, this seems an easy question to answer: yes. However, given finite time and resources, it's sensible to consider whether greater gains could be made by shifting our focus elsewhere, or at least to consider which elements of hazard analysis require attention.

The rate of articles investigating landslide hazard has been rapidly increasing since c.2000, and by the mid-2010s, studies on statistical landslide susceptibility models alone were being published at a rate of over one per week (Reichenbach et al., 2018). The number of different methods available has increased almost exponentially (Fan et al., 2019) despite efforts to develop globally consistent models (Nowicki Jessee et al., 2018). And yet, our model accuracy has not seen anywhere near the same increase. While advanced machine learning approaches typically score marginally better prediction rates than simple statistical models, these improvements are generally minor and nonsignificant (Goetz et al., 2015), with AUROC scores >0.9 quoted for many model types.

In this context, it's clear that a wide variety of modelling approaches already exist that can provide good levels of predictive performance. Yet Reichenbach et al. (2018) demonstrated that few of these studies are rated as 'top quality'. To achieve such a rating, models need to include estimates of model fit, error and predictive performance from independent landslide information – basic elements of any good model. The fact that so few existing landslide susceptibility models contain all these elements is concerning. Consequently, there appears to be far more to be gained from focusing on improving the quality of available landslide models rather than achieving comparatively incremental gains in already good predictive performance.

In addition to improving the quality of our hazard models, there is significant utility in improving the information within susceptibility models. Existing susceptibility models mostly provide information only on the likelihood of any landslide occurring, with no information on the size or type of failure that could occur. However, such variables are critical for landslide runout modelling. As Arrell et al. (2024) showed, we need to understand runout at extremely fine spatial scales if we want to accurately identify what or who might be impacted. Improvements in runout modelling mean such precision is possible at local scales under specific circumstances, when the source zone, volume and movement style, amongst other factors, are all known. However, for earthquakes, when tens to hundreds of thousands of landslides can be triggered across thousands of square kilometres, this remains a significant challenge. While further refinement of hazard models is valuable, the greatest gains may come from improving model quality and integrating landslide size and style predictions rather than developing entirely new susceptibility algorithms. If such models are possible, then existing issues with linking susceptibility and runout models together become easier to address, and combined regional-scale models of landslide initiation and runout hazard for earthquakes become a possibility.

16.6.2 Landslide runout – what to model?

For landslide hazard models to be effective, they need to model the landslide variables that actually result in damage or loss, such as impact pressure, displacement, or velocity. It is therefore critical that the initial hazard analysis considers the causes of damage to people or infrastructure in order to be able to model risk. However, our understanding of physical landslide vulnerability remains uncertain, with a wide variety of intensity metrics being proposed for vulnerability models. The reasons for this variety are complex, but at least partly derive from the myriad ways that landslides can cause damage (Fig. 16.3). If we want to understand the risk posed to buildings from slippage, then our runout models need to accurately model landslide displacement. However, if, in addition, we also want to understand risk from direct impact, then we simultaneously need to model impact pressure, or depth, or velocity. Existing models like RAMMS can consider multiple variables, but applying these at regional scales relevant to earthquakes remains challenging, and for the reasons discussed above, regional-scale susceptibility models rarely contain the information necessary to model such variables accurately.

While developing linked susceptibility and runout models applicable at regional scales would provide a major step forward, it is unlikely to translate directly to a similar step in our understanding of risk without concomitant improvements in our understanding of vulnerability. Improving our ability to rapidly model runout from multiple simultaneous coseismic landslides is critical, but must be linked with ongoing research into physical vulnerability, to ensure that future runout models are able to accurately capture the variables that matter at a spatial accuracy relevant to the assets being considered. For risk to buildings and infrastructure, this will require interdisciplinary research across the computer sciences, engineering geology and civil and structural engineering, while for risk to people, this will also need to include both health science and social science in the broadest sense. In conjunction, we need to improve our understanding of social vulnerability and capacity, and develop approaches that can be meaningfully translated across spatial resolutions. Linking our numerical models of landslide behaviour directly to engineering and social models of infrastructure and human behaviour is essential for understanding how and when coseismic landslides will cause impacts in the future.

16.6.3 Dynamic risk

Perhaps the most challenging aspect of coseismic landslide risk still to be addressed is understanding the temporal variability that results from changing hazard, exposure and vulnerability. Current approaches to calculating risk assume that all three aspects are static, providing static risk estimates. In reality, all three components vary by differing amounts across multiple time scales, from minutes to years to decades and centuries. Consequently, the risk posed by coseismic landslides is highly variable across those time scales, and therefore the existing static view of risk is flawed (Fig. 16.4). To understand and reduce the risk posed by coseismic landslides in the future, we need to develop new tools and methods to account for this temporal variability.

We are beginning to understand how landslide hazard evolves through time, with recent works showing that hazard peaks during earthquake shaking and then undergoes complex changes in the months and years following an earthquake related to ongoing aftershocks, postearthquake rainfall and human activities. However, currently our ability to model this temporally variable hazard is *a posteriori* – while initially this may improve our understanding of the driving factors, ultimately we need to be able to develop *a priori* models. Such efforts will require substantial investment in ongoing observations and recording of landslide behaviour before, during and after future earthquakes, in order to develop multitemporal landslide inventories that can train and test *a priori* models and inform our understanding of the complex mechanics involved in chronic postearthquake landslide hazard.

Similarly, our ability to monitor and predict human mobility is improving rapidly with the spread of technology, particularly mobile phones. However, the heterogeneous and rapidly changing nature of human mobility makes estimating exposure to coseismic landslides complex. Research has shown that the number of people present in a landslide's path can change substantially in just a matter of minutes, meaning that the impacts of both the primary and secondary hazards triggered by an earthquake are highly dependent on the precise timing of the event. Likewise, there is a need to understand

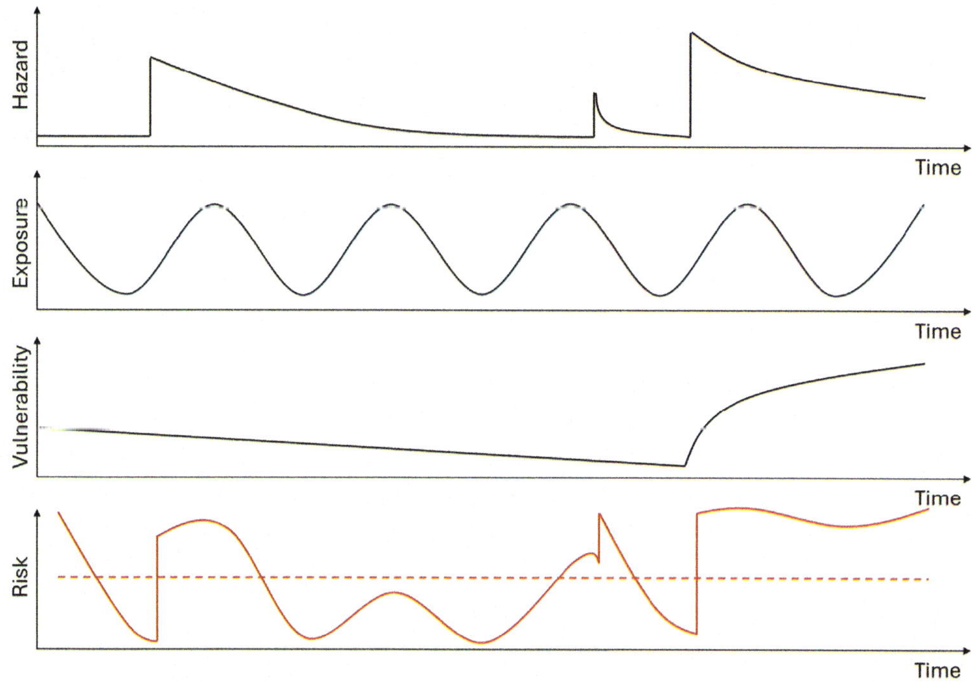

Figure 16.4 Conceptual model of dynamic risk, showing hypothetical temporal variations in hazard, exposure and vulnerability and the consequences for risk, calculated as the product of hazard, exposure and vulnerability. The dashed line in the bottom panel shows a standard 'static' view of risk, showing that such an approach can both over and underestimate the actual risk at different points in time.

human exposure at similarly high spatial resolutions, with changes on the scale of a few metres influencing who is and isn't exposed. Changes in landslide hazard and behaviour over time, both after an earthquake and due to concurrent climate change, will likely also influence human behaviour and mobility, but these effects currently remain unknown.

On top of all this, our vulnerability to coseismic landslides is also dynamic and changes over multiple time scales. The physical vulnerability of our existing buildings and infrastructure is expected to increase over years and decades as components age and decay, while partial damage during an earthquake will increase vulnerability to landslides in the following days to months by decreasing the hazard intensity required to cause further damage. Meanwhile, new buildings and infrastructure would be expected to be less vulnerable to damage as design, construction and siting techniques improve. Developing inventories of our infrastructure and building stocks that contain sufficient details to accurately estimate vulnerability at the time of the event, rather than at the time of construction, will require substantial efforts, as regular data updates are essential. In terms of social vulnerability, at the national and global scale, our vulnerability is generally decreasing over years to decades as living conditions generally improve, but this isn't necessarily reflected at the regional scale (Cutter & Finch, 2008). Equally, social vulnerability can change rapidly as a result of sudden or chronic perturbations, with ongoing effects from climate change and political instability potentially reversing hard-won decreases in vulnerability.

Importantly, if we want to truly understand the temporal variability in coseismic landslide risk, we need to invest in research that understands and quantifies how hazard, exposure and vulnerability change simultaneously. These changes are likely to be highly interlinked, as well as subject to random external pressures, and consideration will need to be given to the various temporal and spatial scales over which these changes occur. Such a challenge is by definition transdisciplinary, requiring multiple disciplines to work together codependently rather than simultaneously but in isolation. Fortunately, this problem is not unique to coseismic landslides, and therefore seeking insights and breakthroughs from other hazard sciences and related fields can meaningfully enhance our collective understanding and practice.

References

Aaron, J., McDougall, S., & Nolde, N. (2019). Two methodologies to calibrate landslide runout models. *Landslides, 16*, 907–920. https://doi.org/10.1007/s10346-018-1116-8.

Allstadt, K. E., Jibson, R. W., Thompson, E. M., Massey, C. I., Wald, D. J., Godt, J. W., & Rengers, F. K. (2018). Improving near-real-time coseismic landslide models: Lessons learned from the 2016 Kaikōura, New Zealand, earthquake. *Bulletin of the Seismological Society of America, 108*(3B), 1649–1664.

Amatya, P., Kirschbaum, D., Stanley, T., & Tanyas, H. (2021). Landslide mapping using object-based image analysis and open source tools. *NASA Technical Reports*. [Available at https://ntrs.nasa.gov/citations/20210000782.

Arrell, K., Rosser, N. J., Kincey, M. E., Robinson, T. R., Horton, P., Densmore, A. L., Oven, K. J., Shrestha, R., & Pujara, D. S. (2024). The dynamic threat from landslides following large continental earthquakes. *PLoS One, 19*(8), e0308444.

Bergstrand, K., Mayer, B., Brumback, B., et al. (2015). Assessing the relationship between social vulnerability and community resilience to hazards. *Social Indicators Research, 122*, 391–409. https://doi.org/10.1007/s11205-014-0698-3.

Bird, J. F., & Bommer, J. J. (2004). Earthquake losses due to ground failure. *Engineering geology, 75*(2), 147–179.

Christen, M., Kowalski, J., & Bartelt, P. (2010). RAMMS: Numerical simulation of dense snow avalanches in three-dimensional terrain. *Cold Regions Science and Technology, 63*(1-2), 1–14.

Corominas, J. (1996). The angle of reach as a mobility index for small and large landslides. *Canadian Geotechnical Journal, 33*(2), 260–271.

Croissant, T., Lague, D., Steer, P., & Davy, P. (2017). Rapid post-seismic landslide evacuation boosted by dynamic river width. *Nature Geoscience, 109*(10), 680–684.

Cutter, S. L., & Finch, C. (2008). Temporal and spatial changes in social vulnerability to natural hazards. *Proceedings of the National Academy of Sciences, 105*(7), 2301–2306. 10.1073/pnas.0710375105.

Cutter, S. L., Boruff, B. J., & Shirley, W. L. (2003). Social vulnerability to environmental hazards. *Social Science Quarterly, 84*(2), 242–261. https://doi.org/10.1111/ssqu.2003.84.issue-2.

Cutter, S. L., Barnes, L., Berry, M., Burton, C., Evans, E., Tate, E., & Webb, J. (2008). A place-based model for understanding community resilience to natural disasters. *Global Environmental Change, 18*(4), 598–606. https://doi.org/10.1016/j.gloenvcha.2008.07.013.

Darling, M. J., Robinson, T. R., Adams, B., Wilson, T. M., & Orchiston, C. (2025). Minutes matter for life safety and risk exposure in Milford Sound, New Zealand. *Scientific Reports, 15*(1), 36459.

Daysh, I., Robinson, T. R., Green, R., & Mason, D. (2025). The role of helicopter sluicing in earthquake response and recovery in Aotearoa New Zealand. *Bulletin of the New Zealand Geotechnical Society, 108*, 66–76.

de Vilder, S., Massey, C., Lukovic, B., Taig, T., & Morgenstern, R. (2022). What drives landslide risk? Disaggregating risk analyses, an example from the Franz Josef Glacier and Fox Glacier valleys, New Zealand. *Natural Hazards Earth Systems Science, 22*, 2289–2316. https://doi.org/10.5194/nhess-22-2289-2022.

Dong, A., Dou, J., Fu, Y., Zhang, R., & Xing, K. (2023). Unraveling the evolution of landslide susceptibility: A systematic review of 30-years of strategic themes and trends. *Geocarto International, 38*(1). https://doi.org/10.1080/10106049.2023.2256308.

Emberson, R., Kirschbaum, D., & Stanley, T. (2020). New global characterisation of landslide exposure. *Natural Hazards and Earth System Sciences, 20*(12), 3413–3424. https://doi.org/10.5194/nhess-20-3413-2020.

Fan, X., Scaringi, G., Korup, O., West, A. J., van Westen, C. J., Tanyas, H., Hovius, N., Hales, T. C., Jibson, R. W., Allstadt, K. E., & Zhang, L. (2019). Earthquake-induced chains of geologic hazards: Patterns, mechanisms, and impacts. *Reviews of Geophysics, 57*(2), 421–503.

Fang, C., et al. (2024). A globally distributed dataset of coseismic landslide mapping via multi-source high-resolution remote sensing images. *Earth System Science Data, 16*, 4817–4842. https://doi.org/10.5194/essd-16-4817-2024.

Froude, M. J., & Petley, D. N. (2018). Global fatal landslide occurrence from 2004 to 2016. *Natural Hazards and Earth System Sciences, 18*, 2161–2181. https://doi.org/10.5194/nhess-18-2161-2018,.

Gallen, S. F., Clark, M. K., Godt, J. W., Roback, K., & Niemi, N. A. (2017). Application and evaluation of a rapid response earthquake-triggered landslide model to the 25 April 2015 Mw 7.8 Gorkha earthquake, Nepal. *Tectonophysics, 714*, 173–187.

Ge, S., Wang, J., & Jiang, C. (2024). Projection of the precipitation-induced landslide risk in China by 2050. *Environmental Research Letters, 19*(2), 024048.

Goetz, J. N., Brenning, A., Petschko, H., & Leopold, P. (2015). Evaluating machine learning and statistical prediction techniques for landslide susceptibility modeling. *Computers & Geosciences, 81*, 1–11.

Gorum, T., Fan, X., van Westen, C. J., Huang, R. Q., Xu, Q., Tang, C., & Wang, G. (2011). Distribution pattern of earthquake-induced landslides triggered by the 12 May 2008 Wenchuan earthquake. *Geomorphology, 133*(3-4), 152–167.

Guillard-Gonçalves, C., Cutter, S. L., Emrich, C. T., & Zêzere, J. L. (2014). Application of social vulnerability index (SoVI) and delineation of natural risk zones in Greater Lisbon, Portugal. *Journal of Risk Research, 18*(5), 651–674. https://doi.org/10.1080/13669877.2014.910689.

Guzzetti, F., Mondini, A. C., Cardinali, M., Fiorucci, F., Santangelo, M., & Chang, K. T. (2012). Landslide inventory maps: New tools for an old problem. *Earth-Science Reviews, 112*(1-2), 42–66.

Guzzetti, F., Reichenbach, P., Cardinali, M., Galli, M., & Ardizzone, F. (2005). Probabilistic landslide hazard assessment at the basin scale. *Geomorphology, 72*(1-4), 272–299.

Haraguchi, M., Nishino, A., Kodaka, A., Allaire, M., Lall, U., Kuei-Hsien, L., Onda, K., Tsubouchi, K., & Kohtake, N. (2022). Human mobility data and analysis for urban resilience: A systematic review. *Environment and Planning B: Urban Analytics and City Science, 49*(5), 1507–1535.

Heim, A. (1932). *Bergsturz und menschenleben*. Fretz & Wasmuth.

Hibert, C., Groult, C., & Malet, J. P. (2025). Seismogenic landslides volume and runout estimation with machine learning. *Landslides, 22*, 2801–2808. https://doi.org/10.1007/s10346-025-02496-w.

Hovius, N., Meunier, P., Lin, C. W., Chen, H., Chen, Y. G., Dadson, S., Horng, M. J., & Lines, M. (2011). Prolonged seismically induced erosion and the mass balance of a large earthquake. *Earth and Planetary Science Letters, 304*, 347–355.

Jibson, R. W. (2007). Regression models for estimating coseismic landslide displacement. *Engineering Geology, 91*(2-4), 209–218.

Jones, J. N., Boulton, S. J., Bennett, G. L., Stokes, M., & Whitworth, M. R. (2021). Temporal variations in landslide distributions following extreme events: Implications for landslide susceptibility modeling. *Journal of Geophysical Research: Earth Surface, 126*(7), e2021JF006067.

Keefer, D. K. (1984). Landslides caused by earthquakes. *Geological Society of America Bulletin, 95*(4), 406–421.

Khan, S., Kirschbaum, D. B., Stanley, T. A., Amatya, P. M., & Emberson, R. A. (2022). Global landslide forecasting system for hazard assessment and situational awareness. *Frontiers in Earth Science, 10*, 878996.

Kincey, M. E., Rosser, N. J., Densmore, A. L., Robinson, T. R., Shrestha, R., Singh Pujara, D., Horton, P., Swirad, Z. M., Oven, K. J., & Arrell, K. (2023). Modelling post-earthquake cascading hazards: Changing patterns of landslide runout following the 2015 Gorkha earthquake, Nepal. *Earth Surface Processes and Landforms, 48*(3), 537–554. https://doi.org/10.1002/esp.5501.

Kincey, M. E., Rosser, N. J., Robinson, T. R., Densmore, A. L., Shrestha, R., Pujara, D. S., Oven, K. J., Williams, J. G., & Swirad, Z. M. (2021). Evolution of coseismic and post-seismic landsliding after the 2015 Mw 7.8 Gorkha earthquake, Nepal. *Journal of Geophysical Research: Earth Surface, 126*(3), e2020JF005803.

Kritikos, T., Robinson, T. R., & Davies, T. R. (2015). Regional coseismic landslide hazard assessment without historical landslide inventories: A new approach. *Journal of Geophysical Research: Earth Surface, 120*(4), 711–729.

Kühnl, M., Sapena, M., Wurm, M., et al. (2023). Multitemporal landslide exposure and vulnerability assessment in Medellín, Colombia. *Nat Hazards, 119*, 883–906. https://doi.org/10.1007/s11069-022-05679-z.

Kumar, S., Sharma, A., & Singh, K. (2024). A Comprehensive Review on Debris Flow Landslide Assessment Using Rapid Mass Movement Simulation (RAMMS). *Geotechnical and Geological Engineering, 42*(7), 5447–5475. 10.1007/s10706-024-02887-1.

Lin, A. F., Zorn, C., Wotherspoon, L., Robinson, T. R., & Pelmard, J. (2025). Prediction of road blockages caused by rainfall-induced landslides. *New Zealand Journal of Geology and Geophysics*, 1–17.

Luo, H. Y., Zhang, L. M., Zhang, L. L., He, J., & Yin, K. S. (2023). Vulnerability of buildings to landslides: The state of the art and future needs. *Earth-Science Reviews, 238*, 104329.

Malamud, B. D., Turcotte, D. L., Guzzetti, F., & Reichenbach, P. (2004). Landslide inventories and their statistical properties. *Earth Surface Processes and Landforms, 29*(6), 687–711.

Manyena, S. B. (2006). The concept of resilience revisited. *Disasters, 30*(4), 433–450.

Marc, O., Hovius, N., & Meunier, P. (2016). The mass balance of earthquakes and earthquake sequences. *Geophysical Research Letters, 43*(8), 3708–3716.

Marc, O., Hovius, N., Meunier, P., Uchida, T., & Hayashi, S. I. (2015). Transient changes of landslide rates after earthquakes. *Geology, 43*(10), 883–886.

Massey, C. I., Townsend, D. T., Lukovic, B., Morgenstern, R., Jones, K., Rosser, B., & de Vilder, S. (2020). Landslides triggered by the MW7. 8 14 November 2016 Kaikōura earthquake: An update. *Landslides, 17*(10), 2401–2408.

McDougall, S., & Hungr, O. (2004). A model for the analysis of rapid landslide motion across three-dimensional terrain. *Canadian Geotechnical Journal, 41*(6), 1084–1097.

Mergili, M., Fischer, J.-T., Krenn, J., & Pudasaini, S. P. (2017). r.avaflow v1, an advanced open-source computational framework for the propagation and interaction of two-phase mass flows. *Geoscientific Model Development, 10*, 553–569. https://doi.org/10.5194/gmd-10-553-2017.

Milledge, D. G., Bellugi, D. G., Watt, J., & Densmore, A. L. (2022). Automated determination of landslide locations after large trigger events: Advantages and disadvantages compared to manual mapping. *Natural Hazards and Earth System Sciences, 22*, 481–508. https://doi.org/10.5194/nhess-22-481-2022.

Milledge, D. G., Densmore, A. L., Bellugi, D., Rosser, N. J., Watt, J., Li, G., & Oven, K. J. (2019). Simple rules to minimise exposure to coseismic landslide hazard. *Natural Hazards and Earth System Sciences, 19*, 837–856. https://doi.org/10.5194/nhess-19-837-2019.

Nam, B. H., Choi, S., Copeland, T., & Kim, Y. J. (2023). Social vulnerability and geohazards: Review and implications. In S. D'Amico, & F. De Pascale (Vol. Eds.), *Geohazards and disaster risk reduction*. Advances in natural and technological hazards research: *vol 51*Cham: Springerhttps://doi.org/10.1007/978-3-031-24541-1_1.

Newmark, N. M. (1965). Effects of earthquakes on dams and embankments. *Géotechnique, 15*(2), 139–160.

Nor Diana, M. I., Muhamad, N., Taha, M. R., Osman, A., & Alam, M. M. (2021). Social vulnerability assessment for landslide hazards in Malaysia: A systematic review study. *Land, 10*(3), 315.

Novellino, A., et al. (2024). Mapping landslides from space: A review. *Landslides, 21*, 1041–1052. https://doi.org/10.1007/s10346-024-02215-x.

Nowicki Jessee, M. A., Hamburger, M. W., Allstadt, K., Wald, D. J., Robeson, S. M., Tanyas, H., Hearne, M., & Thompson, E. M. (2018). A global empirical model for near-real-time assessment of seismically induced landslides. *Journal of Geophysical Research: Earth Surface, 123*(8), 1835–1859.

Pellicani, R., Van Westen, C. J., & Spilotro, G. (2014). Assessing landslide exposure in areas with limited landslide information. *Landslides, 11*, 463–480. https://doi.org/10.1007/s10346-013-0386-4.

Peruzzetto, M., et al. (2020). Operational estimation of landslide runout: Comparison of empirical and numerical methods. *Geosciences, 10*(11), 424. https://doi.org/10.3390/geosciences10110424.

Pollock, W., & Wartman, J. (2020). Human vulnerability to landslides. *GeoHealth, 4*(10) e2020GH000287.

Polwart, M. (2023). *Hazard from post-earthquake landslides: Assessing the temporal and spatial evolution of landslides from the 2016 Kaikōura earthquake* (Unpublished master's thesis). University of Canterbury, p. 93.

Ray, R. L., Lazzari, M., & Olutimehin, T. (2020). Remote sensing approaches and related techniques to map and study landslides. In *Landslides – Investigation and Monitoring*. IntechOpen. https://doi.org/10.5772/intechopen.93681.

Reichenbach, P., Rossi, M., Malamud, B. D., Mihir, M., & Guzzetti, F. (2018). A review of statistically-based landslide susceptibility models. *Earth-Science Reviews, 180*, 60–91.

Robinson, T. R., Davies, T. R., Wilson, T. M., Orchiston, C., & Barth, N. (2016). Evaluation of coseismic landslide hazard on the proposed Haast-Hollyford highway, South Island, New Zealand. *Georisk: Assessment and Management of Risk for Engineered Systems and Geohazards, 10*(2), 146–163.

Robinson, T. R., Davies, T. R. H., Wilson, T. M., & Orchiston, C. (2016). Coseismic landsliding estimates for an Alpine Fault earthquake and the consequences for erosion of the Southern Alps, New Zealand. *Geomorphology, 263*, 71–86.

Robinson, T. R., Rosser, N. J., Davies, T. R., Wilson, T. M., & Orchiston, C. (2018). Near-real-time modeling of landslide impacts to inform rapid response: An example from the 2016 Kaikōura, New Zealand, earthquake. *Bulletin of the Seismological Society of America, 108*(3B), 1665–1682.

Robinson, T. R., Rosser, N. J., Densmore, A. L., Williams, J. G., Kincey, M. E., Benjamin, J., & Bell, H. J. (2017). Rapid post-earthquake modelling of coseismic landslide intensity and distribution for emergency response decision support. *Natural Hazards and Earth System Sciences, 17*(9), 1521–1540.

Rosser, N., Kincey, M., Oven, K., Densmore, A., Robinson, T., Pujara, D. S., Shrestha, R., Smutny, J., Gurung, K., Lama, S., & Dhital, M. R. (2021). Changing significance of landslide hazard and risk after the 2015 Mw 7.8 Gorkha, Nepal earthquake. *Progress in Disaster Science, 10*, 100159.

Saul, G., & Anderson, C. (2019). Kaikoura earthquake 2016: An outline of the geotechnical effort to reopen the coastal highway and railway corridor. In *Earthquake Geotechnical Engineering for Protection and Development of Environment and Constructions*. CRC Press, 4890–4897.

Siders, A. R. (2019). Adaptive capacity to climate change: A synthesis of concepts, methods, and findings in a fragmented field. *Wiley Interdisciplinary Reviews: Climate Change, 10*(3), e573.

Singh, A., Pal, S., Kanungo, D. P., & Pareek, N. (2017). An overview of recent developments in landslide vulnerability assessment – presentation of a new conceptual framework. In M. Mikos, B. Tiwari, Y. Yin, & K. Sassa (Vol. Eds.), Advancing culture of living with landslides. *WLF: 2017*Springerhttps://doi.org/10.1007/978-3-319-53498-5_91.

Sohel, R., Dharanirajan, K., Toushif, J., & Kumar, M. K. (2022). Assessment of social vulnerability of landslides in the Darjeeling district using MCDA-based GIS techniques. *Disaster Advances, 15*(9), 8–15.

Tanyaş, H., van Westen, C. J., Allstadt, K. E., Anna Nowicki Jessee, M., Görüm, T., Jibson, R. W., Godt, J. W., Sato, H. P., Schmitt, R. G., Marc, O., & Hovius, N. (2017). Presentation and analysis of a

worldwide database of earthquake-induced landslide inventories. *Journal of Geophysical Research: Earth Surface, 122*(10), 1991–2015.
Tran, T. X., et al. (2024). A spatial landslide risk assessment based on hazard, vulnerability, exposure, and adaptive capacity. *Sustainability, 16*(21), 9574. https://doi.org/10.3390/su16219574.
United Nations Office for Disaster Risk Reduction (UNDRR). (2017). *Report of the open-ended intergovernmental expert working group on indicators and terminology relating to disaster risk reduction.*
Wilson, R. C., & Keefer, D. K. (1983). Dynamic analysis of a slope failure from the 6 August 1979 Coyote Lake, California, earthquake. *Bulletin of the Seismological Society of America, 73*(3), 863–877.
Woodard, J. B., & Mirus, B. B. (2025). Overcoming the data limitations in landslide susceptibility modelling. *Science Advances, 11*, eadt1541. https://doi.org/10.1126/sciadv.adt1541.
Xiao, Y., Tang, X., Li, Y., Huang, H., & An, B. W. (2022). Social vulnerability assessment of landslide disaster based on improved TOPSIS method: Case study of eleven small towns in China. *Ecological Indicators, 143*, 109316.
Ye, C., et al. (2025). Physically based and data-driven models for landslide susceptibility assessment: principles, applications, and challenges. *Remote Sensing, 17*(13), 2280. https://doi.org/10.3390/rs17132280.
Yin, Y., Wang, F., & Sun, P. (2009). Landslide hazards triggered by the 2008 Wenchuan earthquake, Sichuan, China. *Landslides, 6*, 139–152. https://doi.org/10.1007/s10346-009-0148-5.

Further reading

Modugno, S., Johnson, S. C. M., Borrelli, P., Alam, E., Bezak, N., & Balzter, H. (2022). Analysis of human exposure to landslides with a GIS multiscale approach. *Natural Hazards, 112*(1), 387–412.
Yildiz, A., et al. (2023). Computationally-feasible uncertainty quantification in model-based landslide risk assessment. *Frontiers in Earth Science, 10*. https://doi.org/10.3389/feart.2022.1032438.
Zhao, X., et al. (2023). Application of environmental variables in statistically-based landslide susceptibility mapping: A review. *Frontiers in Earth Science, 11*. https://doi.org/10.3389/feart.2023.1147427.

CHAPTER 17

Summary and future perspectives

Alexander Densmore[1], Tim Davies[2], and Xuanmei Fan[3]
[1]Department of Geography, Durham University, Durham, United Kingdom,
[2]School of Earth and Environment, University of Canterbury, Christchurch, New Zealand,
[3]State Key Laboratory of Geohazard Prevention and Geoenvironment Protection, Chengdu University of Technology, Chengdu, Sichuan, P.R. China

17.1 Introduction

As the preceding chapters make clear, we have made significant advances in our understanding of the geomorphic effects of earthquakes. In a little over a century, as our comprehension of seismology and earthquake physics has advanced, we have, in parallel, moved from nascent observations of surface rupture and liquefaction associated with ground shaking (e.g. Reid, 1910) to a much more holistic, integrated appreciation of the physical, chemical and biological impacts of earthquakes on the Earth's surface, and the ways in which those impacts interact and cascade over extended time periods (see Fan et al., 2019, and references therein). Over that time, we have also begun to recognise the varied ways in which individual earthquakes contribute to regional and even global-scale phenomena, including plate motion, mountain building, basin development, large-scale sediment transfer and global geochemical cycles, again at all time scales. The research summarised in these pages illustrates the wealth of tools, techniques and datasets that are now available to us. These tools allow us to revisit and learn from past events. They also allow us to begin to take a more integrated approach to anticipating the impacts of future earthquakes and to plan more effectively for the effects that they may have on people and societies.

At the same time, there remain some substantial gaps in our knowledge. Some of these stem from our inability to accurately reconstruct the details of past earthquakes so that their broader impacts can be assembled and assessed. As a result, much of our collective progress has come from a fairly small number of well-studied, large historical earthquakes – the events that crop up time and again in the preceding chapters. Other gaps, however, arise from our incomplete understanding of the underlying physical, chemical and biological processes that shape the Earth's surface – in other words, from our inability to grasp the complex dynamic systems that cause landscape change (see Chapter 5). Others arise from our lack of awareness of, or lack of attention to, aspects of the near-surface system that are likely to be important but that are not currently accounted for – such as groundwater dynamics, or the structure and strength of

Geomorphic Impacts of Earthquakes
DOI: https://doi.org/10.1016/B978-0-443-40375-0.00014-1

near-surface materials. Still others arise from the aleatoric uncertainty associated with the diverse range of settings and circumstances in which earthquakes occur. Addressing these gaps is important not just for increasing our understanding of the Earth's surface system, but also for generating ever-better assessments of earthquake hazard and consequent impacts.

In this closing chapter, we identify some promising avenues for future research that might help to address these outstanding knowledge gaps, drawing upon ideas and research directions raised in the preceding chapters.

17.2 Direct measurement of earthquake impacts

Much of our knowledge of the geomorphic impacts of earthquakes is derived not from direct observation but from measurement of a range of proxies – other variables which are related to, or more or less indicative of, the processes that we are interested in. Depending on our understanding of the proxy itself and its causal link to the earthquake, this can result in an incomplete or flawed picture of the impact. We focus here on two examples of this proxy issue: landslide volumes and sediment fluxes.

The volume of sediment mobilised in coseismic and postseismic landslides is a critical indicator of their geomorphic impact. Along with their distribution in space and time, the total landslide volume defines the pattern and rate at which mass can be transferred from hillslopes to the river network and potentially evacuated from the orogen, leading to long-term erosion (Dadson et al., 2004; Hovius et al., 2011; Li et al., 2014; Marc et al., 2019; Pearce & Watson, 1986; Simonett, 1967). It also places a maximum bound on potential fluvial impacts (see Chapter 9) and the magnitude of change in the downstream sediment cascade. The volume of an individual landslide determines its impact at the hillslope scale as well as its potential connectivity to the river network (e.g. Li et al., 2016). Despite this importance, however, the total volume of coseismic and postseismic landslides is typically assessed via scaling relationships that relate landslide area (assessed from aerial or satellite imagery) to volume (e.g. Guzzetti et al., 2009; Larsen et al., 2010). These scaling relationships provide a powerful tool to rapidly estimate the total volume of sediment that has potentially been mobilised in a large earthquake (e.g. Li et al., 2014; Parker et al., 2011; Roback et al., 2018), both at the scale of the entire rupture area and within individual catchments. At the same time, they suffer from some critical drawbacks. The inferred volume is highly dependent on the choice of scaling parameters, which are difficult to know in advance (Larsen et al., 2010), and which are themselves calibrated on relatively limited datasets. This issue could, in principle, be addressed through focused field measurement of landslide areas and volumes after a large earthquake, but that would be complicated by the logistical difficulty of collecting these data in a complex and chaotic postearthquake setting (e.g. Williams et al., 2018) – exactly the time when a scaling relationship is most useful.

Assessment of landslide area is also complicated, both by issues with incomplete visibility of the ground surface over the entire rupture area (Williams et al., 2018) and by amalgamation and overprinting of landslide footprints, which makes distinction and delineation of individual landslides complex (Marc & Hovius, 2015; Fan et al., 2018). Because the landslide area-volume scaling relationship is typically nonlinear, errors in landslide area can propagate into large uncertainties in both individual and total volumes (Li et al., 2014; Marc & Hovius, 2015; see also Chapter 14).

A better way to assess total landslide volume would be to directly measure changes in surface elevation due to material evacuation, transport and deposition – for example, by repeat measurement of high-resolution topography over a sufficiently large area to capture the volume of sediment mobilised in the earthquake. This would require coregistered topographic datasets collected before and after the earthquake, and with sufficiently dense horizontal spacing and vertical resolution to image metre- or submetre scale topographic change. A pioneering study by Bernard et al. (2021) used pre- and postevent LiDAR data over a portion of the area affected by the 2016 M_w 7.8 Kaikōura earthquake in New Zealand to map areas of both negative and positive topographic change, interpreted as due to landslide erosion and deposition, respectively. Their directly measured landslide volume differed by as much as a factor of two from that predicted by globally determined scaling laws (Larsen et al., 2010), illustrating the value of this approach in better constraining both individual and collective sediment volumes (see also Jones et al., 2025). In turn, constraint on sediment volumes gives us insight into the modes and efficiency of sediment input into and transfer through the fluvial system (see Chapter 9). More widespread application of this approach in future large earthquakes will require the systematic collection of high-resolution topographic data at short-enough intervals to allow pre- and post-event differencing, a goal which is increasingly likely to be realised in the next years to decades.

Transfers of landslide sediment into the river system, and consequent downstream impacts on sediment flux and geochemical cycling, represent a second example of indirect measurement of landslide impacts. The pattern and extent of sediment transfers are typically assessed using downstream measurements of sediment discharge in the river network. Given the difficulties of measuring bedload transport rates, especially in steep or high-discharge mountain rivers, these measurements are very often limited to the suspended-sediment load, which is itself often uncertain (Li et al., 2025; Chapter 11). While suspended-sediment discharge measurements are capable of resolving near-instantaneous deliveries of sediment to the river system after a large earthquake (e.g. Wang et al., 2015) and can capture large-scale changes in sediment availability and supply to the river network (e.g. Dadson et al., 2004; Hovius et al., 2011), they are also inherently limited. Coseismic and postseismic landslides may be only partially connected with the river network, constraining their downstream effects on sediment discharge (Li et al., 2016), and sediment may be trapped in short- to medium-term storage along the transport pathway (Fan et al., 2019). Critically, delivery of sediment to downstream sampling stations requires not just supply to the river network but also sufficient water discharge to transport the additional load, leading to potential decoupling or lags between supply in the earthquake and later entrainment, transport and delivery (e.g. Zhang et al., 2019). Finally, while suspended-sediment concentrations may exceed those of bedload, bedload dynamics have much greater effects on river morphology per unit volume than suspended load. These effects conspire to obscure and potentially remove the downstream sediment 'signal' of an earthquake, along with our ability to reconstruct its effects.

As outlined in Chapter 11, new approaches to direct measurement of sediment delivery and transport within the river system have the potential to address these issues. Francis et al. (2022) showed how high-resolution remote sensing data could be combined with field surveys to directly track sediment transfer from hillslopes into the river network after the 2008 Wenchuan earthquake, and demonstrated the value of an integrated sediment budget approach to understand the pathways and fates of coseismic landslide debris. Repeat high-resolution topographic surveys were used by Tunnicliffe et al. (2024) to document the fluvial response to catastrophic sediment delivery after the 2016 Kaikōura earthquake through short-term storage, evacuation and bed armouring. Geochemical tracing techniques allow us to better understand the origin (e.g. Wang et al., 2020) and transport (e.g. West et al., 2014) of earthquake-derived sediments, with the promise of eventually being able to pinpoint sediment sources, transport pathways and both intermediate and long-term sinks on the scale of an earthquake rupture or an entire orogen. More direct measurements of sediment delivery may also help to better illuminate the outstanding question of the time scales required to evacuate earthquake-derived sediment from a mountain range (e.g. Fan et al., 2018; Marc et al., 2019), and how they compare with the return intervals of large earthquakes (see Chapters 9 and 11). In turn, a better understanding of these time scales will help us to define the ways in which sediment and organic carbon are stored and released in these landscapes (see Chapter 11).

17.3 Applications of machine learning

Many of the geomorphic impacts of earthquakes share two inherent characteristics: they are distributed unevenly over the large spatial areas defined by earthquake rupture patterns due to local variability in landscape response, and they are superimposed on a highly dynamic landscape that is being continually modified by surface processes and events that are unrelated to earthquakes. Identifying these impacts thus becomes a problem of identifying a spatially variable parameter against a noisy background – a situation that is ideally suited to pattern recognition using a wide range of machine learning techniques. Increasing accessibility of computational resources and suitable machine learning algorithms, as well as the growing availability of suitable spatial datasets, make this a promising avenue for future research. A number of studies have sought to use machine learning approaches for automated mapping of landslides after earthquakes (e.g. Chen et al., 2023; Fan et al., 2025). Other studies have applied machine learning or deep-learning approaches to recognising and mapping earthquake impacts on infrastructure, such as building damage (e.g. Ghimire et al., 2022; Harirchian et al., 2021; Kuo & Lin, 2025). Other earthquake impacts could be addressed in a similar way. For example, Chevalier (see Chapter 6) highlights the potential returns from using machine learning to map fault traces. When combined with repeat topographic datasets over time, this raises the possibility of generating dynamic fault maps to help visualise how deformation evolves over time during and after an earthquake.

Machine learning approaches may also allow us to recognise, and begin to assess, other controls on geomorphic impacts, including geological, biological and hydrological factors that enable or amplify the processes considered in this book. At the moment, many of these factors are ignored, or their specific influences are subsumed within event-scale inventories, because we lack the tools or capability to isolate and understand their effects. As such, machine learning approaches may allow us to engage with researchers and datasets from cognate fields, such as ecology and hydrology, opening up exciting opportunities for cross-disciplinary collaboration.

17.4 Better pre-event understanding and baseline data

Most of the earthquakes that are mentioned in the preceding chapters occurred in areas with no previous history of large, damaging earthquakes. Some (e.g. the 1999 Chi-Chi and 2015 Gorkha events described in Chapters 13 and 15, and the 1994 Northridge earthquake in California) occurred in areas that were at least known to be earthquake-prone and where there was at least some prior understanding of active faulting. Others, most notably the 2008 Wenchuan earthquake (see Chapter 14), occurred in areas that were poorly understood or where earthquake hazard was not anticipated. Irrespective of the extent to which the events could have been anticipated, the lack of a well-understood earthquake history means that there is, more often than not, little pre-event understanding of what impacts can be expected. It can take months to years for even the initial geomorphic impacts of an earthquake to be adequately assessed and mapped (e.g. Williams et al., 2018), while the dynamic legacy of earthquake impacts can continue to evolve over years to decades or even centuries (e.g. Fan et al., 2019; Kincey et al., 2021; Marc et al., 2019; Wang et al., 2015).

A better baseline understanding of preearthquake conditions, including topography, regolith and vegetation distributions, background rates of landsliding and debris-flow occurrence and fluvial suspended-load and bedload sediment fluxes, would greatly aid the rapid identification and quantification of geomorphic impacts, and would help to constrain the likely 'response time' over which earthquake-triggered changes in processes will decline back to preearthquake levels. Some of these datasets are already available, and more will come over the next few years to decades.

A major step forward is the advent of rapid, large-scale monitoring and visualisation of deformation, which makes it easier to focus attention on parts of the world where future earthquakes are more likely (e.g. England & Jackson, 2011). Continent-scale initiatives such as the European Ground Motion Service (Crosetto et al., 2020) and global-scale programmes such as Looking inside the Continents from Space(Lazecky et al., 2020) raise the prospect of improving our ability to identify active faults and target baseline data collection in ways that have not previously been possible. Why not use these deformation frameworks to task Earth observation satellites with focused campaigns aimed at collating pre-event remote sensing data in advance of the next major earthquake?

17.5 Anticipatory modelling of earthquake impacts

A recurrent issue in previous large earthquakes is our inability to anticipate the extent and duration of earthquake-triggered or secondary hazards, even once the location and broad characteristics of the event are known. This information is critical for informing the immediate response and recovery, during which community groups, emergency services and civil society organisations must grapple with fragmentary answers to fundamental questions, all rendered more difficult by communications outages: Which areas are likely to have been hardest hit, and in what ways? How has access to those areas been affected, and for how long will it be disrupted? How many people and properties are likely to have been affected? How long will secondary hazards (including aftershocks) persist, and over what areas? What are the ongoing threats not just to the affected population but to emergency response or aid workers? Beyond the immediate response, geomorphic impacts can continue to be felt for months or decades (Blagen et al., 2022; Kincey et al., 2022) and can frustrate recovery and eventual reconstruction efforts. How are risks from secondary hazards likely to evolve in space and time (Jones et al., 2021; Rosser et al., 2021)? How can we ensure that reconstruction efforts take those evolving risks into account, to avoid further disruption and loss (Tang et al., 2020)?

It is well-established that the extent and magnitude of secondary hazards are likely to scale with both earthquake magnitude (e.g. Malamud et al., 2004) and hypocentral depth (e.g. Marc et al., 2016). That expectation does not, by itself, provide guidance on the locations or characteristics of those hazards, so it is difficult to anticipate their geomorphic or societal impacts. Our typical approach to this problem is to try to assess the spatial distribution of impacts, either through reconnaissance (e.g. Collins & Jibson, 2015) or mapping. This is a slow process; in the case of the 2008 Wenchuan earthquake, a comprehensive inventory of earthquake-triggered landslides was not published until

3 years after the event (Gorum et al., 2011), and an analysis of the valley-blocking landslide dams that formed a major element of the hazard was published the following year (Fan et al., 2012). After the 2015 Gorkha earthquake, there were nascent efforts to better anticipate the distribution of secondary hazards by extrapolating initial landslide reports (Robinson et al., 2017) or applying a forward model of landslide occurrence (Gallen et al., 2017). While these approaches are promising and were able to anticipate aspects of the coseismic landslide population, neither was able to reproduce the detailed distribution, and neither accounted for the full cascade of earthquake-triggered processes (Fan et al., 2019). Williams et al. (2018) considered the challenges associated with rapid landslide mapping and pointed out the mismatch between the time required to generate robust estimates of earthquake-triggered impacts and the time scales over which information is required for emergency response and initial recovery (see also Datta et al., 2018).

A step forward in our ability to quickly anticipate geomorphic impacts after a large earthquake comes from empirical models, such as the approach outlined by Nowicki Jessee et al. (2018), that relate earthquake shaking to likely patterns of coseismic landslide occurrence. The US Geological Survey's Ground Failure product, part of the PAGER (Prompt Assessment of Global Earthquakes for Response) initiative (Wald et al., 2010), now routinely generates estimates of earthquake-triggered liquefaction and landslide likelihood as well as potential exposure to those hazards for major earthquakes worldwide (Allstadt et al., 2018; 2021; see Chapter 4). Future efforts to anticipate landsliding in particular should seek to exploit the growing availability of instrumental and model-generated data on earthquake shaking after events worldwide. Extension to a wider range of hazards, however, and thus a more comprehensive picture of impacts, will require either an improved process-based understanding of the Earth surface system or, as argued in Chapter 5, a better capacity to understand and simulate emergent system-scale dynamics.

It is also critical to remember the cascading and often compounding nature of earthquake-triggered hazards (Fan et al., 2019), which cautions against a focus just on, for example, coseismic landslide occurrence. Experience in many of the earthquakes summarised here shows that the downstream impacts of this cascade can extend many kilometres beyond the initial rupture area and take weeks, months, or years to be realised (Kincey et al., 2022). Thus in addition to better models of directly triggered earthquake impacts, we also need better ways of tracking and modelling sediment redistribution (see Chapter 7). These advances will allow us to take a more dynamic view of seismic risk, as outlined in Chapter 16, and better plan for the effects of future major earthquakes.

Acknowledgements

The editors would like to express their sincere gratitude to Dr. Niels Hovius for his valuable contributions to the development of this chapter.

References

Allstadt, K. E., Jibson, R. W., Thompson, E. M., Massey, C. I., Wald, D. J., Godt, J. W., & Rengers, F. K. (2018). Improving near-real-time coseismic landslide models: Lessons learned from the 2016 Kaikōura, New Zealand, earthquake. *Bulletin of the Seismological Society of America, 108*, 1649–1664. https://doi.org/10.1785/0120170297.

Allstadt, K. E., Thompson, E. M., Jibson, R. W., Wald, D. J., Hearne, M., Hunter, E. J., Fee, J. M., Schovanec, H., Slosky, D., & Haynie, K. L. (2021). The US Geological survey ground failure product: Near-real-time estimates of earthquake-triggered landslides and liquefaction. *Earthquake Spectra, 38*, 5–36. https://doi.org/10.1177/87552930211032685.

Bernard, T. G., Lague, D., & Steer, P. (2021). Beyond 2D landslide inventories and their rollover: Synoptic 3D inventories and volume from repeat Lidar data. *Earth Surface Dynamics, 9*, 1013–1044. https://doi.org/10.5194/esurf-9-1013-2021.

Blagen, J. R., Davies, T. R. H., Wells, A., & Norton, D. A. (2022). Post-seismic aggradation history of the West Coast, South Island, Aotearoa/New Zealand: Dendrogeomorphological evidence and disaster recovery implications. *Natural Hazards, 114*, 2545–2570. https://doi.org/10.1007/s11069-022-05479-5.

Chen, L., Li, Z., Song, C., Xing, J., Cai, X., Fang, Z., Luo, R., & Li, Z. (2023). Automatic detection of earthquake triggered landslides using Sentinel-1 SAR imagery based on deep learning. *International Journal of Digital Earth, 17*. https://doi.org/10.1080/17538947.2024.2393261.

Collins, B. D., & Jibson, R. W. (2015). *Assessment of existing and potential landslide hazards resulting from the April 25, 2015, Gorkha, Nepal earthquake sequence* (U.S. Geological Survey Open-File Report 2015–1142). U.S. Geological Survey. https://doi.org/10.3133/ofr20151142

Crosetto, M., Solari, L., Balasis-Levinsen, J., Casagli, N., Frei, M., Oyen, A., & Moldestad, D. A. (2020). Ground deformation monitoring at continental scale: The European Ground Motion Service. *International Archives of the Photogrammetry, Remote Sensing and Spatial Information Services, XLIII-B3-2020*. https://doi.org/10.5194/isprs-archives-XLIII-B3-2020-293-2020.

Dadson, S. J., Hovius, N., Chen, H., Dade, W. B., Lin, J. C., Hsu, M. L., Lin, C. W., Horng, M. J., Chen, T. C., Milliman, J., & Stark, C. P. (2004). Earthquake-triggered increase in sediment delivery from an active mountain belt. *Geology, 32*, 733–736. https://doi.org/10.1130/G20639.1.

Datta, A., Sigdel, S., Oven, K. J., Rosser, N., Densmore, A. L., & Rijal, S. (2018). *The role of scientific evidence during the 2015 Nepal earthquake relief efforts* (Overseas Development Institute Report 11044). Overseas Development Institute. https://www.odi.org/publications/11044-role-scientific-evidence-during-2015-nepal-earthquake-relief-efforts

England, P., & Jackson, J. (2011). Uncharted seismic risk. *Nature Geoscience, 4*, 348–349. https://doi.org/10.1038/ngeo1168.

Fan, X., Domènech, G., Scaringi, G., Huang, R., Xu, Q., Hales, T. C., Dai, L., Yang, Q., & Francis, O. (2018). Spatio-temporal evolution of mass wasting after the 2008 M_w 7.9 Wenchuan earthquake revealed by a detailed multi-temporal inventory. *Landslides, 15*, 2325–2341. https://doi.org/10.1007/s10346-018-1054-5.

Fan, X., Scaringi, G., Korup, O., West, A. J., van Westen, C. J., Tanyas, H., Hovius, N., Hales, T. C., Jibson, R. W., Allstadt, K. E., Zhang, L., Evans, S. G., Xu, C., Li, G., Pei, X., Xu, Q., & Huang, R. (2019). Earthquake-induced chains of geologic hazards: Patterns, mechanisms, and impacts. *Reviews of Geophysics, 57*, 421–503. https://doi.org/10.1029/2018RG000626.

Fan, X., van Westen, C. J., Xu, Q., Gorum, T., & Dai, F. (2012). Analysis of landslide dams induced by the 2008 Wenchuan earthquake. *Journal of Asian Earth Sciences, 57*, 25–37. https://doi.org/10.1016/j.jseaes.2012.06.002.

Fan, X., Wang, X., Fang, C., Jansen, J. D., Dai, L., Tanyas, H., Zang, N., Tang, R., Xu, Q., & Huang, R. (2025). Deep learning can predict global earthquake-triggered landslides. *National Science Review, 12*, nwaf179. https://doi.org/10.1093/nsr/nwaf179.

Francis, O., Fan, X., Hales, T. C., Hobley, D., Xu, Q., & Huang, R. (2022). The fate of sediment after a large earthquake. *Journal of Geophysical Research – Earth Surface, 127*, e2021JF006352. https://doi.org/10.1029/2021JF006352.

Gallen, S. F., Clark, M. K., Godt, J. W., Roback, K., & Niemi, N. A. (2017). Application and evaluation of a rapid response earthquake-triggered landslide model to the 25 April 2015 M_w 7.8 Gorkha earthquake, Nepal. *Tectonophysics, 714-715*, 173–187. https://doi.org/10.1016/j.tecto.2016.10.031.

Ghimire, S., Guéguen, P., Giffard-Roisin, S., & Schorlemmer, D. (2022). Testing machine learning models for seismic damage prediction at a regional scale using building-damage dataset compiled after the 2015 Gorkha Nepal earthquake. *Earthquake Spectra, 38*, 2970–2993. https://doi.org/10.1177/87552930221106495.

Gorum, T., Fan, X., van Westen, C. J., Huang, R., Xu, Q., Tang, C., & Wang, G. (2011). Distribution pattern of earthquake-induced landslides triggered by the 12 May 2008 Wenchuan earthquake. *Geomorphology, 133*, 152–167. https://doi.org/10.1016/j.geomorph.2010.12.030.

Guzzetti, F., Ardizzone, F., Cardinali, M., Rossi, M., & Valigi, D. (2009). Landslide volumes and landslide mobilization rates in Umbria, central Italy. *Earth and Planetary Science Letters, 279*, 222–229. https://doi.org/10.1016/j.epsl.2009.01.005.

Harirchian, E., Kumari, V., Jadhav, K., Rasulzade, S., Lahmer, T., & Raj Das, R. (2021). A synthesized study based on machine learning approaches for rapid classifying earthquake damage grades to RC buildings. *Applied Science, 11*, 7540. https://doi.org/10.3390/app11167540.

Hovius, N., Meunier, P., Lin, C. W., Chen, H., Chen, Y. G., Dadson, S., Horng, M. J., & Lines, M. (2011). Prolonged seismically induced erosion and the mass balance of a large earthquake. *Earth and Planetary Science Letters, 304*, 347–355. https://doi.org/10.1016/j.epsl.2011.02.005.

Jones, J. N., Boulton, S. J., Bennett, G. L., Stokes, M., & Whitworth, M. R. Z. (2021). Temporal variations in landslide distributions following extreme events: Implications for landslide susceptibility modelling. *Journal of Geophysical Research – Earth Surface, 126*, e2021JF006067. https://doi.org/10.1029/2021JF006067.

Jones, K. E., Howarth, J. D., Massey, C. I., Luković, B., Sirguey, P., Singeisen, C., Gasston, C., Morgenstern, R., & Ries, W. (2025). An alternative to landslide volume-area scaling relationships: An ensemble approach adopting a difference model to estimate the total volume of landsliding triggered by the 2016 Kaikōura earthquake, New Zealand. *Landslides, 22*, 2219–2236. https://doi.org/10.1007/s10346-025-02479-x.

Kincey, M. E., Rosser, N. J., Densmore, A. L., Robinson, T. R., Shrestha, R., Pujara, D. S., Horton, P., Swirad, Z. M., Oven, K. J., & Arrell, K. (2022). Modelling post-earthquake cascading hazards: Changing patterns of landslide runout following the 2015 Gorkha earthquake, Nepal. *Earth Surface Processes and Landforms, 48*, 537–554. https://doi.org/10.1002/esp.5501.

Kincey, M. E., Rosser, N. J., Robinson, T. R., Densmore, A. L., Shrestha, R., Singh Pujara, D., Oven, K. J., Williams, J. G., & Swirad, Z. M. (2021). Evolution of coseismic and post-seismic landsliding after the 2015 M_w 7.8 Gorkha earthquake, Nepal. *Journal of Geophysical Research – Earth Surface, 126*, e2020JF005803. https://doi.org/10.1029/2020JF005803.

Kuo, W. N., & Lin, S. Y. (2025). Multimodal models for assessing earthquake-induced building damage using metadata and satellite imagery. *Journal of Building Engineering, 111*, 113467. https://doi.org/10.1016/j.jobe.2025.113467.

Larsen, I. J., Montgomery, D. R., & Korup, O. (2010). Landslide erosion controlled by hillslope material. *Nature Geoscience, 3*, 247–251. https://doi.org/10.1038/ngeo776.

Lazecky, M., Spaans, K., González, P. J., Maghsoudi, Y., Morishita, Y., Albino, F., Elliott, J., Greenall, N., Hatton, E., Hooper, A., Juncu, D., McDougall, A., Walters, R. J., Watson, C. S., Weiss, J. R., & Wright, T. J. (2020). LiCSAR: An automatic InSAR tool for measuring and monitoring tectonic and volcanic activity. *Remote Sensing, 12*. https://doi.org/10.3390/rs12152430.

Li, G., West, A. J., Densmore, A. L., Hammond, D. E., Jin, Z., Zhang, F., Wang, J., & Hilton, R. G. (2016). Connectivity of earthquake-triggered landslides with the fluvial network: Implications for landslide sediment transport after the 2008 Wenchuan earthquake. *Journal of Geophysical Research – Earth Surface, 121*, 703–724. https://doi.org/10.1002/2015JF003718.

Li, G., West, A. J., Densmore, A. L., Jin, Z., Parker, R. N., & Hilton, R. G. (2014). Seismic mountain building: Landslides associated with the 2008 Wenchuan earthquake in the context of a generalized model for earthquake volume balance. *Geochemistry, Geophysics, Geosystems, 15*, 833–844. https://doi.org/10.1029/2013GC005067.

Li, G., West, A. J., Jin, Z., Qiu, H., Zhang, F., Wang, J., Hammond, D. E., Densmore, A. L., Hilton, R. G., Dong, S., Fischer, W., & Lamb, M. (2025). Exceptional riverbed sediment flux sustained for ten years after the Wenchuan earthquake. *Nature, 644*, 398–403. https://doi.org/10.1038/s41586-025-09354-8.

Malamud, B. D., Turcotte, D. L., Guzzetti, F., & Reichenbach, P. (2004). Landslides, earthquakes, and erosion. *Earth and Planetary Science Letters, 229*, 45–59. https://doi.org/10.1016/j.epsl.2004.10.018.

Marc, O., & Hovius, N. (2015). Amalgamation in landslide maps: Effects and automatic detection. *Natural Hazards and Earth System Sciences, 15*, 723–733. https://doi.org/10.5194/nhess-15-723-2015.

Marc, O., Behling, R., Andermann, C., Turowski, J. M., Illien, L., Roessner, S., & Hovius, N. (2019). Long-term erosion of the Nepal Himalayas by bedrock landsliding: The role of monsoons, earthquakes, and giant landslides. *Earth Surface Dynamics, 7*, 107–128. https://doi.org/10.5194/esurf-7-107-2019.

Marc, O., Hovius, N., Meunier, P., Gorum, T., & Uchida, T. (2016). A seismologically consistent expression for the total area and volume of earthquake-triggered landsliding. *Journal of Geophysical Research – Earth Surface, 121*, 640–663. https://doi.org/10.1002/2015JF003732.

Nowicki Jessee, M. A., Hamburger, M. W., Allstadt, K., Wald, D. J., Robeson, S. M., Tanyas, H., Hearne, M., & Thompson, E. M. (2018). A global empirical model for near-real-time assessment of seismically induced landslides. *Journal of Geophysical Research – Earth Surface, 123*, 1835–1859. https://doi.org/10.1029/2017JF004494.

Parker, R. N., Densmore, A. L., Rosser, N. J., de Michele, M., Li, Y., Huang, R., Whadcoat, S., & Petley, D. N. (2011). Mass wasting triggered by the 2008 Wenchuan earthquake is greater than orogenic growth. *Nature Geoscience, 4*, 449–452. https://doi.org/10.1038/ngeo1154.

Pearce, A. J., & Watson, A. J. (1986). Effects of earthquake-induced landslides on sediment budget and transport over a 50-yr period. *Geology, 14*, 52–55. https://doi.org/10.1130/0091-7613(1986)14%3C52:EOELOS%3E2.0.CO;2.

Reid, H. F. (1910). The mechanics of the earthquake. In *The California Earthquake of April 18, 1906: Report of the State Investigation Commission*, Vol.2. Carnegie Institution of Washington.

Roback, K., Clark, M. K., West, A. J., Zekkos, D., Li, G., Gallen, S. F., Chamlagain, D., & Godt, J. W. (2018). The size, distribution, and mobility of landslides caused by the 2015 M_w 7.8 Gorkha earthquake, *Nepal. Geomorphology, 301*, 121–138. https://doi.org/10.1016/j.geomorph.2017.01.030.

Robinson, T., Rosser, N. J., Densmore, A. L., Williams, J. G., Kincey, M., Benjamin, J., & Bell, H. J. A. (2017). Rapid post-earthquake modelling of coseismic landslide intensity and distribution for emergency response decision support. *Natural Hazards and Earth System Sciences, 17*, 1521–1540. https://doi.org/10.5194/nhess-17-1521-2017.

Rosser, N., Kincey, M., Oven, K., Densmore, A., Robinson, T., Pujara, D. S., Shrestha, R., Smutny, J., Gurung, K., Lama, S., & Dhital, M. R. (2021). Changing significance of landslide hazard and risk after the 2015 M_w 7.8 Gorkha, Nepal earthquake. *Progress in Disaster Science, 10*. https://doi.org/10.1016/j.pdisas.2021.100159.

Simonett, D. S. (1967). Landslide distribution and earthquakes in the Bewani and Torricelli Mountains, New Guinea – a statistical analysis. In J. N. Jennings, & J. A. Mabbutt (Eds.). *Landform Studies from Australia and New Guinea* (pp. 64–84). Cambridge University Press.

Tang, C., Liu, X., Cai, Y., van Westen, C., Yang, Y., Tang, H., Yang, C., & Tang, C. (2020). Monitoring of the reconstruction process in a high mountainous area affected by a major earthquake and subsequent hazards. *Natural Hazards and Earth System Science, 20*, 1163–1186. https://doi.org/10.5194/nhess-20-1163-2020.

Tunnicliffe, J., Howarth, J., & Massey, C. (2024). Controls on fluvial sediment evacuation following an earthquake-triggered landslide: Observations from LiDAR time series. *Science Advances, 10*. https://doi.org/10.1126/sciadv.adi5560.

Wald, D. J., Jaiswal, K. S., Marano, K. D., Bausch, D. B., & Hearne, M. G. (2010). *PAGER—Rapid assessment of an earthquake's impact* (U.S. Geological Survey Fact Sheet 2010–3036). U.S. Geological Survey. https://doi.org/10.3133/fs20103036

Wang, J., Howarth, J. D., McClymont, E. L., Densmore, A. L., Fitzsimons, S. J., Croissant, T., Groecke, D. R., West, M. D., Harvey, E. L., Frith, N. V., & Hilton, R. G. (2020). Long-term patterns of hillslope erosion by earthquake-induced landslides shape mountain landscapes. *Science Advances, 6*. https://doi.org/10.1126/sciadv.aaz6446.

Wang, J., Jin, Z., Hilton, R. G., Zhang, F., Densmore, A. L., Li, G., & West, A. J. (2015). Controls on fluvial evacuation of sediment from earthquake-triggered landslides. *Geology, 43*, 115–118. https://doi.org/10.1130/G36157.1.

West, A. J., Hetzel, R., Li, G., Jin, Z., Zhang, F., Hilton, R. G., & Densmore, A. L. (2014). Dilution of ^{10}Be in detrital quartz by earthquake-induced landslides: Implications for determining denudation rates and potential to provide insights into landslide sediment dynamics. *Earth and Planetary Science Letters, 396*, 143–153. https://doi.org/10.1016/j.epsl.2014.03.058.

Williams, J. G., Rosser, N. J., Kincey, M. E., Benjamin, J., Oven, K. J., Densmore, A. L., Milledge, D. G., Robinson, T. R., Jordan, C. A., & Dijkstra, T. A. (2018). Satellite-based emergency mapping: Landslides triggered by the 2015 Nepal earthquake. *Natural Hazards and Earth System Sciences, 18*, 185–205. https://doi.org/10.5194/nhess-18-185-2018.

Zhang, F., Jin, Z., West, A. J., An, Z., Hilton, R. G., Wang, J., Li, G., Densmore, A. L., Yu, J., Qiang, X., Sun, Y., Li, L., Gou, L., Xu, Y., Xu, X., Liu, X., Pan, Y., & You, C.-F. (2019). Monsoonal control on a delayed response of sedimentation to the 2008 Wenchuan earthquake. *Science Advances, 5*. https://doi.org/10.1126/sciadv.aav7110.

Index

Note: Page numbers followed by "*f*" and "*t*" refer to figures and tables, respectively.

H

I

P

Q

R

S

Printed and bound by CPI Group (UK) Ltd, Croydon, CR0 4YY
14/09/2026
02216408-0016